Proceedings in Life Sciences

Biophysics and Physiology of Carbon Dioxide

Symposium Held at the
University of Regensburg (FRG)
April 17–20, 1979

Organized and Edited by
C. Bauer G. Gros H. Bartels

With 191 Figures

Springer-Verlag
Berlin Heidelberg New York 1980

Professor Dr. CHRISTIAN BAUER
Institut für Physiologie, Universität Regensburg
Universitätsstraße 31, 8400 Regensburg/FRG

Professor Dr. GEROLF GROS
Institut für Physiologie, Universitätsklinikum Essen
Hufelandstraße 55, 4300 Essen/FRG

Professor Dr. HEINZ BARTELS
Physiologisches Institut, Medizinische Hochschule Hannover
Karl-Wiechert-Allee 9, 3000 Hannover 61/FRG

The organization of the symposium was supported by:
Deutsche Forschungsgemeinschaft,
Freunde der Universität Regensburg and
Bayerisches Staatsministerium für Unterricht und Kultus.

Cover motive: Artistic drawing of the carbonic anhydrase molecule showing the substrates CO_2 and H_2O in position as established by X-ray diffraction analysis. The figure is based on the results of Dr. K. K. Kannan. Mr. S. R. Savant, BARC Bombay has kindly drawn the picture and Mr. Meyerji, BARC has provided the prints.

ISBN-13:978-3-642-67574-4 e-ISBN-13:978-3-642-67572-0
DOI: 10.1007/978-3-642-67572-0

Library of Congress Cataloging in Publication Data. Main entry under title. Biophysics and physiology of carbon dioxide. (Proceedings in life sciences) Bibliography: p. Includes index. 1. Carbon dioxide in the body. 2. Carbon dioxide transport (Physiology). 3. Carbonic anhydrase. I. Bauer, Christian, 1938–. II. Gros, Gerolf, 1942–. III. Bartels, Heinz, 1920–. QP535.C1B56.599.01′2. 80-10093.

Softcover reprint of the hardcover 1st edition 1980

2131/3130-543210

Preface

This volume contains the papers presented at the symposium on *Biophysics and Physiology of Carbon Dioxide* held at Regensburg, April 17–20, 1979. The manuscripts represent the full or even an extended account of the oral presentations. We have decided not to include any part of the discussions which took place after the lectures because this would have led to an undue enlargement of the already substantial volume.

The symposium brought together some 60 scientists of various disciplines including biophysicists, chemists, biochemists, physiologists, pharmacologists, as well as clinicians whose research activities are centered around the various aspects of the reactions and the regulatory role of CO_2 within the body.

In view of the fact that numerous textbooks and Proceedings of Symposia deal expertly with the role of CO_2 in acid-base balance, it was decided not to include this aspect in the present symposium. This holds also for the biochemistry of carboxylation and decarboxylation reactions. Particular emphasis was placed on the following subjects: (1) Chemical reactions of CO_2 in water and facilitated diffusion of CO_2, (2) CO_2 adducts to proteins, in particular hemoglobin, and peptide hormones, (3) structure and function of carbonic anhydrase, (4) CO_2 exchange and carbonic anhydrase activity in respiratory and nonrespiratory systems. Each section contains at least one introductory paper that presents the current knowledge in a more general framework.

Two Symposia were already devoted to CO_2, the "junior partner of oxygen" as Professor F.J.W. Roughton called it in his introductory remarks at the first symposium of this kind held in Haverford, Pennsylvania in 1968 which was organized by R.E. Forster, J.T. Edsall, A.B. Otis, and F.J.W. Roughton [1]. We are delighted to note that quite a few names that already appeared in 1968 are also found in the list of participants of the present conference. The second meeting on the subject of CO_2 was organized by G. Nahas and K.E. Schaefer [2]. It took

---

1 CO_2: Chemical, biochemical and biophysical aspects. Forster RE, Edsall JT, Otis AB, Roughton FJW (eds) (1969) NASA publication SP-188

2 Carbon dioxide and metabolic regulations. Nahas G, Schaefer KE (eds) (1974) Springer, New York Berlin Heidelberg

place in Monaco in 1973 and dealt mainly with the role of CO_2 in the regulation of acid-base balance and organ functions.

The wealth of knowledge which has been accumulated since 1968 convinced us that it would be desirable to have a conference at which the recent developments in the fields indicated above were taken into account. It is the advantage of such gatherings that a publication emerges which in its entirety can be considered as an up-to-date report on the state of affairs, written by a devoted study group.

The symposium was sponsored by the International Union of Physiological Sciences (IUPS) and we are grateful to IUPS and to Dr. E. Agostoni, presently Head of the Commission for Respiratory Physiology of IUPS, for this sponsorship.

We wish to thank the Deutsche Forschungsgemeinschaft very much for its generous financial support. Without this support the symposium could not have taken place. Additional funding by the Freunde der Universität Regensburg and the Bayerisches Staatsministerium für Unterricht und Kultus is gratefully acknowledged. Furthermore we should like to thank Miss G. Wischka for her secretarial assistance and also the officials of the town of Regensburg who provided a very pleasant framework.

With great pleasure we acknowledge the readiness of the Springer-Verlag, Heidelberg, to publish the Proceedings of the symposium in full length and for doing it so swiftly and in such a generous presentation.

Regensburg, Essen, Hannover, February 1980

CHRISTIAN BAUER
GEROLF GROS
HEINZ BARTELS

Contents

Chemical Reactions of CO_2 in Water
Chairman: J.PIIPER

Chemical Reactions of CO_2 in Water
W. KNOCHE (With 3 Figures) 3

Facilitated Diffusion of CO_2
Chairman: J. PIIPER

Facilitation of CO_2 Through Layers with a Spatial Distribution of Carbonic Anhydrase
J.S. SCHULTZ (With 7 Figures) 15

Facilitated Diffusion of CO_2 and Carbonic Anhydrase Activity
J.A. QUINN and T.L. DONALDSON (With 8 Figures) 23

Mechanisms of Facilitated CO_2 and H^+ Diffusion in Protein Solutions
G. GROS, H. GROS, D. LAVALETTE, B. AMAND, and F. POCHON (With 6 Figures). 36

The Effect of Diffusion Potentials on Facilitated CO_2 Diffusion
J.H. MELDON (With 4 Figures) 49

Evidence for Facilitated Transport of CO_2 in Muscle Tissue
P. SCHEID, T. KAWASHIRO, and J. PIIPER (With 5 Figures) . . 58

Carbamate Reactions of Proteins
Chairman: R.E. FORSTER

The Binding of CO_2 to Human Deoxyhemoglobin: An X-Ray Study Using Low-Salt Crystals
A. ARNONE, P.H. ROGERS, and P.D. BRILEY (With 3 Figures) 67

The Determination of CO_2 Bound to Hemoglobin as Carbamate
M. PERRELLA and L. ROSSI-BERNARDI (With 3 Figures) . . . 75

Phylogeny and Physiological Significance of Oxygen-Linked CO_2 Binding in Hemoglobin
C. BAUER 84

The Carbamate Reaction with Proteins: Observation by ^{13}C-NMR and Evaluation of Structural Consequences
F.R.N. GURD, J.B. MATTHEW, R.J. WITTEBORT, J.S. MORROW, and S.H. FRIEND (With 3 Figures) 89

The Measurement of the CO_2 Hb Binding Curve of Human Hemoglobin by ^{13}C-NMR Spectrometry
O. BURKHARD and W.K.R. BARNIKOL (With 2 Figures) 102

Formation and Physiological Role of Carbamino Adducts of Angiotensin, Bradykinin, and S-Methylglucagon
R.S. GURD, R.D. ENGLAND, D.F. HAYES, R.J. WITTEBORT, and T.M. ROTHGEB (With 4 Figures) 106

Interaction Between Hemoglobins, CO_2 and Anions
R. BAUMANN (With 5 Figures) 114

The Influence of CO_2 on the First Binding Constant of Oxygen to Human Hemoglobin A_0
C. POYART, E. BURSAUX, and P. MATZKE (With 5 Figures) . 122

Carbonic Anhydrase – Structure and Function
Chairmen: S. LINDSKOG and J.E. COLEMAN

Current Concepts of the Mechanism of Action of Carbonic Anhydrase
J.E. COLEMAN (With 10 Figures) 133

Some Structural Aspects of Carbonic Anhydrase
I. BERTINI, C. LUCHINAT, and A. SCOZZAFAVA (With 2 Figures) 151

Primary Structure of Carbonic Anhydrases
L.E. HENDERSON (With 1 Figure) 154

Evolution of Structure and Function in the Carbonic Anhydrase Isozymes of Mammals
R.E. TASHIAN, D. HEWETT-EMMETT, S.K. STROUP, M. GOODMAN, and Y.-S.L. YU (With 1 Figure) 165

Carbonic Anhydrase Activity in Intact Red Cells as Measured by Means of ^{18}O Exchange Between CO_2 and Water
R.E. FORSTER and N. ITADA (With 3 Figures) 177

Crystal Structure of Carbonic Anhydrase
K.K. KANNAN (With 15 Figures) 184

Kinetics and Mechanistic Implications of CO_2 Hydration Activity of Human Erythrocyte Carbonic Anhydrases
R.G. KHALIFAH (With 8 Figures) 206

Models for Metal Ion Function and Evolution of the Catalytic Step in Carbonic Anhydrase
P. WOOLLEY (With 4 Figures) 216

Consequences of Metal Ion Substitutions in Human Carbonic Anhydrase B. A Carbon-13 Nuclear Magnetic Resonance Comparative Study
R.G. KHALIFAH and P.J. MORLEY (With 3 Figures) 226

Rate-Limiting Steps in the Catalytic Action of Carbonic Anhydrase
S. LINDSKOG (With 2 Figures) 230

The pH-Independence of Carbonic Anhydrase Activity: Apparent pK_a Due to Inhibition by HSO_4^-
S.H. KOENIG, R.D. BROWN, III, and G.S. JACOB (With 4 Figures) 238

Role of Buffer in Catalysis of the Hydration of CO_2 by Carbonic Anhydrase
D.N. SILVERMAN, C.K. TU, and G.C. WYNNS (With 3 Figures) 254

^{35}Cl and ^{81}Br Nuclear Magnetic Resonance Studies of Carbonic Anhydrase
R.L. WARD (With 7 Figures) 262

NMR-Studies of Carbonic Anhydrase
J.M. PESANDO and R.K. GUPTA (With 2 Figures) 273

CO_2 Exchange and Carbonic Anhydrase in Respiratory Systems
Chairmen: H.H. LOESCHCKE and E.D. CRANDALL

Cl^- and HCO_3^- Movements Across the Red Blood Cell Membrane
R.E. FORSTER, A.L. OBAID, E.D. CRANDALL, and N. ITADA (With 3 Figures) 285

The Rate Limiting Event in the Depletion of ^{18}O from CO_2 in Red Cell Suspensions
D.N. SILVERMAN and C.K. TU (With 5 Figures) 291

Kinetics of CO_2 Uptake by Red Cells
R.A.B. HOLLAND (With 3 Figures) 299

Intravascular End-Capillary CO_2-H^+ Disequilibrium
E.D. CRANDALL, A. BIDANI, and J.E. O'BRASKY
(With 2 Figures) 306

Changes in pH and P_{CO_2} with Time in Pulmonary Post-Capillary Blood in Cats
J. PONTE and M.J. PURVES (With 4 Figures) 315

Kinetics of Acid-Base Parameters of Blood After Respiratory and Metabolic Disturbances
A. LUTTMANN, K. MÜCKENHOFF, and H.H. LOESCHCKE
(With 6 Figures) 321

Carbonic Anhydrase in Lung Tissue
R.A. KLOCKE (With 3 Figures). 331

Pulmonary Capillary Carbonic Anhydrase
R.M. EFFROS, L. SHAPIRO, P. SILVERMAN, and M. LIEBER
(With 3 Figures) 339

Can CO_2 Be Actively Eliminated by the Lungs?
G.H. GURTNER and R.J. TRAYSTMAN (With 2 Figures) 343

Experimental Results on Gas-Blood CO_2 Equilibrium in Lungs
J. PIIPER, M. MEYER, and P. SCHEID 351

Intracapillary CO_2 Gradients
R.M. EFFROS 353

Contribution of Red Cell-Plasma H^+ Disequilibrium to Alveolar-Blood P_{CO_2} Differences During Rebreathing
E.D. CRANDALL and A. BIDANI (With 2 Figures). 356

Diffusion Limitation in Alveolar-Capillary CO_2 Transfer in Human Lungs: Experimental Evidence from Rebreathing Equilibration
J. PIIPER, M. MEYER, and P. SCHEID 359

Physiological Consequences of Carbonic Anhydrase Inhibition During Rest and Exercise, and on the Bohr Effect
T.H. MAREN and E.R. SWENSON (With 3 Figures). 366

CO_2 Transport and Carbonic Anhydrase in Placenta, Muscle and Calcium-Mobilizing Tissues. Localization of Carbonic Anhydrase
Chairman: C. ALBERS

Placental CO_2 Exchange
E.P. HILL and L.D. LONGO (With 6 Figures) 379

Transfer of Bicarbonate in a Hemochorial Placenta
W. MOLL, E. KASTENDIECK, and H. GIRARD (With 8 Figures) 390

Recent Developments in Carbonic Anhydrase Localization: Occurrence of Carbonic Anhydrase in Calcium-Mobilizing Tissues
C.V. GAY 399

Arguments for and Against the Histochemical Carbonic Anhydrase Test
O. LEDER (With 1 Figure) 406

Activity and Properties of Carbonic Anhydrase in Striated Muscle
G. GROS, W. SIFFERT, and A. SCHMID (With 3 Figures) 410

CO_2 Transport and Carbonic Anhydrase in Kidney and Brain
Chairman: C. ALBERS

Significance of Carbonic Anhydrase for HCO_3^- Absorption and H^+ Secretion in Renal Tubules
E. FRÖMTER (With 2 Figures) 419

Formation and Ionic Composition of Cerebrospinal Fluid: The Role of Carbonic Anhydrase
T.H. MAREN and B.P. VOGH (With 6 Figures) 426

Autoregulation of CSF H^+ Activity
G.H. GURTNER, H.M. MAKHZOUMI, and R.J. TRAYSTMAN (With 2 Figures) 436

Transients and Steady State of Chloride-Bicarbonate Relationships of Brain Extra-Cellular Fluid
H.H. LOESCHCKE and H.R. AHMAD (With 4 Figures) 439

Subject Index 449

List of Participants 455

Contributors

You will find the addresses at the beginning of the respective contribution

AHMAD, H.R. 439
AMAND, B. 36
ARNONE, A. 67
BARNIKOL, W.K.R. 102
BAUER, C. 84
BAUMANN, R. 114
BERTINI, I. 151
BIDANI, A. 306, 356
BRILEY, P.D. 67
BROWN, R.D. 238
BURKHARD, O. 102
BURSAUX, E. 122
COLEMAN, J.E. 133
CRANDALL, E.D. 285, 306, 356
DONALDSON, T.L. 23
EFFROS, R.M. 339, 353
ENGLAND, R.D. 106
FORSTER, R.E. 177, 285
FRIEND, S.H. 89
FRÖMTER, E. 419
GAY, C.V. 399
GIRARD, H. 390
GOODMAN, M. 165
GROS, G. 36, 410
GROS, H. 36
GUPTA, R.K. 273
GURD, F.R.N. 89
GURD, R.S. 106
GURTNER, G.H. 343, 436
HAYES, D.F. 106
HENDERSON, L.E. 154
HEWETT-EMMETT, D. 165
HILL, E.P. 379
HOLLAND, R.A.B. 299
ITADA, N. 177, 285
JACOB, G.S. 238
KANNAN, K.K. 184
KASTENDIECK, E. 390
KAWASHIRO, T. 58
KHALIFAH, R.G. 206, 226
KLOCKE, R.A. 331
KNOCHE, W. 3
KOENIG, S.H. 238
LAVALETTE, D. 36
LEDER, O. 406
LIEBER, M. 339
LINDSKOG, S. 230
LOESCHCKE, H.H. 321, 439
LONGO, L.D. 379
LUCHINAT, C. 151
LUTTMANN, A. 321
MAKHZOUMI, H.M. 436
MAREN, T.H. 366, 426
MATTHEW, J.B. 89
MATZKE, P. 122
MELDON, J.H. 49
MEYER, M. 351, 359
MOLL, W. 390
MORLEY, P.J. 226
MORROW, J.S. 89
MÜCKENHOFF, K. 321
OBAID, A.L. 285
O'BRASKY, J.E. 306
PERRELLA, M. 75
PESANDO, J.M. 273
PIIPER, J. 58, 351, 359
POCHON, F. 36
PONTE, J. 315
POYART, C. 122
PURVES, M.J. 315
QUINN, J.A. 23
ROGERS, P.H. 67

ROSSI-BERNARDI, L. 75
ROTHGEB, T.M. 106
SCHEID, P. 58, 351, 359
SCHMID, A. 410
SCHULTZ, J.S. 15
SCOZZAFAVA, A. 151
SHAPIRO, L. 339
SIFFERT, W. 410
SILVERMAN, D.N. 254, 291, 339
STROUP, S.K. 165
SWENSON, E.R. 366
TASHIAN, R.E. 165
TRAYSTMAN, R.J. 343, 436
TU, C.K. 254, 291
VOGH, B.P. 426
WARD, R.L. 262
WITTEBORT, R.J. 89, 106
WOOLLEY, P. 216
WYNNS, G.C. 254
YU, Y.-S.L. 165

Chemical Reactions of CO_2 in Water

Chairman: J. PIIPER

Chemical Reactions of CO_2 in Water

W. KNOCHE [1]

1 Introduction

The gas CO_2 is quite soluble in water in which more than 99% exists as the dissolved gas and less than 1% as carbonic acid H_2CO_3, which partly dissociates to give H^+, HCO_3^-, and CO_3^{2-}. The subsequent contributions deal with the central role CO_2 plays in the living organism, whereas here we restrict ourselves to the reaction between CO_2 and water.

CO_2 contributes 0.033% to the composition of the atmosphere, which means that its concentration in air at 10°C is about $1.4 \cdot 10^{-5}$ mol dm^{-3}. At this temperature its solubility in water is 0.064 mol per kg water, at a partial pressure of 1 atm CO_2. At such low pressure the solubility follows Henry's law, i.e., the concentration of a gas in solution is directly proportional to the partial pressure of the gas in the vapor phase in equilibrium with that solution. Therefore $2.1 \cdot 10^{-5}$ mol dm^{-3} CO_2 are dissolved in water at the partial pressure of CO_2 in air ($3.3 \cdot 10^{-4}$ atm) and CO_2 is nearly evenly distributed between air and water. In that respect it differs much from other gases, e.g., the absorption coefficient A given as the ratio (mol/l in solution)/(mol/l in gas phase) is A = 0.049 for O_2 and A = 0.024 for N_2. The solubility and absorption coefficients of CO_2 in water for different temperatures are given in Table 1. The concentration of CO_2 in blood is about $1.3 \cdot 10^{-3}$ mol dm^{-3} corresponding to a partial pressure of about 0.05 atm CO_2.

Table 1. Solubility Q = mol CO_2 dissolved per kg H_2O at a partial pressure of 1 atm CO_2, and absorption coefficient A = (mol CO_2 dissolved per l water)/(mol CO_2/l in the gas phase); (Harnad and Davies, 1943)

T°C	Q mol/kg	A
0	0.0770	1.725
10	0.0640	1.247
20	0.0393	0.942
30	0.0302	0.750
40	0.0241	0.615
50	0.0198	0.519

1 Max-Planck-Institut für biophysikalische Chemie, 3400 Göttingen/FRG

2 Equilibrium of the System CO_2 /Water

Since carbon dioxide can hydrate and twice dissociate in water, the reaction scheme may be written as

$$CO_2 + H_2O \underset{k'_1}{\overset{k_1}{\rightleftharpoons}} H_2CO_3 \underset{k'_2}{\overset{k_2}{\rightleftharpoons}} HCO_3^- + H^+ \rightleftharpoons CO_3^{2-} + 2H^+ \tag{1}$$

The equilibrium is determined by the three equilibrium constants

$$K_h = \frac{[H_2CO_3]}{[CO_2]} \tag{2}$$

$$K_a = \frac{[HCO_3^-][H^+] \cdot f_\pm^2}{[H_2CO_3]} \tag{3}$$

$$K_{a2} = \frac{[CO_3^{2-}][H^+] \cdot f_{2-} f_+}{[HCO_3^-] f_-} \tag{4}$$

and by the total concentration of CO_2 present in solution

$$c_{tot} = [CO_2] + [H_2CO_3] + [HCO_3^-] + [CO_3^{2-}]. \tag{5}$$

f_i are activity coefficients, the activities of the uncharged species CO_2 and H_2CO_3 are taken to be equal to their concentrations, and the activity of water is unity. c_{tot} can directly be determined by experimental measurement. The hydrogen-ion activity is obtained from the pH of the solution and conductivity measurements give information about the concentrations of the charged species H^+, HCO_3^-, and CO_3^{2-}. However, since it is not possible to differentiate between CO_2 and H_2CO_3, only the "over-all" first ionisation constant

$$K_1 = \frac{[H^+][HCO_3^-] f_\pm^2}{[CO_2] + [H_2CO_3]} \tag{6}$$

can be obtained which is related to K_a and K_h by

$$K_1 = \frac{K_a \cdot K_h}{1 + K_h} \approx K_a \cdot K_h \tag{7}$$

since $K_h \ll 1$ as already stated in the introduction. The values of the dissociation constants at 25°C are $K_1 = 4.45 \cdot 10^{-7}$ mol dm^{-3} and $K_{a2} = 4.70 \cdot 10^{-11}$ mol dm^{-3} Blood has a pH of about 7.4, and therefore the second dissociation step does not play a role in physiological systems and in the following we consider only the first two reaction steps in scheme (1). Kinetic methods enable K_1 to be split into K_h and K_a as will be discussed below.

Carbon dioxide functions as a pH buffer in blood. A non-volatile compound buffers best if it is at 50% in its basic form and at 50% in its acidic form, whereas for the system CO_2 in blood it is $[HCO_3^-] : ([CO_2] + [H_2CO_3]) \approx 20 : 1$ (This ratio is estimated from K_1, pH, and $f_\pm^2 \approx 0.6$ for an ionic strength of 0.16 in blood). Nevertheless optimal buffer conditions are realized since the small concentration of acid $[CO_2] + [H_2CO_3]$ in the blood is maintained by exchange with the enormous CO_2 reservoir of the environment.

3 Kinetics of the Hydration of CO_2

From kinetic measurements the rate constants k_1 of the hydration and k_1' for the dehydration reactions can be obtained. The hydration constant is then given by

$$K_h = k_1/k_1' \tag{8}$$

and K_a is determined by Eq. (7). Several authors have determined these constants, and the results have been summarized by Kern (1960), Edsall (1968), and Patel et al. (1973). We applied the pressure-jump relaxation technique for a redetermination, since the values obtained for k_1, K_a, and K_h by different authors differ by more than a factor of two.

3.1 Pressure-Jump-Technique

In a pressure-jump relaxation experiment the pressure applied to a system in which a chemical equilibrium is established is suddenly changed. The equilibrium constant depends on pressure according to

$$\left(\frac{\partial \ln K}{\partial P}\right)_T = -\frac{\Delta V^0}{RT} \tag{9}$$

where ΔV^0 is the volume of reaction. Therefore the change in pressure is accompanied by a change in equilibrium constant and in equilibrium concentrations and the reaction has to proceed in one direction in order to obtain the new equilibrium position. The reactions proceeding can be followed if they are slow compared with the time needed for the pressure change. In a pressure jump experiment the concentrations reach their new equilibrium values exponentially, since the equilibria are only slightly shifted. The time constant of the exponential is called the relaxation time and depends on rate constants and equilibrium concentrations of the reaction studied.

In Fig. 1 the change of proton concentration is shown schematically after a sudden pressure drop at time $t = 0$ applied to an aqueous solution of CO_2 (the second dissociation is not considered). In a reaction which is too fast to be observed with this rechnique the equilibrium between carbonic acid and protons and bicarbonate ions is established causing a change in proton concentration of δ_1 $[H^+]$. In a second much slower reaction step the equilibrium between CO_2 and H_2CO_3 is reached.

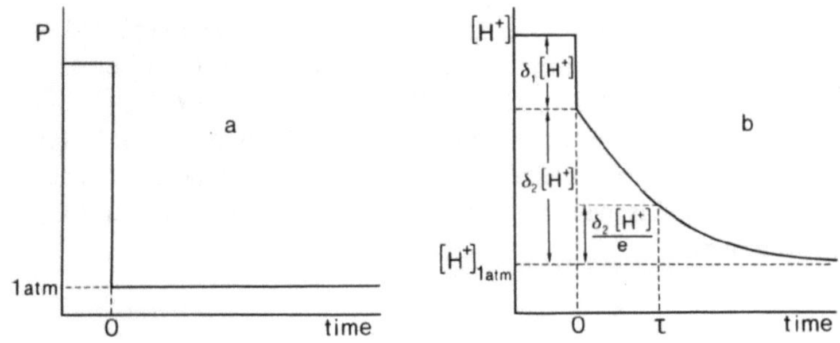

Fig. 1. Change of pressure a and of proton concentration b at a pressure jump experiment applied to an aqueous solution of CO_2

Since in this time region H_2CO_3 is in equilibrium with H^+ and HCO_3^-, also $[H^+]$ changes during the slow reaction and it is given by

$$[H^+] = [H^+]_{1\ atm} + \delta_2\ [H^+] \exp(-t/\tau). \tag{10}$$

$[H^+]$ can be followed conductometrically.

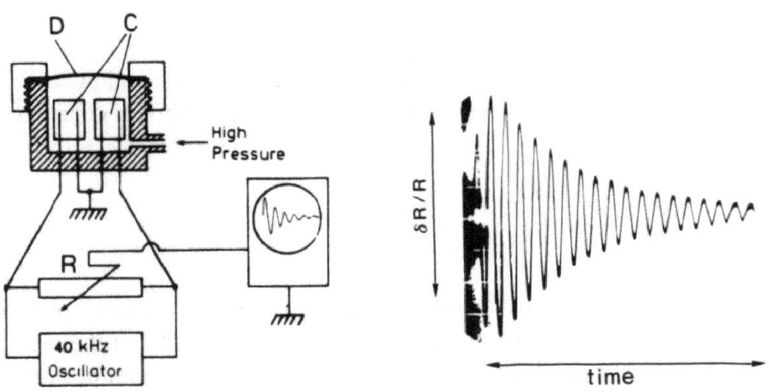

Fig. 2. Pressure-jump apparatus (schematic)

An apparatus which can be used for this pressure-jump experiment is shown schematically in Fig. 2 (Knoche and Wiese, 1973). Inside the autoclave (where the pressure can be changed rapidly) two identical conductivity cells C are mounted. One of these is filled with the solution to be examined and the second with a solution having the same electrical conductivity but showing no relaxation. By comparing the changes in resistance of the two cells, disturbances caused by temperature fluctuations are largely cancelled. The cells are closed with soft polyethylene membranes. The autoclave is filled with a liquid (e.g., water) and closed with a rupture disc (D). Together with the potentiometer the two cells form a Wheatstone bridge operated at a frequency of 40 kHz.

The bridge is tuned at the ambient pressure of 1 atm. Then the pressure is increased slowly by pumping more liquid into the autoclave. Thereby the equilibrium of the reaction studied is shifted and the conductivity bridge is unbalanced. At about 100 atm the rupture disc bursts spontaneously and the pressure in the cells drops to 1 atm within about 50 μs. This time interval is needed for the bursting of the metal membrane. The fast pressure change generates a voltage peak across a piezoclectric condenser which is used to trigger the oscilloscope. The trace on the oscilloscope screen shows how the investigated solution regains equilibrium at the ambient pressure of 1 atm. The relaxation oscillograms as seen in Fig. 2b shows the exponential decay of an ac signal. The shortest relaxation time measurable with this technique is about 50 μs. There are no limitations to long times and relaxation times of 20 s have been measured conveniently.

3.2 Relaxation Time for the Hydration of CO_2

Neglecting the second dissociation step, the reaction scheme is given as (see Eigen et al., 1961)

$$\begin{array}{ccc} CO_2 + H_2O & \underset{k_1'}{\overset{k_1}{\rightleftharpoons}} & H_2CO_3 \\ {}_{k_3'}\nwarrow\!\!\searrow^{k_3} & & {}^{k_2}\swarrow\!\!\nearrow_{k_2'} \\ & H^+ + HCO_3^- & \end{array} \tag{11}$$

At pH > 8 the reaction of CO_2 with OH^- has to be considered too which, however, does not contribute to the rate of the reaction in the pH range of this study (Sirs, 1958). The rate law for the hydration/dehydration reaction is

$$\frac{d\,[CO_2]}{dt} = -k_1\,[CO_2] + k_1'\,[H_2CO_3] - k_3\,[CO_2] + k_3'\,[H^+]\,[HCO_3^-] \tag{12}$$

(The activity of water is unity).

On the pressure-jump time scale the dissociation equilibrium is established, i.e.,

$$[HCO_3^-] \cdot [H^+] = K_a\,[H_2CO_3]. \tag{13}$$

Activity coefficients have not to be taken into account, since measurements were performed using very dilute solutions. The solutions were prepared by dissolving CO_2 in water. Therefore

$$[H^+] = [HCO_3^-]. \tag{14}$$

In a pressure jump experiment the total concentration of CO_2 stays constant, i.e.,

$$[CO_2] + [H_2CO_3] + [HCO_3^-] = \text{const} \tag{15}$$

combining Eqs. (12) to (15) yields Eq. (10) with

$$1/\tau = k_1 + k_1'\,\frac{2\,[H^+]}{K_a + 2\,[H^+]} + k_3 + k_3'\,\frac{2\,[H^+]\,K_a}{K_a + 2\,[H^+]} \tag{16}$$

[for details of the derivation of Eq. (16) see e.g., Strehlow and Knoche, 1977]

By inserting

$$k_3' = k_3 / K_1 \tag{17}$$

and Eqs. (7) and (8) into Eq. (16) we get

$$1/\tau = (k_1 + k_3) \left[1 + \frac{2\,[H^+]}{K_1 + 2\,[H^+]\,K_h}\right] \tag{18}$$

K_1 is known from potentiometric measurements and $k_1 + k_3$ and K_h have to be obtained from the kinetic measurements. Edsall has chosen the term k_{CO_2} for the sum of the rate constants $k_1 + k_3$.

The carbon dioxide solutions were prepared by bubbling CO_2 through triply distilled water and then diluting to the desired concentration. The proton concentration was determined by measuring the conductance of the solutions which is

$$\kappa = \lambda_H\,[H^+] + \lambda_{HCO_3}\,[HCO_3^-] \tag{19}$$

with $\lambda_H = 349.8$ and $\lambda_{HCO_3} = 44.5\ cm^2\ ohm^{-1}\ mol^{-1}$ at 25°C, and by applying Eq. (14)

$$\kappa = (\lambda_H + \lambda_{HCO_3})\,[H^+] \tag{20}$$

The concentration of CO_2 chosen was sufficiently high so that the conductance due to the dissociation of water and to impurities does not seriously affect the results. On the other hand the highest concentration of H^+ was about 10^{-4} mol dm^{-3} so that the equivalent ionic conductances λ can be taken to be equal to those at infinite dilution. The conductance was measured in the pressure jump cell so that losses of CO_2, which are difficult to avoid when handlung the solution, did not influence the determinations of concentration.

In the concentration range studied the "one" in the brackets of Eq. (18) may be neglected since $K_1 \ll [H^+]$ and $K_h \ll 1$. Thus we get

$$\tau = \frac{1}{k_{H_2CO_3}} + \frac{K_a}{k_{H_2CO_3} \cdot 2\,[H^+]} \tag{21}$$

with

$$k_{H_2CO_3} = k_{CO_2}/K_h = k_1' + K_a k_3' \tag{22}$$

From the plot in Fig. 3 we get at 25°C $k_{H_2CO_3} = (28 \pm 3)\ s^{-1}$ and $K_a = (3.2 \pm 0.4) \cdot 10^{-4}$ mol dm^{-3}. K_h is calculated from Eq. (7) to $K_h = (1.4 \pm 0.2) \cdot 10^{-5}$ and k_{CO_2} from Eq. (22) to $k_{CO_2} = (0.040 \pm 0.003)\ s^{-1}$. Since the concentration range shown in Fig. 3 is rather limited, a measurement was performed at very low concen-

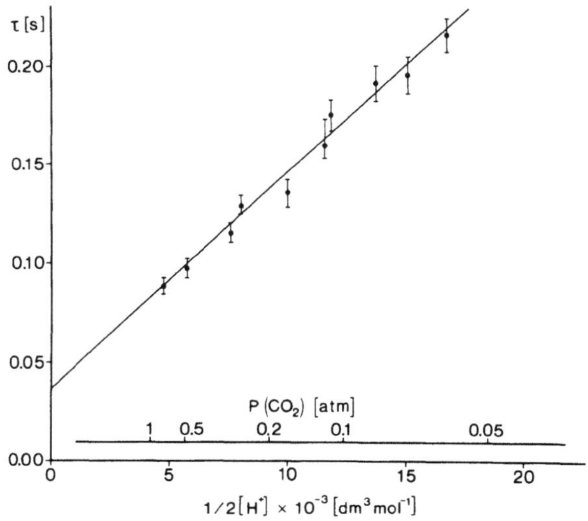

Fig. 3. Relaxation time vs. $1/2 \cdot [H^+]$ at 25°C according to Eq. (21)

tration of CO_2 where we obtained $\tau = 2$ s at 25°C for $[H^+] \approx 3 \cdot 10^{-6}$ mol dm^{-3} in agreement with reaction scheme (11) and the values obtained for $k_{H_2CO_3}$ and K_a. However, we did not include this point into the numerical evaluation because of the relatively large inaccuracy of the determination of the proton concentration.

Discussion

$k_{H_2CO_3}$ and K_a are obtained according to Eq. (21) from slope and intercept of the straight line in Fig. 3. K_h is calculated from these constants and the well-known equilibrium constant K_1, and finally Eq. (22) yields $k_{CO_2} = (0.040 \pm 0.003)$ s^{-1} in agreement with values obtained at 25°C by other authors in recent years: (0.0358 ± 0.0015) s^{-1} (Ho and Sturtevant, 1963); (0.0375 ± 0.002) s^{-1} (Gibbons and Edsall, 1963); 0.039 s^{-1} (Kernohan, 1965); (0.037 ± 0.002) s^{-1} (Khalifah, 1971). The hydration rate does not depend on ionic strength since only uncharged species are involved.

Our value for the rate constant of dehydration, $k_{H_2CO_3} = (28 \pm 3)$ s^{-1} agrees with the values (25.5 ± 1.5) s^{-1} (Rossi-Bernardi and Berger, 1968), and 29.4 s^{-1} (Patel et al., 1973). However, these results are in contradiction to values for $k_{H_2CO_3}$ of 13.7 s^{-1} at ionic strength I = 0.055 (Gibbons and Edsall, 1963); 18 s^{-1} for $I \to 0$ (Magid and Turnbeck, 1968), and 31 s^{-1} at 37°C and I = 0.15 (Swenson and Maren, 1978). However, these last three authors measured the ratio $k_{H_2CO_3}/K_a$ and calculated $k_{H_2CO_3}$ using the value of $K_a = 1.72 \cdot 10^{-4}$ mol dm^{-3} as determined by Wissbrun et al. (1954). If we apply $K_a = 3.2 \cdot 10^{-4}$ mol dm^{-3} as obtained in this study, then we get $k_{H_2CO_3} = 33.5$ s^{-1} from the value of Magid and Turnbeck, which is close to our value. The difference between the results of Magid and Turnbeck and those of Gibbons and Edsall can be explained by the different ionic strength at which the results were obtained; the former refers to I = 0 and the latter to I = 0.055.

Thus the discrepancy in the values reported for $k_{H_2CO_3}$ may be due to the use of an erroneous value of K_a.

The hydration of CO_2 may be compared with the hydration of other carbonyl compounds which have also been studied using the pressure jump relaxation technique (Buschmann and Knoche, 1980). Rate and equilibrium constants for a few compounds are shown in Table 2. The reaction rates are mainly determined by the charge density at the carbonylic carbon atom. A strong acid catalysis is observed for the hydration of carbonyl compounds which could not be detected in the case of CO_2. The hydration constant of CO_2 is very small compared to the hydration constants of the other carbonyls. This may reflect the fact that the collinear structure of the CO_2 is unique among all these compounds.

Table 2. Rate and equilibrium constants for the hydration of carbonyl compounds (Buschmann and Knoche, 1979)

CO_2	$k_1\ s^{-1}$	K_h
Propionaldehyde	0.021	1.24
Pyruvic acid	0.30	1.86
α-Keto glutaric acid	0.13	1.13
Choracetone	0.07	0.09
1,1,1 Trifluoro-acetone	91	35

The dissociation of carbonic acid is a protolytic reaction which occurs extremely rapidly, i.e., its rate is determined purely by the diffusion of the reacting species. In our case the proton and the carbonic ion are attracted by Coulombic interaction and the rate constant can be calculated by (Debye, 1942)

$$k_2' = \frac{4\pi N z_A z_B e_0{}^2}{\epsilon kT} \cdot \frac{D_A + D_B}{\exp\left(\frac{z_A z_B e_0{}^2}{\epsilon akT}\right) - 1} \tag{23}$$

N is Avogadro's number, z_A and z_B the valencies of the ions, e_0 the elementary charge, ϵ the dielectric constant, k Boltzmann's constant, D_A and D_B the diffusion coefficients of the ions, and a the distance of closest approach of the ions. For $z_A z_B = -1$ and $a \approx 0.4$ nm we obtain $k_2' \approx 6.5 \cdot 10^{-10} dm^3\ mol^{-1}\ s^{-1}$.

This constant has also been determined applying the elctrical field jump relaxation technique (where the electric field is changed suddenly) yielding $k_2' = 4.7 \cdot 10^{-10}$ $dm^3\ mol^{-1}\ s^{-1}$ (Eigen, 1963) in good agreement with the theoretical value. k_2 is calculated from k_2' and K_a.

The results for the hydration of CO_2 as obtained by chemical relaxation techniques are summarized in Table 3. A splitting of k_{CO_2} into k_1 and k_3 and $k_{H_2CO_3}$ into k_1' and k_3' is not possible from the kinetic measurements reported here.

Table 3. Rate and equilibrium constants for the reaction of CO_2 with water at 25°C and ionic strength I = 0

k_{CO_2}	$(0.040 \pm 0.003)\ s^{-1}$
$k_{H_2CO_3}$	$(28 \pm 3)\ s^{-1}$
K_h	$(1.4 \pm 0.2) \cdot 10^{-3}$
k_2	$1.5 \cdot 10^7\ s^{-1}$
k_2'	$4.7 \cdot 10^{10}\ dm^3\ mol^{-1}\ s^{-1}$
K_a	$(3.2 \pm 0.4) \cdot 10^{-4}\ mol\ dm^{-3}$

References

Buschmann HJ, Knoche W (1979) The reversible hydration of carbonyl compounds in aqueous solution. Ber Bunsenges Phys Chem (in press)

Debye P (1942) Reaction rates in ionic solutions. Trans Electrochem Soc 82: 265–272

Edsall JT (1968) Carbon dioxide, carbonic acid, and bicarbonate ion: Physical properties and kinetics of interconversion I. In: CO_2, chemical, biochemical and physiological aspects. Nasa SP 188, Washington DC, p 15–27

Eigen M (1963) Protonenübertragung, Säure-Base-Katalyse und enzymatische Hydrolyse, Teil 1: Elementarvorgänge. Angew Chem 75: 489–508

Eigen M, Kustin K, Maass G (1961) Die Geschwindigkeit der Hydratation von SO_2 in wäßriger Lösung. Z Phys Chem NF 30: 130–136

Gibbons BH, Edsall JT (1963) Rate of hydration of carbon dioxide and dehydration of carbonic acid at 25°C. J Biol Chem 238: 3502–3507

Harnad HS, Davies R (1943) The ionization constant of carbonic acid in water and the solubility of carbon dioxide in water and aqueous salt solutions from 0° to 50°C. J Am Chem Soc 65: 2030–2037

Ho C, Sturtevant JM (1963) The kinetics of the hydratation of carbon dioxide at 25°C. J Biol Chem 238: 3499–3501

Kern DM (1960) The hydration of carbon dioxide. J Chem 37: 14–23

Kernohan JC (1965) The pH-activity curve of bovine carbonic anhydrase and its relationship to the inhibition of the enzyme by anions. Biochim Biophys Acta 96: 304–317

Khalifah RG (1971) The carbon dioxide hydration activity of carbonic anhydrase. I. Stop-flow kinetic studies of the native human isoenzym B and C. J Biol Chem 246: 2561–2573

Knoche W, Wiese G (1973) An improved apparatus for pressure-jump relaxation measurements. Chem Instrum 5: 91–98

Magid E, Turnbeck BO (1968) The rates of the spontaneous hydration of CO_2 and the reciprocal reaction in neutral aqueous solutions between 0° and 38°C. Biochim Biophys Acta 165: 515–524

Patel RC, Boe RJ, Atkinson G (1973) The CO_2-water system, Part I. J Sol Chem 2: 367–372

Rossi-Bernardi L, Berger RL (1968) The rapid measurements of pH by the glass electrode. The kinetics of dehydration of carbonic acid at 25°C and 37°C. J Biol Chem 243: 1297–1302

Sirs JA (1958) Electrometric stopped-flow measurements of rapid reactions in solution, Part 1. Conductivity measurements. Trans Faraday Soc 54: 201–206

Sirs JA (1958) Electrometric stopped-flow measurements of rapid reactions in solution, Part 2. Glass electrode pH measurements. Trans Faraday Soc 54: 207–212

Strehlow H, Knoche W (1977) Fundamentals of chemical relaxation. Verlag Chemie, Weinheim

Swenson ER, Maren T (1978) A quantitative analysis of CO_2 transport at rest and during maximal exercise. Respir Physiol 35: 129–159

Wissbrun KF, Frenck DM, Patterson A, Jr (1954) The true ionization constant of carbonic acid in aqueous solution from 5° to 45°C. J Phys Chem 58: 693–695

Facilitated Diffusion of CO_2

Chairman: J. PIIPER

Facilitation of CO_2 Through Layers with a Spatial Distribution of Carbonic Anhydrase

J.S. SCHULTZ [1]

1 Introduction

In previous papers (Suchdeo and Schultz, 1974; Schultz et al., 1974), we have shown that one of the major parameters that govern facilitated diffusion in menbranes is the Damköhler number for the system, kL^2/D. Calculation techniques were developed to estimate the facilitation of CO_2 through bicarbonate solutions with and without carbonic anhydrase. These experimental and theoretical studies primarily considered films with homogeneous composition. In this paper, we extend these approaches to films with a spatial distribution of carbonic anhydrase. Selegny et al. (1971), have performed some experiments of this type and Donaldson and Quinn (1975), have developed analytical solutions for isotopic tracer flux thorugh layered systems.

2 Theory

For purposes of calculating CO_2 transport through bicarbonate films, the reactions can be represented as follows (Suchdeo and Schultz, 1974):

$$CO_2 + H_2O \underset{k_{-1}}{\overset{k_1}{\rightleftharpoons}} H_2CO_3 \qquad (pH \leqslant 8)$$

$$CO_2 + OH^- \underset{k_{-2}}{\overset{k_2}{\rightleftharpoons}} HCO_3^- \qquad (pH \geqslant 10)$$

$$\frac{(H^+)(HCO_3^-)}{(H_2CO_3)} = K_3$$

$$\frac{(H^+)(CO_3^=)}{(HCO_3^-)} = K_4$$

$$(H^+)(OH^-) = K_6$$

1 Department of Chemical Engineering, University of Michigan, Ann Arbor, Michigan 48109/ USA

and with carbonic anhydrase,

$$CO_2 + H_2O + E \overset{K_{ms}}{\rightleftharpoons} X \underset{k_p'}{\overset{k_p}{\rightleftharpoons}} HCO_3^- + H^+ + E$$

$(E = E^-, \text{ or } EH)$

$$E^- + H^+ \overset{K_H}{\rightleftharpoons} EH$$

$$E + HCO_3^- \overset{K_{mi}}{\rightleftharpoons} EHCO_3^-$$

Taken together, these kinetic mechanisms lead to the following diffusion-reaction equations.

$$\frac{d^2\overline{C}_1}{dx_2} = \alpha_1{}^2\,\phi_1 + \alpha_e{}^2\,\phi_e \tag{1}$$

where $\alpha_1{}^2 = \dfrac{k_1 L^2}{D_1}$ and $\alpha_e{}^2 = \dfrac{k_p E_0 L^2}{D_1}$

and for the noncatalyzed reaction

$$\phi_1 = (\overline{C}_1 - \frac{mC_3{}^2}{1 - C_3})\left[1 + n(\frac{1 - C_3}{C_3})\right] \tag{2}$$

while for carbonic anhydrase catalyzed reaction

$$\phi_e = (\overline{C}_1 - \frac{mC_3{}^2}{1 - C_3})\,/\,(1 + \overline{C}_1\,\frac{(CO_2)_{x=0}}{K_{ms}} + C_3\,\frac{C_6^\circ}{K_{mi}})\,(1 + \frac{C_3}{(1 - C_3)}\,\frac{2K_4}{K_H}) \tag{3}$$

where $\overline{C}_1 = (CO_2)\,/\,(CO_2)_{x=0};\ m = \dfrac{2C_6^\circ}{(KH)\,P_0}$

$$n = \frac{k_2 K_2}{2k_1 K_4}\ ;\ C_6^\circ = NaHCO_3;\ C_3 = \frac{(HCO_3^-)}{(NaHCO_3)}$$

L = layer thickness; D_1 = diffusivity of CO_2

Typical theoretical and experimental results (Suchdeo and Schultz, 1974) are shown in Figs. 1 and 2.

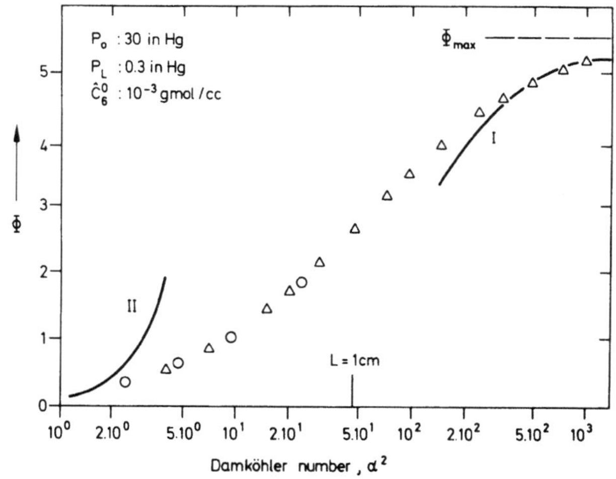

Fig. 1. Comparison of theoretical methods for estimating the diffusion of carbon dioxide across bicarbonate films. (From Suchdeo and Schultz, 1974). The facilitation factor, Φ, defined as

$$\Phi = \left[\frac{CO_2 \text{ flux with carrier}}{CO_2 \text{ flux, no carrier}} - 1\right]$$ is plotted against the Damköhler number α^2. The figure also illustrates the effect of membrane thickness, L, and reaction rate constants k_1 and k_pE_0. See Eq. (1) for definition of Damköhler numbers. Φ_{max} is maximum facilitation factor in the limit of chemical reaction equilibrium throughout the membrane. ○ Quasilinearization – numerical solution. △ Kreuzer-Hoofd approx. – Limiting solutions. *I* Near-equilibrium. *II* Near-diffusion

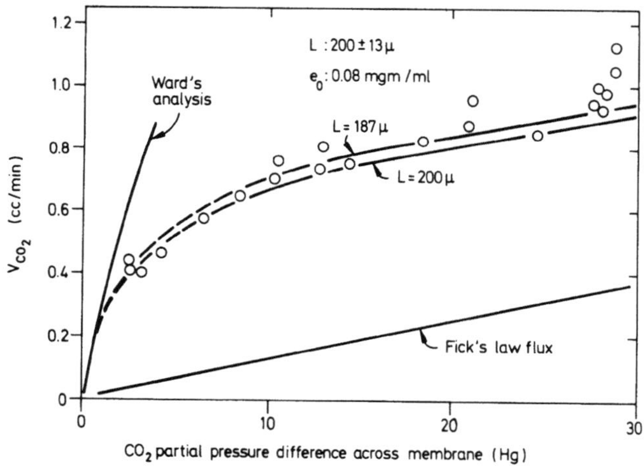

Fig. 2. Comparison of theoretical methods for calculating facilitated carbon dioxide flux and experimental measurements for the case of bicarbonate solutions with carbonic anhydrase. *Solid lines* through the points calculated by a modified Kreuzer-Hoffd procedure (Suchdeo and Schultz, 1974)

Figure 1 shows the effect of the Damköhler numer, $\alpha_1{}^2$, on facilitation $(\Phi = \frac{\text{flux of } CO_2 \text{ with carrier}}{\text{flux of } CO_2\text{, no carrier}} - 1)$. In the absence of carbonic anhydrase, only the first term ($\alpha_1{}^2\ \phi_1$) of the kinetic expressions in Eq. (1) is required. In particular, it is noteworthy that facilitation varies from zero (diffusion regime) to maximum (equilibrium regime) over three orders of magnitude in the Damköhler parameter. Thus, to change a given physical system that is in the diffusion regime to one in which facilitation dominates, requires a change in kinetic constants over several orders of magnitude. For the case of CO_2 transport, this dramatic increase in kinetics can be achieved by carbonic anhydrase.

For example, for a 200 μ film, CO_2, diffusion is approximately represented by the lower "Fick's Law" solid line in Fig. 2. Whereas, in the presence of carbonic anhydrase, the CO_2 flux is increased 3–10 times. These curves were calculated (Suchdeo and Schultz, 1974) using the second term, $\alpha_e{}^2\ \phi_e$, for the catalyzed reaction alone.

Both these examples show the importance of the Damköhler parameter on the extent of facilitation of CO_2 in bicarbonate solutions. Yet translating these results to physiological systems requires the consideration of another factor; namely, the spatial distribution of carbonic anhydrase. In the several tissues where CO_2 transport is important, e.g., lung, muscle, kidney, erythrocyte, carbonic anhydrase is localized to a particular region of what may be considered the CO_2 transfer pathway.

In order to assess the importance of CO_2 spatial distribution, we undertook some theoretical-numerical calculations using the quasi-linearization approach of Bellman that we had validated earlier (Suchdeo and Schultz, 1971). The system consisted of a 200 μ film saturated with 1 N $NaHCO_3$ solution, and with an imposed CO_2 concentration driving force of 50 mm Hg on one side of the film to 0.5 mm Hg on the other side. For calculational purposes, the film was divided into 32 elements, and the diffusion reaction Eqs. (1–3) linearized at each point. In order to simulate spatial distributions of the enzyme, the enzyme related kinetic expression ($\alpha_e^2\ \phi_e$) was included in the computational algorithm for selected points. Computations for groups of points representing from 1/8 to the entire film were selected as shown in the shaded rectangular areas in Fig. 3 and Fig. 4.

3 Results

The effect of carbonic anhydrase on increasing CO_2 transport is shown in the upper curve of Fig. 3. Over the range of enzyme concentrations studied, about a twofold increase in flux was calculated where the enzyme is uniformly distributed across the film. However, the effectiveness of the enzyme on increasing the flux is drastically reduced where there are regions of the film which do not contain carbonic anhydrase as shown in the lower curves of Figs. 3 and 4. Even in cases where the nonenzyme region was only 1/4 of the film thickness, a considerable reduction in flux is seen. The calculations show that the absence of carbonic anhydrase at either interfacial boundary causes a significant decrease in flux. In fact, increasing the enzyme-free fraction across the membrane shows relatively little additional reduction in flux (Fig. 5).

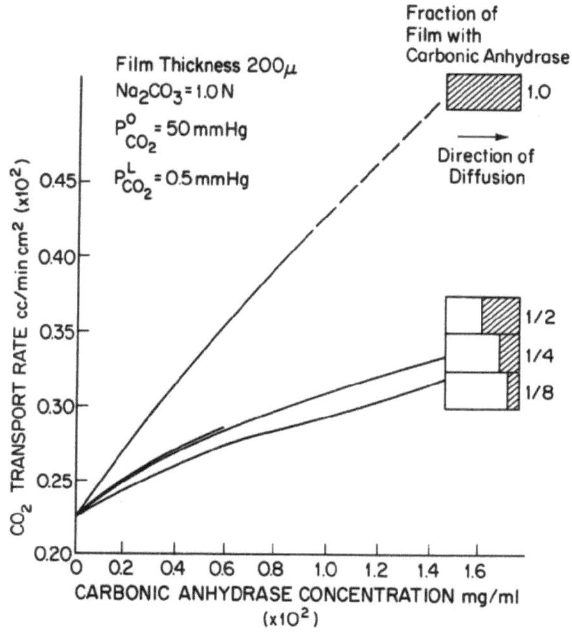

Fig. 3. Effect of spatial distribution of carbonic anhydrase within a film on the facilitated transport of carbon dioxide. Cases illustrating the effect of various fractions of the membrane containing carbonic anhydrase at various concentrations on the *downstream* side are shown. Calculations made by the quasilinearization method

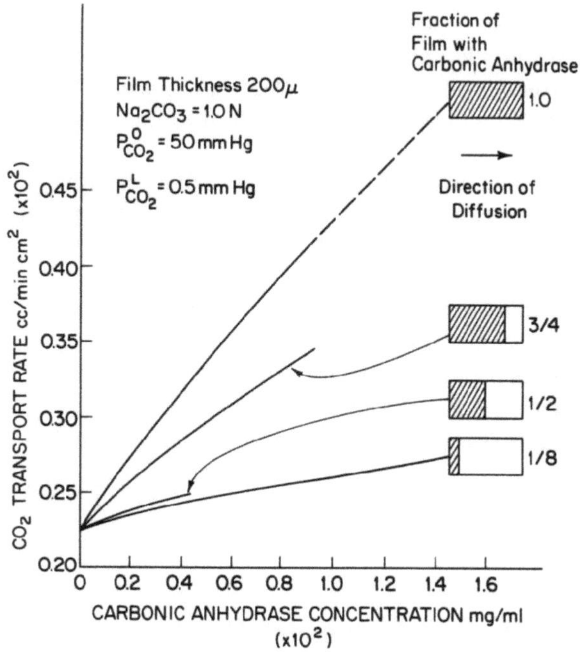

Fig. 4. Effect of spatial distribution of carbonic anhydrase within a film on the facilitated transport of carbon dioxide. Cases illustrating the effect of various fractions of the membrane containing carbon anhydrase at various concentrations on the *upstream* side are shown. Calculations made by the quasilinearization method

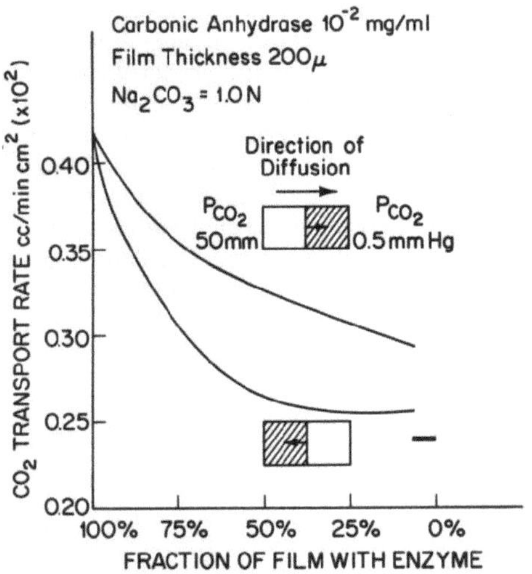

Fig. 5. Effect of relative spatial distribution of carbonic anhydrase in a membrane on facilitated transport of carbon dioxide, at a given concentration of carbonic anhydrase

If carbonic anhydrase is spatially distributed across the film, there are significant differences in the CO_2 concentration profiles depending on the direction of diffusion as related to the enzyme depleted regions. Fig. 3 shows the effect of having the enzyme on the downstream section of the membrane, whereas Fig. 4 shows the opposite condition, i.e., enzyme on the upstream section. It is quite apparent that carbonic anhydrase is more effective in providing CO_2 transport when it is on the downstream section of the film. This result might be considered unexpected since the slow step in facilitated CO_2 transport is the hydration reaction which occurs in the forward sense on the upstream portion of the film. Yet, if the so-called "back-pressure" effect is recalled, the importance of enzyme on the downstream side of the film can be rationalized.

Fig. 6 shows a reduction in the measured flux of CO_2 across a film when the downstream pressure increases. An increase in the apparent downstream CO_2 partial pressure reduces the driving force for HCO_3^- across the film, which in turn has a primary effect on flux across the film.

This asymmetry in facilitated transport due to nonuniformity in membrane structure is a general phenomenon (Schultz, 1971).

The effect of spatial variations in membrane enzyme on CO_2 concentrations profiles within the membrane is shown in Fig. 7. One obvious trend is that the symmetry of the concentration profiles is distorted when the enzyme is not uniformly distributed.

4 Conclusions

Carbonic anhydrase is most effective in promoting facilitated CO_2 transport when it is uniformly dispersed throughout the diffusion pathway. Therefore, tissues which have a spatial distribution of carbonic anhydrase will have a depressed facilitated flux

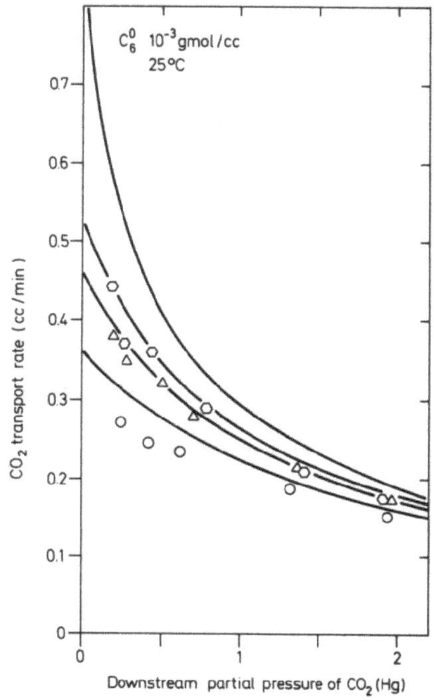

Fig. 6. Effect of downstream (back) partial pressure of carbon dioxide on facilitated diffusion of carbon dioxide across membranes containing a uniform distribution of carbonic anhydrase across the membrane. *Upper solid curve* is expected flux in the limit of high enzyme levels (equilibrium limit). Enzyme concentration mg/ml ○0.08, △0.32, ○ 0.64. Thickness 0.1 cm, P_0 28.3 in Hg. (Model calculations, Kreuzer-Hoofd method)

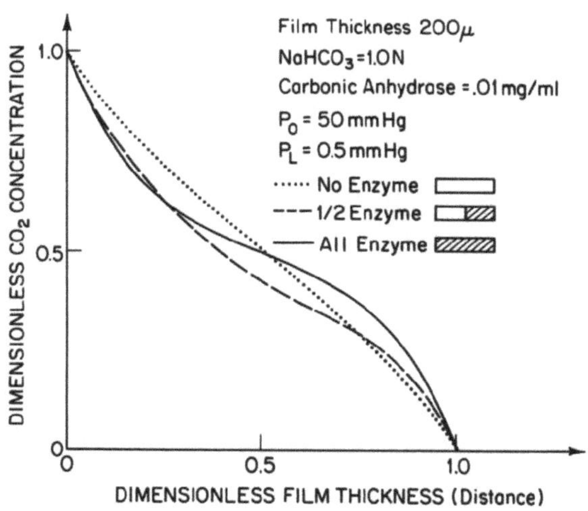

Fig. 7. Concentration profiles of carbon dioxide across membranes with different spatial distributions of carbonic anhydrase

as compared to a similar situation when the enzyme is uniformly dispersed. The flux will be asymmetric in that the effectiveness of the enzyme is somewhat greater when the enzyme is at the downstream end of the diffusion pathway. Although not treated in these calculations, an enzyme distribution in which both interfaces had enzyme with the center of the membrane enzymeless would probably behave like a film with a uniform distribution. This can be asserted based on previously published results (Suchdeo and Schultz, 1974) which show that the central portion of a facilitated diffusion membrane is effectively at equilibrium anyway and, thus, the enzyme in this portion of the film would have relatively little effect.

Acknowledgments. The previous work of S. Suchdeo was significantly drawn upon for this paper. B. Carnahan and S. Mansouri helped in the computer analysis. The work was partially supported by a grant from NSF No. 78-13316.

References

Donaldson TL, Quinn JA (1975) Carbon dioxide transport through enzymatically active synthetic membranes. Chem Eng Sci 30: 103

Schultz JS (1971) Passive asymmetric transport through biological membranes. Biophys J 11: 924–943

Schultz JS, Goddard JD, Suchdeo SR (1974) Facilitated transport via carrier mediated diffusion in membranes. Am Inst Chem Eng J 20: 417–445

Selegny E, Brown G, Thomas D (1971) Enzymatically active model membranes. Physiol Veg 9: 25

Suchdeo SR, Schultz JS (1971) Mathematical approaches to the analysis of facilitated transport phenomenon in membranes. Chem Eng Prog Symp Ser 67: 165–173

Suchdeo SR, Schultz JS (1974) Mass transfer of CO_2 across membranes: facilitation in the presence of bicarbonate ion and the enzyme carbonic anhydrose. Biochem Biophys Acta 352: 412–440

Facilitated Diffusion of CO_2 and Carbonic Anhydrase Activity

J.A. QUINN [1] and T.L. DONALDSON [2]

1 Introduction

If the rate of transport of a reactive permeant through an immobilized liquid layer is facilitated by reversible chemical reactions catalyzed by an enzyme contained within the layer, then measurement of the permeant flux can be used to determine kinetic properties of the enzyme. As opposed to conventional kinetic methods, the quantity measured is a steady-state flux rather than a transient change in concentration. Equations governing the flux can be solved analytically for first-order kinetics and kinetic parameters can be determined from straight-line plots of the data. In this study of CO_2 transport and carbonic anhydrase (CA) activity, the flux was measured using tagged $^{14}CO_2$. The results indicate a constant specific activity for CA over an 80-fold range in concentration extending to 4 mg/ml. Aspects of this study have previously been discussed in Donaldson (1974) and Donaldson and Quinn (1974, 1975).

2 Facilitated Transport of CO_2

Facilitated transport through a membrane is a simultaneous reaction and diffusion process in which the normal Fickian diffusion of a permeant is augmented by the parallel diffusion of products of a reaction between the permeant and membrane components. For the facilitated transport of CO_2 in an aqueous membrane, the reactions involved are (Edsall, 1969)

$$\begin{aligned} CO_2 + H_2O &\rightleftharpoons HCO_3^- + H^+ \\ CO_2 + OH^- &\rightleftharpoons HCO_3^- \end{aligned} \qquad (1)$$

The bicarbonate ion diffuses in parallel with the dissolved CO_2 and can augment the total CO_2 flux considerably under proper conditions. The relatively slow conversion of CO_2 to bicarbonate and the reverse reaction are catalyzed by CA. The maximum flux augmentation is limited, however, by chemical reaction equilibrium, and is thus related to the equilibrium constant

1 Department of Chemical and Biochemical Engineering, University of Pennsylvania, Philadelphia, Pennsylvania 19104/USA

2 Department of Chemical Engineering, University of Rochester, Rochester, New York 14627/USA

$$K_{EQ} = \frac{[HCO_3^-][H^+]}{[CO_2]} \quad (2)$$

The facilitated transport of CO_2 has been discussed in detail in reports by Longmuir et al. (1966), Ward and Robb (1967), Enns (1967), Suchdeo and Schultz (1974), Meldon (1973), Otto and Quinn (1971), Donaldson and Quinn (1974, 1975), Smith et al. (1977).

3 Experimental Method

The facilitated transport of tracer $^{14}CO_2$ through various types of membranes was measured in a diffusion cell equipped with a gas flow proportional counter and a Geiger-Müller tube mounted in the upper and lower chambers, respectively. The diffusion cell was similar to the cell of Enns (1967), and the complete apparatus is described more fully elsewhere (Donaldson and Quinn, 1975). A schematic drawing of the apparatus is shown in Fig. 1. We examined the facilitated transport of CO_2 in two membrane configurations featuring an immobilized liquid membrane. In the first, the bicarbonate solution was sandwiched as a thin layer between highly permeable polymer films. In these membranes we measured the facilitated flux due to the uncatalyzed reactions, as well as the catalytic effect of CA in homogeneous solution and bound to the supporting polymer films. In the second set of experiments, analogous measurements were made using Millipore filters impregnated with bicarbonate-enzyme solution and cross-linked protein membranes containing enzyme and bicarbonate. The results cover a wide range of enzyme concentrations in both types of membranes. Schematic representations of the various membranes are shown in Fig. 2.

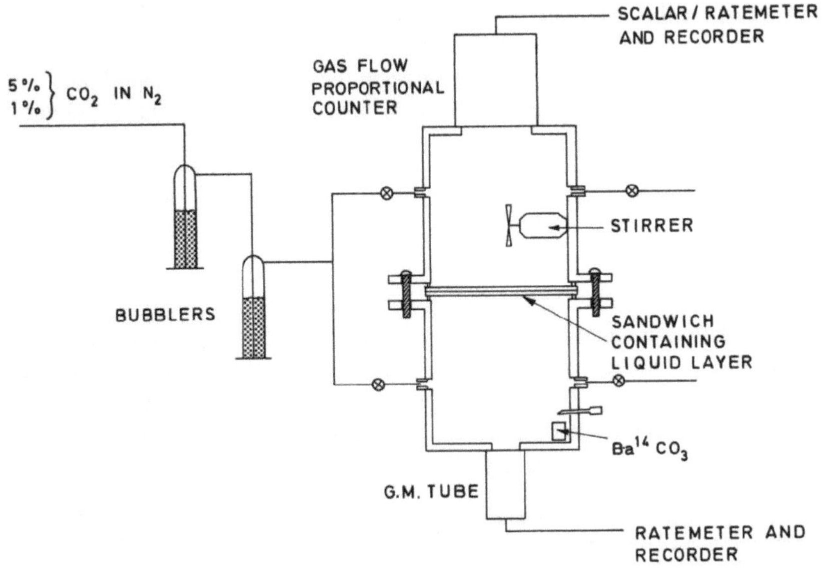

Fig. 1. Diffusion cell for tracer $^{14}CO_2$ diffusion through membranes

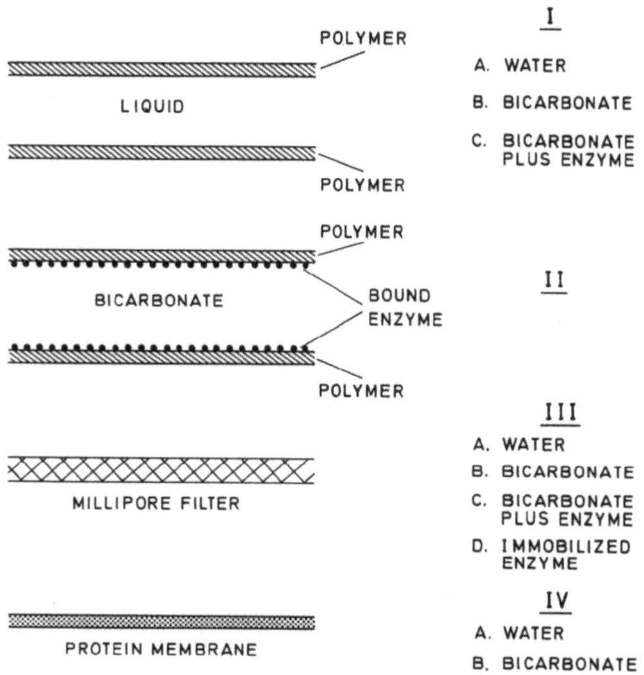

Fig. 2. The four types of immobilized liquid membranes investigated

The sandwich containing the thin liquid membrane, impregnated Millipore filter membrane, or protein membrane was placed in the diffusion cell and both chambers were purged with a gas mixture of 1% or 5% CO_2 in N_2 for approximately 30 min. There were no pH gradients across the membranes in these experiments, since both sides of the membranes were exposed to the same partial pressure of CO_2.

The bottom chamber was then closed off and a small amount of $^{14}CO_2$ (approximately 25 μCi) was generated by releasing a drop of HCl through a hypodermic needle into a cup containing $Ba^{14}CO_3$. The loss of $^{14}CO_2$ from the lower chamber by diffusion into the upper chamber was negligible during the experiment, and consequently the count rate in the lower chamber remained constant after the initial transient generation period.

The purge in the top chamber was continued for approximately another half-hour until the $^{14}CO_2$ count rate in the bottom chamber became constant. The top chamber was next sealed off and the rate of accumulation of $^{14}CO_2$ in the top chamber was recorded. The concentration of $^{14}CO_2$ in the bottom chamber was less than 1% of the CO_2 in the purge gas and consequently did not significantly alter the total CO_2 concentration. No supporting buffers or other ions were used; the enzyme substrate-product system, CO_2-HCO_3^-, served as a buffer. Bovine CA was obtained from Worthington Biochemical Corp. and used without further purification. The activity was determined by standard techniques (Wilbur and Anderson, 1948) and the activity specified by Worthington was verified. The enzyme purity was estimated to be 75% by ultraviolet absorption and atomic absorption for Zn (Donaldson, 1974).

3.1 Polymer-Liquid Membrane Sandwich

Two types of liquid membrane were studied. The first was the polymer-liquid sandwich similar to that described by Otto and Quinn (1971). The liquid membrane was held between two thin silicone co-polymer films, each approximately 35 μm thick, which were supported by Plexiglas discs with open holes for gas transfer. The polymer films were sealed to the Plexiglas and were reusable for a number of transport experiments. The thickness of the liquid membrane was 775 μm and was determined by the thickness of a stainless steel spacer plate placed between the polymer films; the immobilized liquid film was entrapped within holes in the stainless steel plate.

3.2 Millipore Filter Membranes

Millipore filter membranes (Type AA, 0.8 μm pore size) were soaked in the solution to be studied for at least 4 h prior to the transport experiments. The membranes were stored in the bicarbonate or bicarbonate/enzyme solutions and refrigerated at 6°C between experiments. Carbonic anhydrase was also immobilized within the matrix of Millipore filters, using a modification of the glutaraldehyde crosslinking technique of Broun et al. (1973). Filters were soaked for several hours in pH 7 phosphate buffer containing 4 mg/ml bovine serum albumin, 0.1 mg/ml CA, and 0.4% glutaraldehyde and then placed on glass plates at – 5°C overnight for completion of the cross-linking process. The filters were then rinsed, handled, and stored refrigerated in bicarbonate in the same manner as the other Millipore membranes and the enzyme-active polymer films.

4 Mathematical Model

At steady state the transport of each species in the membrane is governed by an equation of the form

$$D_i \frac{d^2 [i]}{dx^2} = R_i \tag{3}$$

where x is the coordinate in the direction of transport, D_i is the diffusivity of species i, the brackets refer to molar concentrations, and R_i is the net rate of depletion of species i by chemical reaction. This system of equations is solved subject to the boundary conditions of constant CO_2 concentrations at the membrane boundaries and no flux of bicarbonate ion at the membrane boundaries (Otto and Quinn, 1971; Donaldson and Quinn, 1975). The facilitation effect is expressed as the total flux of CO_2 through the membrane normalized with respect to the ordinary Fickian flux of molecular CO_2. The analytical expression for this ratio of fluxes is

$$\psi = \frac{1 + F}{1 + (F/\phi)\tanh\phi} \tag{4}$$

$$\text{where } F = \frac{D_{HCO_3^-} K_{EQ}}{D_{CO_2} [H^+]} \quad \text{and} \quad \phi = \frac{1}{2}\sqrt{\frac{k L^2}{D_{CO_2}} \left(\frac{1 + F}{F}\right)} \tag{5}$$

in which $D_{HCO_3^-}$ is the bicarbonate diffusivity, L is the diffusional path length (or membrane thickness), and k is the combined pseudo-first-order rate constant for the CO_2 hydration reactions in Eq. (1). The corresponding pseudo-first-order enzymatic rate constant is designated k_{enz}. As noted earlier, the pH is constant throughout the membrane under our experimental conditions, and hence the reaction rates of the tagged $^{14}CO_2$ species are first-order; and since $[CO_2] \gg [^{14}CO_2]$, the total CO_2 concentration is uniform throughout the membrane and the equilibrium expression in Eq. (2) is applicable. An alternate, more useful form of $K_{EQ}/[H^+]$ is its equivalent from the equilibrium expression, $[HCO_3^-] / [CO_2]$.

5 Results

All experiments were carried out at ambient temperature, 23° ± 1°C, and average values for D_{CO_2}, $D_{HCO_3^-}$, and CO_2 solubility of 1.85×10^{-5} cm^2/s, 1.07×10^{-5} cm^2/s, and 3.6×10^{-2} mol/cm^3 atm, respectively, were used (Donaldson and Quinn, 1975). No corrections were made for ionic strength effects; the ionic strength was 0.05 M or less, except for one series of experiments at 0.10 M bicarbonate.

5.1 No Facilitation

The transport of $^{14}CO_2$ was measured through polymer-liquid sandwiches and Millipore filters contaning acidified water of pH < 4. There is no capacitance for bicarbonate at this pH and the total transport is by simple molecular CO_2 diffusion alone. The membrane resistance for the polymer-liquid sandwiches agreed well with the predicted value based on the liquid membrane thickness and the permeability of CO_2 in water, and the permeability of CO_2 in the polymer (Donaldson, 1974). Otto and Quinn (1971) have also discussed diffusion through similar pure water membranes.

The effective diffusional path length through the Millipore filters can be estimated from the diffusion of tracer $^{14}CO_2$ through acidified water. An average path lenth of 245 μm (corrected for 82% porosity) was obtained for filters having a nominal thickness of 160 μm; the tortuosity is therefore about 1.53.

5.2 Uncatalyzed Reaction

The facilitation due to the uncatalyzed reaction was studied in the polymer-liquid sandwich with 0.01 M $KHCO_3$ and 5% CO_2. The pH was about 7, so the dominant reaction was the combination of CO_2 and H_2O. The average flux ratio, ψ, for a number of runs was 1.66. Assuming a value of 0.58 for $D_{HCO_3^-}/D_{CO_2}$, the rate constant k calculated from this result and Eq. (4) is 0.0351 s^{-1} at 23°C. Literature values for this rate constant vary considerably, and are reviewed by Edsall (1969) and Garg and Maren (1972). Edsall's own value is reported as 0.0375 s^{-1} at 25°C, and Maren's recent value is 0.036 s^{-1} at 25°C. Our result is consistent with these values when the 2°C temperature difference is considered.

5.3 Enzyme-Catalyzed Reaction

The flux ratio measured in liquid membranes and in Millipore filter membranes is shown in Fig. 3 as a function of bicarbonate concentration for a wide range of enzyme concentrations. The flux ratio is roughly independent of the enzyme concentration and the diffusional path length, or in other words, independent of the modulus ϕ. This result suggests that the reaction equilibrium limit has been reached. The broken line represents the approximate upper limit of the data based on Eq. (4) as $\phi \to \infty$.

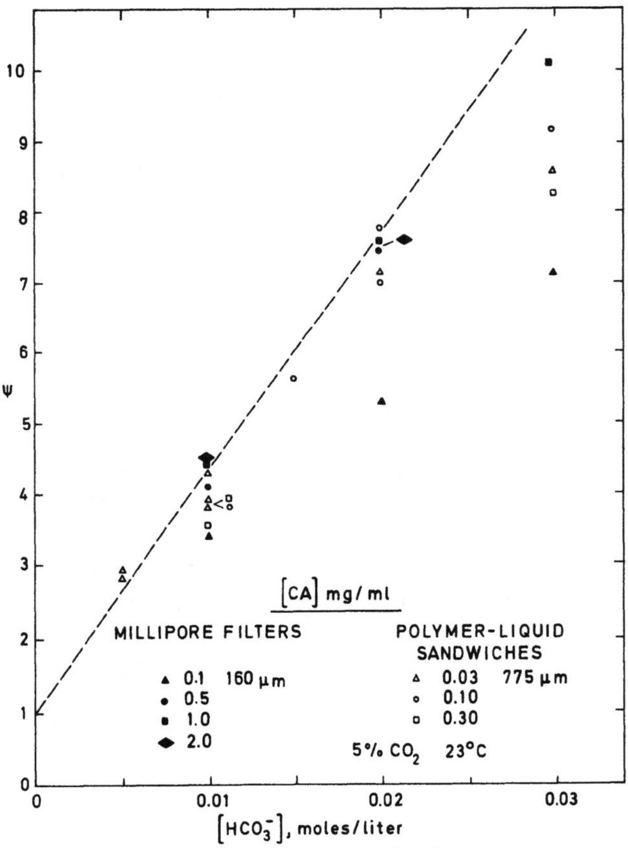

Fig. 3. The flux ratio of CO_2 measured in thin liquid membranes and Millipore filter membranes at several enzyme concentrations. The *broken line* is an approximation to the upper limit of the data corresponding to Eq. (4) as $\phi \to \infty$

No kinetic rate constants can be obtained from the data in Fig. 3 if the transport occurred at reaction equilibrium. To obtain rate data it was necessary to go to higher values of the flux ratio where the transport was reaction-rate limited. The data obtained with the Millipore filters and 0.1 mg/ml CA are already apparently reaction-rate limited to some extent. These experiments were extended to larger values of F (and consequently higher equilibrium limits) by lowering the CO_2 partial pressure from 5% to 1%. The increase in F was carried out in this manner rather than by increasing the bicarbonate concentration in order to minimize the variation in ionic strength.

The flux ratio in Millipore filters soaked in solutions containing 0.10 mg/ml CA is shown in Fig. 4 as a function of the parameter F. The solid line illustrates the trend of the data, and the broken line represents the reaction equilibrium upper limit ($\psi = 1 + F$ in the limit of large ϕ). The data in Fig. 4 fall well below the thermodynamic equilibrium limit, and the transport is therefore reaction-rate limited. The value of ψ is approaching the saturation plateau predicted by Eq. (4) at large F.

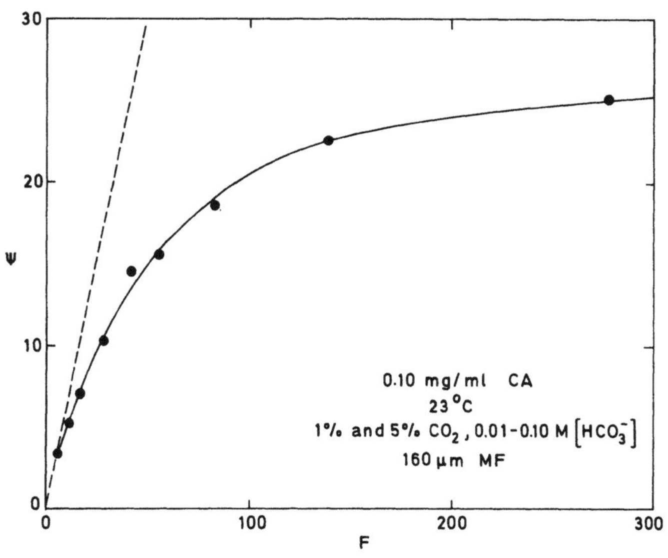

Fig. 4. Flux ratio of CO_2 measured in Millipore filter membranes at constant enzyme concentration showing reaction-rate-limited transport. The *broken line* is the reaction equilibrium limit. The *solid line* is simply a curve drawn through the points

Equation (4) is nonlinear in both F and ϕ, but it can be cast into a more nearly linear form by taking the reciprocal of both sides:

$$\frac{1}{\psi} = \frac{1}{1+F} + \left(\frac{F}{1+F}\right)\frac{\tanh \phi}{\phi} \qquad (6)$$

The parameter F is 25 or greater for a portion of the data in Fig. 4 and thus $F \gg 1$. The pH over this range is such that the enzyme activity is nearly maximal and constant. These data are plotted in the double reciprocal form of Eq. (6) in Fig. 5. The remainder of the data at lower F values cannot be used for this plot because they do not satisfy the criterion $F \gg 1$ and because of the variation of the enzyme activity with pH. The value of the modulus ϕ determined from the intercept in Fig. 5 is 29.4 and corresponds to a rate constant $k_{enz} = 110\ s^{-1}$. The half-time of the enzymatic reaction is thus about 7 ms, and the uncatalyzed reaction with a half-time of about 12 s is clearly negligible. The turnover number can now be calculated using a K_M of 12 mM (Kernohan, 1965), and an enzyme purity of 75%. The calculated turnover number is $5.3 \times 10^5\ s^{-1}$, which may be compared with Kernohan's value of $10 \times 10^5\ s^{-1}$ for the bovine enzyme measured by stopped-flow methods. The factor-of-two difference is most likely a result of bicarbonate inhibition (see discussion below).

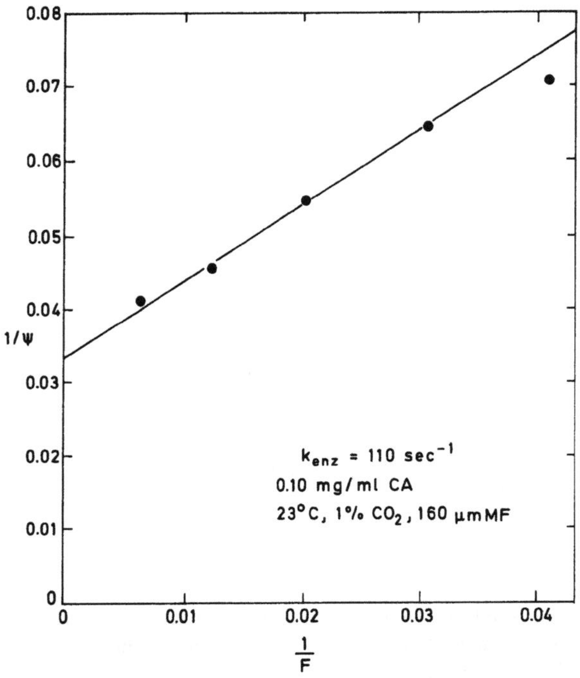

Fig. 5. Double reciprocal plot of large F data from Fig. 4. The *line* is the limiting linear form of Eq. (6)

Values of K_M cannot be obtained independently from the present transport data. The usual method for determining the Michaelis constant is to measure the reaction rate as a function of substrate concentration $[CO_2]$. With the tracer system, K_M could be determined in principle from a plot of $1/k_{enz}$ vs. $[CO_2]$. The Michaelis constant for the bovine enzyme has been thoroughly investigated by Kernohan (1964, 1965) and found to be 12 mM at 25°C and independent of pH.

5.4 pH Dependence of Enzyme Activity

The pH-dependent activity of CA is well known, and corresponds to the addition of a proton to the basic form of an ionizing group at low pH which inactivates the enzyme for the hydration reaction (Kernohan, 1964; DeVoe and Kistiakowsky, 1961; Roughton and Booth, 1946). This pH-dependent activity can be deduced from the data in Fig. 4. The pH for each point was calculated from the known bicarbonate and CO_2 concentration and the equilibrium relationship in Eq. (2) and the equilibrium constant K_{EQ} was corrected for ionic strength effects (Harned and Bonner, 1945). The flux ratios were converted to k_{enz} values, which were plotted as a function of pH. The sigmoidal result is shown in Fig. 6. The maximum k_{enz}, of course, is approximately 100 s^{-1}, which was also determined from the intercept in Fig. 5.

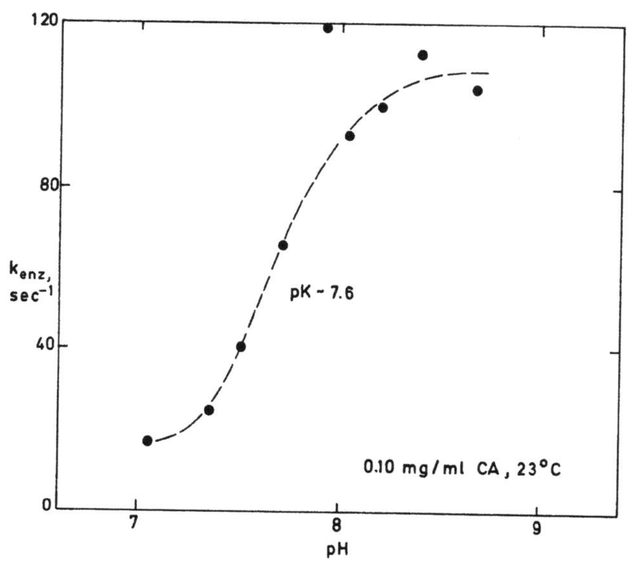

Fig. 6. Dependence of enzyme activity on pH

The apparent pK for the pH-dependent activity is about 7.6 under these experimental conditions. This value is somewhat higher than the apparent pK of 6.9 reported by Kernohan (1965). However, Kernohan's value was measured in the presence of chloride and no bicarbonate. The difference in apparent pK is probably due to the absence of chloride and the presence of bicarbonate in our membrane reaction system. Both of these ions are noncompetitive inhibitors, and are more effective at low pH than at high pH (Kernohan, 1965; Roughton and Booth, 1946). Unfortunately, a quantitative measure of the bicarbonate inhibition cannot be deduced from the data in Fig. 4 since these data represent a range of bicarbonate concentrations from 0.01 to 0.10 M in addition to the variation in pH. These transport experiments were not designed specifically to investigate the pH-activity relationship of the enzyme, but we wish to point out that the tracer transport technique is capable of revealing the pH-dependent activity and that the results are consistent with what is known about the enzyme. It is interesting to note that Lindskog (1966) measured an apparent pK of 7.7 for the absorption spectrum change of Co(II)-carbonic anhydrase with pH in a solution containing only CO_2 and 0.10 M bicarbonate. In the presence of Cl^- (concentration unspecified), the pK was 7.1 (Lindskog, 1963).

5.5 High Enzyme Concentration

The transport results over an 80-fold range in bulk solution enzyme concentration are shown in Fig. 7. These data were obtained at constant 1% CO_2 and 50 mM bicarbonate, and a pH of 8.4. The enzyme solutions were prepared by dilution from a common stock solution. The solid line is simply a curve through the data points.

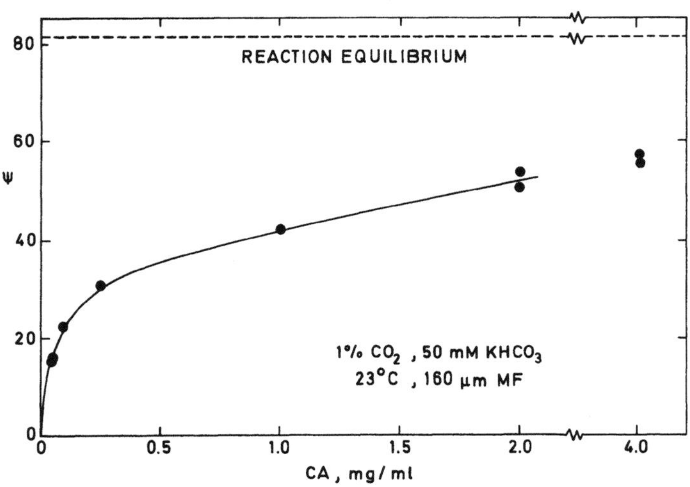

Fig. 7. Flux ratio of CO_2 measured in Millipore filter membranes as a function of enzyme concentration

The data are well below the equilibrium limit of 80, and the enzyme kinetic parameters can be deduced by replotting the data in the double reciprocal form in Fig. 8. The parameter ϕ is sufficiently large that $\tanh \phi = 1$, and since the rate constant k_{enz} is proportional to enzyme concentration, the proper parameter for the abscissa is $[CA]^{-1/2}$. The linearity of the data at higher enzyme concentrations indicates that the enzyme specific activity is constant over the concentration range 0.1 to 4.0 mg/ml. The data at 0.05 mg/ml CA (at 4.5 on the abscissa) are below the line due to adsorption of CA on the Millipore filters (Donaldson and Quinn, 1974). The turnover number can be determined from the slope of the straight line. Assuming $K_M = 12$ mM and 75% enzyme purity, the turnover number is $5.5 \times 10^5\ s^{-1}$, in excellent agreement with $5.3 \times 10^5\ s^{-1}$ determined from the 0.1 mg/ml data of Fig. 4.

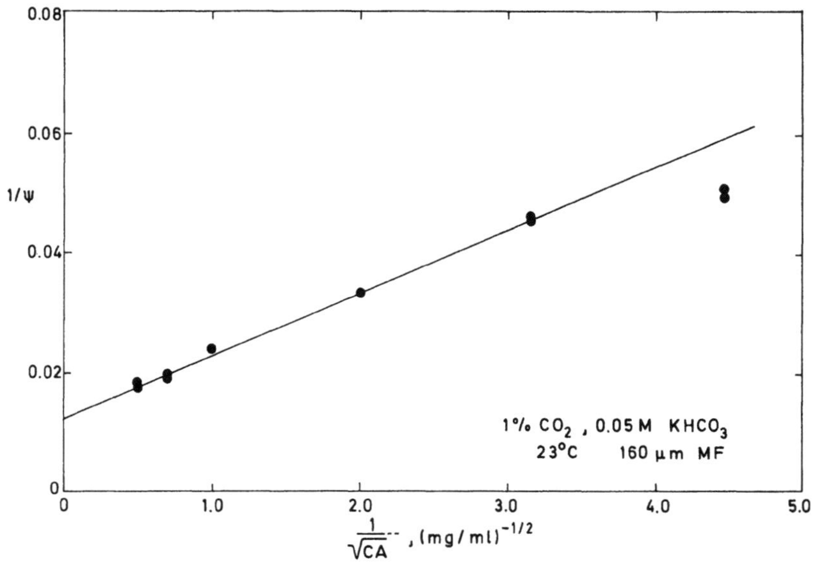

Fig. 8. Double reciprocal plot of data from Fig. 7. The points at 4.5 on the abscissa fall below the line as a result of enzyme adsorption on the Millipore filters

5.6 Bicarbonate Inhibition

It is generally believed that there is little bicarbonate binding as a substrate for the dehydration reaction, but that bicarbonate is a noncompetitive inhibitor which combines with the enzyme at an anion-binding site (Kernohan, 1964). The extent of bicarbonate inhibition is reduced in the presence of Cl^- and other anions (which also inhibit) and it is believed that the bicarbonate is displaced from the anion-binding site by chloride and other anions.

The transport data of Figs. 7 and 8 over a range of enzyme concentrations were obtained at constant CO_2 and bicarbonate concentrations, and the bicarbonate inhibition effect can be estimated. The activity of the inhibited enzyme is expressed by

$$\frac{k_{enz}}{k^0_{enz}} = \frac{1}{1 + [HCO_3^-]/K_I} \tag{7}$$

for a noncompetitive inhibitor, where k^0_{enz} is the uninhibited activity and K_I is the bicarbonate binding (inhibition) constant. Based on Kernohan's turnover number of $10 \times 10^5\ s^{-1}$ with no inhibition and our turnover number of $5.5 \times 10^5\ s^{-1}$ in 50 mM bicarbonate at pH 8.4, the bicarbonate inhibition constant is about 60 mM. Kernohan (1965) observed a reduction of 40% in the enzyme activity in the presence of 80 mM bicarbonate in pyrophosphate buffers at pH values from 8.2 to 8.6, and Maren (1967) reported a 50% inhibition with 50 mM bicarbonate in a barbital buffer system in which the enzymatic hydration rate was independent of pH. These inhibition effects suggest that the binding constant is about 50–100 mM, in good agreement with our estimate of 60 mM. Extensive comparison and modeling of the anion inhibition is difficult since the inhibition effects are generally pH-dependent for the bovine enzyme, but again our results are consistent with other bicarbonate inhibition observation.

Acknowledgment. This investigation was supported in part by a grant from the National Science Foundation.

References

Broun G, Thomas D., Selegny E (1972) Structured bienzymatical models formed by sequential enzymes bound into artificial supports: active glucose transport effect. J Membr Biol 8: 313–332

Broun G, Thomas D, Gellf G, Domurado D, Berjonneau AM, Guillon C (1973) New methods for binding enzyme molecules into a water insoluble matrix: Properties after insolubilization. Biotechnol Bioeng 25: 359–375

DeVoe H, Kistiakowsky GB (1961) The enzyme kinetics of carbonic anhydrase from bovine and human erythrocytes. J Am Chem Soc 83: 274–280

Donaldson TL (1974) Carbon dioxide transport and carbonic anhydrase kinetics: Reaction and diffusion studies with enzyme-active membranes. Ph D Diss, University of Pennsylvania

Donaldson TL, Quinn JA (1974) Kinetic constants determined from membrane transport measurements: carbonic anhydrase activity at high concentrations. Proc Natl Acad Sci USA 71: 4995–4999

Donaldson TL, Quinn JA (1975) Carbon dioxide transport through enzymatically active synthetic membranes. Chem Eng. Sci 30: 103–115

Edsall JT (1969) Carbon dioxide, carbonic acid, and bicarbonate ion: Physical properties and kinetics of inter-conversion. In: Forster RE, Edsall JT, Otis AB, Roughton FJW (eds) CO_2: Chemical, biochemical, and physiological aspects. NASA SP-188, Washington DC, pp 15–27

Enns T (1967) Facilitation by carbonic anhydrase of carbon dioxide transport. Science 155: 44–47

Garg LC, Maren TH (1972) The rate of hydration of carbon dioxide and dehydration of carbonic acid at 37°. Biochim Biophys Acta 261: 70

Harned HS, Bonner, FT (1945) The first ionization of carbonic acid in aqueous solutions of sodium chloride. J Am Chem Soc 67: 1026–1031

Kernohan JC (1964) The activity of bovine carbonic anhydrase in imidazole buffers. Biochim Biophys Acta 81: 346–356

Kernohan JC (1965) The pH-activity curve of bovine carbonic anhydrase and its relationship to the inhibition of the enzyme by anions. Biochim Biophys Acta 96: 304–317

Lindskog S (1963) Effects of pH and inhibitors on some properties related to metal binding in bovine carbonic anhydrase. J Biol Chem 238: 945–951

Lindskog S (1966) Interaction of Co(II)-carbonic anhydrase with anions. Biochemistry 5: 2641–2646

Longmuir IS, Forster RE, Woo CY (1966) Diffusion of carbon dioxide through thin layers of solution. Nature (London) 209: 393–394

Maren TH (1967) Carbonic anhydrase: chemistry, physiology, and inhibition. Physiol Rev 47: 595–781

Meldon JH (1973) Reaction-enhanced mass transfer in thin liquid films. Ph D Diss, Massachusetts Institute of Technology

Otto NC, Quinn JA (1971) The facilitated transport of carbon dioxide through bicarbonate solutions. Chem Eng Sci 26: 949–961

Roughton FJW, Booth VH (1946) The effect of substrate concentration, pH, and other factors upon the activity of carbonic anhydrase. Biochem J 40: 419–430

Smith DR, Lander RJ, Quinn JA (1977) Carrier-mediated transport in synthetic membranes. In: Li, NN (ed) Recent developments in separation science. CRC Press, pp 225–241

Suchdeo SR, Schultz JS (1974) The permeability of gases through reacting solutions: The carbon dioxide-bicarbonate membrane system. Chem Eng Sci 29: 13–23

Ward WJ, Robb WL (1967) Carbon dioxide-oxygen separation: Facilitated transport of carbon dioxide across a liquid film. Science 156: 1481–1484

Wilbur KM, Anderson NG (1948) Electrometric and colorimetric determination of carbonic anhydrase. J Biol Chem 176: 147–154

Mechanisms of Facilitated CO_2 and H^+ Diffusion in Protein Solutions [1]

G. GROS, H. GROS, D. LAVALETTE, B. AMAND, and F. POCHON [2]

Facilitated diffusion of CO_2 under steady-state conditions is based on two simultaneous processes (a) diffusion of HCO_3^- and (b) an equivalent H^+ flux that is brought about by a facilitated H^+ diffusion. We have previously shown that, in this latter process, proteins and other buffers can function as proton carriers in such a powerful way that facilitated H^+ fluxes at neutral pH values are 1000–10,000 times greater than the fluxes of free protons (Gros and Moll, 1974; Gros et al., 1976). Inside cells, where soluble proteins and other mobile buffers are present, proton fluxes should almost exclusively occur by such a facilitated transport process. Our main aim in this study was to understand the mechanisms underlying these facilitated proton fluxes, and we have used the measurement of facilitated CO_2 diffusion as a tool to obtain quantitative information on proton transport in protein solutions.

In the following, we would like to show that facilitated proton transport can be based on two modes of molecular motion of the carrier molecule: it can occur by translational diffusion and by rotational diffusion of the protein. To show this, we have measured facilitated CO_2 and H^+ fluxes in solutions of various proteins and other buffers. We have then calculated the protein diffusivities that have to be postulated to explain the measured proton fluxes, and compared the postulated protein diffusivities with experimentally determined translational and rotational diffusion coefficients of these proteins. While translational protein diffusion led to facilitated proton transport with all buffers studied, the role of rotational diffusion seems to depend on the molecular size of the carrier molecule: rotation can result in proton transport when the buffer molecule is large (e.g., in the case of earthworm hemoglobin), it does not contribute to facilitated proton flux when the buffer molecule is small (e.g., in the case of inorganic phosphate).

1 Materials and Methods

1.1 Preparation of Solutions

Sperm whale myoglobin was purchased from Serva (Heidelberg), dissolved in 0.15 M NaCl, and centrifuged to remove small amounts of precipitated material. The supernatant was further purified by gel filtration on a Sephadex G75 column (Pharmacia,

1 Supported by the Deutsche Forschungsgemeinschaft under Gr 489/2
2 Institut für Physiologie, Universität Regensburg/FRG, and the Institut Curie, Section de Biologie, 91405 Orsay/France

Freiburg i.Br.) and extensive dialysis against 0.15 M NaCl. Human hemoglobin was prepared by water hemolysis of washed red cells. Ghosts were removed by centrifugation at 20,000 *g* in the presence of 0.15 M NaCl. The supernatant was also dialyzed against 0.15 M NaCl. Hemoglobin of the earthworm (*Lumbricus terrestris*) was prepared from live earthworms as described previously (Gros, 1978). All protein solutions were adjusted to pH 7.6 at $pCO_2 = 0$ and 25°C. Ionic strength was 0.15. Bovine carbonic anhydrase was added at a concentration (0.1 g%) sufficient to eliminate the CO_2 hydration reaction as a limiting step in the facilitated diffusion of CO_2.

1.2 Measurement of Facilitated CO_2 Diffusion

CO_2 fluxes were measured under steady-state conditions across thin layers of solution using an experimental set-up described previously (Gros and Moll, 1971; Gros and Moll, 1974). The layers were prepared by soaking 160 μm thick Millipore filters (type HA; Millipore Corp., Bedford, Mass.) in the solution under investigation. Temperature was 25°C. Facilitated diffusion was usually quantitated by comparing the CO_2 diffusion coefficients (= total flux of CO_2 per concentration difference of dissolved CO_2) obtained in a low CO_2 partial pressure range (pCO_2 values at the boundaries of the layers about 2 and 3 torr, respectively) with the CO_2 diffusion coefficients obtained in a high pCO_2 range (boundary pCO_2 values 200 and 600 torr, respectively). Figure 1 illustrates, for the case of a 33 g% solution of human hemoglobin, how the CO_2 diffusion coefficient, D_{CO_2}, varies with the average pCO_2 within the layer. Above an average pCO_2 of 100 torr, D_{CO_2} is independent of pCO_2.

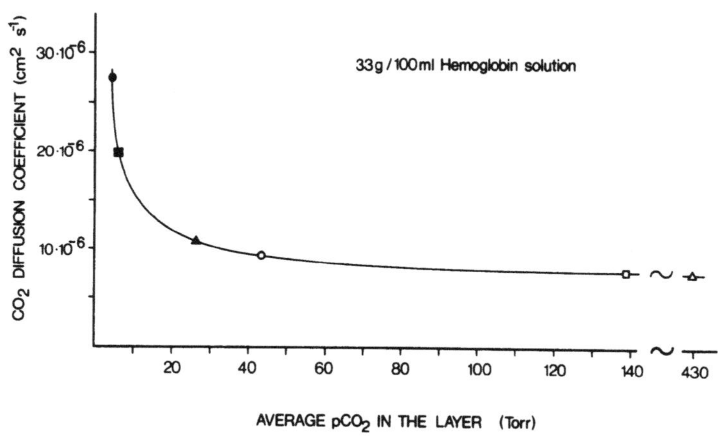

Fig. 1. CO_2 diffusion coefficient in $cm^2 s^{-1}$ in a 33 g% solution of human hemoglobin as a function of the average pCO_2 prevailing in the diffusion layer. The average pCO_2 was calculated as the arithmetic mean of the pCO_2 values at the boundaries of the layer. Temperature 22°C, pH ($pCO_2 = 0$) ≈ 7.6, ionic strength 0.13. Carbonic anhydrase concentration was that of human erythrocytes. The following boundary pCO_2 values were employed: 5.5/3.4 torr (●), 7.8/4.0 torr (■), 37.5/15.0 torr (▲), 63.8/22.8 torr (○), 211/66.1 torr (□) and 665/187 torr (△)

Here, no significant facilitation occurs, and D_{CO_2} represents free diffusion of dissolved CO_2 only (Gros and Moll, 1971). Lowering pCO_2 below 100 torr, on the other hand, results in an increasing rise in the value of the CO_2 diffusion coefficient, indicating an increasing contribution of facilitated CO_2 diffusion to the total CO_2 flux. Facilitated CO_2 diffusion, therefore, was estimated from the difference of the CO_2 diffusion coefficient measured at low pCO_2 and that determined at high pCO_2.

1.3 Translational Diffusion Coefficients of Proteins

Translational diffusion coefficients, D_{trans}, were taken from the literature. All data were obtained from measurements with millipore filters, and conditions of pH and ionic strength were similar to those used in the CO_2 diffusion experiments. Part of the values had to be corrected to 25°C.

1.4 Rotational Diffusion Coefficient of Earthworm Hemoglobin

Earthworm carbonmonoxy hemoglobin (HbCO) was photodissociated by a short pulse of polarized light at 530 nm. The decay of the resulting dichroism was observed at 445 nm using an apparatus described previously (Lavalette et al., 1977). The recombination of CO with hemoglobin being negligible on the time-scale of that decay, the latter was solely due to rotational protein diffusion, and the rotational diffusion constant could be calculated from its time constant. Earthworm hemoglobin concentration was 0.2g%, pH was 7, the buffer was 0.15 M NaCl and 0.1 M phosphate, and temperature was 25°C. The details of the method will be published elsewhere (Gros et al., 1980).

2 Results and Discussion

2.1 Facilitated CO_2 and H^+ Diffusion in Earthworm Hemoglobin Solutions

In this section, we describe the experimental results together with a quantitative analysis that we obtained for solutions of earthworm hemoglobin. The extracellular hemoglobin of the earthworm is one of the largest known proteins: it has a molecular weight of $3.7 \cdot 10^6$ and a molecular diameter of about 280 Å, and is of approximately spherical shape (Levin, 1963; Vinogradov et al., 1977; and this study).

Figure 2 shows the CO_2 diffusion coefficients in earthworm hemoglobin solutions plotted as a function of earthworm hemoglobin concentration. The two curves represent measurements in two CO_2 partial pressure ranges: the lower curve was obtained at boundary pCO_2 values of 200 and 600 torr, the upper curve at pCO_2 values at the boundaries of 2 and 3 torr. The difference between the curves indicates that facilitated CO_2 diffusion occurs in the low pCO_2 range. It is present at all protein concentrations studied. At an earthworm hemoglobin concentration of 26 g%, for example, facilitated diffusion is about equal to free diffusion of CO_2.

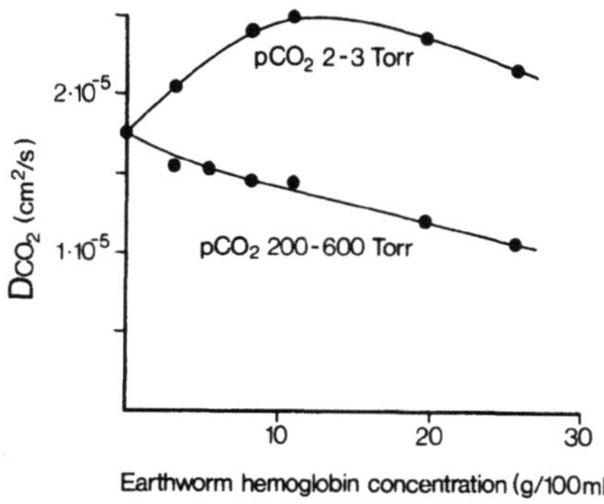

Fig. 2. CO_2 diffusion coefficient in $cm^2 s^{-1}$, D_{CO_2}, in earthworm hemoglobin solutions as a function of earthworm hemoglobin concentration in g%. The measurements were performed in a low pCO_2 range (*upper curve* boundary pCO_2 values about 2 and 3 torr, respectively) and in a high pCO_2 range (*lower curve* boundary pCO_2 values 200 and 600 torr, respectively). Temperature 25°C, pH (pCO_2 = 0) = 7.6, ionic strength 0.15. The standard errors of the CO_2 diffusion coefficients range from 0.5% to 2.5% (n = 6–10). The difference between the two curves indicates the facilitation of CO_2 diffusion that occurs at low CO_2 partial pressures

What is the physicochemical basis of facilitated CO_2 diffusion in earthworm hemoglobin solutions? We have previously proposed that facilitated CO_2 diffusion can occur by diffusion of HCO_3^- and a concomitant H^+ transport by translational diffusion of protein. In the case of a 30 g% serum albumin solution, this mechanism can quantitatively account for the measured facilitated CO_2 fluxes (Gros and Moll, 1974). In earthworm hemoglobin solutions a facilitation similar to that in albumin solutions occurs, even though earthworm hemoglobin is a much larger molecule than albumin (M.W. 67,000) and thus should diffuse much more slowly. The question arises, therefore, whether translational protein diffusion is the sole basis of facilitated CO_2 and H^+ diffusion in earthworm hemoglobin solutions. To answer this question, we shall in the following derive a simplified mathematical formulation of facilitated diffusion by translational protein diffusion, use it to calculate the translational diffusion coefficients that are necessary to explain the facilitated CO_2 fluxes measured in earthworm hemoglobin solutions, and then compare the postulated diffusivities with the directly measured translational diffusion coefficients of earthworm hemoglobin.

2.1.1 Facilitated CO_2 and H^+ Transport by Translational Protein Diffusion

In this section, an equation is derived that describes the total (free + facilitated) flux of CO_2, F_{tot}, as a function of the translational diffusion coefficient of the protein, D_{trans}. Using this equation, the values of D_{trans} were estimated that are needed to account for the measured CO_2 fluxes.

The flux of dissolved CO_2, F_{CO_2}, is expressed by Fick's first diffusion law:

$$F_{CO_2} = - D_{CO_2} \cdot \frac{d\,[CO_2]}{dx}, \qquad (1)$$

where D_{CO_2} is the CO_2 diffusion coefficient measured in a high pCO_2 range, $[CO_2]$ is the concentration of dissolved CO_2, and x is the diffusion path. All other fluxes are also expressed in terms of Fick's equation, even though we are now dealing with diffusion of ions. This is permissible because under the present conditions of ionic strength diffusion potentials are small (Meldon et al., 1978) and their effect on ionic fluxes can be neglected (Gros and Moll, 1974; Gros, 1976). The flux of bicarbonate, $F_{HCO_3^-}$, is then given by:

$$F_{HCO_3^-} = -D_{HCO_3^-} \cdot \frac{d\,[HCO_3^-]}{dx}, \tag{2}$$

where $D_{HCO_3^-}$ is the diffusion coefficient of HCO_3^-, and $[HCO_3^-]$ is the concentration of bicarbonate. The flux of protons is given by the product of the translational diffusion coefficient of the protein, D_{trans}, and the concentration gradient of buffered protons:

$$F_{H^+} = D_{trans} \cdot [Prot] \cdot \beta \cdot \frac{dpH}{dx}, \tag{3}$$

where [Prot] is the concentration of the protein, β its buffer capacity, and dpH/dx the pH gradient in the layer. In Eq. (3), the flux of free H^+ is neglected since it is three orders of magnitude smaller than the protein facilitated H^+ flux.

The total (free + facilitated) flux of CO_2 is given by the sum of the fluxes of dissolved CO_2 and of HCO_3^-:

$$F_{tot} = F_{CO_2} + F_{HCO_3^-}, \tag{4}$$

and the fluxes of HCO_3^- and H^+ have to be of equal size:

$$F_{HCO_3^-} = F_{H^+}. \tag{5}$$

Equation (5) satisfies the condition that the entire process of facilitated diffusion be electrically silent. As carbonic anhydrase is present in excess, CO_2 and HCO_3^- are assumed to be in chemical equilibrium everywhere in the layer:

$$\frac{[H^+]\,[HCO_3^-]}{[CO_2]} = K_1'. \tag{6}$$

By transformation and rearrangement of Eqs. (1)–(6) we obtain the following expression for F_{tot}:

$$F_{tot} = \left[-D_{CO_2} - \frac{D_{trans}\,[Prot]\,\beta}{\dfrac{D_{trans}\,[Prot]\,\beta\,[CO_2]}{C_1 - D_{CO_2}\,[CO_2] - xF_{tot}} + 2.303\,[CO_2]} \right] \frac{d\,[CO_2]}{dx} \tag{7}$$

where C_1 is a constant.

Equation (7) was integrated numerically for the following boundary conditions: the CO_2 concentrations at the boundaries, $x = 0$ and $x = l$, are those of the CO_2 diffusion experiment considered; the average concentration of buffered protons in the layer minus that at $pCO_2 = 0$ is equal to the average HCO_3^- concentration in the layer: $\frac{1}{l}\,[\mathrm{Prot}]\,\beta \int_{x=0}^{l} dpH - [\mathrm{Prot}]\,\beta\, pH_{(pCO_2=0)} = \frac{1}{l} \int_{x=0}^{l} d\,[HCO_3^-]$. The equation was solved in such a way that the protein diffusion coefficient was obtained, which has to be postulated to explain the measured facilitated CO_2 fluxes.

The results are shown in Fig. 3, where the diffusion coefficient of earthworm hemoglobin is plotted versus earthworm hemoglobin concentration. The upper curve (●) represents the diffusion coefficients that one must postulate to explain the measured facilitated fluxes. These values may be compared to the directly measured translational diffusion coefficients of earthworm hemoglobin that are shown on the lower curve (□) It is apparent from Fig. 3 that the postulated protein diffusivities are significantly greater than the translational diffusivities at all protein concentrations studied. Thus, translational diffusion can only explain part of the measured facilitated proton fluxes. At a concentration of 3 g% about 60% of the total flux seems to be due to translational transport, at 26 g% only about 1/4 of the total flux can be explained by translational diffusion.

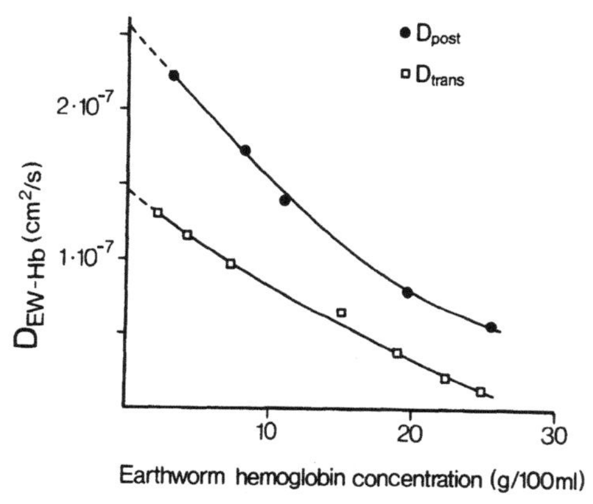

Fig. 3. Diffusion coefficient of earthworm hemoglobin in $cm^2 s^{-1}$, $D_{EW\text{-}Hb}$, as a function of earthworm hemoglobin concentration in g%. □ indicate the directly measured translational diffusion coefficients of earthworm hemoglobin, D_{trans} (quoted from Gros, 1978); ● represent the protein diffusivities that have to be postulated to explain the facilitated CO_2 fluxes measured in earthworm hemoglobin solutions (see Fig. 2). Temperature 25°C (the D_{trans} values were measured at 21°C and corrected to 25°C), pH ≈ 7.3, ionic strength 0.15. The standard errors of D_{trans} range from 1% to 6% (n = 9–20). Extrapolation to zero protein concentration (see *dashed portions of the curves*) gives limiting values of $1.46 \cdot 10^{-7}\ cm^2 s^{-1}$ for D_{trans}, and $2.55 \cdot 10^{-7}\ cm^2 s^{-1}$ for D_{post}

2.1.2 Proton Transport by Rotational Protein Diffusion

Several years ago Moll (1962) and Wyman (1966) discussed rotational protein diffusion as a possible mechanism of facilitated oxygen diffusion in hemoglobin solutions. Wyman showed that rotation cannot play a significant role because the chemical reaction is too slow for the oxygen to be bound to and released from the hemoglobin during the short time of, say, half a rotation of the hemoglobin molecule. However, the velocity of protonation is several orders of magnitude faster than that of oxygenation, and it thus appeared possible that rotational diffusion does contribute to facilitated proton transport. We asked now whether a combination of rotational and translational diffusion of earthworm hemoglobin can quantitatively explain our experimental data. We therefore derived an equation that describes facilitated proton transport by rotational diffusion, and we measured the rotational diffusion coefficient of earthworm hemoglobin.

Proton flux by rotational diffusion of a spherical protein is given by Gros (1976) and Gros et al. (1980):

$$F_{H^+} = \frac{\pi}{4}\, r^2\, \Theta\, [\text{Prot}]\, \beta\, \frac{dpH}{dx}\,, \tag{8}$$

where r is the molecular radius of the protein, and Θ its rotational diffusion coefficient. The total proton flux mediated by translation and rotation is then given by the sum of Eqs. (3) and (8):

$$F_{H^+} = \left(D_{trans} + \frac{\pi}{4}\, r^2\, \Theta\right) [\text{Prot}]\, \beta\, \frac{dpH}{dx}\,, \tag{9}$$

or, using an apparent transport coefficient, $\bar{D}_{rot\,+\,trans}$, that describes the effects of rotation as well as translation, we can write:

$$F_{H^+} = \bar{D}_{rot\,+\,trans}\, [\text{Prot}]\, \beta\, \frac{dpH}{dx} \tag{10}$$

We now obtained a direct estimate of $\bar{D}_{rot\,+\,trans}$ in the following manner. We determined Θ for earthworm hemoglobin at 25°C and very low protein concentration ($c \to 0$) with the method described above. From the value we found, $\Theta = 6.94 \cdot 10^4 s^{-1}$, we calculate the expression $\pi/4\, r^2\, \Theta$ to be $1.05 \cdot 10^{-7}\ cm^2 s^{-1}$ (the inserted value of r, 138 Å, is derived from Θ using $\Theta = kT/8\pi\eta r^3$). By extrapolation of the directly measured translational diffusivities to zero protein concentration (see lower curve in Fig. 3) we estimate the limiting value of the translational diffusion coefficient of earthworm hemoglobin to be $1.46 \cdot 10^{-7}\ cm^2 s^{-1}$. The apparent transport coefficient at $c \to 0$ is then obtained as the sum of $\pi/4\, r^2\, \Theta$ and D_{trans} to be:

$$\bar{D}_{rot\,+\,trans} = 2.51 \cdot 10^{-7}\ cm^2 s^{-1}.$$

Unfortunately, it was not possible to determine Θ at higher protein concentrations, but we can compare this directly determined value of the apparent transport coefficient with the postulated value, when we extrapolate the postulated diffusivities to zero protein concentration. This is shown by the dashed portion of the upper curve in Fig. 3 and yields for $c \rightarrow 0$:

$$D_{post} \approx 2.6 \cdot 10^{-7}\ cm^2 s^{-1}.$$

Postulated and directly determined values agree within a few percent, and we conclude therefore that, at least at low protein concentrations, facilitated proton transport in earthworm hemoglobin solutions can be quantitatively explained by rotational together with translational diffusion of the protein.

2.2 Facilitated CO_2 and H^+ Diffusion in Other Protein and Buffer Solutions

Can all proteins and buffers transport protons by rotational diffusion, or is this phenomenon restricted to very large carrier molecules such as earthworm hemoglobin? To study this question, we measured facilitated diffusion in solutions of phosphate (M.W. ~100), myoglobin (M.W. 17,000), and human hemoglobin (M.W. 65,000).

The results are shown in Figs. 4 a–c. There is clear evidence for facilitated diffusion of CO_2 in all these systems: In the case of phosphate the CO_2 diffusion coefficient (measured at boundary pCO_2 values of ~ 30 and 60 torr) is significantly greater in the presence than in the absence of carbonic anhydrase (Fig. 4a; data quoted from Gros et al., 1976). For myoglobin (Fig. 4b) and hemoglobin (Fig. 4c), facilitated diffusion is demonstrated by the much larger CO_2 diffusivities found at low boundary pCO_2 values as compared to those measured at high pCO_2 values.

We have now analyzed the experimental data of Fig. 4 in the same fashion as described above for earthworm hemoglobin. In the case of the smallest buffer molecule employed, phosphate, the measured facilitated fluxes can be fully explained by translational buffer diffusion alone; this has been shown previously using a theory somewhat more elaborate than that given above (Gros et al., 1976). Rotational diffusion of phosphate thus does not appear to lead to a facilitation of proton transport. The same conclusion holds for facilitated proton transport in solutions of the second smallest molecule, myoglobin. This is shown in Fig. 5a, where the diffusion coefficient of myoglobin is plotted versus myoglobin concentration. The closed circles represent the values postulated from facilitated CO_2 diffusion (Fig. 4b), the open squares are the translational diffusion coefficients reported by Riveros-Moreno and Wittenberg (1972). It is apparent that within the experimental scatter there is no difference between postulated and translational myoglobin diffusivities. As in the case of phosphate, rotational diffusion seems to play no role for facilitated proton transport in myoglobin solutions.

The results obtained for the second largest molecule, human hemoglobin, are shown in Fig. 5b. The open squares represent the translational diffusion coefficients, the closed circles represent the protein diffusion coefficients postulated from facilitated CO_2 fluxes (Fig. 4c). At low hemoglobin concentrations, 5–10 g%, postulated and translational diffusivities are almost identical. Here, the contribution of rotational diffusion seems to be also very small. At a concentration of 36.5 g%, however, postulated and translational diffusion coefficients are significantly different. This may be

CO$_2$ DIFFUSION IN SOLUTIONS OF INORGANIC PHOSPHATE

Phosphate concentration	66 mM	705 mM
D_{CO_2} (cm²/s) (with 0.2% carbonic anhydrase)	$3.9 \cdot 10^{-5}$	$6.7 \cdot 10^{-5}$
D_{CO_2} (cm²/s) (without carbonic anhydrase)	$1.6 \cdot 10^{-5}$	$1.3 \cdot 10^{-5}$

a

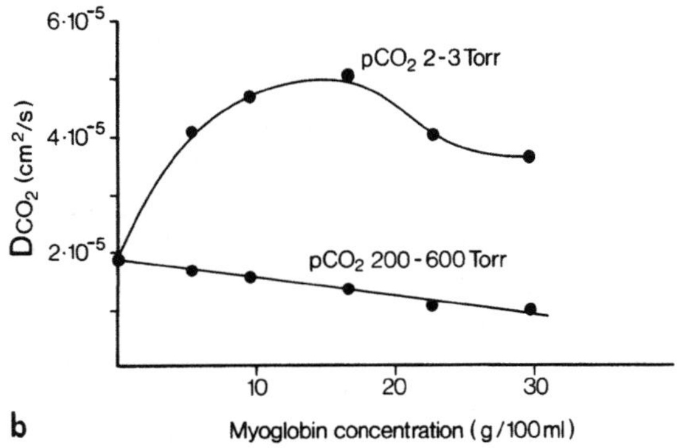

b

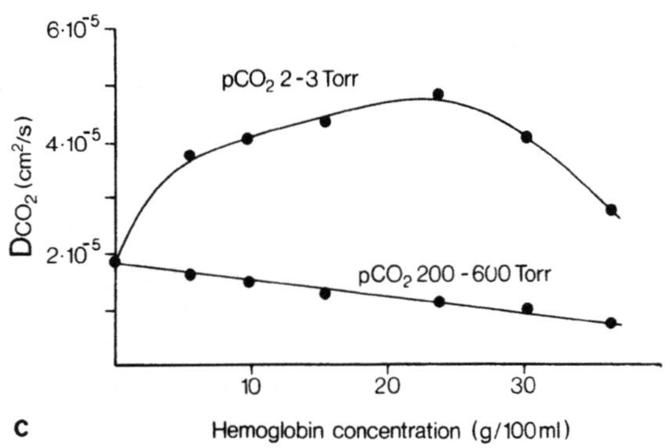

c

Fig. 4 a–c. Demonstration of facilitated CO_2 diffusion in various buffer solutions. **a** CO_2 diffusion coefficients, D_{CO_2}, with and without carbonic anhydrase in 66 mM and 705 mM phosphate solutions. Temperature 22–25°C, pH 7.0 in the 66 mM and 7.3 in the 705 mM solution, boundary pCO_2 values 57/26 torr in the 66 mM and 56/30 torr in the 705 mM phosphate solution. The CO_2 diffusion coefficients obtained in the absence of carbonic anhydrase represented free CO_2 diffusion only, the increased values obtained after addition of carbonic anhydrase represent free and facilitated CO_2 diffusion. (Data quoted from Gros et al., 1976; with permission from the Rockefeller University Press). **b** CO_2 diffusion coefficients in myoglobin solutions as a function of myoglobin concentration. Temperature 25°C, pH ($pCO_2 = 0$) = 7.6, ionic strength 0.15. The difference between the curve obtained in a low pCO_2 range and that obtained in a high pCO_2 range is a measure of the facilitation of CO_2 diffusion in the low pCO_2 range. **c** CO_2 diffusion coefficients in solutions of human hemoglobin, measured in a low and a high pCO_2 range, as a function of hemoglobin concentration. Temperature 25°C, pH ($pCO_2 = 0$) = 7.6, ionic strength 0.15. As in the case of earthworm hemoglobin and myoglobin, only free diffusion of CO_2 occurs in the high pCO_2 range

seen more clearly in Table 1: At a hemoglobin concentration of 36.5 g% the postulated diffusion coefficient is twice the translational value, i.e., only one half of the measured proton flux can be explained by translational diffusion, and rotational diffusion appears to be important. At a hemoglobin concentration of 9.8 g%, on the other hand, there is only a difference of 8% between postulated and translational diffusion coefficients.

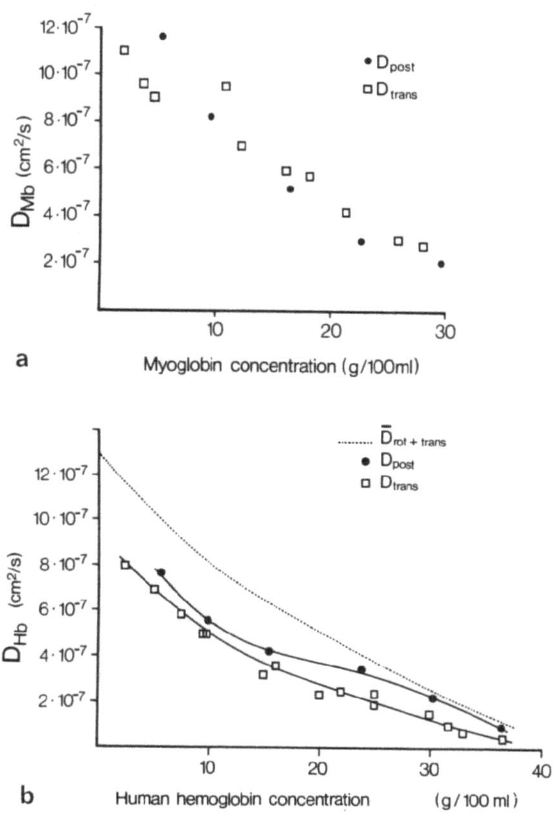

Fig. 5. a,b. Translational protein diffusion coefficients, D_{trans}, versus protein diffusion coefficients postulated from facilitated CO_2 diffusion measurements, D_{post}. **a** Myoglobin diffusion coefficients as a function of myoblogin concentration. Translational diffusivities (□) are from Riveros-Moreno and Wittenberg (1972) and have been dorrected to 25 °C. **b** Diffusion coefficients of human hemoglobin as a function of hemoglobin concentration. Translational diffusivities (□) are those of Moll (1966), Keller et al. (1971), and Gros (1978), and have been corrected to 25 °C. The values of $\bar{D}_{rot + trans}$, shown by the *dotted line*, were calculated according to Eqs. (9) and (10) using the translational diffusion coefficients shown and the rotational diffusion coefficients of human hemoglobin, corrected to 25 °C, as reported by Lindstrom et al. (1976)

Table 1. Translational and postulated diffusion coefficients of human hemoglobin in solutions with hemoglobin concentrations of 9.8 and 36.5 g%. Given are the mean values of the diffusion coefficients and the standard errors in percent of the mean. At a concentration of 9.8 g% about 90% of the measured facilitated CO_2 flux can be explained by translational diffusion, at 36.5g% only 50% of the facilitated flux is due to translation

	Human hemoglobin concentration	
	9.8 g/100 ml	36.5 g/100 ml
D_{trans} $(cm^2 s^{-1} \cdot 10^{-7})$	5.1 (± 1.8%)	0.51 (± 2.9%)
D_{post} $(cm^2 s^{-1} \cdot 10^{-7})$	5.5 (± 2.2%)	1.00 (± 1.2%)

Summarizing the experimental results, we may conclude that the two smallest buffer molecules employed, phosphate and myoglobin, can transport protons by translational diffusion only. The largest molecule, earthworm hemoglobin, transports protons by translational and rotational diffusion at all protein concentrations studied. For human hemoglobin, the second largest molecule, rotational diffusion plays a role at high protein concentration, but is unimportant at lower concentrations.

2.3 Facilitated Transport by Rotational Diffusion and Molecular Size of the Carrier

In this section, we shall make an attempt to give a qualitative interpretation of these results. Figure 6 gives a schematic representation of the process of facilitated proton transport by rotational protein diffusion: At the side of low pH, a proton is bound by an unprotonated buffering group of the protein molecule 1. When this protein molecule performs half a rotation about its axis perpendicular to the plane of the figure, the buffering group becomes exposed to a more alkaline pH, the proton may dissociate and will then be accepted by protein molecule 2. The chemical reaction that takes place is a proton exchange between two protein molecules and is shown in the lower part of Fig. 6: the protonated buffering group of molecule 1 (B_1H^+) reacts with an unprotonated buffering group of molecule 2 (B_2) to give a protonated group on molecule 2 (B_2H^+) and an unprotonated group on molecule 1 (B_1). By successively repeated steps of chemical reaction and protein rotation, a facilitated flux of protons across a layer of protein solution is established. This mechanism can only work when, at a given rotational velocity of the protein, the association of protons is fast enough to occur during the short time the proton binding site spends on the more acid side of the molecule. Similarly, the dissociation reaction has to be fast enough to occur during the time the binding site stays on the side which is exposed to a more alkaline pH. Since large molecules rotate more slowly than small molecules, protons have a better chance of jumping on (at the low pH side) and jumping off (at the high pH side) when the protein is large than when it is small. For a given reaction velocity, therefore, we may expect facilitated transport by rotation to be more efficient with large than with small carrier molecules.

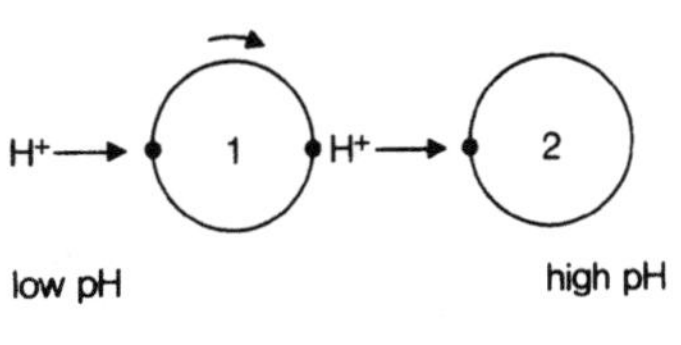

Chemical Reaction:

$$B_1H^+ + B_2 \rightleftharpoons B_2H^+ + B_1$$

Fig. 6. Schematic representation of facilitated proton transport by rotational protein diffusion. A proton is bound by an unprotonated buffering group of protein molecule 1 at the side of low pH. When the protein rotates as indicated by the *arrow*, the buffering group becomes exposed to a more alkaline pH, and the proton may be transferred to an unprotonated group on protein molecule 2. The chemical reaction that takes place, proton exchange, is indicated in the lower part of the figure. By successively repeating the steps of chemical reaction and rotation the proton can be transported through the protein solution

The present data are consistent with this view. This is illustrated in Table 2, which gives a comparison of rotational and chemical relaxation times for the four buffer molecules studied. It should be noted that the chemical relaxation times of Table 2 represent rough estimates based on the assumption of a forward velocity

Table 2. Observed proton flux by rotation as a fraction of the maximal proton flux by rotation, chemical relaxation time, rotational relaxation time, and molecular radius of the buffers used in this study: phosphate, myoglobin, human hemoglobin, and earthworm hemoglobin. The chemical relaxation times are calculated for the concentrations indicated assuming the forward velocity constant of the proton exchange reaction (see Fig. 6) to be 10^9 l/mol/s, and assuming that the experimentally determined buffer capacities of the proteins are represented by buffering groups whose pK is identical to the pH existing in the layer. The rotational relaxation times (= 3 x rotational correlation times) given are limiting values that are valid for $c \rightarrow 0$ (values for human and earthworm hemoglobin from own measurements, for myoglobin from Anderson et al., 1970). (Only a minor difference is expected between τ_{rot} at protein concentrations of 0 and 5g%). The figure for a 36.5 g% hemoglobin solution was taken from Lindstrom et al. (1976). The "observed" proton flux by rotation was estimated as difference between the measured facilitated H^+ flux and the H^+ flux expected from translational diffusion. The maximal proton flux by rotation is that expected when the proton exchange reaction is fast enough to be not a limiting step in proton transfer by rotation; it is given by Eq. (8)

		Molecular radius	Rotational relaxation time τ_{rot}	Chemical relaxation time τ_{chem}	Observed H^+ flux due to rotation/ Maximal H^+ flux by rotation
Phosphate	(66 mM)	~ 3 Å	~ 0.1 ns	30 ns	0
Myoglobin	(5 g%)	18 Å	23 ns	170 ns	0
Human hemoglobin	(5 g%)	31 Å	120 ns	150 ns	0.2
	(36.5 g%)		510 ns	20 ns	0.7
Earthworm hemoglobin	(5 g%)	138 Å	7200 ns	140 ns	1.0

constant for the proton exchange reaction of 10^9 l mol^{-1} s^{-1}. In Table 2, the buffers are listed in the order of their molecular size, together with their radius, rotational and chemical relaxation time, and the ratio of the "observed" H^+ flux due to rotation over the "maximal" H^+ flux by rotation. This latter quantity was calculated as that part of the measured facilitated H^+ flux that cannot be attributed to translational diffusion, divided by the maximal flux due to rotational diffusion as it is predicted by Eq. (8). This ratio should be 1, i.e., the proton flux by rotation should be maximal, when the protonation reaction is fast compared to the rotation of the protein. Table 2 shows that for the smallest buffer, phosphate (molecular radius $r \approx 3$ Å), the ratio is zero; in other words, rotation does not lead to a transport of protons. Chemical relaxation (τ_{chem} = 30 ns) is 300 times slower than rotational relaxation (τ_{rot} = 0.1 ns) in this case. Here, the proton apparently has no chance of getting on and off the carrier molecule fast enough. Similarly, for the second smallest molecule, myoglobin (r = 18 Å), the chemical relaxation is seven times slower than the rotational relaxation, and no proton flux by rotation occurs. For human hemoglobin at a concentration of 5 g% the chemical relaxation is almost identical to the rotational relaxation, and we find 20% of the maximally possible proton transport by rotation. As the concentration of human hemoglobin is raised from 5 to 36.5 g%, the chemical relaxation becomes faster and the rotational relaxation becomes slower (Lindstrom et al., 1976), so that the chemical relaxation is now 25 times faster than the rotational relaxation; this is associated with an increase in proton flux by rotation to 70% of its maximally possible value. (The same information is given in Fig. 5b: The dotted line represents $\bar{D}_{rot + trans}$ over the entire hemoglobin concentration range studied. It can be seen that the diffusivities postulated from facilitated diffusion, $D_{\underline{post}}$, are close to $\bar{D}_{rot + trans}$ at high hemoglobin concentrations, but much smaller than $\bar{D}_{rot + trans}$ at low concentrations.) In the case of earthworm hemoglobin, the chemical relaxation is 50 times faster than the rotational relaxation, and here we observe the maximal value for proton transport by rotational diffusion.

We conclude that all buffers studied can sustain facilitated proton transport by translational diffusion in 160 μm thick layers. Proton transport by rotational buffer diffusion, however, can only occur when the chemical relaxation of the proton exchange reaction is faster than, or at least about equal to, the rotational relaxation of the buffer.

Acknowledgment. G.G. is deeply indebted to Professor W. Moll for his continuing and stimulating interest in this study.

References

Anderson SR, Brunori M, Weber G (1970) Fluorescence studies of aplysia and sperm whale apomyoglobins. Biochemistry 9: 4723–4729

Gros G (1976) Die erleichterte Diffusion von CO_2 in Proteinlösungen. Habilitationsschrift, Universität Regensburg

Gros G (1978) Concentration dependence of the self-diffusion of human and *Lumbricus terrestris* hemoglobin. Biophys J 22: 453–468

Gros G, Moll W (1971) The diffusion of carbon dioxide in erythrocytes and hemoglobin solutions. Pflügers Arch 324: 249–266

Gros G, Moll W (1974) Facilitated diffusion of CO_2 across albumin solutions. J Gen Physiol 64: 356–371

Gros G, Moll W, Hoppe H, Gros H (1976) Proton transport by phosphate diffusion – a mechanism of facilitated CO_2 transfer. J Gen Physiol 67: 773–790

Gros G, Lavalette D, Moll W, Amand B, Pochon F, Gros H (1980) Manuscript in preparation

Keller KH, Canales ER, Yum SI (1971) Tracer and mutual diffusion coefficients of proteins. J Phys Chem 75: 379–387

Lavalette D, Amand D, Pochon F (1977) Rotational relaxation of 70S ribosomes by a depolarization method using triplet probes. Proc Natl Acad Sci USA 74: 1407–1411

Levin Ö (1963) Electron microscope observations on some 60s erythrocruorins and their split products. J Mol Biol 6: 95–101

Lindstrom TR, Koenig SH, Boussios T, Bertles JG (1976) Intermolecular interactions of oxygenated sickle hemoglobin molecules in cells and cell-free solutions. Biophys J 16: 679–689

Meldon JH, de Koning J, Stroeve P (1978) Electrical potentials induced by CO_2 gradients in protein solutions and their role in CO_2 transport. Bioelectrochem Bioenerg 5: 77–87

Moll W (1962) Die Carrier-Funktion des Hämoglobins beim Sauerstoff-Transport im Erythrocyten. Pflügers Arch 275: 412–419

Moll W (1966) The diffusion coefficient of haemoglobin. Respir Physiol 1: 357–365

Riveros-Moreno V, Wittenberg JB (1972) The self-diffusion coefficients of myoglobin and hemoglobin in concentrated solutions. J Biol Chem 247: 895–901

Vinogradov SN, Shlom JM, Hall BC, Muzukami H (1977) The dissociation of *Lumbricus terrestris* hemoglobin: a model of its subunit structure. Biochim Biophys Acta 492: 136–155

Wyman J (1966) Facilitated diffusion and the possible role of myoglobin as a transport mechanism. J Biol Chem 241: 115–121

The Effect of Diffusion Potentials on Facilitated CO_2 Diffusion

J.H. MELDON [1]

1 Introduction

Bicarbonate-facilitated CO_2 transport is, as described elsewhere in this section, of long-standing biological and industrial interest. It can be said to date back at least half a century to A.V. Hill's (1928) comment on August Krogh's (1918–1919) measurements of tissue O_2 and CO_2 permeabilities: "It seems rather curious that the larger molecule should diffuse faster."

However, the appreciable research effort that has greatly clarified the mechanism of enhanced CO_2 transfer only began with the work of Longmuir et al. (1966), who measured CO_2 diffusion across thin layers of buffer solution and lysed and whole blood. Their work – along with that of Enns (1967). Ward and Robb (1967) and others since – firmly established that enhancement is promoted by catalysis of the intrinsically slow CO_2 hydration reaction leading to bicarbonate formation.

The extent of bicarbonate formation is limited, as well, by the buildup of hydrogen ions. Hence more alkaline solutions are better CO_2 transporters, as has repeatedly been shown. By the same token, buffers – which consume hydrogen ions generated by the dissociation of carbonic acid – also promote bicarbonate concentration gradients and, therefore, diffusion (Meldon, 1973; Meldon et al., 1977).

The buffer, as demonstrated by Gros et al. (1976), must, however, be diffusible. An immobilized buffer will, in fact, eradicate bicarbonate-facilitated CO_2 diffusion. The reason is the diffusion potential caused by the lesser mobility of the buffer compared to that of bicarbonate ion. This effect, which is particularly great in the case of protein solutions, is the subject of the discussion that follows.

2 The Diffusion Potential

The significance of the diffusion potential in electrolyte transport is treated in the classical paper by Vinograd and McBain (1941). They demonstrated experimentally the coupling of the diffusion of any one ionic species to the diffusion of all others present; and explained their results with the theories of Nernst, Planck and Arrhenius.

The unidirectional flux, N, of any species j is given by:

$$N_j = -D_j\left(\frac{dC_j}{dx} + z_jC_j\ \frac{F}{RT}\ \frac{dV}{dx}\right) \tag{1}$$

1 Dept. of Chemical Engineering, Tufts University, Medford, Mass. 02155/USA

where x is the spatial coordinate, F is Faraday's and R the gas constant, T is absolute temperature and V is the electrical potential; D denotes diffusivity, C concentration and z charge.

In an electrically floating solution, the current must everywhere be zero, i.e.:

$$\Sigma z_j N_j = 0 \tag{2}$$

and – except in liquid layers of thickness approaching that of biological membranes (100 Å) – electroneutrality can also be presumed to prevail locally; thus:

$$\Sigma z_j C_j = 0 \tag{3}$$

From inspection of these three equations it can be inferred that Eqs. (2) and (3) can only be made consistent if either (i) all the D_j are equal, or (ii) dV/dx is not equal to zero. Since ion diffusion coefficients do in fact vary appreciably from one another, it follows that dV/dx – the diffusion potential – will generally attain some positive or negative value.

As expressed in Eq. (1), the potential affects the diffusion of all ionic species. Moreover, according to the signs of z_j, dC_j/dx and dV/dx, its effect upon a particular species may be to accelerate or slow down its diffusion. Thus Vinograd and McBain showed circumstances under which the flux of certain species is slowed almost to a halt, and others in which hydrogen ion, the fastest of all ions, behaves as if its mobility is greater again by 40%.

In a solution of a simple 1:1 electrolyte like HCl, Eqs. (1)–(3) can be combined to show that the anion (a) and cation (c) diffuse together as a neutral salt of diffusivity equal to $2D_aD_c/(D_a + D_c)$. Thus, H^+ and Cl^-, with intrinsic diffusivities which differ by an order of magnitude, each diffuse with an effective diffusivity approximately twice the intrinsic value for chloride.

Analogous effects prevail in the case of CO_2 transport in aqueous media. The analysis is complicated by (i) simultaneous chemical reaction, (ii) the Nernstian distribution, within the liquid layer, of nonreactive ions like Na^+ or Cl^-, and (iii) the non-constancy of ionic charge which characterizes titratable proteins. Nonetheless, mathematical solution is still possible, and agreement with experiment is excellent.

3 Analysis of CO_2 Transport

Of particular physiological interest is the diffusion of CO_2 in moderately alkaline media. Such systems involve the following set of reversible chemical reactions, which are intrinsic to aqueous solutions.

$$CO_2 + H_2O \rightleftharpoons H_2CO_3 \rightleftharpoons HCO_3^- + H^+ \tag{I}$$

$$CO_2 + OH^- \rightleftharpoons HCO_3^- \tag{II}$$

$$HCO_3^- \rightleftharpoons CO_3^{2-} + H^+ \tag{III}$$

$$H_2O \rightleftharpoons OH^- + H^+ \tag{IV}$$

In the presence of weak acid HX, these are further coupled to the buffer reaction:

$$HX^z \rightleftharpoons X^{z-1} + H^+ \qquad (V)$$

where z is the charge of the protonated form of the weak acid; e.g., for phosphate buffer at physiological pH, X is HPO_4 and z is minus one.

In the special case of a protein (Pr) buffer, reaction V may be written as:

$$Pr^z \rightleftharpoons Pr^{z-1} + H^+ \qquad (VI)$$

where z, which is now the charge of a particular protein molecule, may exhibit a range of integral values at any one pH, as a result of the simulataneous equilibria of many amino acid ionizations.

When a gradient in the partial pressure of CO_2 – pCO_2 – is imposed across a layer of moderately alkaline solution containing a protein, the reactions proceed in the manner illustrated in Fig. 1. All species tend to diffuse from the side where their concentrations are higher – as denoted by larger type – to where they are lower.

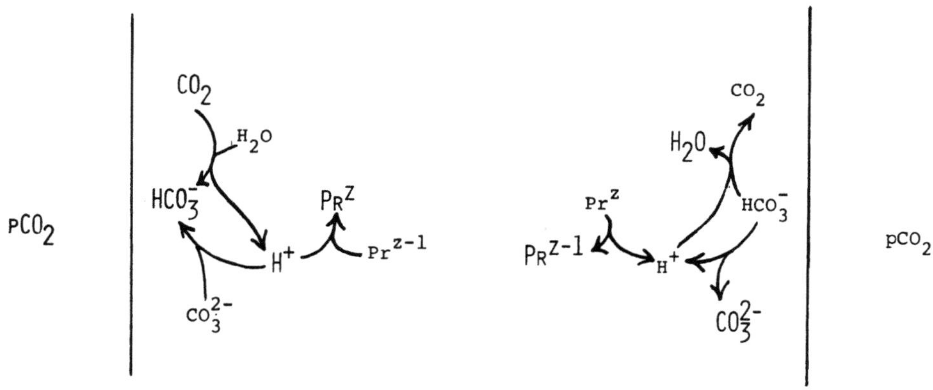

Fig. 1. Directions of the important chemical reactions on the two sides of a liquid film exposed to a gradient in CO_2 partial pressure. ***Larger type*** denotes the larger concentration of a given species

Consider, first, the situation in the absence of protein or any other extraneous buffer. Bicarbonate diffuses cocurrently with CO_2, while carbonate diffuses countercurrently at a rate half that of bicarbonate if there is to be no current (the fluxes of hydrogen and hydroxyl ions being negligible). Since the diffusivity of CO_3^{2-} is substantially less than that of HCO_3^{2-} (more than 20% less at infinite dilution), the high pCO_2 (upstream) side becomes positive relative to the downstream side of the liquid film.

Because of the potential difference, nonreactive cations like Na^+ or K^+ accumulate downstream, while anions like Cl^-accumulate upstream. This lowers the upstream pH and raises the downstream pH. That, in turn, lowers the bicarbonate concentration gradient, and hence its flux.

Furthermore, the electrically driven portion of the HCO_3^- flux – as expressed by the second term in Eq. (1) – is counter to the total HCO_3^- flux. In short, the potential adversely affects bicarbonate-facilitated CO_2 transport, both by decreasing its concentration gradient, and by causing countercurrent electrical migration. The net result, as theoretically estimated by Meldon et al. (1979), is a decrease in CO_2 flux facilitation in dilute solutions by as much as 30%, compared with a hypothetical case in which the carbonate diffusion coefficient is set equal to that of bicarbonate.

Let us next consider the case in which a simple weak acid buffer like phosphate is present. Thus in Fig. 1, replace Pr^{z-1} by X^{z-1}, and Pr^z by X^z. The buffer binds hydrogen ions released by CO_2 hydration – thus promoting bicarbonate formation – on the upstream side, and releases hydrogen ions – thus promoting CO_2 formation – on the downstream side. The effect on net CO_2 transfer in dilute solutions (Meldon et al., 1977) is shown in Fig. 2 for a number of inorganic weak acids including arsenious, the anion of which further promotes CO_2 transfer by catalyzing CO_2 hydration. The effect of noncatalytic buffers maximizes when the weak acid pK is in the range of pH values prevailing in solution, i.E., when buffering is maximal.

As demonstrated by Gros et al. (1976), the effect of a weak acid depends not on its pK alone, but also on its mobility. The appreciable CO_2 flux enhancement in a solution of inorganic phosphate is all but eliminated when the simple buffer is replaced by slow-moving cellulose phosphate. Gros and coworkers concluded that a buffer, to be effective, must be mobile.

If the buffer is much less diffusible than bicarbonate ion, a diffusion potential is established which is of the same sign as but much greater magnitude than that in the buffer-free system discussed above. However, Gros et al. (1976) choose to interpret the effect in somewhat different fashion, i.e., in terms of "buffer-facilitated proton transport", and this author is not entirely comfortable with such an approach.

When Gros et al. say that "under the present experimental conditions there are three prerequisites for facilitation", one of them being "the presence of mobile buffers acting as proton carriers", one is tempted to conclude that without any buffer there could be no facilitation. That, of course, is contradicted by the large body of data exhibiting appreciable facilitation of CO_2 transport in simple carbonate/bicarbonate solutions discussed earlier in this section. Apparently, by "under the present conditions" Gros and coworkers imply pH $\approx$ 7. It is true that in the latter regime there is a very low concentration of CO_3^{2-} – the only viable counterion to HCO_3^- in the absence of buffer – and, therefore, facilitation is minimal.

In addition, the expression "proton transport" should be used with care, since CO_2 is the only species of which there is net transport in the systems of interest here. Reconsider, for example, the buffer-free CO_3^{2-}/HCO_3^- solutions discussed earlier. Referring to Fig. 1, it should be clear that the absence of net proton transport is, in such cases, ensured by the diffusion of water to the high-pCO_2 side of the film. The same must be true of buffer-containing systems as well.

For these reasons this author prefers to analyze such phenomena in terms of electrical charges and diffusion potentials. However, it is acknowledged that, since protons are the medium of charge transfer here, a rigorous analysis of diffusion and species conservation of protons must also shed light on the mechanism of facilitated CO_2 transport.

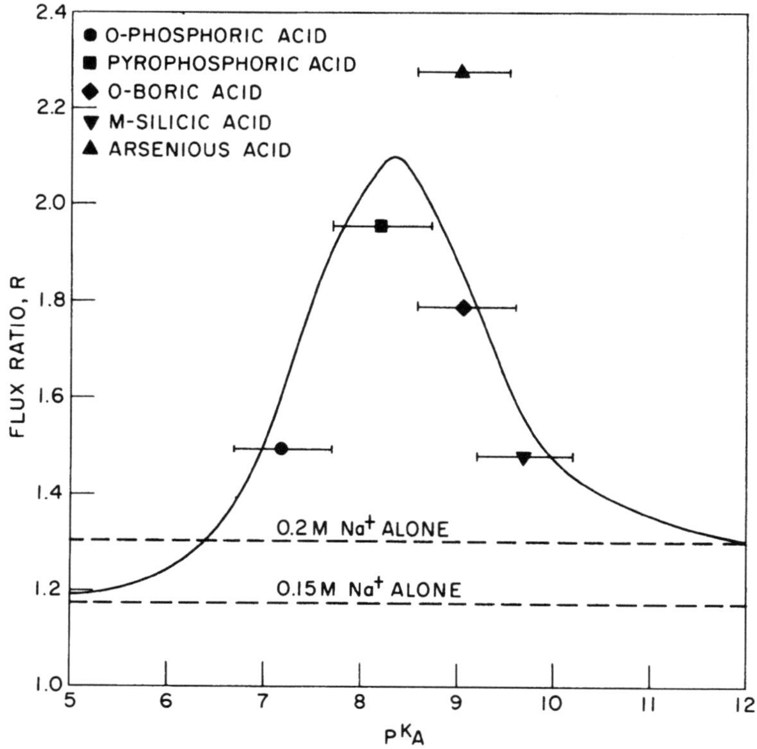

Fig. 2. The degree of steady-state CO_2 flux enhancement through a 1 1/2 mm-thick film of solution, separated by two silicone membranes from gas phases of pCO_2 = 90 and 10 mm Hg, respectively. R is the ratio of the flux to that measured with a water layer between the membranes. pK_A is the pK of the weak acids listed in the upper right corner, which are added to a 0.2 M NaOH solution. Reprinted from Meldon et al. (1977), with permission of the authors. See the original for further details

Let us, finally, consider CO_2 transport in systems containing proteins like albumin and hemoglobin. Because the diffusivities of the latter species are much less than that of bicarbonate, diffusion potentials are greatest here. At moderatley alkaline pH the carbonate ion concentration, as mentioned above, is very low. The protein must then be the counter-ion to bicarbonate which ensures zero electrical current.

The electrical field promotes net charge carriage by protein which is equal and opposite to that carried by bicarbonate ion. Referring to Eq. (1), the smaller the protein's average charge, $\overline{z_{Pr}}$, the greater the field must be. Thus, the potential difference is expected to maximize when the pH range in solution bounds the protein's isoelectric point, $\overline{pH_I}$. This is indeed found to be the case (see below).

Combination of Eq. (1)–(3) yields:

$$-\frac{dV}{dx} = -\frac{RT}{F}\left(\frac{D_B - D_{Pr}}{D_B\, C_B + D_{Pr}\,(\overline{z_{Pr}^2}\, C_{Pr} + \Sigma\, z_i^2\, C_i)}\right)\frac{dC_B}{dx} \tag{4}$$

where B denotes bicarbonate and i denotes reactively intert ions like Na^+, K^+ or Cl^-. The term $\overline{z_{Pr}^2}$ is the average squared protein charge, which can be estimated by the equation of Linderstrøm-Lang (1943):

$$\overline{z_{Pr}^2} = (\overline{z_{Pr}})^2 + \frac{\beta}{2.303} \tag{5}$$

and the average protein charge can be estimated using:

$$\overline{z_{Pr}} = -\beta(\mathrm{pH} - \mathrm{pH_I}) \tag{6}$$

i.e., β is the negative average slope of the protein titration curve.

Equation (4) indicates that the potential difference increases not only with an increase in the difference between D_B and D_{Pr}, and a decrease in the protein charge, but also with an increase in the size of the bicarbonate ion concentration gradient and a decrease in the ionic strength. The overall effects have been confirmed by measurements of actual potential differences (DeKoning et al., 1978; Meldon et al., 1978).

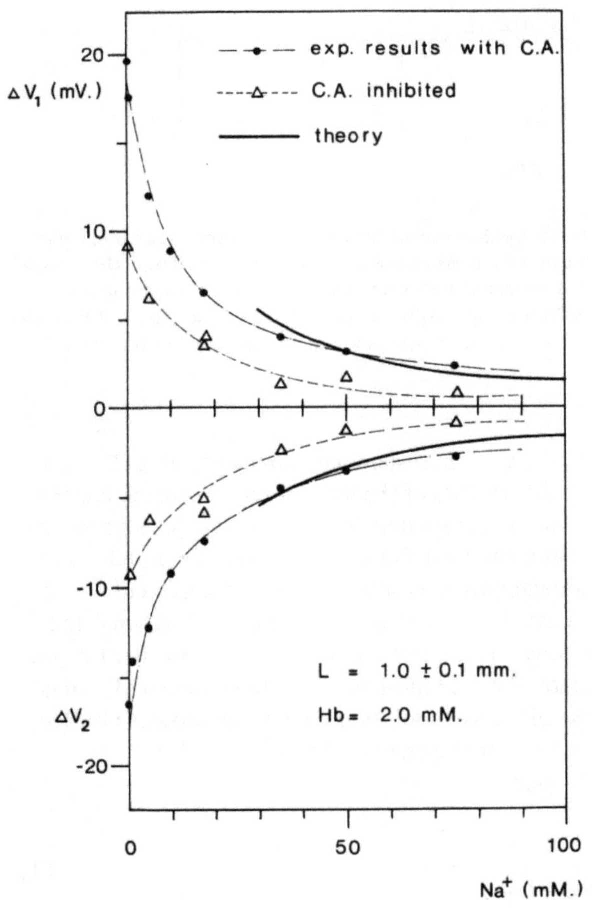

Fig. 3. Electrical potentials measured across 1 mm-thick layers of 2 mM bovine Hb solutions with active (*circles*) and poisoned (*triangles*) carbonic anhydrase. The *solid curve* is derived from theory. Na^+ denotes the added concentration of $NaHCO_3$. ΔV_1 and ΔV_2 are the results for the high pCO_2 gas respectively above and below the liquid firm. Published previously in part by Meldon et al. (1978). See the original for further details

Figure 3 presents the effect of varying $NaHCO_3$ concentration on the potentials measured across 1-mm thick films of 2 mM bovine hemoglobin solutions, with and without active carbonic anhydrase (C.A.). The boundary pCO_2 values were fixed at 38 and 2 mm Hg, respectively, in order to maximize dC_B/dx, and hence the electrical signal. Also included are theoretical values calculated on the assumptions of (i) local reaction equilibrium (i.E., active C.A.), and (ii) $pH \gg pH_I$ everywhere in the film.

The lower values measured in the presence of poisoned C.A. are consistent with the associated diminishing of dC_B/dx. The decreasing signals with increasing alkali concentrations are due to (i) the increasingly negative protein charge (which is close to zero in the presence of minimal alkali) and (ii) the ionic strength effect, which together overcome an increasing dC_B/dx.

Figure 4 presents experimental results for varying bovine hemoglobin concentrations. These further confirm the effect of decreasing alkali. In addition, the potential also increases as the protein concentration increases, as a result of an increase in dC_B/dx and, presumably, a decrease in D_{Pr}/D_B as well.

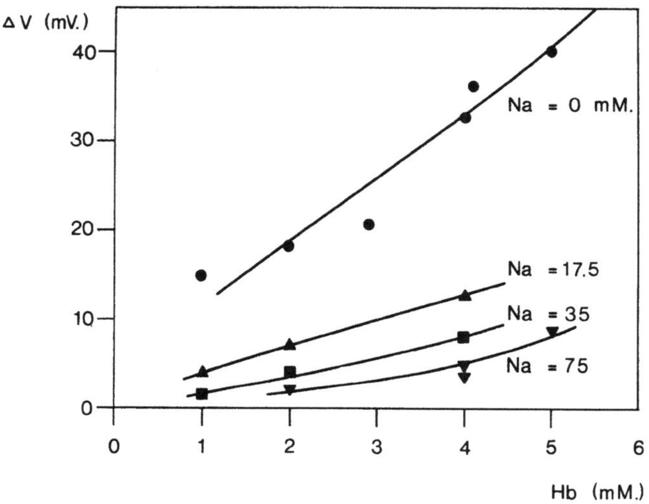

Fig. 4. Electrical potentials measured with a range of bovine Hb and $NaHCO_3$ concentrations and active carbonic anhydrase. The high pCO_2 gas stream was below the liquid layer. Reprinted from Meldon et al. (1978) with permission of the authors. See the original for further details

The theoretical analysis employed to predict the potential difference (Meldon et al., 1978) necessarily calculates the bicarbonate concentration gradient across the film as well. Hence it can also be used to predict bicarbonate-facilitated CO_2 transport. For the bovine hemoglobin systems above, bicarbonate diffusion enhances CO_2 transport by a factor, E, between 1 and 2, as opposed to a factor, E_0, of the order of 10 if $D_{Pr} = D_B$.

Table 1 lists similar effects caclculated (Meldon, 1975) for the 30% albumin solutions studied by Gros and Moll (1974), for which the boundary pCO_2 values were close to 7 and 4 mm Hg, respectively. Here the theoretical potential difference is less than 1 millivolt because (i) the pCO_2 gradient, and hence dC_B/dx, is small; and (ii) the isoelectric point of albumin is near 5, while that of hemoglobin is near 7; thus, in moderately alkaline solutions the albumin charge is highly negative. While the diffusion potential is relatively small, its effect is comparable. This is because the albumin charge causes it to accumulate appreciably – and, therefore, to lower the pH and, in turn, C_B – on the high pCO_2 side of the film. The predicted CO_2 flux enhancement, E, is gain found to be much less than E_0, in quantitative agreement with theory.

Table 1. Simulated results of Gros and Moll (1974)

30% albumin solution			
C_M, mM [a]	26	90	225
E_{Exp}	1.2	1.9	6.4
E_{Theory}	1.3	2.2	7.1
E_0	1.3	11.4	35.3
ΔV_{Theory}, mV	0.364	0.187	0.042

a C_M denotes calculated value of $\Sigma z_i C_i$

Diffusion potential effects are, in conclusion, significant. In protein systems the effects are particularly profound, leading to relatively low facilitation of CO_2 transport. Further work, however, is required to determine their significance in vivo.

References

DeKoning J, Stroeve P, Meldon JH (1978) Electrical potentials during CO_2 transport in hemoglobin solutions. Adv Exp Med Biol 94: 183–188

Enns T (1967) Facilitation by carbonic anhydrase of carbon dioxide transport. Science 155: 44–47

Gros G, Moll W (1974) Facilitated diffusion of CO_2 across albumin solutions. J Gen Physiol 64: 356–371

Gros G, Moll W, Hoppe H, Gros H (1976) Proton transport by phosphate diffusion – a mechanism of facilitated CO_2 transfer. J Gen Physiol 67: 773–790

Hill AV (1928) The diffusion of oxygen and lactic acid through tissues. Proc R Soc London B 104: 39–96

Krogh A (1918–19) The rate of diffusion of gases through animal tissues, with some remarks on the coefficient of invasion. J Physiol (London) 52: 391–408

Linderstrøm-Lang K, quoted in: Cohn EJ, Edsall JT (1943) Proteins, amino acids, and peptides. Reinhold, New York, p 462

Longmuir IS, Forster RE, Woo C-Y (1966) Diffusion of carbon dioxide through thin layers of solution. Nature (London) 209: 393–394

Meldon JH (1973) Reaction-enhanced mass transfer in thin liquid films. Ph D Thesis MIT, Cambridge Mass

Meldon JH (1975) The theoretical effect of diffusion potentials on carbon dioxide transport in protein solutions. Proc 5th Int Biophys Cong, Copenhagen, Abstr 413

Meldon JH, Smith KA, Colton CK (1977) The effect of weak acids upon the transport of carbon dioxide in alkaline solutions. Chem Eng Sci 32: 939–950

Meldon JH, De Koning J, Stroeve P (1978) Electrical potentials induced by CO_2 gradients in protein solutions and their role in CO_2 transport. Bioelectrochem Bioenerg 5: 77–87

Meldon JH, Smith KA, Colton CK (1979) An analysis of electrical effects induced by carbon dioxide transport in alkaline solutions. Recent Dev Sep Sci 5: 1–10

Vinograd JR, McBain JW (1941) Diffusion of electrolytes and of the ions in their mixtures. J Am Chem Soc 63: 2008–2015

Ward AJ, Robb WL (1967) Carbon dioxide-oxygen separation: facilitated transport of carbon dioxide across a liquid film. Science 156: 1481–1484

Evidence for Facilitated Transport of CO_2 in Muscle Tissue

P. SCHEID, T. KAWASHIRO, and J. PIIPER [1]

1 Introduction

There is convincing evidence that molecular diffusion may not be the only mechanism by which CO_2 is transported through liquid media. In fact, in buffered solutions, the movement of HCO_3^- and H^+ ions, coupled with mobile buffers, may enhance net CO_2 transfer substantially above the value expected for simple molecular diffusion (cf. Gros et al., 1976). The magnitude of this facilitation depends, amongst other variables, on the P_{CO_2} in the solution, being particularly pronounced at low levels of P_{CO_2}, and on the presence of carbonic anhydrase.

For muscle tissue, CO_2 facilitation has not been reported, and in fact it was argued that such enhancement would be of disadvantage for the respiring cells (Roughton, 1935). We have studied the diffusivity of CO_2 in the intact skeletal muscle, using techniques elaborated earlier (Kawashiro et al., 1975), and found a significant dependence of Krogh's diffusion constant for CO_2 upon P_{CO_2}. This paper reviews the material which we have reported in full detail (Kawashiro and Scheid, 1976).

2 Principle of Measurement

The principle for measuring Krogh's diffusion constant for CO_2 (and for O_2) in excised, respiring muscle is shown in Fig. 1. The muscle separates as a thin sheet two chambers. A large chamber is open and continuously flushed with pre-conditioned gases, the P_{CO_2} of which is constant (P_0); a small chamber can be flushed to establish steady state conditions at an initial P_{CO_2} [P_1 (0)], and is then closed, and the variation in time of this P_{CO_2} [P_1 (t)] is recorded continuously with an ordinary CO_2 electrode, used for measurement of blood gases.

In general, CO_2 transfer will result in changing P_1, until an equilibrium is reached when no CO_2 will cross the tissue surface bordering upon the closed chamber, and all CO_2 metabolically produced will leave the tissue via its surface to the open chamber. The tissue P_{CO_2} profile in this situation is depicted in Fig. 1, and the P_{CO_2} difference across the tissue in equilibrium, ΔP_{eq}, is related to the metabolic rate of CO_2 production, m, Krogh's diffusion constant, K, and the muscle thickness, l, as shown in Fig. 1.

---

1 Abteilung Physiologie, Max-Planck-Institut für Experimentelle Medizin, 3400 Göttingen/FRG

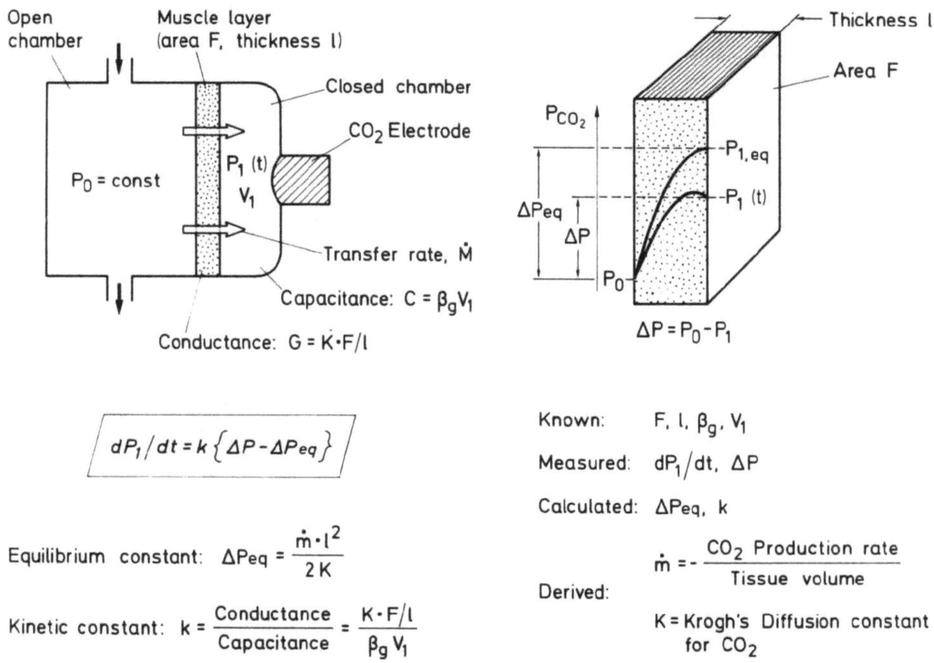

Fig. 1. Principle of measuring Krogh's diffusion constant, K, and metabolic activity, $\dot{m}$, in a respiring thin muscle sheet

When the P_{CO_2} difference across the tissue, ΔP, differs from ΔP_{eq}, P_1 will change at a rate that is proportional to the difference ($\Delta P - \Delta P_{eq}$), the rate constant, k, being related to K and l, and to tissue area, F, to chamber volume, V_1, and to the capacitance coefficient of the gas phase, β_g (cf. Piiper et al., 1971).

A plot of dP_1/dt against $\Delta P = P_1 - P_0$ allows to obtain both k and ΔP_{eq}, and, with known geometrical parameters, to calculate K and m. In the present study such plots were created at various levels of P_0, and thus of mean tissue P_{CO_2}, P_t, which enabled us to investigate relations between K (and $\dot{m}$) and P_t.

3 Methods

We used rat skeletal muscle (abdominal transverse and internal oblique), excised from animals immediately after death by an overdose of ether anesthesia. The muscle sheet (l ranging from 300 to 400 μm; F = 2.0 cm^2) was mounted to the small chamber (closed later; V_1 = 0.43 cm^3) which was attached to the open chamber. The whole apparatus with the CO_2 electrode (Radiometer) was thermostated at 37°C, the respective P_{CO_2} values being P_0 and P_1. The O_2 level was high enough to ensure sufficient O_2 supply.

After steady-state conditions had been attained, the small chamber was closed and $dP_1/\dot{d}t$ obtained from the first 6-min period. P_0 was varied between 0 and 140 torr (only one level being tested per experiment), and P_1 was varied randomly in a range of about ± 7 torr around the equilibrium value. P_t was calculated as the average from each P_{CO_2} profile (Kawashiro et al., 1976).

The thickness of the tissue was obtained at the end of the experiments by weighing (Kawashiro and Scheid, 1975).

4 Results

Figure 2 shows a typical plot of $dP_1/\dot{d}t$ against ΔP obtained in one experimental tissue. P_0 in this case was 14 torr, and $dP_1/\dot{d}t$ was obtained from 5 measurements at 3 levels of P_1. The calculated results for K_{CO_2} and $\dot{m}_{CO_2}$ are indicated.

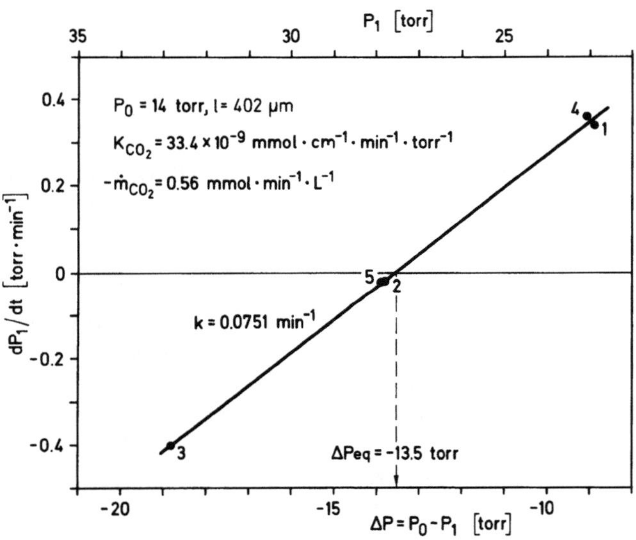

Fig. 2. Results obtained in one muscle tissue

Figure 3 shows a plot, comprising all results in 58 experiments, of K_{CO_2} against mean tissue P_{CO_2}, P_{tCO_2}. K_{CO_2} declines with P_{tCO_2}, reaching an apparently asymptotic value, the approach to which can be fitted by a single exponential, as shown in Fig. 3.

Metabolic CO_2 production rate, $\dot{m}_{CO_2}$, was apparently independent of P_{tCO_2}, suggesting no interference of tissue P_{CO_2} with metabolic activity.

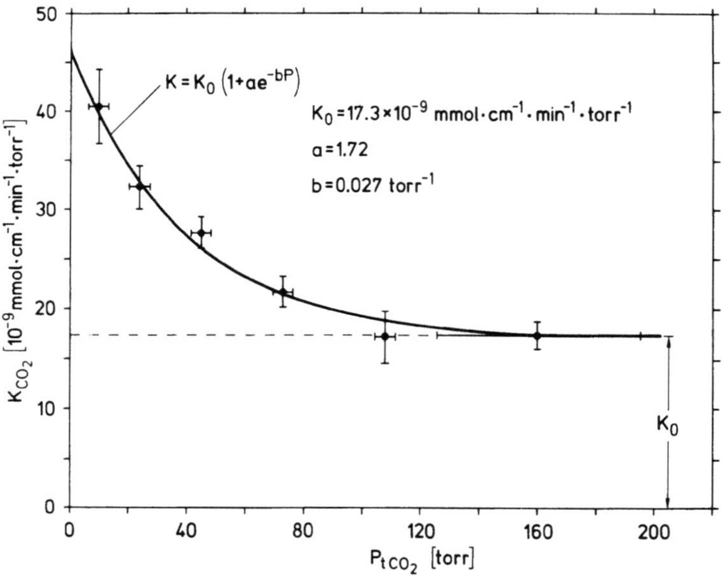

Fig. 3. Plot of Krogh's diffusion constant for CO_2, K_{CO_2}, against mean tissue P_{CO_2}, P_{tCO_2}. *Points* reflect results (± S.E.) obtained in 58 experiments. The *solid line* is an exponential function, fitted to the points by the least-square method. K_0, asymptotic value in this curve

5 Discussion

5.1 Facilitated Transport of CO_2

The results may be interpreted to show that CO_2 transfer through the muscle sheet is facilitated, particularly at lower levels of P_{CO_2}. Figure 4 compares K values for CO_2 at different levels of P_{CO_2} with the value for O_2 in the same preparation (Kawashiro et al., 1975). For high P_{CO_2}, the ratio K_{CO_2}/K_{O_2} approaches the water value, suggesting that the asymptotic K_{CO_2} value represents molecular CO_2 diffusion. It must then be concluded that at low levels of P_{CO_2}, net transfer of CO_2 is enhanced by some mechanism.

	$\dot{m}$ [mmol/(min·l)]	K [10^{-9} mmol·cm^{-1}·min^{-1}·torr^{-1}]			
		Muscle			Water
		P_{CO_2} = 0	P_{CO_2} = 40 torr	P_{CO_2} > 160 torr	
CO_2	0.74	47.1	27.4	17.7	47.2
O_2	0.87	1.31	1.31	1.31	2.76
CO_2/O_2	0.85	36	21	14	17

Fig. 4. Krogh's diffusion constant, K, and metabolic rate, $\dot{m}$, for CO_2 (this study) and for O_2 (Kawashiro et al., 1975) at some selected levels of tissue P_{CO_2}. For discussion see text

5.2 Mechanism of CO_2 Facilitation

For buffered solutions (cf. Gros et al., 1976) CO_2 facilitation occurs via net transport of HCO_3^-, and the following features are required for a significant contribution of this mechanism to CO_2 transfer:

a) Gradient of HCO_3^- across the tissue;
b) Mobile buffers for net transfer of H^+ ions;
c) Carbonic anhydrase.

In our preparation, a HCO_3^- gradient exists, notably in the low P_{CO_2} range. Mobile buffers, e.g. phosphate, are present in the intracellular space of the muscle. The presence of carbonic anhydrase in skeletal muscle has recently been suggested from direct measurements (cf. Chap. by Gros, this vol.).

5.3 Significance of CO_2 Facilitation

It has been shown (Suchdeo and Schultz, 1974) that the magnitude of CO_2 facilitation diminishes with the path length HCO_3^- travels between formation from CO_2 and recombination. It may thus appear that our results, obtained in a muscle sheet comprising about 10 layers of muscle fibers, overestimate the importance of CO_2 facilitation in the transport of metabolically produced CO_2 to the surrounding blood vessels. However, we believe that the facilitation measured by us is representative for the physiological CO_2 transport, as the mobile buffers required for facilitation are unlikely to cross the cell boundaries. It is then likely that formation of HCO_3^- and recombination have to take place in each cell, such that the path length to be considered is the diameter of the muscle fiber rather than the entire thickness of the membrane.

We have estimated the importance of facilitation for intracellular CO_2 transport under physiological conditions. To this end, we have calculated P_{CO_2} profiles in a solid cylinder in the absence of facilitation (K_{CO_2} constant and equal to the asymptotic value) or in its presence (K_{CO_2} depending on tissue P_{CO_2} as in Fig. 3). The results are summarized in Fig. 5 which shows that with CO_2 facilitation, P_{CO_2} is lower anywhere in the cylinder than with simple diffusion alone. The reduction in the maximum tissue P_{CO_2} (in the center of the cylinder) is about the same at the arterial and venous ends and during rest and exercise. However, particularly during exercise, CO_2 facilitation may become of significance, as in this case the reduction in tissue P_{CO_2} (at the center of the cylinder) is of the order of 1 torr and thus not negligible.

6 Conclusion

CO_2 facilitation appears to exist in skeletal muscle of the rat, and it constitutes an advantage for CO_2 release from muscle, particularly during exercise.

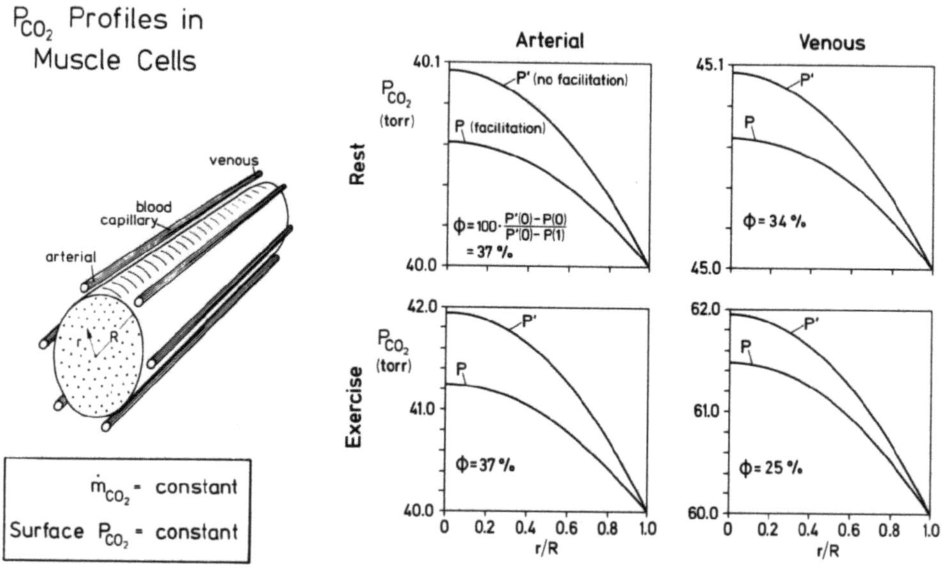

Fig. 5. Effect of facilitated CO_2 transport for P_{CO_2} profiles in a solid cylinder. Metabolic CO_2 production in the cylinder is homogeneously distributed throughout; P_{CO_2} at the surface of the cylinder is constant, at arterial or venous levels. See text for details

References

Gros G, Moll W, Hoppe H, Gros H (1976) Proton transport by phosphate diffusion – a mechanism of facilitated CO_2 transfer. J Gen Physiol 67: 773–790

Kawashiro T, Scheid P (1976) Measurement of Krogh's diffusion constant of CO_2 in respiring muscle at various CO_2 levels: evidence for facilitated diffusion. Pflügers Arch 362: 127–133

Kawashiro T, Nüsse W, Scheid P (1975) Determination of diffusivity of oxygen and carbon dioxide in respiring tissue: results in rat skeletal muscle. Pflügers Arch 359: 231–251

Piiper J, Dejours P, Haab P, Rahn H (1971) Concepts and basic quantities in gas exchange physiology. Respir Physiol 13: 292–304

Roughton FJW (1935) Recent work on carbon dioxide transport by the blood. Physiol Rev 15: 241–296

Suchdeo SR, Schultz JS (1974) Mass transfer of CO_2 across membranes: facilitation in the presence of bicarbonate ion and the enzyme carbonic anhydrase. Biochem Biophys Acta 352: 412–440

Carbamate Reactions of Proteins

Chairman: R. E. FORSTER

The Binding of CO_2 to Human Deoxyhemoglobin: An X-Ray Study Using Low-Salt Crystals

A. ARNONE, P.H. ROGERS, and P.D. BRILEY [1]

1 Introduction

The sites for the oxygen-linked formation of carbamate groups in hemoglobin were determined in 1969 by Kilmarin and Rossi-Bernardi [8]. They showed that if the NH_2-terminal amino groups of both the α and the β subunits are blocked by reaction with cyanate, linkage between CO_2 binding and O_2 binding is no longer observed. Subsequent studies in a number of laboratories have shown that the NH_2-terminal amino groups of the β chains are responsible for most (about 80%) of the oxygen-linked binding of CO_2 [5, 9, 10]. However, since the presence of 2,3-diphosphoglycerate reduces the amount of CO_2 bound to deoxyhemoglobin [5, 9, 10], the fraction of carbamate groups associated with the NH_2-termini of the α chains is probably increased in vivo.

Carbamate formation is governed by the following equilibria.

1. $Hb\text{-}NH_3^+ \rightleftharpoons Hb\text{-}NH_2 + H^+$
2. $Hb\text{-}NH_2 + CO_2 \rightleftharpoons Hb\text{-}NHCOO^- + H^+$

Since carbamate formation is favored in deoxyhemoglobin relative to oxyhemoglobin, it follows that the pK values of the NH_2-terminal amino groups must be lower in deoxyhemoglobin and/or the carbamate groups must be preferentially stabilized in deoxyhemoglobin. In the case of the α chains, the pK value of the NH_2-terminal amino groups [7] is acutally greater in deoxyhemoglobin than in liganded hemoglobin (7.79 versus 6.95, respectively). For the β chains, the pK value of the NH_2-terminal amino groups [7] is only slightly less in deoxyhemoglobin (6.84) versus liganded hemoglobin (7.05). Therefore, it is reasonable to expect that the carbamate groups of the α and β chains should be stabilized by favorable noncovalent bonds in deoxyhemoglobin. The aim of our work is to use X-ray crystallographic analysis to determine directly the stereochemistry of carbamate interactions in deoxyhemoglobin.

A previous X-ray study found evidence for carbamte formation at the β chain NH_2-termini as well as an unexpected binding site for CO_2 in a hydrophobic hole near the β heme group [1]. However, the hemoglobin crystals used in that study were initially grown from solutions of concentrated ammonium sulfate and then transferred for one day to a solution composed of 70% methypentanediol with CO_2 (1 atm)/bicarbonate buffer at pH 7.4. Therefore, the conditions of this experiment

1 The University of Iowa, Department of Biochemistry, Iowa City, Iowa 52242/USA

(i.e., high CO_2 pressure, high organic solvent concentration, and an undetermined amount of sulfate) may have exaggerated a minor effect, the binding of CO_2 in a hydrophobic hole to the rear of the β heme group, and inhibited a major effect, the reaction of CO_2 with the NH_2-terminal groups of the α chains.

We have now carried out another X-ray study of carbamino hemoglobin using crystals of human deoxyhemoglobin grown from solutions of polyethylene glycol [15] to achieve an environment which is much closer to physiological conditions. Difference electron density analysis using these crystals clearly shows formation of carbamate groups at the NH_2-termini of both the β chains and the α chains, but does not show any other CO_2 binding sites. The carbamate group at Val 1β is oriented so that it can form an intrasubunit salt-bridge with the positively charged side chain of Lys 82 β. In contrast, the carbamate group at Val 1 α forms an intrasubunit hydrogen bond with the hydroxyl group of Ser 131 α. These carbamate-protein interactions appear to be related to the mode of inorganic anion binding to human deoxyhemoglobin.

2 Materials and Methods

2.1 Low-Salt Crystallization of Human Deoxyhemoglobin

Hemoglobin was prepared from freshly drawn blood by washing the red cells with 0.9% NaCl, lysing with water, and (after centrifuging to remove the membranes) gel filtering through a G25 Sephadex column equilibrated with 0.01 M ammonium phosphate at pH 7.2 (see reference [13] for more details). The hemoglobin was then crystallized in the deoxy state from solutions of polyethylene glycol (PEG) according to the method of Ward et al. [15] except that the mother liquor also contained 0.10 KC and 6.2 mM ferrous chloride. Specifically, the solutions which yield the best crystals have the following composition; 10 mg/ml human hemoglobin, 10 mM potassium phosphate, 100 mM potassium chloride, 8% to 9% weight PEG (molecular weight 6000), and 6.2 mM ferrous chloride. The final mixture, which has a pH near 7.0, was prepared in a glove box (filled with nitrogen) by mixing stock solutions of the individual components.

Crystals with dimensions exceeding 1.0 mm grow in about a month. The crystals are orthorhombic with space group symmetry $P2_1 2_1 2$ and unit cell dimensions a = 96.8 Å, b = 98.9 Å, and c = 65.8 Å. These values are very similar to those reported by Ward et al. [15]: those being 97.1 Å, 99.4 Å, and 66.0 Å, respectively.

2.2 Crystal Soaking Conditions

The CO_2 soaking solution has the following composition: 95 mM potassium chloride, 12% by weight PEG, and CO_2 (100 mm Hg)/bicarbonate buffer at pH 7.5. Large well-formed crystals of deoxyhemoglobin were soaked in 1.0 ml of this solution for

1 day prior to data collection. Other crystals were soaked in a similar solution in which the CO_2/bicarbonate buffer was replaced by 50 mM bis Tris so that the pH is 7.5. Oxygen was eliminated from the soaking solutions by bubbling with CO_2 (100 mm Hg)/N_2 or pure N_2.

Depending on the soaking conditions (i.e., with or without CO_2), crystals were mounted in quartz capillary tubes under an atmosphere of CO_2 (100 mm Hg)/N_2 or 1 atm N_2.

2.3 Data Collection and Reduction

The diffraction data were collected to a resolution of 4.4 Å on an Enraf-Nonius CAD4 diffractometer using the omega scan mode. Degradation due to radiation damage never exceeded 10% as determined by repeated measurement of four standard reflections. An empirical correction for radiation damage was incorporated into the crystal scaling corrections as described previously [14]. Absorption was corrected for by the method of North et al. [12].

2.4 Phasing and Difference Map Calculation

Structure factor phases (to a resolution of 4.4 Å) were first determined by the method of isomorphous replacement using a single heavy-atom derivative. The heavy-atom derivative was easily prepared by reacting human hemoglobin with 1 equivalent of p-chloromercuribenzoate per $\alpha\beta$ dimer according to the published procedure of Muirhead et al. [11]. Since only one heavy-atom derivative was used, anomalous dispersion data were required to resolve the phasing ambiguity for the noncentric reflections.

The experimentally determined phases were then used to calculate an anomalous difference Fourier map of native human deoxyhemoglobin. This map clearly revealed the locations of the heme iron atoms which were then further refined using the anomalous dispersion data and a phasing and refinement computer program written by Michael Rossman and extensively modified by Ten Eyck [14]. Knowing the iron atom coordinates allowed us to position an atomic model of human deoxyhemoglobin [6] in the unit cell and, therefore, to calculate structure factor phases. We repeated this procedure once more by using the calculated phases in the refinement of the iron atom positions and then calculating new phases from the repositioned atomic model.

The difference electron density map was computed using the difference amplitudes ($|F|_{CO_2} - |F|_{noCO_2}$) together with the calculated phases, and then symmetry avaraged about the molecular dyad that relates the two equivalent $\alpha\beta$ dimers [11].

3 Results and Discussion

3.1 Carbamate Interactions at the NH_2-Termini of the β Chains

A common way to interpret difference electron density maps is to superimpose the same regions of the difference electron density with the electron density of the native protein. In Fig. 1a and b we show such a superposition in the region of the NH_2-termini of the β chains. Figure 1a shows this region of the 2.5 Å native electron density map as determined from high-salt crystals (i.e., crystals grown from 2.3 M ammonium sulfate). In Fig. 1b sections of the CO_2 difference electron density map (as determined from the low-salt PEG crystals) have been superimposed on the native map.

The only significant features in this region of the CO_2 difference map are two intense positive peaks centered immediately adjacent to the native electron density of the NH_2-termini of the β chains. These positive peaks clearly represent the binding of CO_2 to the NH_2-terminal amino groups, presumably as carbamate.

From the coordinates of these peaks, carbamate groups were positioned on an atomic model of human deoxyhemoglobin. This model building study shows that the carbamate groups can easily form intrasubunit salt-bridges with Lys 82 β. It seems less likely that the carbamate groups can form intersubunit bonds with Lys 82 or His 143 on the opposite β chain, but this point will have to be checked when higher resolution studies are completed.

Since no other significant difference electron density peaks are observed in this region of the hemoglobin molecule, we can conclude that CO_2 binding does not result in major changes in β chain tertiary structure. Similarly, there is no evidence for anion displacement. In the previous X-ray study [1] with high-salt crystals which were soaked in 70% methylpentanediol and CO_2 (1 atm)/bicarbonate, negative difference electron density peaks occurred at the positions of the features labeled XO_4^- in the native electron density map (see Fig. 1a). XO_4^- is a sulfate or phosphate anion which apparently remains bound to each β chain after soaking in the methylpentanediol solution, but is displaced by the formation of carbamate. Since no large negative difference peaks appear in our current difference map, we must conclude that under the conditions of this experiment (i.e., 100 mM chloride at pH 7.5) very little chloride is bound in the immediate vicinity of the NH_2-terminal amino groups of the β chains.

3.2 Carbamate Interactions at the NH_2-Termini of the α Chains

The stereochemistry of the NH_2-terminal region of each α chain is shown in Fig. 2. Recent X-ray studies have shown that there are two anion binding sites associated with the NH_2-terminal amino groups of the α chains [2–4]. Specifically, it has been determined that selenate and sulfate (and possibly phosphate) bind between the α-amino group of Val 1 α and the hydroxyl group of Ser 131 on the same α chain, as well as between Val 1 α and the guanidinium side chain of Arg 141 on the opposite α chain.

The native electron density (in the high-salt crystals) of the region near the NH_2-terminal amino groups of the α chains is shown in Fig. 3a. This region includes Val 1α

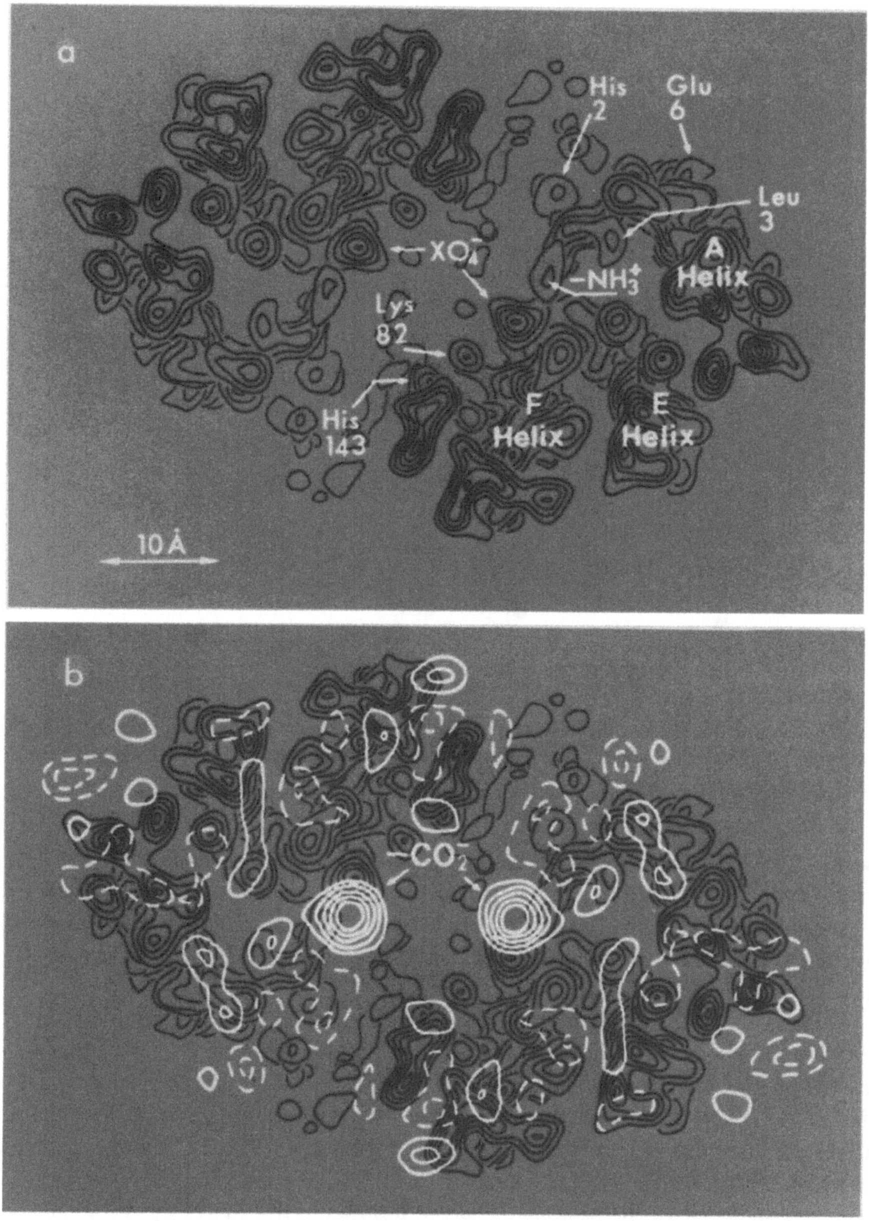

Fig. 1a, b. Fourier maps showing the interaction of CO_2 with the NH_2-terminal amino groups of the β chains. **a** Composite of two-dimensional sections of the native electron density map as seen in "high-salt" crystals at 2.5 Å resolution. The map consists of 9 sections spaced 1 Å apart, − 15 Å to − 23 Å below the center of the tetramer. All features occur in pairs (one on each β subunit) related by the molecular two-fold axis of symmetry. With the exception of XO_4^-, an inorganic anion (see text), only one member of each pair has been labeled. **b** Section − 18 of the CO_2 difference electron density map at 4.4 Å resolution (*white contours*) superimposed on the native protein electron density. Positive difference electron density is marked by *solid white contours*, negative difference density by *dashed white contours*. The diffraction data used to calculate the difference map was collected with "low-salt" crystals (see text)

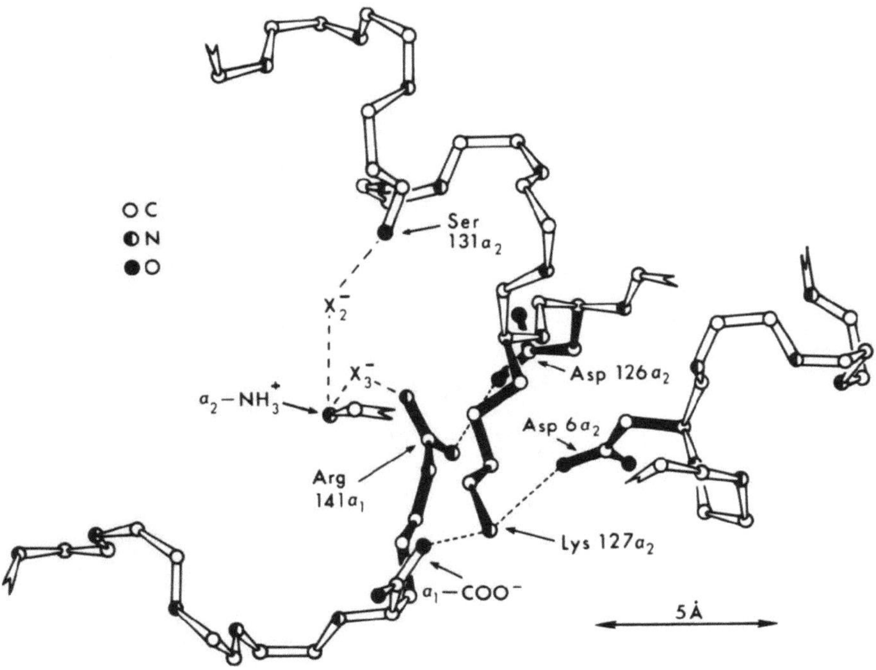

Fig. 2. Sketch showing the environment of the NH_2-terminal amino group of the α_2 subunit and the anion binding sites, X_2^- and X_3^-. For clarity, the peptide backbone between the first and fifth residues has been deleted

as well as two low-level features, labeled X_1 and X_2^-. X_1 is probably a water molecule since it cannot be replaced by selenate [2]. X_2^-, on the other hand, can be replaced by selenate [2], and therefore probably represents a sulfate anion which is bound between the α-amino group of Val 1 α and the hydroxyl group of Ser 131 α (not shown in Fig. 3a) in the high-salt crystals.

In Fig. 3b the CO_2 difference electron density map has been superimposed on the native electron density shown in Fig. 3a. As expected, positive difference peaks show the presence of bound CO_2 adjacent to the NH_2-terminal amino groups of the α chains. Moreover, these peaks are less intense than their counterparts at the β chain NH_2-termini, and this is consistent with solution studies which show that the β chains have a higher affinity for CO_2 [5, 9, 10]. The α chain carbamate groups, however, are not oriented toward the positively charged residues in the area (Lys 127 α and Arg 141 α), but toward the hydroxyl group of Ser 131 α on the same α chain. *That is, the carbamate group of each α chain occupies the X_2^- binding site and, by forming a hydrogen bond with Ser 131 α, stabilizes the deoxyhemoglobin structure.*

Competition between sulfate and CO_2 for site X_2^- could explain why carbamate formation was not observed in the previous X-ray study [1]. Similarly, we must conclude that since carbamate formation is observed in the present study, chloride anions must not bind to X_2^- with *high occupancy* at a chloride concentration of 100 mM and pH 7.5.

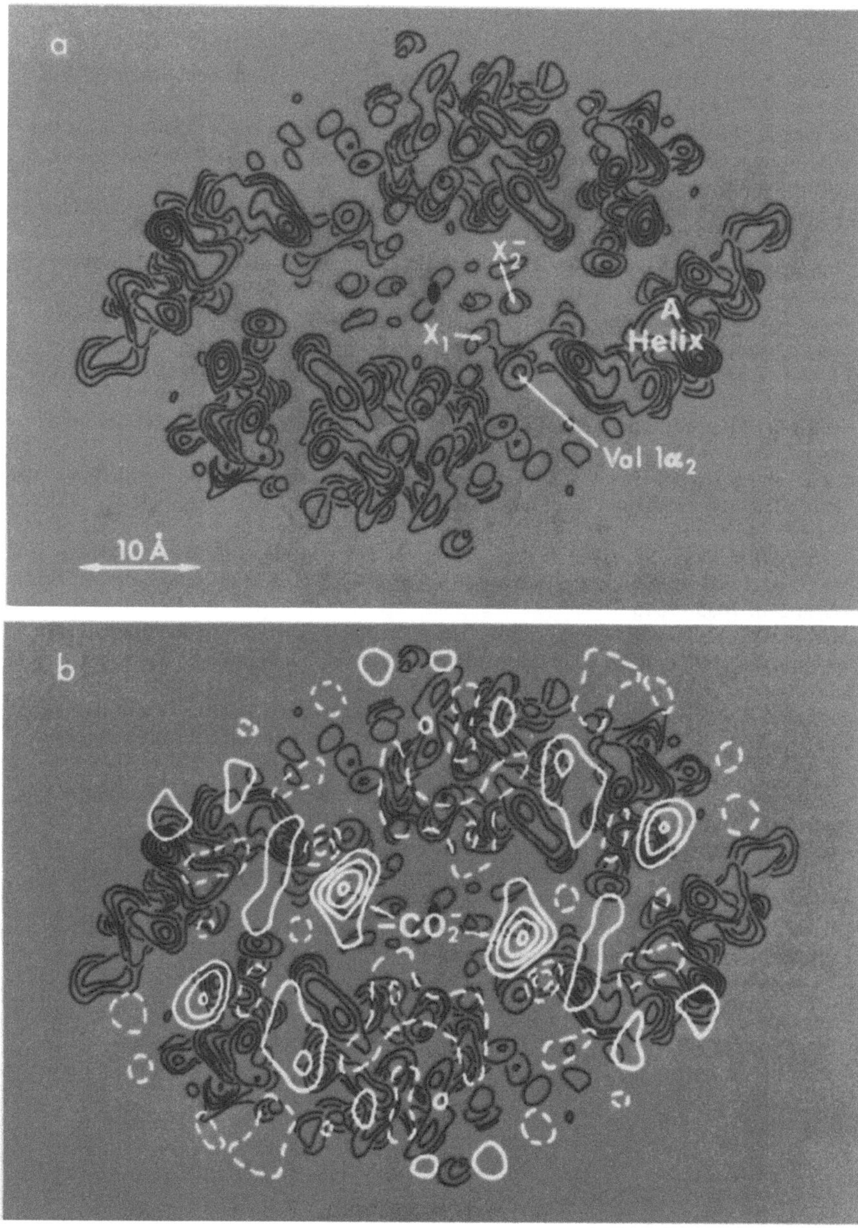

Fig. 3a, b. Fourier maps showing the interaction of CO_2 with the NH_2-terminal amino groups of the α chains. **a** Native electron density as in Fig. 1a. The map consists of 6 sections, 16 Å to 21 Å above the center of the tetramer. Features X_1 and X_2^- are described in the text. **b** Section + 19 of the CO_2 difference map (*white contours*) superimposed on the native protein electron density

References

1. Arnone A (1974) X-ray studies of the interaction of CO_2 with human deoxyhemoglobin. Nature 247: 143–145
2. Arnone A, Williams DL, Jr (1977) Crystallographic evidence for anion binding sites at the NH_2-termini of the α subunits of human deoxyhemoglobin. In: Labie D, Poyart C, Rosa S (eds) Interactions moléculaires de l'hemoglobine. INSERM, Paris, pp 15–22
3. Arnone A, O'Donnell S, Schuster T (1976) X-ray diffraction studies of carbamylated human deoxyhemoglobin. Fed Proc Fed Am Soc Exp Biol 35: 1604
4. Arnone A, Benesch RE, Benesch R (1977) Structure of human deoxyhemoglobin specifically modified with pyridoxal compounds. J Mol Biol 115: 627–642
5. Bauer C, Baumann R, Engels V, Pacyna B (1975) The carbon dioxide affinity of various human hemoglobins. J Biol Chem 250: 2173–2176
6. Fermi G (1975) Three-dimensional fourier synthesis of human deoxyhemoglobin at 2.5 Å resolution: refinement of the atomic model. J Mol Biol 97: 237–256
7. Garner MH, Bogardt RA, Jr, Gurd FRN (1975) Determination of the pK values of the α-amino groups of human hemoglobin. J Biol Chem 250: 4398–4404
8. Kilmartin JV, Rossi-Bernardi L (1969) Inhibition of CO_2 combination and reduction of the Bohr effect in haemoglobin chemically modified at its α-amino groups. Nature 222: 1243–1246
9. Kilmartin JV, Fogg J, Luzzana M, Rossi-Bernardi L (1973) Role of the α-amino groups of the α and β chains of human hemoglobin in oxygen-linked binding of carbon dioxide. Biol Chem 248: 7039–7043
10. Matthew JB, Morrow JS, Wittebort RD, Gurd FRN (1977) Quantitative determination of carbino adducts of α and β chains in human adult hemoglobin in presence and absence of carbon monoxide and 2,3-diphosphoglycerate. J Biol Chem 252: 2234–2244
11. Muirhead H, Cox JM, Mazzarella L, Perutz MF (1967) Structure and function of haemoglobin III, a three-dimensional fourier synthesis of human deoxyhemoglobin in 5.5 Å resolution. J Mol Biol 28: 117–156
12. North ACT, Philips DC, Mathews FS (1968) A semi-empirical method of adsorption correction. Acta Crystallogr Sect A 24: 351–355
13. Perutz MF (1968) Preparation of hemoglobin crystals. J Cryst Growth 2: 54–56
14. Ten Eyck LF, Arnone A (1976) Three-dimensional fourier synthesis of human deoxyhemoglobin at 2.5 Å resolution. J Mol Biol 100: 3–11
15. Ward KB, Wishner BC, Lattman EE, Lover WE (1975) Structure of deoxyhemoglobin A crystals grown from polyethylene glycol solutions. J Mol Biol 98: 161–177

The Determination of CO_2 Bound to Hemoglobin as Carbamate

M. PERRELLA [1] and L. ROSSI-BERNARDI [2]

1 Introduction

The investigation of the reaction between CO_2 and hemoglobin has proved important for a clearer understanding of the physiological functions of human blood. The elucidation of the mechanism of CO_2 binding to the four N-terminal groups of human hemoglobin has also provided a better insight into the structure-function relationship of this protein. Unfortunately, the methods so far developed for the determination of CO_2 bound to hemoglobin as carbaminohemoglobin (carbhemoglobin) are not simple to apply, and this may well explain why only a few papers have appeared in the last few years on this problem.

Since a complete description of all analytical steps and the instrumentation used to determine carbhemoglobin has never been published, this paper gives a complete and up-to-date description of the original technique, which we have developed and used extensively in several previous studies on the reaction of CO_2 with human and bovine hemoglobin. The method described is a derivation of that originally developed by Faurholt [1] for the determination of the CO_2 adducts of simple amines, and later applied by Ferguson and Roughton [2] to the study of carbhemoglobin. Both methods made use of barium hydroxide to remove all CO_2 in solution exept for that forming carbhemoglobin, but unfortunately the analytical accuracy has proved, in the presence of hemoglobin, rather low.

Since it was found by Ferguson and Roughton that the presence of hemoglobin somehow prevented the complete precipitation of barium carbonate from solution, Rossi-Bernardi et al. [9] and Perrella et al. [4, 7] proposed to replace barium hydroxide with a suspension of anion exchanger resin. The introduction of the ion-exchange step and several other improvements have resulted in a four-fold increase in the accuracy of the method for carbhemoglobin determination over the original Ferguson and Roughton procedure, allowing its successful application to the physicochemical characterization of the interaction of CO_2 with normal [4] and chemically modified human hemoglobin [6]. A summary of the method has appeared elsewhere [7]. The method herein described can be directly applied to normal human hemoglobin. Comments are made on its possible extension to the study of other hemoglobins.

---

1 Cattedra di Enzimologia, Universitá di Milano, Via G. Celoria 2, 20133 Milan/Italy
2 Cattedra di Chimica Biologica, Universitá di Milano, Ospedale San Raffaele, Via Olgettina 6o, 20090 Segrate (Milan)/Italy

2 Description of the Method

2.1 The Equilibrium Reactions of CO_2 in Aqueous Solutions of Hemoglobin

The reactions of CO_2 in an aqueous solution containing amino groups of amines or proteins (R-NH, where R may be hemoglobin) conform to the following scheme:

$$CO_2 + H_2O \rightleftharpoons H_2CO_3 \rightleftharpoons H^+ + HCO_3^- \ldots\ldots K_1 \qquad (1)$$

$$CO_2 + OH^- \rightleftharpoons HCO_3^- \ldots\ldots\ldots\ldots\ldots\ldots K_{OH^-} \qquad (2)$$

$$HCO_3^- \rightleftharpoons CO_3^= + H^+ \ldots\ldots\ldots\ldots\ldots\ldots K_2 \qquad (3)$$

$$CO_2 + R-NH_2 \rightleftharpoons R-NHCOO^- + H^+ \ldots\ldots K_c \qquad (4)$$

$$R-NH_3^+ \rightleftharpoons R-NH_2 + H^+ \ldots\ldots\ldots\ldots K_z \qquad (5)$$

K_c is the fundamental carbamate constant for a particular amino group, and K_z is its ionization constant. The undissociated carbamic acid species R-NHCOOH can be neglected at pH > 6.8 [10]. In hemoglobin, only two kinds of group bind CO_2 carbamino-wise, i.e., the ϵ-amino groups of lysine, $pK_z \sim 10$ [11], and the four terminal α-amino groups, $pK_z \sim 7.5$ [3].

2.2 Equilibration of the Hemoglobin Solution

Samples (2 to 3 ml) of purified and concentrated solutions of deoxy or carboxyhemoglobin (10 to 18 mM in heme) in 0.1 M KCl, containing also acetazolamide (1 mg/ml) or ethoxolamide (0.2 mg/ml), are equilibrated with a gas phase containing CO_2 in a double-walled glass tonometer (Fig. 1A). Temperature control (37°C ± 0.1) is obtained by circulating water in the tonometer from a thermostated bath. The tonometer is mounted on a standard Radiometer frame (Fig. 1A), which allows rapid vibration and mixing of the viscous hemoglobin solution. The equilibrating gas mixtures saturated with water at 37°C (Fig. 1F) stream continuously over the sample. This apparatus allows 99% deoxygenation of concentrated (up to 250 g/l) normal hemoglobin in about 30 min. Complete CO_2 equilibrium with a gas phase containing CO_2 can be obtained in about 15 min.

The pH of the equilibrated sample is measured by the use of a Radiometer glass microelectrode (type G297/G2). The pH of the hemoglobin sample can be adjusted to a programmed value above pH 7 by the addition of a few microliters of 1 or 2 M $KHCO_3$. The total CO_2 content of the equilibrated sample can be measured according to Peters and Van Slyke [8].

2.3 Chemical Entrapment of the Free Dissolved CO_2

The hemoglobin solution, equilibrated with a known P_{CO_2}, is transferred anaerobically from the tonometer to one of the two syringes shown in Fig. 1B and mixed with a triethylamino-HCl buffer (TEA-HCl) contained in the other syringe, in a simplified, hand-driven, continuous-flow apparatus (Fig. 1D). The buffer is 0.3 M in amine.

Fig. 1. The complete setup used for the determination of carbamate compounds of human hemoglobin. *A* thermostated glass vessel used to equilibrate hemoglobin with a gas phase. *B* thermostated tuberculin syringes. *C* thermostated vessel. *D* a plunger to mix the content of the two syringes by hand. *M* mixing chamber. *E, F* thermostated gas humidifying vessels

The HCl concentration (0.05 to 0.1 M) is adjusted so that the pH after mixing of the solution in the mixing chamber (Fig. 1 M) is in the range 11 to 11.5. The syringes used are two tuberculin syringes provided with thermostating jackets for temperature control. The TEA-HCl buffer is kept at 0° to 1°C and the hemoglobin solution at the temperature of equilibration. The hemoglobin-containing syringe is also fitted with a Teflon plunger.

The mixture from M (Fig. 1) (at pH about 11) is led into a 1-mm bore, 10 cm-long plastic tubing to ensure complete mixing of reagents within a few ms and from there delivered into the double-walled glass chamber (Fig. 1C), which is thermostated at 0° to 1°C. The mixture is then left for 5 s, during which time the free, dissolved CO_2 reacts mainly with the newly deprotonated ϵ-amino groups of lysine residues of hemoglobin and disappears from solution. The rather low pH ($\sim$ 11), which is obtained a few ms after mixing, ensures that free CO_2 is taken up almost exclusively through reaction (4) and not through reaction (2). It has been shown (4) that if the pH after mixing in M is about 12, a small but significant amount of the dissolved CO_2 will react through reactions (1) or (2). The effect of carbonic anhydrase on reaction (1) is inhibited by the inhibitors added to the hemoglobin solution.

2.4 Separation of $HbCO_2$ from HCO_3^- and $CO_3^=$

Next, 0.3 to 0.5 ml of ice-cold 1 N KOH and 2 ml of a suspension of Bio-Rad Ag 1x8 resin (in the OH^- form) are added in rapid succession, thus raising the pH to 12 to 12.5. The purpose of this step is (i) to remove any HCO_3^- or $CO_3^=$ by ion exchange with OH^-, leaving $HbCO_2$ as the only form of CO_2 in solution, and (ii) to stabilize $HbCO_2$, which at this pH only slowly dissociates to carbonate [1].

The resin suspension is prepared by addition of two volumes of water to one volume of packed resin. The addition of KOH may be unnecessary, since the resin releases basic groups upon aging, so that following its addition the pH of the suspension is 12 to 13. The magnet in C is not effective in stirring the thick suspension which results from the addition of the resin. The nitrogen stream coming from E (Fig. 1) and used to exclude air from C is therefore bypassed and allowed to bubble in the liquid. Gas bubbling is continued for 70 to 80 s before collecting about 2.5 ml of resin-free sample for gasometric analysis. Filtration of the resin is accomplished by drawing the sample into the analytical burette by suction through a 7-cm long piece of plastic tubing filled with a nylon net. The time needed for sample collection is 30 to 50 s. The overall elapsed time from mixing to sample collection is usually 150 ± 10 s. The amount of resin and the stirring time needed to remove carbonate ions exhaustively from the resulting mixture in C are determined by replacing the hemoglobin solution with a carbonate-bicarbonate buffer. For human hemoglobin studies, the composition of this solution (typically 100 mM total CO_2, 0.1 M KCl, and practically zero P_{CO_2}) is chosen to approximate the behavior of a deoxyhemoglobin solution equilibrated with a 22% CO_2 gas phase at pH 7.4, 37°C. By determining the CO_2 content of the carbonate-bicarbonate solution collected from C (Fig. 1) at different times and resin concentrations, it was found that a 30 to 40-s reaction time and 2 ml of resin suspension diluted as described above are sufficient to remove all the HCO_3^- and $CO_3^=$ ions. For safety, this time interval should be routinely doubled.

2.5 CO_2 Analysis

Total CO_2 determinations can be carried out according to Peters and Van Slyke [8] in a standard Van Slyke apparatus. For the 2-ml samples collected from C and used for the determination of $HbCO_2$, readings are taken at the 0.5-ml mark. Owing to the high pH of the solution, acidification to pH 3.5 is carried out by use of 1.2 N lactic acid. The presence of the volatile triethylamine buffer makes it necessary to avoid the use of strong alkali for CO_2 absorption, since at pH ~ 12 a significant release of the amine occurs. Thus, the alkali ordinarily used to absorb CO_2 is replaced with a glycine-NaOH buffer (2.5 M Gly–1.5 N NaOH, pH 9). The blank for the CO_2 analysis is obtained by replacing the hemoglobin solution in syringe B (Fig. 1) with an equal volume of an 0.1 M solution of bicarbonate in 0.1 M KCl. Under the conditions described above for the CO_2 analysis, the blank value should be about 7 mm Hg. Higher values indicate that the amount of added exchanger resin is not sufficient to remove all the bicarbonate or that the resin has a reduced capacity.

2.6 Determination of Hemoglobin Concentration

During the ion-exchange process in C, the hemoglobin concentration of the original sample changes to an unknown extent. Only the accurate evaluation of this dilution factor allows the correct estimation of the sum of CO_2 dissolved plus $HbCO_2$ in the hemoglobin sample under investigation. Since human, unlike bovine, hemoglobin is denatured at alkaline pH, the cyanmethemoglobin method of Drabkin cannot be safely used. The determination of human hemoglobin concentration, which was exposed to alkaline pH, is carried out by diluting the protein in 0.1 N KOH and reading its absorbance in a Beckman Acta III spectrophotometer at 390 nm. Separate experiments with hemoglobin of known concentration under identical conditions have shown that absorbance is linear with protein concentration, at least up to 0.8 absorbance units, which is the highest value ever read. Absorbance decreases slightly (3% to 5%) with time, but a plateau is reached after 7 to 8 h and is maintained for at least 20 h. To make an allowance for differences in rate of denaturation of different samples (for instance, deoxyhemoglobin is more easily denatured than HbCO), the concentration of hemoglobin sampled from C (Fig. 1) is calculated assuming as a standard an approximately equal amount of protein from the original hemoglobin sample, whose concentration is in turn obtained by the cyanmethemoglobin method of Drabkin ($\epsilon\ \frac{1\ \text{mM Fe}}{1\ \text{cm}} = 11.05$ at 540 nm). The standard is then diluted, and a parallel reading at 390 nm is taken after the same time of denaturation. If measurements are in triplicate and readings are taken after 15 h standing at 20° to 25°C, the reproducibility of the measurements is about 1%.

2.7 Calculation of Hemoglobin Carbamate

Gasometric analysis of the total CO_2 of the diluted hemoglobin sample obtained from C (Fig. 1) directly gives the sum of $HbCO_2$ and CO_2, since HCO_3^- and $CO_3^=$ have been removed in the ion exchange process in C. The sum of $HbCO_2$ plus CO_2 in the original hemoglobin sample is obtained by multiplying the gasometric value by a factor which takes into account the dilution undergone by the hemoglobin. The dilution factor is simply the ratio between hemoglobin concentration in the syringe shown in Fig. 1B and that in C. A precise estimation of Q', the solubility coefficient of CO_2 in the hemoglobin solutions used, is thus needed to subtract this quantity from the total carbamate obtained from the gasometric analysis and obtain the true $HbCO_2$. The fractional saturation of CO_2 per hemoglobin tetramer is given by $\phi = [HbCO_2]/[Hb]$, where [Hb] represents the concentration of tetrameric hemoglobin.

3 Tests Designed for the Assessment of the Optimal Conditions for Each Step in the Procedure

It is important that the pH after the first mixing step be higher than 11. This is clearly seen in Fig. 2, which shows total carbamate (i.e., original carbhemoglobin at equilibrium plus dissolved CO_2) vs. the pH of the first mixing. Only at about and over

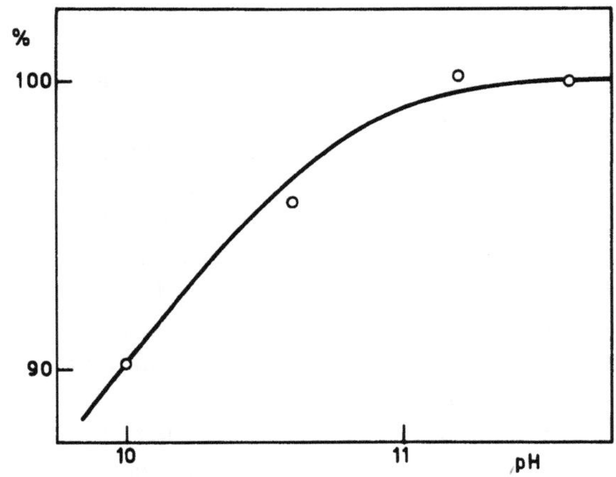

Fig. 2. Carbamates of human hemoglobin (expressed as percentage of the value of pH 12) vs. the pH value of the first mixing

pH 11 is all the dissolved CO_2 removed by reaction (4) between CO_2 and the newly deprotonated ϵ-amino groups of the protein. The pH of the hemoglobin-resin suspension should be kept at 12.5. Since at constant buffer concentration the pH after mixing will be a function of hemoglobin concentration and of the pH and P_{CO_2} of the sample, the buffer-HCl concentration contained in one of the two syringes (Fig. 1B) is adjusted for each experimental $HbCO_2$ determination to give a pH of 11.0 to 11.6.

To check the stability of the carbamate in the pH range 11.2 to 13.5, samples of deoxyhemoglobin at constant protein concentrations were equilibrated at pH 7.4 with a CO_2-N_2 gas mixture with a P_{CO_2} of 120 mm Hg. Under these conditions the concentration of carbamate is close to the maximum value ever measured in our studies of human hemoglobin. Figure 3a shows the decrease in the amount of carbamate vs. time of exposure at pH 11.5. Since the hemoglobin-buffer mixture was never left longer than 5 s at pH 11.5, the loss of carbamate through reaction (4) can be considered negligible during the first stage of the double-mixing procedure.

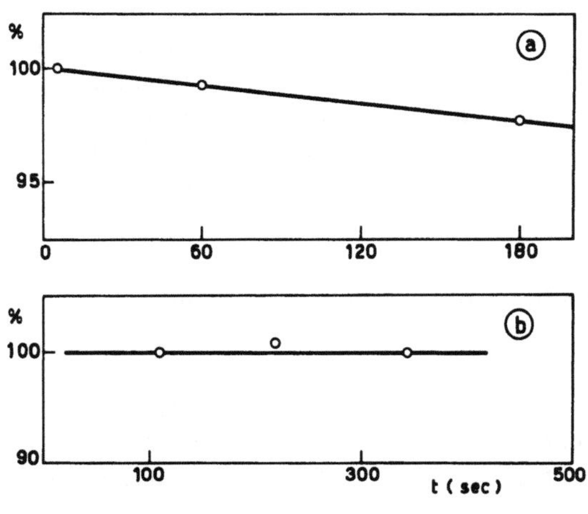

Fig. 3. a Decrease in the amount of carbamate (expressed as percentage of the value found after 5 s) vs. time, at pH 11.5. **b** Stability of carbamates at pH 12.5 (expressed as percentage of the value found after 10 s)

Figure 3b shows that the same holds true if the pH is raised from 11.5 to 12.5 by the addition of 1 N KOH, and even if the alkaline hemoglobin-buffer solution is left at this pH for up to 3 to 4 min before adding the resin suspension.

Tests have been carried out to determine if any denatured hemoglobin is bound by the resin in C: 1.5 ml HbO_2 were added to a 5-ml volumetric flask, followed by 1.5 ml of TEA-HCl buffer and Bio-Rad AG 1x8 suspension, and 1 N KOH was then added to volume. The mixture, at pH ~ 13, was stirred in the cold for 3 min, and a sample was taken for absorbance measurement at 390 nm after dilution in 0.1 N KOH. Similarly, 1.5 ml of the same HbO_2 sample was treated with the same amount of TEA-HCl (at pH ~ 7), Bio-Rad resin and water to volume (final pH ~ 7.5). A sample of this solution when diluted in 0.1 N KOH had the same absorbance as the sample exposed to the resin at alkaline pH.

4 An Absolute Test of Validity of the Method for Carbamate Determination

This test is based on the determination of the solubility coefficient, Q', of CO_2 in concentrated solutions of hemoglobin at 37°C containing 0.1 M KCl and has been described and discussed elsewhere [4]. In summary, it consists in the determination of Q' by two independent methods. (a) The total CO_2 content of a solution of carboxyhemoglobin equilibrated with a known P_{CO_2} at pH 4.6 is determined gasometrically. Under these conditions no carbhemoglobin is present, and the total CO_2 differs from the dissolved CO_2 only because it contains a small amount of bicarbonate, which, however, can be calculated from the Henderson-Hasselbach relationship and the pH. (b) The dissolved CO_2 in a hemoglobin solution of the same concentration in protein and salt as the solution used for the experiment described above, equilibrated also with the same P_{CO_2} but at pH 6.2, is determined as carbhemoglobin by the procedure described in this paper. At pH 6.2, no significant amount of carbhemoglobin is formed at equilibrium with either deoxy or carboxyhemoglobin, and the only species of CO_2 in solution besides free, dissolved CO_2 is bicarbonate. The latter is removed by the resin during the procedure, and the dissolved CO_2 is estimated as carbhemoglobin. The two methods, applied under different conditions of P_{CO_2}, have yielded values of Q' in close agreement (within 2%), thus validating the whole procedure on an absolute basis.

5 Applicability of the Method to the Determination of Carbhemoglobin Content of Hemoglobins Other Than Normal Human Hemoglobin

The possibility of extending carbhemoglobin studies by the method herein described to abnormal human hemoglobins, or to hemoglobins of other species, depends on the stability of the protein in alkaline solution and the amount of available material. The method can be applied whenever no precipitation of protein occurs in alkaline solution. Some hemoglobins may not be denatured by the exposure to pH 12.5 for 4 or 5 min at 1°C. Thus their concentration could be measured directly by the Drabkin method, making the procedure of complete denaturation in 0.1 N KOH unnecessary.

The following considerations should be indicative of the minimum amount of protein required for carbamate studies by the present method. Let it be assumed that the error in reading the gas pressure at the 0.5-ml mark of the Van Slyke apparatus is ± 1 mm Hg. If the total carbhemoglobin plus dissolved CO_2 concentration at equilibrium is 2 to 3 mM and 2 ml of alkaline protein solution are used for the gasometric analysis, the error in their estimation is 2% to 3% (the dilution undergone by the original hemoglobin solution is approximately threefold). These circumstances occur when the carbhemoglobin content of a sample of 14 mM human deoxyhemoglobin equilibrated at pH 7.4 with a P_{CO_2} 14 mm Hg is determined. The total volume of hemoglobin required would be about 1.3 ml, which corresponds to 300 mg of protein. This volume includes the amount of solution required for pH and concentration measurements and the amount lost in the dead volumes of the equipment (about 0.3 ml). When the free, dissolved CO_2 plus carbhemoglobin amounts to 10 to 11 mM, as in the case of the same sample cited above of deoxyhemoglobin equilibrated with a $P_{CO_2} \sim 120$ mm Hg, about 150 mg of protein would be required to carry out the analysis with the same error as in the previous example.

It is thus clear that a significant reduction in the amount of protein required can only be obtained by a suitable modification of Van Slyke's technique used to determine the total CO_2 content of the sample from C (Fig. 1). It is likely that this reduction, which is important to extend the applicability of the present method to the study of the CO_2 reactions of other hemoglobins, can only be obtained by the use of the infrared method developed by Van Kempen [12] for the estimation of total CO_2 in solution.

Acknowledgment. The work reported in this paper has been supported by a grant from the Consiglio Nazionale delle Ricerche, Rome.

References

1. Faurholt C (1925) Etudes sur les solutions aqueouses de carbamates et de carbonates. J Chim Phys 22: 1–44
2. Ferguson JKW, Roughton FJW (1934) The chemical relationships and physiological importance of carbamino compounds of CO_2 with haemoglobin. J Physiol (London) 83: 87–102
3. Kilmartin JV, Rossi-Bernardi L (1971) The binding of carbon dioxide by horse haemoglobin. Biochem J 124: 31–45
4. Perrella M, Bresciani D, Rossi-Bernardi L (1975a) The binding of CO_2 to human haemoglobin. J Biol Chem 250: 5413–5418
5. Perrella M, Bresciani D, Rossi-Bernardi L (1975b) Supplemental material to: The binding of CO_2 to human hemoglobin. J Biol Chem Document No 74-M-851
6. Perrella M, Kilmartin JV, Fogg J, Rossi-Bernardi L (1975c) Identification of the high and low affinity CO_2-binding sites of human haemoglobin. Nature (London) 256: 759–761
7. Perrella M, Rossi-Bernardi L, Roughton FJW (1972) The carbamate equilibrium between CO_2 and bovine haemoglobin at 25°C. In: Rorth M, Astrup P (eds) Oxygen affinity of haemoglobin and red cell acid-base status. IV Alfred Benzon Symposium. Munksgaard, Copenhagen, pp 177–204
8. Peters JP, Van Slyke DD (1932) Quantitative clinical chemistry. Williams and Wilkins, Baltimore, pp 281–286

9. Rossi-Bernardi L, Pace M, Roughton FJW, Van Kempen L (1969) The estimation of haemoglobin-CO_2 compounds by gel filtration and ion-exchange chromatography. In: Forster RE, Edsall JT, Otis AB, Roughton FJW (eds) CO_2: chemical, biochemical and physiological aspects. US Government Printing Office, NASA No SP-188, Washington, pp 65–71
10. Roughton FJW, Rossi-Bernardi L (1969) pK and the rate of dissociation of glycylglycine-carbamic acid molecule. In: Forster RE, Edsall JT, Otis AB, Roughton FJW (eds) CO_2: chemical, biochemical and physiological aspects. US Government Printing Office, NASA no. SP-188, Washington, pp 41–44
11. Tanford C., Nozaky Y (1966) Titration of the histidyl and α-amino groups of hemoglobin. J Biol Chem 241: 2832–2839
12. Van Kempen LHJ (1972) Estimation of free and hemoglobin-bound CO_2. Ph D Thesis, University of Nijmegen

Phylogeny and Physiological Significance of Oxygen-Linked CO_2 Binding in Hemoglobin

C. BAUER[1]

The most conservative class of vertebrates are the jawless Cyclostomes. Two living families belong to this class: the Myxinidae (hagfish) and the Petromyzonidae (lampreys). Even though fossilized fragments of more recent members of this class were not preserved because of the absence of hard tissue in Cyclostomes, it is quite likely that their predecessors are the Ostracoderms from the Ordovician strata that are about 400 million years old (Young, 1962). These more or less armored fish resemble the more recent Cyclostomes in three important ways, namely the absence of (a) lower jaws (b) pectoral fins and (c) an ossified internal skeleton (Heintz, 1963). If one accepts the view that morphological, physiological, and biochemical evolution of vertebrates are interconnected even though they may proceed at different rates (Wilson et al., 1977) it follows that an animal species that has changed relatively little with time during phylogenetic evolution with respect to morphology may also have retained certain functional characteristics. For example, the complete absence of nerves regulating the heart beat in the Pacific hagfish or the lack of tubules in the kidney of this animal (Jensen, 1966) may be regarded as findings in favor of this argument. Likewise, the occurrence of monomeric hemoglobin in lampreys and hagfish would point toward an archaic form of life represented by these animals since in all other, non-cyclostome vertebrates hemoglobin contains four hemes per molecule. Nonetheless, some components of hagfish hemoglobin show a tendency to form aggregates which results in an apparent heme-heme interaction (Bannai et al., 1972). Furthermore, the hemolysate of the Atlantic hagfish (*Myxine glutinosa*) exhibits a significant decrease in oxygen affinity upon addition of CO_2 whilst high concentrations of NaCl or organic phosphates are without effect (Bauer et al., 1975). At pH 7.3 for example, addition of CO_2 equivalent to a pCO_2 of 40 torr leads to an increase in oxygen half saturation pressure (P_{50}) by a factor of 1.5. Similarly, the Bohr effect when expressed as $\Delta \log P_{50}/\Delta pH$ is -0.07 if pH is changed by adding acid or base, but rises to -0.17 when the change in pH is brought about by changing pCO_2 at constant bicarbonate concentration. These results suggest that the reciprocal linkage between oxygen and carbon dioxide binding appeared quite early in the evolution of hemoglobin function.

However, a clear-cut molecular interpretation of the effect of CO_2 binding on oxygen affinity and Bohr effect of hagfish hemoglobin cannot yet be advanced. The fact that the parameter n in the Hill equation increases with increasing pCO_2 and that furthermore three isolated hemoglobin components did not show any change of the oxygen affinity upon addition of CO_2 would be in keeping with the interpretation

1 Physiologisches Institut der Universität Regensburg, Postfach 397, 8400 Regensburg/FRG

that the presence of CO_2 leads to an increased tendency of hagfish hemolysate to aggregate whereby the aggregated molecules have a lower oxygen affinity than the parent monomers. An alternative explanation is that the tertiary structure of deoxy-hemoglobin is slightly altered upon binding of CO_2. If such changes of the tertiary structure occur, they may be too small to be detected by X-ray analysis (Arnone, 1974). In any event, experiments on the reactivity of -SH groups in hemoglobin in the absence and presence of CO_2 give indications of structural changes of human and sheep hemoglobin induced by CO_2 binding (Baumann, 1974).

The hemoglobin from cartilage and bony fish are all composed of two unlike pairs of globin chains, i.e., they are heterotetramers like the hemoglobins of all higher vertebrates or Gnathostomata. However, little is known on the formation of oxygen-linked carbamate compounds in these hemoglobins. Farmer (1979) studied the effect of CO_2 on oxygen affinity and Bohr effect in two water-breathing (*Leiostomus xanthurus, Brachyplatystoma* sp.) and two air-breathing (*Hoplosternum littorale, Lepidosiren paradoxa*) fish species. She finds that in general the effect of CO_2 on oxygen affinity is smaller than in human hemoglobin which could be due to the fact that the α-amino termini of the α-chain, i.e., a CO_2-binding site are blocked in some of these hemoglobins (Gillen and Riggs, 1973; Farmer, 1979). Therefore, this study did not indicate that air-breathing fish have hemoglobins with CO_2-binding properties that are different from those of water-breathing fish, despite the fact that the CO_2 concentration in the blood of air-breathers is very likely to be higher than that of water-breathers.

Hemoglobins from birds and adult mammals in the absence of organic phosphates are, within the same order of magnitude, equally susceptible to CO_2 with respect to oxygen affinity and Bohr effect. It was demonstrated that the formation of oxygen-linked carbamate when estimated from the difference in Bohr effect induced by CO_2 and fixed acids in chicken hemoglobin is in very good agreement with more direct measurements on human hemoglobin under similar conditions (Bauer, 1970; Weingarten et al., 1978). However, if one considers the contribution of oxygen-linked carbamate compounds to the CO_2 exchange during the respiratory cycle under physiological conditions, some important quantitative differences come to light. For example, the marked difference between the "CO_2 Bohr effect" and the "fixed acid Bohr effect" which is due to the formation of oxygen-linked carbamate compounds is completely abolished in chicken hemoglobin in the presence of inositol hexaphosphate which is an analog of the physiological phosphate compound in bird erythrocytes, inositol pentaphosphate (Weingarten et al., 1978). Therefore, under physiological conditions, there is very little, if any, formation of oxygen-linked carbamate in chicken and probably also in duck and geese blood (Baumann and Baumann, 1977; Petschow et al., 1977; Meyer et al., 1978).

In human blood, i.e., in the presence of physiological concentrations of 2,3-diphosphoglycerate (DPG) the formation of oxygen-linked carbamino compounds is much reduced, but not abolished, as in bird blood. This antagonism is due to the fact that CO_2 and DPG share binding sites at the β-chains of human adult hemoglobin (Bauer, 1969). For example, the fraction of oxygen-linked carbamate ($-\Delta HbCO_2/\Delta HbO_2$) estimated at 37°C drops from 0.20 in the absence of DPG to 0.08 in the presence of physiological concentrations of DPG (Bauer and Schröder, 1972; Klocke,

1973; Bursaux et al., 1979). Therefore, the contribution of carbamino compounds to the CO_2 exchange in human blood is not very considerable and amounts to about 10% (Bauer and Schröder, 1972; Klocke, 1973; Bursaux et al., 1979). Under conditions of heavy muscular exercise this fraction may increase to about 15%–20% because the higher temperatures in the working muscle would favor carbamate formation especially at low oxygen saturations in the blood (Hlastala et al., 1977).

Since human fetal hemoglobin binds DPG less firmly than the human adult pigment, one would expect a more important formation of oxygen-linked carbamate in fetal blood than in adult blood. Earlier estimates gave a value for $-\Delta HbCO_2/\Delta HbO_2$ of 0.117 (Bauer and Schröder, 1972) whilst Bursaux et al. (1979) found a value of 0.068. This relatively low figure is in accord with the observation that the ionization constant of the N-terminal α-amino groups of the fetal γ-chains is much lower than of the adult β-chains (Gros and Bauer, 1978) which leads to an intrinsically lower CO_2 affinity in fetal compared to adult deoxyhemoglobin. The contribution of oxygen-linked carbamate for the CO_2 exchange across the placenta is therefore likely to be lower than originally estimated (Bauer and Schröder, 1972), amounting to about 10% (Bursaux et al., 1979).

In mammals like sheep and other ruminants that lack significant concentrations of DPG in their erythrocytes, it seems possible that oxygen-linked carbamino compounds are more important for CO_2 exchange in lungs and tissues because there is no competition between the binding of phosphates and CO_2. However, as in human fetal hemoglobin, the CO_2 affinity both for bovine and for sheep hemoglobin is not very large, which results in a 15% contribution of oxygen-linked carbamino compounds to CO_2 exchange in both species (Ferguson and Roughton, 1934; Baumann et al., 1975).

In summarizing the results on the physiological significance of carbamate compounds in the carbon dioxide exchange during the respiratory cycle it becomes apparent that in mammals 10%–20% of the CO_2 given off in the lungs or taken up by the tissues is due to the linkage between oxygen binding and carbamate formation. In birds there seems to be no oxylabile carbamate at all, at least in the presence of the physiological phosphate inositol pentaphosphate or its analog inositol hexaphosphate.

There is one order of animals, however, that actually represents distant relatives of the birds, namely the crocodiles, in which oxygen-linked CO_2 binding has a much more pronounced influence on the oxygen affinity of hemoglobin than in any other vertebrate studied so far. At 20°C and pH 7.3 P_{50} increases by a factor of 8 upon addition of CO_2 equivalent to a pCO_2 of 40 torr (Bauer and Jelkmann, 1977). This pronounced effect of CO_2 on the oxygen affinity of crocodile hemoglobin which is quantitatively similar to that exerted by inositol hexaphosphate or inositol pentaphosphate on the oxygen affinity of bird hemoglobin (Baumann and Baumann, 1977; Bauer et al., 1978) is not only observed in dilute hemoglobin solutions but also in whole blood both at 25° and 37°C (Jelkmann and Bauer, unpublished experiments). It should be noted in this connection that none of the organic phosphate esters that normally govern the oxygen affinity of the blood in mammals, birds, fish, or amphibians has any influence on the P_{50} of crocodile hemoglobin.

The large CO_2 effect in this pigment could be brought about by (i) a direct inhibition of oxygen binding by gaseous CO_2, e.g., in the heme pocket, (ii) by a particularly strong and specific effect of HCO_3^- and (iii) by an unusually potent linkage between oxygen binding and formation of carbamino compounds ($HbCO_2$). This last possibility can be safely excluded because both direct measurements of $HbCO_2$ in deoxygenated and heme-liganded crocodile hemoglobin (Perrella, Bauer, Mosca, Rossi-Bernardi, unpublished experiments) and of the proton-liberation associated with carbamate formation (Gros and Bauer, unpublished experiments) gave no indication of oxylabile carbamate binding as was originally supposed (Bauer and Jelkmann, 1977).

In order to discriminate tentatively between possibilities (i) and (ii), titration curves were performed of deoxy- and carbonmonoxy-hemoglobin in the absence and presence of CO_2. The reasoning was that if a charged, buffering molecule like HCO_3^- were released upon binding of carbon monoxide this should result in a decrease in charge between carbonmonoxy and deoxyhemoglobin, i.e., of the number of Bohr protons (ΔZ) released upon CO binding. On the other hand, if molecular CO_2 was bound in an oxygen-linked fashion one would not expect a significant change of the Bohr effect unless there was a pronounced change in the ionization constant of some residues that titrate in the physiological pH region. Appropriate experiments revealed that at pH 7.3 and 37°C, ΔZ per heme was 0.37 in the absence of CO_2 and $\sim$ 0.07 in the presence of CO_2. This result indicates that per mol of oxygen bound about 0.3 mol of HCO_3^- are liberated upon CO_2 binding (Perrella, Bauer, Mosca, Rossi-Bernardi, unpublished experiments).

This figure agrees with the difference in total CO_2 content between deoxygenated and CO-liganded hemoglobin solutions in a comparable pH range (pH 7.1 – pH 7.45) which amounts to 0.31–0.36 mol of CO_2 per molecule CO bound to the heme. However, more experiments also with different techniques comprising X-ray crystallography and NMR measurements with $^{13}CO_2$ are needed before the nature of the strong O_2–CO_2 interaction can be established with certainty.

References

Arnone A (1974) X-ray studies of the interaction of CO_2 with human deoxyhemoglobin. Nature (London) 247: 143–145

Bannai S, Sugita Y, Yoneyama Y (1972) Studies on hemoglobin from the Hagfish *Eptatretus burgeri*. J Biol Chem 247: 505–510

Bauer C (1969) Antagonistic influence of CO_2 and 2,3-diphosphoglycerate on the Bohr effect of human hemoglobin. Life Sci 8: 1041–1046

Bauer C (1970) Reduction of the carbon dioxide affinity of human haemoglobin solutions by 2,3-diphosphoglycerate. Respir Physiol 10: 10–19

Bauer C, Jelkmann W (1977) Carbon dioxide governs the oxygen affinity of crocodile blood. Nature (London) 269: 825–827

Bauer C, Schröder E (1972) Carbamino compounds of haemoglobin in human adult and fetal blood. J Physiol (London) 227: 457–471

Bauer C, Engels U, Paléus S (1975) Oxygen binding to haemoglobins of the primitive vertebrate *Myxine glutinosa L.* Nature (London) 256: 66–68

Bauer C, Jelkmann W, Rollema H (1978) Mechanisms controlling the oxygen affinity of bird and reptile blood. A comparision between the functional properties of chicken and crocodile hemoglobin. In: Piiper J (ed) Respiratory function in birds, adult and embryonic. Springer, Berlin Heidelberg New York, pp 61–66

Baumann FH, Baumann R (1977) A comparative study of oxygen transport in bird blood. Respir Physiol 31: 333–343
Baumann R (1974) Effect of CO_2 binding to haemoglobin on the reactivity of the SH groups at position β93. Biochem Biophys Res Commun 61: 845–851
Baumann R, Bauer C, Haller EA (1975) Oxygen-linked CO_2 transport in sheep blood. Am J Physiol 229: 334–339
Bursaux E, Poyart C, Guesnon P, Teisseire B, Herigault R, Bohn B (1979) Comparative effects of CO_2 on the affinity for O_2 fetal and adult erythrocytes. Pflügers Arch 378: 197–203
Farmer M (1979) The transition from water to air-breathing: effects of CO_2 on hemoglobin function. Comp Biochem Physiol 62 A: 109–114
Ferguson JKW, Roughton FJW (1934) The chemical relationships and physiological importance of carbamino compounds of CO_2 with haemoglobin. J Physiol (London) 83: 87–102
Gillen RG, Riggs A (1973) Structure and function of the isolated hemoglobins of the American eel, ***Anquilla rostrata.*** J Biol Chem 248: 1961–1969
Gros G, Bauer C (1978) High pK value of the N-terminal amino groups of the γ-chain causes low CO_2 binding of human fetal hemoglobin. Biochem Biophys Res Commun 80: 56–62
Heintz A (1963) Phylogenetic aspects of myxinoids. In: Brodal A, Fänge R (eds) The biology of myxine. Universitetsforlaget, Oslo, pp 9–21
Hlastala MP, Woodson RD, Wranne B (1977) Influence of temperature on hemoglobin-ligand interaction in whole blood. J Appl Physiol 43: 545–550
Jensen D (1966) The hagfish. Sci Am 214: 82–90
Klocke RA (1973) Mechanism and kinetics of the Haldane effect in human erythrocytes. J Appl Physiol 35: 673–681
Meyer M, Holle JP, Scheid P (1978) Bohr effect induced by CO_2 and fixed acid at various levels of O_2 saturation in duck blood. Pflügers Arch 376: 237–240
Petschow D, Würdinger I, Baumann R, Duhm J, Braunitzer G, Bauer C (1977) Causes of high blood O_2 affinity of animals living at high altitude. J Appl Physiol 42: 139–143
Weingarten JP, Rollema HS, Bauer C, Scheid P (1978) Effects of inositol hexaphosphate on the Bohr effect induced by CO_2 and fixed acids in chicken hemoglobin. Pflügers Arch 377: 135–141
Wilson AC, Carlson SS, White TJ (1977) Biochemical evolution. Annu Rev Biochem 46: 573–639
Young JZ (1962) The life of vertebrates, 2nd ed. Oxford University Press, New York Oxford

The Carbamate Reaction with Proteins: Observation by ^{13}C-NMR and Evaluation of Structural Consequences [1]

F.R.N. GURD [2], J.B. MATTHEW [3], R.J. WITTEBORT [4], J.S. MORROW [5], and S.H. FRIEND [2]

1 General

1.1 The Carbamate Reaction

The classical studies of Bohr et al. (1904) and Christiansen et al. (1914) led to the first recognition of the general nature and importance of carbamate formation from the reaction of CO_2 with amino groups (Henriques, 1928). The general importance of the reaction has been stressed by Neuberg et al. (1955) and Edsall and Wyman (1958), and its importance in CO_2 homeostasis by Hastings (1970), Morrow et al. (1974b), Morrow and Gurd (1975), Wittebort et al. (1978), and Gurd et al. (1979). Since the carbamino derivatives are rarely the dominant form of the total CO_2 present in a system, they have been intensively studied in relatively few cases whose interest was recognized early (Stadie and O'Brien, 1937; Roughton, 1964).

The reaction of a primary amine with CO_2 is like that with cyanic acid or water in which the nucleophile attacks the electron-poor carbon (Caplow, 1968) to form the carbamic acid derivative

$$
\begin{array}{lll}
 & \text{R-NH}_2 + CO_2 & \rightleftharpoons \text{R-NH-CO}_2\text{H} \\
 & \quad\quad\quad + & \quad\quad \rightleftharpoons \\
H^+ \rightleftharpoons\ K_z & \quad\ \ H_2O & \text{R-NHCO}_2^- + H^+ \\
 & \quad\ \ \rightleftharpoons & \\
\text{R-NH}_3^+ & \quad\ \ H_2CO_3 & \\
 & \quad\ \ \rightleftharpoons & \\
 & \quad\ \ H^+ + HCO_3^- & \\
 & \quad\ \ \rightleftharpoons & \\
 & \quad\ \ H^+ + CO_3^{=} &
\end{array}
\qquad (1)
$$

The carbamic acid, R-NH-CO_2H, is a strong enough acid (pK ~ 5.0) that under physiological conditions in which the carbamate equilibrium is important its dissociation

---

1 Supported by NIH Grants HL-05556, HL-14680 and GM-1046. This is the 111th paper in a series dealing with coordination complexes and catalytic properties of proteins and related substances

2 Department of Chemistry and Medical Sciences Program, Indiana University, Bloomington, Indiana 47405/USA

3 Department of Molecular Biophysics and Biochemistry, Yale University/USA

4 National Magnet Laboratory, Massachusetts Institute of Technology/USA

5 Department of Pathology, Yale University/USA

can be taken as complete. Hence, an overall equilibrium constant K_c, defined as $[R\text{-}NHCOO^-][H^+]/[R\text{-}NH_2][CO_2]$ is used. The use of K_z for the dissociation constant of the amine in this field goes back to the study of carbamino formation with the zwitterionic form of glycine by Stadie and O'Brien (1936). The Eq. set (1) includes the formation of bicarbonate according to an overall constant K_1', equal to $[HCO_3^-][H^+]/[CO_2]$, and the further dissociation of bicarbonate, expressed by K_2'. In this discussion $[CO_2]$ represents the free molecule and not the total carbonates. Note that the reversible reactions of hydration and amination of CO_2 have relatively slow steps compared with the protonation and deprotonation processes.

The mol fraction of amine present as carbamate, $[RNHCO_2^-]/[RNHCO_2^-] + [RNH_2] + [RNH_3^+]$, is expressed as Z, as follows (Kilmartin and Rossi-Bernardi, 1973; Morrow et al., 1974a)

$$Z = \frac{K_c K_z [CO_2]}{K_c K_z [CO_2] + K_z [H^+] + [H^+]^2} \tag{2}$$

Equation (2) is usually recast in terms of total carbonates rather than $[CO_2]$ (Morrow et al., 1974a). Equation (2) indicates that a plot of Z vs. pH will be bell-shaped because the concentration of free amine will be limiting at low pH and of CO_2 at high pH (Morrow et al., 1974a). The equilibria described in Eq. set (1) are all sensitive to variation in ionic strength and to substitution of D_2O for H_2O in the solvent (Morrow et al., 1974a).

1.2 Measurements by ^{13}C-NMR

The use of ^{13}C-NMR to quantitative carbamate formation in hemoglobin was introduced by Matwiyoff and Needham (1972) and Morrow et al. (1973). CO_2 enriched in ^{13}C is equilibrated with the protein or peptide at a given pH, P_{CO_2}, ionic strength and temperature. Since the ^{13}C of the carbamate, bicarbonate-carbonate and CO_2 is not directly protonated, the spectra are accumulated without proton decoupling. The absence of a directly bonded hydrogen atom minimizes the efficient dipolar relaxation process, thereby favoring a small natural line width and making detection more effective. A careful study of factors that might affect the accuracy of quantitation has been published (Morrow et al., 1974a), as well as a treatment to test limits of exchange rates between carbamino and free CO_2 forms (Matthew et al., 1977).

The chemical shift position of the carbamino adduct lies upfield of the amide and carboxyl carbons in a protein ^{13}C-NMR spectrum. It is also distinguishable normally from the single resonance of the bicarbonate-carbonate component. The chemical shift positions of the bicarbonate and carbonate species differ, but are in the fast exchange limit in NMR terms which produces a single resonance observed at the weighted average of the constituent chemical shift positions. Conversely, this resonance position will provide an independent check on the pH of the experiment, in the same way and with the same theoretical limitations as in the parallel use of ^{31}P-NMR for intramolecular pH measurements (Ogawa et al., 1978). Because of the slow conversion rates between bicarbonate and CO_2, the CO_2 resonance is separately observable in an upfield position at the lower part of the pH range used to study carbamino formation. Conversely, in the presence of carbonic anhydrase the CO_2 and bicarbonate-carbonate resonances are markedly affected (Morrow et al., 1973). The bicarbonate component is potentially an important anion in protein binding (Gurd et al., 1974).

1.3 Carbamate Reaction with Proteins

The carbamate reaction with proteins can be studied by ^{13}C-MNR introducing the isotopically enriched form $^{13}CO_2$. By comparison with the time over which the NMR measurements must be accumulated, the rate of equilibration of the label throughout the total carbonate components is fast enough to allow accurate analysis of the equilibria. The natural abundance of ^{13}C is 1.1%, so that striking amplification of the signal is obtainable with enriched CO_2 and carbonates. In our work we have seldom resorted to the maximum enrichment obtainable. The carbamate carbon incorporated into the protein is not significantly influenced by other magnetically active nuclei unless a paramagnetic center is present. Furthermore, the relatively slow relaxation of this unprotonated ^{13}C nucleus means, in general NMR terms, that the resonance is narrow and relatively easily distinguished. Since the carbonyl groups in the protein itself are similar in these respects, their total resonance integral may be used as an internal reference (Morrow et al., 1974a). Because of the relatively narrow resonances observed with the carbamino carbon nucleus it is possible with standard techniques to make useful observations even in relatively large proteins of approximately 100,000 daltons (O'Leary et al., 1979). Up to the present, little difficulty has been encountered in distinguishing carbamino carbon resonances from other chemical forms (Morrow et al., 1974a; Matthew et al., 1977; O'Leary et al., 1979).

2 Carbamate Formation in Hemoglobin

2.1 Observations by ^{13}C-NMR

In this section we shall be concerned with carbamino formation in normal adult human hemoglobin, a topic of great inherent interest and the most appropriate case at this time for a discussion of detailed results and their chemical and physiological evaluation. Figure 1 (Matthew et al., 1977) shows representative spectra of (A) deoxyhemoglobin and (B) carbon monoxide hemoglobin equilibrated with $^{13}CO_2$ at various values of pH and containing total carbonates and NaCl as indicated in the legend. The broad envelope of resonances extending from 15 to 25 ppm represents the 328 carbonyl carbon atoms at natural abundance per $\alpha\beta$ dimer. The large resonance at 32.8 ppm represents the enriched bicarbonate-carbonate components in rapid exchange. The $^{13}CO_2$ itself, in slow exchange, occurs at 68.38 ppm and is not shown. The resonance at 28.4 ppm seen in both panels is assigned to carbamino derivatives of ϵ-amino groups (Morrow et al., 1974a; Morrow et al., 1976; Matthew et al., 1977).

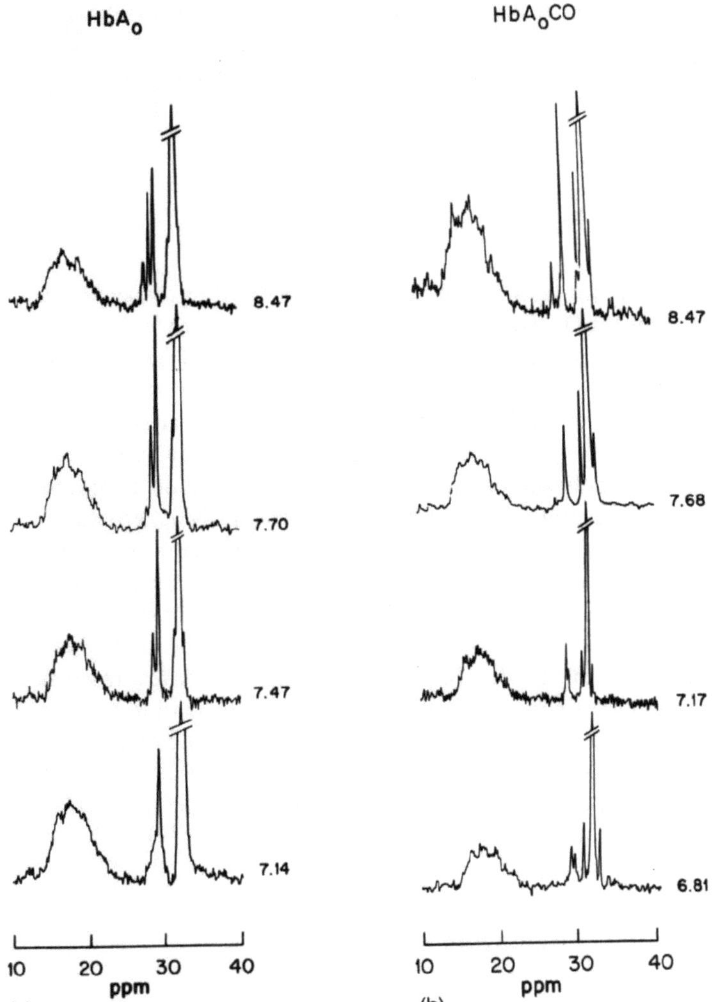

Fig. 1. Shown here are 25.2 MHz ^{13}NMR spectra of human adult hemoglobin A_0 equilibrated at various pH values with $^{13}CO_2$. **a** deoxyhemoglobin; **b** carbon monoxide hemoglobin. The pH values are listed at the *right* of each spectrum. The parts per million scale on the abscissa is referred to external CS_2. Total carbonates were 35 to 57 mM, and $^{13}CO_2$ (mol fraction $^{13}CO_2$ 0.82 to 0.92) was equilibrated at pressures of 4 to 374 torr. The hemoglobin concentration was usually 11.0 to 16.6 mM expressed as heme concentration in 0.05 M NaCl. The bicarbonate-carbonate resonance appears near 33 ppm, usually with spinning side bands in evidence. Measurements were obtained between 29° and 31°C

On the basis of specific blockage of the Val-1α and Val-1β amino groups by cyanate (Williams et al., 1975; Morrow et al., 1976), it is possible to show that in the deoxyhemoglobin the carbamino derivatives of these two amino sites are represented by the resonances at 29.2 and 29.8 ppm, whereas in the liganded hemoglobin they both occur at 29.8 ppm. There is good reason to associate the difference with local structural differences in the vicinity of Val-1α in the two quaternary structures (Matthew et al., 1977). Note, however, that the well-established quaternary change in the general vicinity of the Val-1β groups is not reflected in the carbamino resonance positions.

From careful measurements of the resonance intensities as a function of pH, involving numerous observations of the types illustrated in Fig. 1, it is possible to determine pK_z and pK_c values as summarized in Table 1. From this table it is clear that the carbamino formation is stronger (lower pK_c) for Val-1β than Val-1α in the deoxy state, and that pKc rises for both groups in the liganded state. These results signify the well-known linkage between the homotropic heme ligand effector and the heterotropic effector CO_2 (Haldane Effect), and show that the effect is inherently more marked for Val-1β than for Val-1α. The pK_z values, conversely, illustrate the longer-established dominant role of Val-1α (Kilmartin and Rossi-Bernardi, 1973; Garner et al., 1975) in the action of H^+ as a heterotropic effector (Bohr Effect).

2.2 Functional Significance

The significance of the change in pK_c in passing from the deoxy to the liganded hemoglobin forms can only be appreciated fully in conjunction with the change in pK_z and the variations in P_{CO_2} associated functionally with the oxygen binding cycle. These changes are illustrated in the four web plots of Fig. 2. The form of the web plot was introduced for a similar purpose to illustrate the possible role of carbamino formation in homeostatic regulation of respiratory exchange (Morrow et al., 1974b). The plots in Fig. 2 bring out the differences in carbamino formation for both Val-1α and Val-1β, but show clearly the quantitatively far greater effect for Val-1β. The roles of the Val-1α and Val-1β groups as the pH is varied are summarized in Fig. 3 for the Haldane and Bohr effects. Note that pK_z for Val-1β undergoes little change in passing between the quaternary states, and so this group has been omitted from Fig. 3B.

The full Bohr effect goes beyond the contribution of Val-1α (Kilmartin and Rossi-Bernardi, 1973). Until recently a very prominent place was accorded to His-146β (Kilmartin et al., 1973; Matthew et al., 1977); however, results at more nearly physiological ionic strength indicate less of a role for this particular group (O'Donnell et al., 1978; Matthew et al., 1979b). There seems little question, however, that Val-1α represents the most important group of several contributing significantly to the alkaline Bohr effect (Matthew et al., 1979b).

2.3 Comparison with Other Methods

The ^{13}C-NMR method of observation of carbamate derivatives in hemoglobin has the advantage of a direct method of observation applicable over a range of conditions. The required instrumentation is now relatively accessible. Because of the identity of the chemical shifts of the Val-1α and Val-1β adducts in the carbon monoxide or oxyhemoglobin structures (Morrow et al., 1976), it is necessary to distinguish between

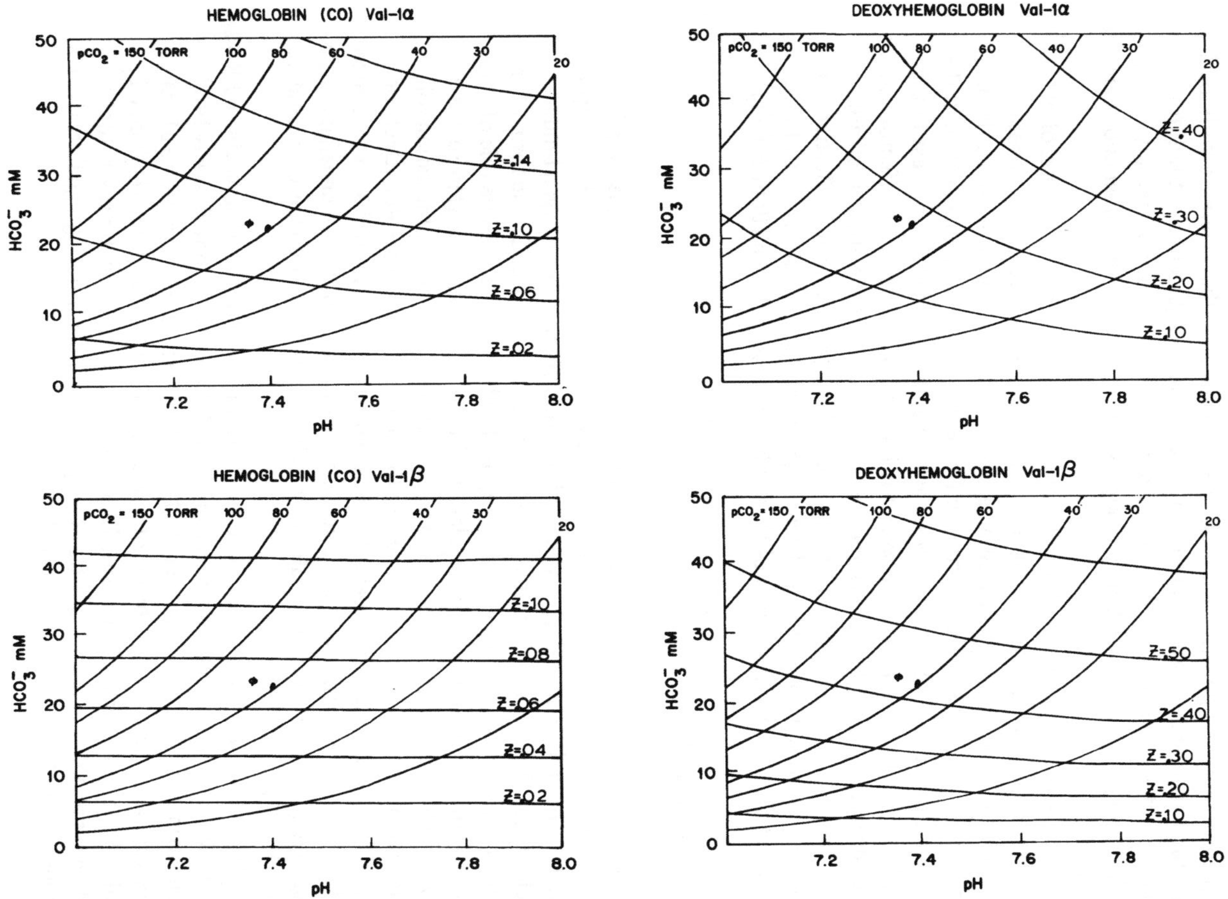

Fig. 2. Relation between pH, P_{CO_2}, bicarbonate concentration, and mol fraction, Z, of carbamate for Val-1α and Val-1β in carbon monoxide and deoxyhemoglobins. The *upward sloping lines from left to right* are P_{CO_2} isobars and the *downward sloping lines* are at the indicated constant Z values. θ and ϕ indicate the conditions corresponding to arterial and venous blood, respectively

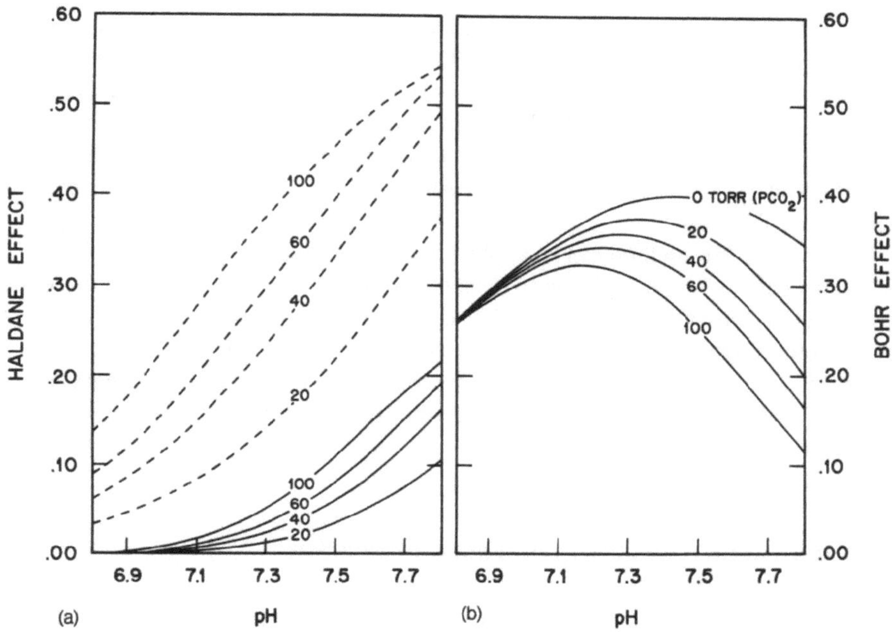

Fig. 3. a the change in mol fraction of carbamate, Z, in going from carbon monoxide to deoxyhemoglobin is shown for Val-1β (- - -) and Val-1α (——) at the indicated P_{CO_2} values. **b** the Bohr effect, ΔH^+, due to Val-1α is shown for the indicated P_{CO_2} values

them by specific blockage of one or the other α-amino group by some means such as the cyanate reaction (Williams et al., 1975; Morrow et al., 1976). The reliability of this procedure is indicated by the self-consistency of the results (Morrow et al., 1976; Matthew et al., 1977). In terms of difficulty and the demands for preparing considerable supplies of the blocked material, the ^{13}C-NMR method has some advantages over gasometric methods (Roughton and Rossi-Bernardi, 1969). It is gratifying that these two quite different techniques provide similar results where comparisons can be made (Kilmartin and Rossi-Bernardi, 1973; Morrow et al., 1976).

All the ^{13}C-NMR results drawn on here and in our earlier reports were made on the fully deoxygenated, the fully formed carbon monoxide derivative, or, more rarely, the fully oxygenated derivative. Possibly because of limitations of sensitivity, we have been unable to detect any evidence of distinct behavior in intermediate mixtures of various proportions of deoxy and carbon monoxide hemoglobin (see the important results of Poyart, Bursaux and Matzke, this vol.), but only the expected summations of individual contributions corresponding to the two dominant quaternary structures (J.B. Matthew, unpublished results).

The NMR technique has been useful in allowing sufficiently close analysis of the pH dependence of the stability of the Val-1β carbamino derivative to show a modest, pH-dependent stabilization in the liganded tetramer, even though the net stabilization of this derivative is so poor that it is difficult to detect by other means (Matthew et al., 1977). The technique is also useful in distinguishing directly the carbamino adducts from other chemical forms that may bind to the protein. It is flexible enough to be applicable to smaller (Gurd et al., 1979) and somewhat larger molecules (O'Leary et al., 1979) to test their ability to form carbamino derivatives. Because of a special characteristic of the NMR method it is possible under some conditions to define limits on the rates of the processes involved at both the Val-1α and Val-1β sites (Matthew et al., 1977) that fit within experimental errors with overall estimates (Forster et al., 1968).

The ^{13}C-NMR method can be applied to other hemoglobins (Morrow, 1974; Morrow et al., 1974c) in addition to hemoglobin A and observations can be made on the intact erythrocyte (Matwiyoff and Needham, 1972; Morrow et al., 1973). The resourceful, classical studies of Bauer and coworkers (Bauer et al., 1975; Gros and Bauer, 1978) place beyond doubt the above-mentioned resonance assignments determined by selective chemical blocking, and point the direction of future analyses of carbamino interactions in hemoglobins.

3 Electrostatic Aspects of Carbamino Formation

The preferential formation of carbamino derivatives of the deoxy relative to the liganded form of hemoglobin, primarily involving the Val-1β sites, is responsible for the role of CO_2 as a heterotropic effector. The Val-1α sites, conversely, are responsible to a large degree for the role of H^+ as a second heterotropic effector. The specialization of the terminal valine residues is achieved in part by the differences in pK_c values shown in Table 1, and in larger part by sensitivity of the pK_z of the Val-1α residues to the quaternary state of the protein. Preferential occupancy of Val-1α by a proton in the deoxy state precludes the formation of the carbamino derivative. The two effectors are thus independently effective and their contributions can be additive.

Table 1. Carbamino formation data for human hemoglobin A

	Chemical [a] shift	pK_z	pK_z [b]	pK_c
			Deoxyhemoglobin	
Val-1β	29.8	6.91 ± 0.32	6.84 ± 0.12	4.64 ± 0.08
Val-1α	29.2	7.83 ± 0.19	7.79 ± 0.10	4.89 ± 0.10
			Carbon Monoxide Hemoglobin	
Val-1β [c]	29.8	– [d]	7.05 ± 0.05	5.8 [d]
Val-1α [e]	29.8	7.16 ± 0.36	6.95 ± 0.13	5.5 ± 0.08

a Referenced to external CS_2
b For comparison values of pK_z determined by Garner et al. are included
c Carbamylated α chain NH_2-terminal hemoglobin derivative
d Carbonmonoxy β chain data does not fit two-parameter pK_c, pK_z function
e Carbamylated β chain NH_2-terminal hemoglobin derivative

The pK_c values shown in Table 1 for the deoxyhemoglobin form are similar to what are found with small molecules, whereas those for both Val-1β and Val-1α in the liganded form show strong destabilization of the carbamino derivatives (Matthew et al., 1977). The reason for the strong destabilization of the carbamino derivatives in the liganded hemoglobin form is not known. It has been shown that the rate of decomposition of the Val-1α carbamino adduct is more rapid than that of the Val-1β adduct (Matthew et al., 1977), and a similar process in more exaggerated form very likely applies to both Val-1α and Val-1β in the liganded state. Although a high resolution structure of the carbon monoxide derivative of human hemoglobin A will shortly be available (J.M. Baldwin, personal communication), at present it is possible to deal best with the electrostatic properties of the deoxyhemoglobin in providing a suitable site for carbamino formation at Val-1β.

The electrostatic interactions in human deoxyhemoglobin have been analyzed by a procedure that takes into account the charge separation and other parameters of the Tanford-Kirkwood theory (Tanford and Kirkwood, 1957) and, in addition, the degree of solvent accessibility of each charged site (Lee and Richards, 1971). The extended treatment was developed and tested with myoglobins of several species (Shire et al., 1974a, b, 1975; Matthew et al., 1978; Botelho et al., 1978; Friend and Gurd, 1979a, b).

The experience with myoglobin (Botelho et al., 1978; Friend and Gurd, 1979a, b) and hemoglobin (Matthew et al., 1979 a, b) leads to the prediction from the extended theory that wide variations will occur in specific electrostatic interactions between groups and in the overall electrostatic effects experienced by a given group, depending on the variations in both distances of separation and degrees of exposure of each group to solvent. Because certain structural factors (Gros and Bauer, 1978) cause the pK_{int} for both Val-1α and Val-1β to be low, the dissociation of both these groups can reach into the physiologically important pH range (Matthew et al., 1979a, b).

The application to hemoglobin (Matthew, 1978; Matthew et al., 1979a, b) shows that the net sum of electrostatic interactions between Val-1α and all other charged sites in deoxyhemoglobin accounts at pH 7.6 for 4% of the total electrostatic stabilization of the molecule, whereas for Val-1β the corresponding net sum comes to less than 1%. In the transition between the deoxy and liganded structures it is the more strongly interacting Val-1α that undergoes the greater change in pK_z (Table 1) which allows it to act as the most prominent Bohr group. By contrast, the Val-1β site which is more isolated in an electrostatic sense can undergo relatively easily the functional variation in net charge from positive through zero to negative with carbamate formation. In terms of distances of separation from other charged sites, Val-1α falls within 7 Å of three other charged sites of which 1 (Arg-141 α) is within 5.5 Å; Val-1β, on the other hand, falls within 7 Å of no other charged site. As Arnone, Rogers, and Riley (this vol.) have shown by X-ray studies, the carbamate derivative of Val-1β is oriented so that it falls within an effective distance for electrostatic interaction with the ε-amino group of Lys-82 β. The electrostatic stabilization of the derivative, as opposed to the free Val-1β amino group, is therefore significant. It is interesting that Val-1 of sperm whale myoglobin, which forms readily a carbamino derivative (Morrow et al., 1974a), also is relatively isolated electrostatically (Friend and Gurd, 1979a, b).

The chemical free energy of carbamino formation, which includes the substantial effect of dissociating the relatively strong carbamic acid, is sufficient to drive the formation reaction and allow the carbamino reaction to play its significant role as effector of the quaternary transition.

Chloride ion binding contributes separately to the electrostatic constellations of the deoxy and liganded hemoglobin structures (Arnone et al., 1976; O'Donnell et al., 1978; Hayashi et al., 1974; Tucker and Perutz, 1977; Chiancone et al., 1976; Wiechelman et al., 1978; Rollema et al., 1975; Matthew et al., 1979b). The binding of chloride ion close to Val-1α (Arnone et al., 1976; O'Donnell et al., 1978) contributes significantly at physiological concentrations to the electrostatic environment of that group; in a sense this anion binding represents a competing factor with carbamino formation as well as a contributing factor to the change of pK on formation of the liganded derivative. Partly because of the anion binding to the hemoglobin, the pK values of Val-1β and Val-1α are affected by ionic strength (Matthew et al., 1979b). For example, the effective pK of Val-1β will decrease with increasing ionic strength, a factor that will increase the availability of the unprotonated, reactant form [equation set (1)] and so increase Z.

4 Regulatory Aspects

In summary, it is the changes in both pK_c and pK_z values which define the separate functional roles of Val-1α and Val-1β in the Bohr and Haldane effects, respectively. First, without the large increase in pK_c for Val-1α, the accompanying decrease in pK_z necessary for the Bohr effect would lead to an enhancement of carbamino formation at this site in the liganded hemoglobin. This consequence for the α subunit would run counter to the allosteric effector role of CO_2 with respect to the β subunit. Secondly, the maintenance of a low level of carbamino formation with Val-1α under both liganded and unliganded conditions allows this site to serve its function as essentially a Bohr group without significant interference.

The carbamino interactions in hemoglobin illustrate a mechanism for a regulatory role for CO_2. This has been explored as reported in an accompanying paper in this volume (Gurd et al., 1979) for peptides of high biological activity. Another regulatory system of great interest is ribulosebisphosphate carboxylase from *Rhodospirillum rubrum*, an enzyme that is activated by CO_2 and which catalyzes the first step in CO_2 fixation as well as the oxidation of the ribulosebisphosphate. O'Leary et al. (1979) present strong evidence from ^{13}C-NMR techniques similar to those discussed above that the CO_2 activation of this key regulatory enzyme in photosynthesis involves formation of a carbamate with an enzyme amino group, probably an ε-amino group of relatively low pK.

Acknowledgments. We acknowledge with pleasure the advice and encouragement of Professors A.B. Hastings and J.T. Edsall.

References

Arnone A, O'Donnell S, Schuster T (1976) X-ray diffraction studies of carbamylated human deoxyhemoglobin. Fed Proc 35: 1604

Bauer C, Baumann R, Engels U, Pacyna B (1975) The carbon dioxide affinity of various human hemoglobins. J Biol Chem 250: 2173–2176

Bohr C, Hasselbalch KA, Krogh A (1904) Über einen in biologischer Beziehung wichtigen Einfluß, den die Kohlensäurespannung des Blutes auf dessen Sauerstoffbindung übt. Scand Arch Physiol 16: 402–412

Botelho LH, Friend SH, Matthew JB, Lehman LD, Hanania GIH, Gurd FRN (1978) Proton nuclear magnetic resonance study of histidine ionizations in myoglobins of various species. The comparison of observed and computed pK values. Biochemistry 17: 5197–5205

Caplow M (1968) Kinetics of carbamate formation and breakdown. J Am Chem Soc 90: 6795–6803

Chiancone E, Norne JE, Forsen S, Mansouri A, Winterhalter KH (1976) Anion binding to proteins. NMR quadrupole relaxation study of chloride binding to various human hemoglobins. FEBS Lett 63: 309–312

Christiansen J, Douglas CC, Haldane JS (1914) The absorption and dissociation of carbon dioxide by human blood. J Physiol (London) 48: 244–277

Edsall JT, Wyman J (1958) Biophysical chemistry, vol I. Academic Press, London New York, pp 1–26

Forster RE, Constantine HP, Craw MR, Rotman HH, Klocke RA (1968) Reactions of CO_2 with human hemoglobin solution. J Biol Chem 243: 3317–3326

Friend SH, Gurd FRN (1979a) Electrostatic stabilization in myoglobin: pH dependence of summed electrostatic contributions. Biochemistry 18: 4612–4619

Friend SH, Gurd FRN (1979b) Electrostatic stabilization in myoglobin: Interactive free energies between individual sites. Biochemistry 18: 4620–4630

Garner MH, Bogardt RA, Jr, Gurd FRN (1975) Determination of the pK values for the α-amino groups of human hemoglobin. J Biol Chem 250: 4398–4404

Gros G, Bauer C (1978) High pK value of the N-terminal amino group of the γ-chain causes low CO_2 binding of human fetal hemoglobin. Biochem Biophys Res Commun 80: 56–62

Gurd FRN, Morrow JS, Keim P, Visscher RB, Marshall RC (1974) ^{13}C NMR studies of the interaction of Hb and carbonic anhydrase with $^{13}CO_2$. Adv Exp Biol Med 48: 109–124

Gurd RS, England RD, Hayes DF, Wittebort RJ, Rothgeb TM (1979) Formation and physiological role of carbamino adducts of angiotensin, bradykinin, and S-methylglucagon. This volume

Hastings AB (1970) A biochemist's anabasis. Annu Rev Biochem 39: 1–24

Hayashi A, Suzuki T, Stamatoxannopoulos G (1974) Electrophoretic and functional abnormalities of haemoglobin tacoma β 30 (B12) Arg → Ser. Biochim Biophys Acta 351: 453–456

Henriques OM (1928) Die Bindungsweise des Kohlendioxyds im Blute. Biochem Z 200: 1–24

Kilmartin JV, Rossi-Bernardi L (1973) Interaction of hemoglobin with hydrogen ions, carbon dioxide, and organic phosphates. Physiol Rev 53: 836–890

Kilmartin JV, Breen JJ, Roberts GCK, Ho C (1973) Direct measurement of the pK values of an alkaline Bohr group in human hemoglobin. Proc Natl Acad Sci USA 70: 1246–1249

Lee B, Richards FM (1971) The interpretation of protein structures: Estimation of static accessibility. J Mol Biol 55: 379–400

Matthew JB (1978) Hemoglobin interactions with hydrogen ion, carbon dioxide, and 2,3-diphosphoglycerate: A study by ^{13}C nuclear magnetic resonance and electrostatic theory. Ph D Thesis, Indiana University

Matthew JB, Morrow JS, Wittebort RJ, Gurd FRN (1977) Quantitative determination of carbamino adducts of α and β chains in human adult hemoglobin in the presence and absence of carbon monoxide and 2,3-diphosphoglycerate. J Biol Chem 252: 2234–2244

Matthew JB, Friend SH, Botelho LH, Lehman LD, Hanania GIH, Gurd FRN (1978) Discrete charge calculations of potentiometric titrations for globular proteins: Sperm whale myoglobin, hemoglobin alpha chain, cytochrome c. Biochem Biophys Res Commun 81: 416–421

Matthew JB, Hanania GIH, Gurd FRN (1979a) Electrostatic effects in hemoglobin: Hydrogen ion equilibria in deoxy- and oxy- human hemoglobin A. Biochemistry 18: 1919–1928

Matthew JB, Hanania GIH, Gurd FRN (1979b) Electrostatic effects in hemoglobin: Bohr effect and ionic strength dependence of individual groups. Biochemistry 18: 1928–1936

Matwiyoff NA, Needham TE (1972) Carbon-13 NMR spectroscopy of red blood cell suspensions. Biochem Biophys Res Commun 49: 1158–1164

Morrow JS (1974) The interaction of carbon dioxide with amino acids, peptides, and hemoglobin: A study by ^{13}C nuclear magnetic resonance. Ph D Thesis, Indiana University

Morrow JS, Gurd FRN (1975) Nuclear magnetic resonance studies of hemoglobin: functional state correlations and isotopic enrichment strategies. CRC Crit Rev Biochem 4: 221–287

Morrow JS, Keim P, Visscher RB, Marshall RC, Gurd FRN (1973) The interaction of [^{13}C] carbon dioxide and bicarbonate with human hemoglobin preparations (carbamino/NMR/carbonic anhydrase/relaxations). Proc Natl Acad Sci 70: 1414–1418

Morrow JS, Keim P, Gurd FRN (1974a) CO_2 adducts of certain amino acids, peptides, and sperm whale myoglobin studied by carbon 13 and proton nuclear magnetic resonance. J Biol Chem 249: 7484–7494

Morrow JS, Gurd RS, Gurd FRN (1974b) The chemical basis and possible role of carbamino homeostatic mechanisms. In: Blout ER, Bovey FA, Goodman M, Lotan N (eds) Peptides, polypeptides and proteins. Wiley, New York, pp 594–604

Morrow JS, Wittebort RJ, Gurd FRN (1974c) Ligand-dependent aggregation of chicken hemoglobin A_I. Biochem Biophys Res Commun 60: 1058–1065

Morrow JS, Matthew JB, Wittebort RJ, Gurd FRN (1976) Carbon 13 resonances of $^{13}CO_2$ carbamino adducts of α and β chains in human adult hemoglobin. J Biol Chem 251: 477–484

Neuberg C, Grauer A, Kreidl M (1955) The basis of the carbamate reaction. Arch Biochem Biophys 58: 169–177

O'Donnell S, Mandaro R, Schuster T (1978) Mechanism of the alkaline Bohr effect at Val α1. Biophys J 21: 201a

Ogawa S, Rottenberg H, Brown TR, Shulman RG, Castillo CL, Glynn P (1978) High-resolution ^{31}P nuclear magnetic resonance study of rat liver mitochondria. Proc Natl Acad Sci USA 75: 1796–1800

O'Leary MH, Jaworski RJ, Hartman FC (1979) ^{13}C nuclear magnetic resonance study of the CO_2 activation of ribulosebisphosphate carboxylase from *Rhodospirillum rubrum*. Proc Natl Acad Sci USA 76: 673–675

Rollema HS, de Bruin SH, Janssen LHM, van Os GAJ (1975) The effect of potassium chloride on the Bohr effect of human hemoglobin. J Biol Chem 250: 1333–1339

Roughton FJW (1964) Transport of oxygen and carbon dioxide. In: Fenn WE, Rann H (eds) Handbook of physiology, Sect 3: Respiration, vol I. American Physiological Society, Washington DC, p 767

Roughton FJW, Rossi-Bernardi L (1969) Carbon dioxide: Chemical, biochemical and physiological aspects. NASA SP 188, Washington DC, pp 41–45

Shire SJ, Hanania GIH, Gurd FRN (1974a) Electrostatic effects in myoglobin. Hydrogen ion equilibria in sperm whale ferrimyoglobin. Biochemistry 13: 2967–2974

Shire SJ, Hanania GIH, Gurd FRN (1974b) Electrostitic effects in myoglobin. pH and ionic strength variations of ionization equilibria for inidivual groups in sperm whale ferrimyoglobin. Biochemistry 13: 2974–2979

Shire SJ, Hanania GIH, Gurd FRN (1975) Electrostatic effects in myoglobin. Application of the modified Tanford-Kirkwood theory to myoglobins from horse, California grey whale, Harbor seal, and California sea lion. Biochemistry 14: 1352–1358

Stadie WC, O'Brien H (1936) The carbamate equilibrium: I. J Biol Chem 112: 723–758

Stadie WC, O'Brien H (1937) The carbamate equilibrium: II. J Biol Chem 117: 439–470

Tanford C, Kirkwood JG (1957) Theory of protein titration curves. J Am Chem Soc 79: 5333–5339

Tucker PW, Perutz MF (1977) Mechanism of charge compensation and impairment of co-operative functions in Hb tacoma. J Mol Biol 114: 415–420

Wiechelman KJ, Fox J, McCurdy PR, Ho C (1978) Proton nuclear magnetic resonance studies of Mb providence: A residue involved in anion binding. Biochemistry 17: 791, 795
Williams RC, Jr, Chung LL, Schuster TM (1975) Preparation and purification of carbamylated intermediates of human hemoglobin. Biophys Res Commun 62: 118–128
Wittebort RJ, Hayes DF, Rothgeb TM, Gurd RS (1978) The quantitation of carbamino adduct formation of angiotensin II and bradykinin. Biophys J 24: 765–778

The Measurement of the CO_2 Hb Binding Curve of Human Hemoglobin by ^{13}C-NMR Spectrometry

O. BURKHARD and W.K.R. BARNIKOL [1]

It is well known that under physiological conditions CO_2 binds to the four terminal amino groups of hemoglobin (Hb) (Kilmartin and Rossi-Bernardi, 1971). The carbamino compounds which are formed by this reaction are involved in the Haldane effect.

If we keep the concentration of the hydrogen ions, the ionic strength and the concentration of Hb constant, we can apply a simplified model based on the following reaction:

$$\mathrm{Hb} + \mathrm{CO_2} \rightleftharpoons \mathrm{HbCO_2} \tag{1}$$

Carbamino Hb is abbreviated by $HbCO_2$. For reaction (1) we can define an effective equilibrium constant λ:

$$\lambda = \frac{[\mathrm{HbCO_2}]}{[\mathrm{Hb}]\,[\mathrm{CO_2}]} \tag{2}$$

$[HbCO_2]$, $[Hb]$, and $[CO_2]$ are the concentration of $HbCO_2$, Hb, and CO_2. The CO_2 saturation of Hb (Z) is defined in the usual manner. By using Eq. (2) we obtain:

$$Z = \frac{\lambda\,[\mathrm{CO_2}]}{1 + \lambda\,[\mathrm{CO_2}]} \tag{3}$$

By using the well-known Hill plot we are able to obtain further information on the mechanism of the CO_2 binding to Hb, especially to examine whether the generally assumed simple theory outlined above is correct or not. In the Hill plot $\log [Z/(1-Z)]$ is plotted versus $\log P_{CO_2}$ (P_{CO_2}: partial pressure of CO_2). Usually we will obtain a straight line, and the slope of this line (n) is a very important parameter.

If n has the value 1, CO_2 binds to hemoglobin according to the simple model. If the four different sites of Hb do not interact and have different equilibrium constants n becomes less than 1. If n is considerably greater than 1, the four amino groups interact among themselves and bind CO_2 in a cooperative manner.

For our experiments we used ^{13}C-NMR spectrometry. Desoxygenated solutions of human Hb were equilibrated with the proper amount of N_2 and CO_2 containing 90% ^{13}C-isotope. The pH of the solutions was adjusted by titration. The Hb concen-

1 Physiologisches Institut der Universität Mainz, Saarstr. 21, 6500 Mainz/FRG

tration was about 11 g/100 ml. The NMR measurements were mainly done with a Varian XL 100 NMR spectrometer. Before and after each NMR measurement pH, P_{CO_2}, and PO_2 were controlled.

The NMR peaks were identified according to the results of Morrow et al., who had shown in a series of recent papers that the NMR signals which are obtained in the range of 28–30 p.p.m. (relative to CS_2) are caused by the carbamino derivative of Hb (Morrow et al., 1973, 1974, 1976; Matthew et al., 1977).

To do quantitative analysis we used the method of internal standard. This means that we have applied the following equation:

$$\frac{[HbCO_2]}{[HbCO_2] + [Hb]} = K \frac{I}{I_{Hb}} \Rightarrow Z = \frac{I/I_{Hb}}{(I/I_{Hb})_{max}} \tag{4}$$

I is the intensity of the carbamino signal and I_{Hb} the intensity of the amide signal. To prove the validity of this method, we have analyzed quantitatively mixtures of known composition in the same manner. The experimental results show clearly that the method of internal standard is working.

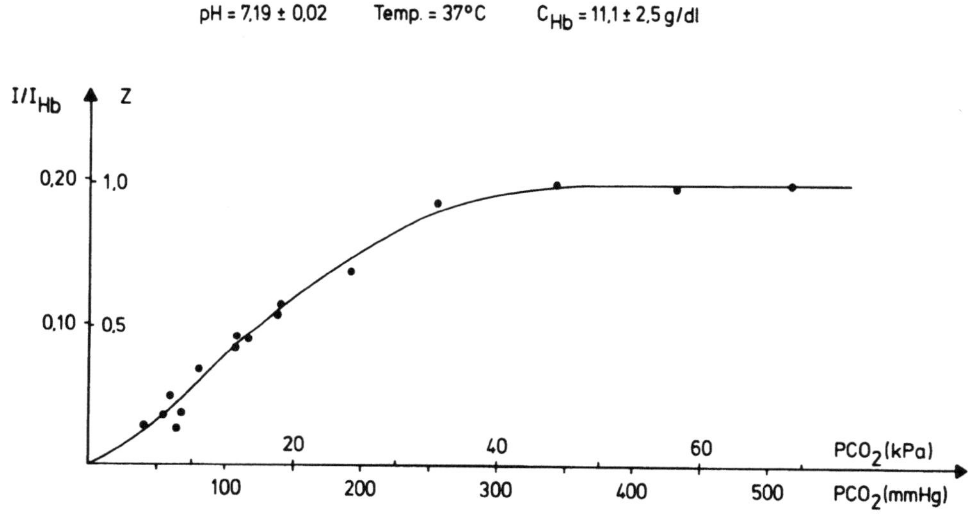

Fig. 1. CO_2 Hb binding curve of desoxygenated hemoglobin

In Fig. 1 the CO_2 binding curve we have measured is given, we have a S-shaped curve. The P_{CO_2} at 50% saturation is approximately 16 kPa. Above a P_{CO_2} of 50 kPa a real saturation behavior is observed.

Figure 2 shows the Hill plot of our data. The slope of the Hill plot is 2.1, which is in contrast to the results of former investigators. For example Perrella et al. (1972) reported an n-value of 0.9. But Perrella et al. only measured the CO_2 Hb binding curve

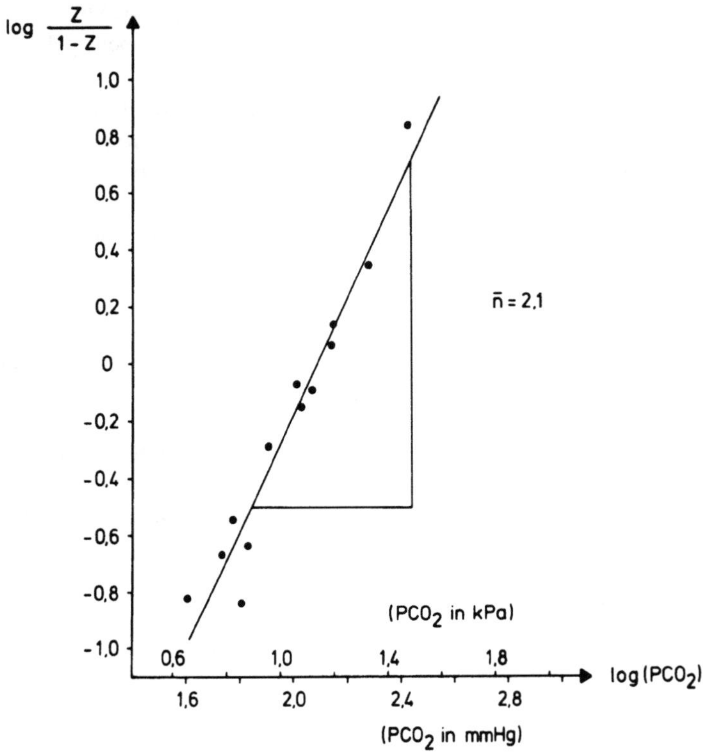

Fig. 2. Hill plot of the CO_2 Hb binding curve

at low CO_2 saturation. We have found during a study on the O_2 Hb binding curve that especially at low O_2 saturation the n-value is greatly reduced (Barnikol and Burkhard, 1978). In our opinion one obtains only reliable n-values from the whole CO_2 Hb binding curve. It should be emphasized that by evaluating our data according to the Hill plot no model is involved.

Our results are of considerable theoretical interest, because, as was stated above, an n-value greater than 1 gives evidence that the binding sites of a polymer interact among themselves. The conclusion of our experiments is that CO_2, like O_2, binds to Hb in a cooperative manner.

Acknowledgments. We wish to thank Dr. Pook and Mr. Pryss of the firm Ch. Boehringer/Ingelheim for carrying through the NMR measurements.

References

Barnikol WKR, Burkhard O (1978) Die Abhängigkeit des Hill'schen Parameters n von der O_2-Sättigung des Hämoglobins. Die Feinstruktur der O_2-Hb-Bindungskurve. Der direkte Nachweis für mindestens drei Zustände des Hb-Moleküls mit unterschiedlicher O_2-Affinität. Pflügers Arch Suppl 373: R 45

Kilmartin JV, Rossi-Bernardi L (1971) The binding of carbon dioxide by horse hemoglobin. Biochem J 124: 31–45

Matthew JB, Morrow JS, Wittebort RJ, Gurd FRN (1977) Quantitative determination of carbamino adducts of α and β chains in human adult hemoglobin in presence and absence of carbon monoacide and 2,3-disphosphoglycerate. J Biol Chem 252: 2234–2244

Morrow JS, Keim P, Visscher RB, Marschall RC, Gurd FRN (1973) Interactions of $^{13}CO_2$ and bicarbonate with human hemoglobin preparations. Proc Natl Acad Sci USA 70: 1414–1418

Morrow JS, Keim P, Gurd FRN (1974) CO_2 adducts of certain amino acids, peptids and sperm whale myoglobin studied by carbon 13 and proton nuclear magnetic resonance. J Biol Chem 249: 7484–7494

Morrow JS, Matthew JB, Wittebort RJ, Gurd FRN (1976) Carbon 13 resonances of $^{13}CO_2$ carbamino adducts of α and β chains in human adult hemoglobin. J Biol Chem 251: 477–484

Perrella M, Rossi-Bernardi L, Roughton FJW (1972) The carbamate equilibrium between CO_2 and bovine haemoglobin at 25 °C. In: Rørth M, Astrup P (eds) Oxygen affinity of hemoglobin and red cell acid base status. Munksgaard, Copenhagen, pp 177–204

Formation and Physiological Role of Carbamino Adducts of Angiotensin, Bradykinin, and S-Methylglucagon [1]

R.S. GURD, R.D. ENGLAND, D.F. HAYES, R.J. WITTEBORT, and T.M. ROTHGEB [2]

1 Introduction

Quantitation of the binding of carbon dioxide to the α-amino group of the β-subunits of hemoglobin by ^{13}C-nuclear magnetic resonance techniques has enhanced our understanding of the function of hemoglobin [3, 5]. We have further extended the investigation of the role of carbon dioxide binding to peptide hormones [10, 12]. Such binding represents a nonenzymatic, reversible, post-translational modification. If an intact amino terminal determines the biological fate or activity of any particular peptide hormone, then the addition of CO_2 could provide regulation for the endocrine system of which that peptide is a component. Possible mechanisms include a charge alteration, addition of bulk, or modification of metal binding [6].

We initially selected the vasoactive peptides angiotensin and bradykinin for study since they control vasomotor tone and hence tissue perfusion [12]. The resultant of their combined actions in turn affects local P_{CO_2} and pH. Additionally we have studied glucagon, a somewhat larger peptide which is present in greater concentrations than normal in clinical conditions characterized by metabolic acidosis [10]. Hyperglucagonemia is believed by some to be responsible in part for the complications of diabetes mellitus [7, 11].

To establish a regulatory role for carbon dioxide binding to these peptide hormones several conditions must be satisfied. First, binding must be shown to occur with physiological concentrations of H^+ and CO_2. Second, the extent of this binding should vary with physiological or pathological perturbations of pH and P_{CO_2}. Third, the binding should alter some subsequent biochemical phenomenon and fourth, this biochemical difference must result in a physiological perturbation in the intact organism.

This report summarizes our progress in demonstrating that carbon dioxide binding satisfies the first three of these criteria, suggesting that carbamino mechanisms may well be involved in physiological or pathological processes.

---

1 Supported by NIH Grants RO1 AM-21121, HL-05556 and GM-1046 and the Insurance Medical Scientist Scholarship Fund Sponsored by American United Life. This is the 112th paper in a series dealing with coordination complexes and catalytic properties of proteins and related substances

2 Department of Chemistry and Medical Sciences Program, Indiana University, Bloomington, Indiana, 47405/USA

2 Theory and Methods

Detailed procedures have been published elsewhere [3, 5, 6, 10, 12]. In essence, two constants, K_z, the proton dissociation constant for the amino group and K_c, the association constant for the binding of CO_2 to the nonprotonated form of the amine determine the extent of CO_2 binding to any amine for all values of pH and P_{CO_2}. Values for K_z have been determined by ^{13}C NMR techniques. The value for K_c for each peptide has been calculated after first measuring the value of Z, the mol fraction of peptide in the adduct form determined by equilibrating the peptide with $^{13}CO_2$ under known conditions of pH, total amine, total carbamate, and ^{13}C enrichment.

Angiotensin and bradykinin could be studied directly. However, the limited solubility of native glucagon precluded NMR studies at physiological pH. Therefore, S-methylglucagon was chosen as a model for glucagon. Prior studies in our laboratory have shown that methylation of the single methionine residue -27 of the 29-membered peptide yields a less active but soluble agonist of glucagon [9]. Our unpublished results indicate that its diminished activity correlates with diminished binding but at elevated hormone concentrations near maximum adenylate cyclase activity occurs. ^{13}C NMR spectra of this S-methyl derivative confirm its flexibility, circular dichroism studies indicate a lack of concentration-dependent conformation change characteristic of glucagon, and K_z values are similar to those measured for histidinamide [10]. These characteristics indicate a lack of intermolecular aggregation and conformation change. They permit the thermodynamic constants determined on the N-terminal region of S-methylglucagon to be used for those of native hormone and give justification for our use of S-methylglucagon as a working model for native glucagon in its predominant physiological form for the purpose of these studies [10].

Enzymatic analyses of aminopeptidase M activity were performed at pH 7.35–7.40 at ionic strength ranging from 0.155 to 0.300 at various total carbonates. Both leucyl-p-nitro anilide cleavage and angiotensin II cleavage were studied.

3 Results and Discussion

3.1 Physical Constants

Figure 1 shows a typical titration. In this case two resonances give similar pK_z values for angiotensin II. Similar titrations of each peptide were performed [10, 12].

Figure 2 shows a ^{13}C NMR spectrum at 25.2 MHz of S-methylglucagon equilibrated with $^{13}CO_2$ at various pH values. With pH values increasing from 7.4 to 8.9, the area of the carbamate resonance (peak 1) rises. At higher pH values than shown, this area decreases describing the expected bell-shaped curve [6]. The values of pK_z and pK_c which we have determined for each peptide are summarized in Table 1.

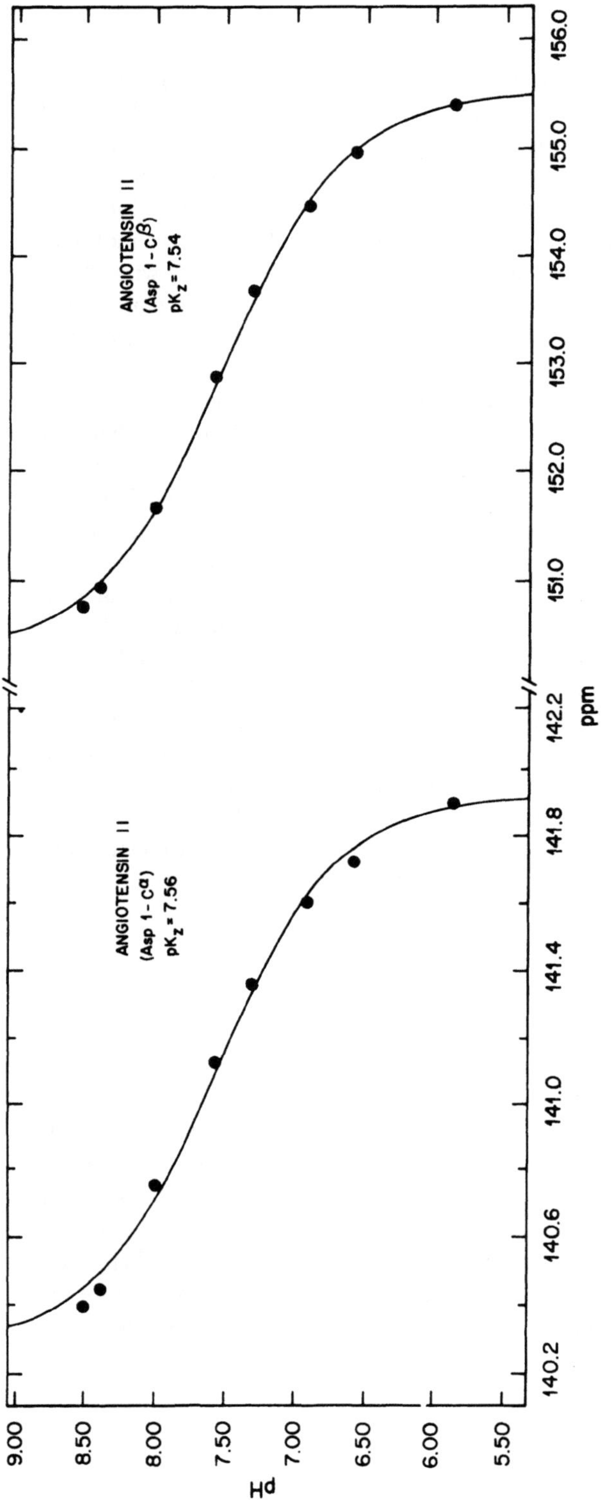

Fig. 1. The ^{13}C NMR determination of K_Z of the amino terminal of angiotensin II. Experimental points are given. The theoretical curve is the Henderson-Hasselbalch function using the determined pK. Both titrations yield K_Z values within experimental error [12]

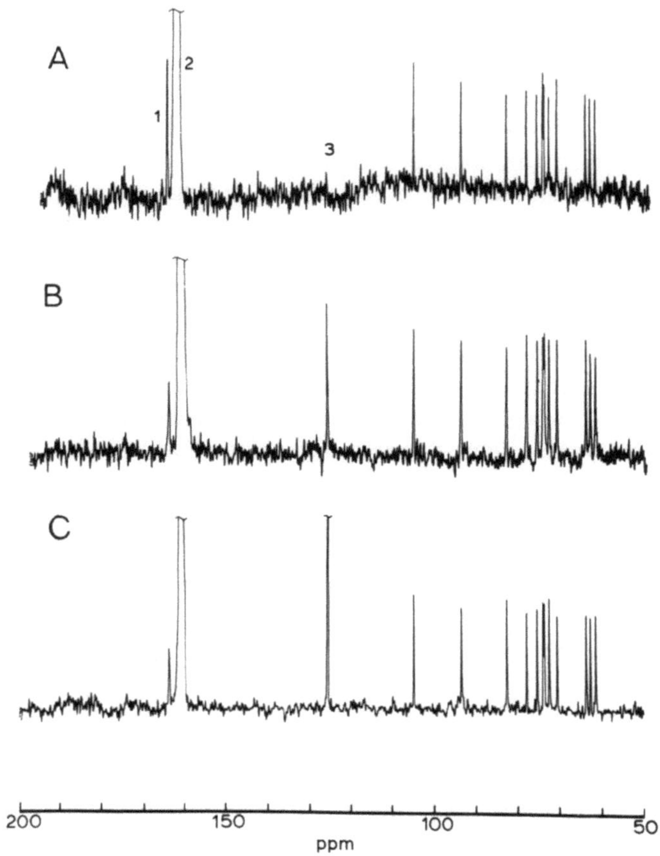

Fig. 2. The ^{13}C NMR spectrum of S-methylglucagon equilibrated with $^{13}CO_2$ at pH values of 8.9 (*A*) 7.7 (*B*) and 7.4 (*C*). Peaks *1, 2,* and *3* result from ^{13}C-enriched carbamate, bicarbonate and CO_2, respectively. Additional resonances are from sucrose added as a reference for quantitation [10]

Table 1. Summary of experimental constants determining the extent of carbamino adduct formation

Peptide	pK_Z [a]	pK_C [a]
Angiotensin II α-amino	7.55 ± 0.05 [b]	4.80 ± 0.10
Bradykinin α-amino	7.14 ± 0.05 [c]	5.02 ± 0.10
S-Methylglucagon α-amino	7.23 ± 0.05 [d]	4.66 ± 0.11

a -log K_Z and -log K_C where K_Z and K_C are the equilibrium constants for the proton dissociation from, and CO_2 addition to, the α-amino terminal, respectively [10, 12]

b Determined at an ionic strength of 0.15 at 30°C

c Determined at an ionic strength of 0.01 at 30°C

d Determined at an ionic strength of 0.02 at 28°C

3.2 Physiological Consideration

A series of curves of constant mol fraction, Z, may be superimposed upon standard pH-bicarbonate-P_{CO_2} diagrams. A typical web plot is shown for S-methylglucagon in Fig. 3. Curves of constant Z slope downward from left to right and values of Z which might be encountered in acid-base disturbance, as indicated by A, C, X, and Y, are seen to differ from the normal value of Z shown at B. From this figure it is readily appreciated that the extent of adduct formation differs minimally with perturbations of pH and P_{CO_2} characteristic of respiratory-induced acid-base imbalance (X and Y), whereas a near tenfold variation of Z occurs with conditions more typical of metabolic perturbations in acid-base balance (A and C). The web plots for the three peptides are remarkably similar with some variation in the slopes of Z as a consequence of pK_z values. Significantly, at lower pH values more relevant for intracellular conditions, marked changes in P_{CO_2} would be required to elicit even small changes in Z. Data for each peptide with typical acid-base perturbations are given in Table 2.

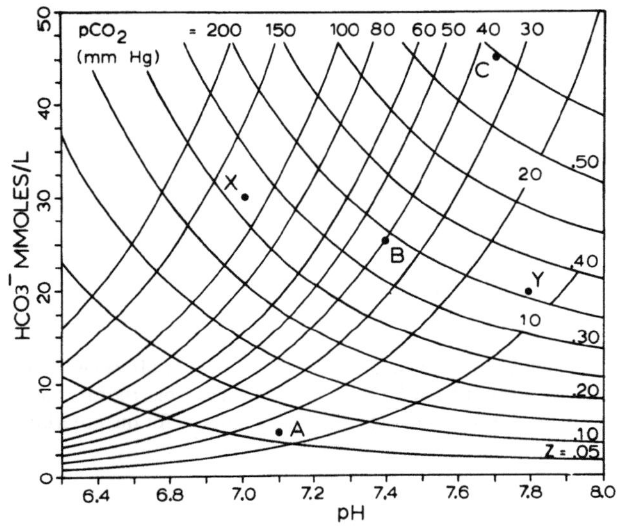

Fig. 3. The web plot of S-methylglucagon, Point *B* represents the mol fraction of carbamino adduct at normal values of pH, P_{CO_2}, and HCO_3^- of plasma. *X* and *Y* correspond to conditions of respiratory acidosis and alkalosis while *A* and *C* correspond to those of metabolic acidosis and alkalosis [10]

Enzymic Degradation. The role of the amino terminal for the biological effectiveness of each hormone differs, a phenomenon which allows specificity of the effect of changing mole fractions of adduct. For example, angiotensin II which produces a direct vasopressor effect by receptor interaction is converted by aminopeptidase activity to des-Asp AII (AIII) which principally mediates receptors controlling aldosterone secretion [2]. In contrast, bradykinin is degraded by dipeptidyl carboxypeptidase activity. Bradykinin may be generated from lysylbradykinin in vivo but the

Table 2. Mol fraction of carbamino adduct form of angiotensin II (AII), bradykinin (BK), and S-methylglucagon (SMG) at varying conditions of pH, P_{CO_2} and HCO_3^-. Calculated from the constants of Table 1

Physiological state	pH	P_{CO_2} (mm Hg)	HCO_3^- (mM)	Z-AII	Z-BK	Z-SMG
Normal	7.4	40	24	0.17	0.16	0.34
Respiratory acidosis	7.0	115	27	0.11	0.12	0.27
Respiratory alkalosis	7.8	13	22	0.21	0.16	0.36
Metabolic acidosis	7.1	15	4	0.02	0.03	0.07
Metabolic alkalosis	7.6	50	45	0.32	0.26	0.54

action of each on receptors is comparable [1]. Figure 4 shows the inhibition obtained when angiotensin II is subjected to aminopeptidase M degradation as a model for angiotensinase when conditions are conducive to formation of CO_2 adducts. That adduct formation is restricted to the substrate or that other species are not involved is unclear since kinetic analyses of this reaction with the model substrate leucyl-p-nitro anilide appear quite complex [4].

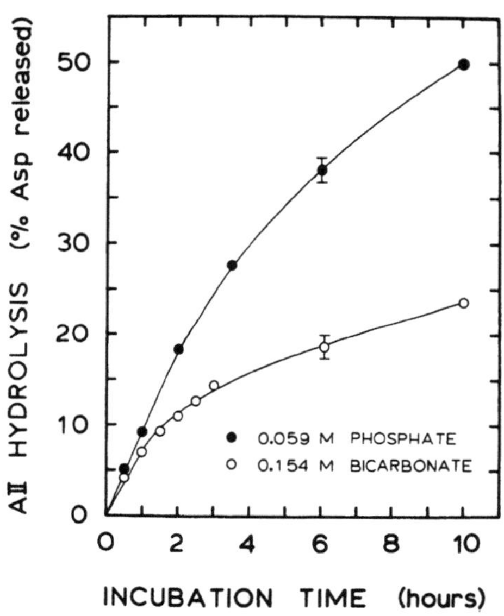

Fig. 4. The time course of aminopeptidase M cleavage of Ile[5]-angiotensin II. Hydrolysis was carried out at 37°C with 15 μg enzyme and 2.2 mg substrate at pH 7.4. Aspartate was quantitated by amino acid analysis and the results were normalized by the use of glycine as an internal standard

Receptor Occupancy. For glucagon and S-methylglucagon the principal role of the His-1 residue is to activate adenylate cyclase once the hormone is bound to its receptor. The following equilibria pertain to that binding when CO_2 is present.

$$\begin{array}{lccl} & CO_2 & & \\ & + & K & \\ R\ + & G & \rightleftharpoons & G \cdot R \\ \lambda & \upharpoonleft\downharpoonright & K' & \\ R\ + & CO_2 \cdot G & \rightleftharpoons & CO_2 \cdot G \cdot R \end{array}$$

Here K and K′ are apparent association constants for the binding of glucagon (G) to receptor (R) and λ is the effective pH-dependent association constant for binding of CO_2 to hormone. For adduct formation to be sensed by the cell two criteria are important: both G and $CO_2 \cdot G$ must bind with comparable affinity ($K \cong K'$) and subsequent metabolic events, say adenylate cyclase activation, must differ. From the above equilibria the following equation may be derived [8]:

$$\frac{[G \cdot R]}{[CO_2 \cdot G \cdot R]} = \frac{K}{K'} \cdot \frac{1}{\lambda\,[CO_2]}$$

Estimates of K/K′ may be used with calculated values of λ [$\lambda = K_cK_z/K_z\,(H^+) + (H^+)^2$] to indicate the relative occupancy of receptors by the native and adduct hormone at various P_{CO_2} and pH.

Table 3 summarizes the theoretical occupancy of glucagon receptors by carbaminoglucagon at several values of K/K′ with conditions comparable to those shown in Fig. 3. From the table one can see (1) that the closer K/K′ is to unity, the greater the occupancy by carbaminoglucagon rather than by native glucagon, (2) that the occupancy varies with perturbations of acid-base balance and (3) that the variation in occupancy is greater with metabolic than with respiratory perturbations. Carbon dioxide binding may be uniquely important for the ultimate activity of extended flexible hormones such as glucagon for which specialization of the amino terminal is directed toward activation and the binding region of hormone to receptor is sufficiently distant so that binding affinities are unaffected by adduct formation. Similar considerations of simultaneous equilibria suggest that carbamino hormone formation is unlikely to perturb the hormon-receptor binding process per se [12].

Table 3. Theoretical occupancy of glucagon receptors by carbamino glucagon with typical acid-base disturbance and variable relative affinity (K/K′)

Physiological state	pH	P_{CO_2}	K/K′ = 1	K/K′ = 2	K/K′ = 5
Normal	7.4	40	0.34	0.21	0.10
Respiratory acidosis	7.0	115	0.26	0.15	0.07
Respiratory alkalosis	7.8	12	0.36	0.22	0.10
Metabolic acidosis	7.1	15	0.06	0.03	0.01
Metabolic alkalosis	7.7	35	0.54	0.37	0.19

4 Conclusions

Three peptide hormones of major physiological importance have been shown to form carbamino adducts at physiological concentrations of H^+ and CO_2. The extent of carbamino formation varies widely with conditions characteristic of metabolic perturbations of acid-base balance. Specificity of response to these adducts depends upon the role of the amino terminal for the biological effectiveness of the hormone. A model enzyme for that which degrades angiotensin II to angiotensin III exhibits diminished activity with conditions where adduct formation is favored. Glucagon may be representative of a particular group of hormones whose receptors bind native hormone and its adduct form to an equivalent extent, yet subsequent activation of adenylate cyclase is extensively modified. Our studies suggest a regulatory role for carbamino adduct formation to peptide hormones as well as to other proteins whose biological roles and fates are determined by their α-amino terminals. Other extracellular proteins and peptides with blocked amino terminals may also be affected if chemical properties or conformational constraints place the proton dissociation constants of ε-amino groups into the range described here [3].

References

1. Colman RW (1974) Formation of human plasma kinin. N Engl J Med 291: 509–515
2. Freeman RH, Davis JO, Lohmeier TE, Spielman WS (1977) [Des-Asp^1] angiotensin II: mediator of the renin-angiotensin system? Fed Proc 36: 1766–1770
3. Gurd FRN, Matthew JB, Wittebort RJ, Morrow JS, Friend SH (1979) The carbamate reaction with proteins: observation by ^{13}C-NMR and evaluation of structural consequences. Proc Symp Biophys Physiol Carbon Dioxide, Regensberg, W Germany, April 17–20, this vol
4. Hayes DF (1977) Inhibition of aminopeptidase M by the carbon dioxide-bicarbonate-carbonate buffer system. Master Thesis, Indiana University, Pt 1
5. Matthew JB, Morrow JS, Wittebort RJ, Gurd FRN (1977) Quantitative determination of carbamino adducts of α and β chains in human adult hemoglobin in presence and absence of carbon monoxide and 2,3-diphosphoglycerate. J Biol Chem 252: 2234–2244
6. Morrow JS, Keim P, Gurd FRN (1974) CO_2 adducts of certain amino acids, peptides, and sperm whale myoglobin by carbon 13 and proton nuclear magnetic resonance. J Biol Chem 249: 7484–7494
7. Raskin P, Unger RH (1978) Hyperglucagonemia and its suppression: importance in the metabolic control of diabetes. N Engl J Med 299: 433–436
8. Rothgeb TM (1978) Methylation and demethylation of the methionyl residues of sperm whale myoglobin and glucagon. Ph D Thesis, Indiana University
9. Rothgeb TM, Jones BN, Hayes DF, Gurd RS (1977) Methylation of glucagon, characterization of the sulfonium derivative, and regeneration of the native covalent structure. Biochemistry 16: 5813–5818
10. Rothgeb TM, England RD, Jones BN, Gurd RS (1978) Physical characterization of S-methylglucagon and quantitation of carbamino adduct formation. Biochemistry 17: 4564–4571
11. Unger RH, Orci L (1975) The essential role of glucagon in the pathogenesis of the endogenous hyperglycemia of diabetes mellitus. Lancet 1: 14–16
12. Wittebort RJ, Hayes DF, Rothgeb TM, Gurd RS (1978) The quantitation of carbamino adduct formation of angiotensin II and bradykinin. Biophys J 24: 765–778

Interaction Between Hemoglobins, CO_2 and Anions

R. BAUMANN [1]

1 Introduction

The role of oxygen-linked carbamate formation in the regulation of hemoglobin function under physiological conditions is largely dependent on the interaction between CO_2 and other allosteric effectors, notably the organic phosphates. The α-amino groups of the α- and β-chain N-termini are the established binding sites for carbon dioxide in the physiological pH range [14] and quantitative estimates of the contributions of both sets of sites have been made for human hemoglobin, where the β-chains play the dominant role in oxygen-linked CO_2 binding [6, 15, 18]. However, direct competition between 2,3-diphosphoglycerate and CO_2 for the β-chain N-terminal amino group results in a minor role of carbamate formation, both with respect to oxygen-linked CO_2 transport as well as regulation of oxygen affinity [5, 6] of human hemoglobin, and such may be the case in all hemoglobins which have organic phosphates as dominant allosteric effectors under physiological conditions. Among mammalian hemoglobins, there are two groups where oxygen affinity in vivo is regulated independent of organic phosphates, i.e., bovine and cervidae hemoglobins and the hemoglobins of the felidae, all of which have substitutions or deletions at the organic phosphate binding site [11, 21].

The two hemoglobins of the domestic cat are of particular interest, because of the unusually high incidence of a β-chain N-terminally acetylated variant, hemoglobin B, which constitutes between 5% to 50% of total hemoglobin [20]. Both cat hemoglobins have His β2 substituted by phenylalanine [21], and the B hemoglobin differs in primary structure at three sites from Hb A : β1 Gly → N-acetylser; β144 Lys → Arg; β139 Asn → Ser. The substitutions at sites crucial for CO_2 and organic phosphate binding makes these hemoglobins interesting models for comparative studies of CO_2-binding as well as of binding of inorganic anions such as chloride, which aside from CO_2 and protons constitute the only potentially important physiological effector molecules.

2 Materials and Methods

Cat hemoglobins A and B were prepared as described previously [7] and stored as CO-hemoglobin in liquid nitrogen until use. The isolated fractions were analyzed for

1 Physiologisches Institut, Medizinische Hochschule, Karl-Wiechert-Allee 9, 3000 Hannover/FRG

purity by isoelectric focusing on polyacrylamide gel. Organic and inorganic phosphates were removed by passing the respective hemoglobin samples over a 1.5 x 90 cm column of Sephadex G-20, equilibrated with 0.05 mol Tris HCl pH 8.5 in 0.16 mol Cl^-, and subsequently over a 2.5 x 40 cm column of Sephadex G-20 equilibrated with 0.1 mol NaCl.

Concentration and dialysis of the hemoglobin solutions were done in an Amicon U.F. cell with PM 10 membrane.

Total carbon dioxide content of hemoglobin solutions equilibrated with N_2/CO_2 or O_2/CO_2 mixtures was measured in 0.1 ml samples with a Hewlett Pachard GC 5750 and automatic integrator unit 3380 A. Prior to the measurements CO was removed from hemoglobin by exposure to strong light. The completion of this procedure was checked by measuring the oxygen capacity of the sample with a Lex O_2Con (Lexington Instruments, Waltham Mass.). Hemoglobin concentration for these measurements was adjusted to 200 g/l. Methemoglobin did not exceed 7%. Oxygen equilibrium curves were done either with the method of Benesch et al. [9] or as described in [8].

3 Results and Discussion

At low concentration of salt the oxygen affinity of cat hemoglobin B is lower than that of cat hemoglobin A (Fig. 1). The primary mechanism adapting the oxygen affinity of both hemoglobins, so that variation in the hemoglobin A/B ratio does not affect the in vivo oxygen affinity, is their different reactivity towards CO_2 and chloride (Fig. 1). Since both the action of CO_2 as well as of chloride is stronger on HbA than on HbB, the oxygen affinities of both hemoglobins are the same in the presence of physiological concentrations of H^+, CO_2, and chloride. In a similar manner the Bohr effect of both hemoglobins is adjusted (Table 1) in the physiological pH range.

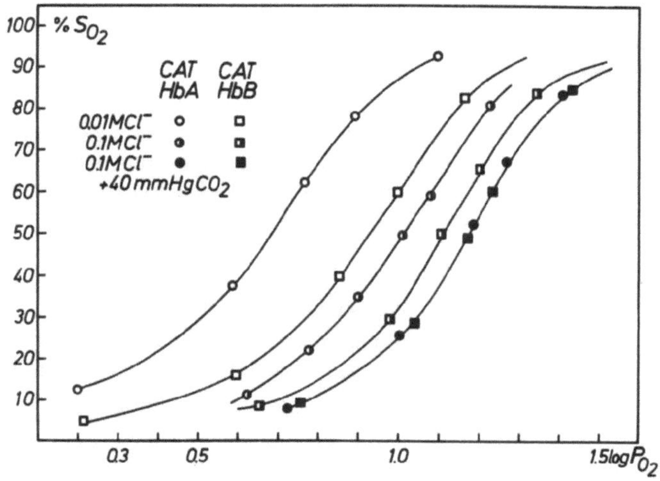

Fig. 1. Adjustment of the oxygen affinities of cat HbA and HbB through CO_2 and chloride pH 7,3, 20°C. (After Baumann and Haller [7])

Table 1. Effect of Cl^- and CO_2 on the Bor effect of cat HbA and HbB

$\Delta \log P_{50}/\Delta pH$ (pH 7.0–7.5)	HbA	HbB
0.01 M Cl^-	– 0.60	– 0.61
0.1 M Cl^-	– 0.77	– 0.64
0.1 M Cl^- + 40 torr CO_2	– 0.48	– 0.51

There is no conceivable competition between oxygen-linked CO_2 and chloride binding up to a chloride concentration of 0.5 mol in both hemoglobins. This is in keeping with observations of Bauer and Jung [4] on sheep hemoglobins A and B, who also found no competition between chloride and CO_2 for these hemoglobins. On the other hand, inorganic phosphate (P_i) reduces oxygen-linked CO_2-binding to both cat hemoglobins and a qualitatively similar behavior is observed for human hemoglobin (Figs. 2 and 3).

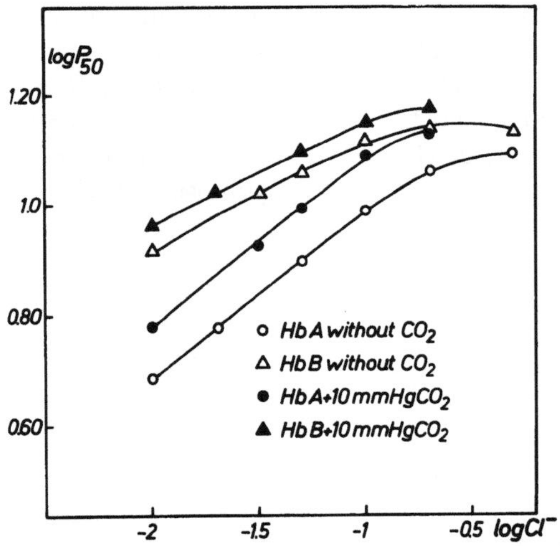

Fig. 2. Influence of CO_2 and chloride on the oxygen affinity of cat HbA and HbB pH 7.3 20°C. (After Bauman and Haller [7])

From Fig. 3 it can also be seen that the oxygen-linked binding of CO_2 to cat hemoglobin A which has free N-termini is reduced compared to human adult hemoglobin. The effect of CO_2 on hemoglobin B, where only the α-chains are free to bind CO_2, is of the same magnitude as observed with comparable human hemoglobin variants HbA_{Ic} and HbF_{Ic} [6].

The results obtained from CO_2 binding curves (Figs. 4, 5 and Table 2), show that oxygen-linked binding to the cat hemoglobin A β-chain is only a third of what is found for human hemoglobin under the same conditions.

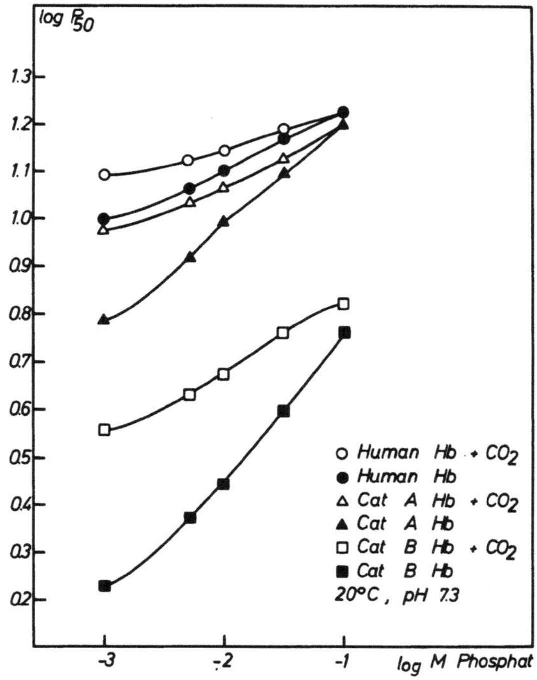

Fig. 3. Influence of inorganic phosphate on the oxygen affinity of human and cat hemoglobin in the absence and presence of 40 mm Hg (5.3 kPA) CO_2

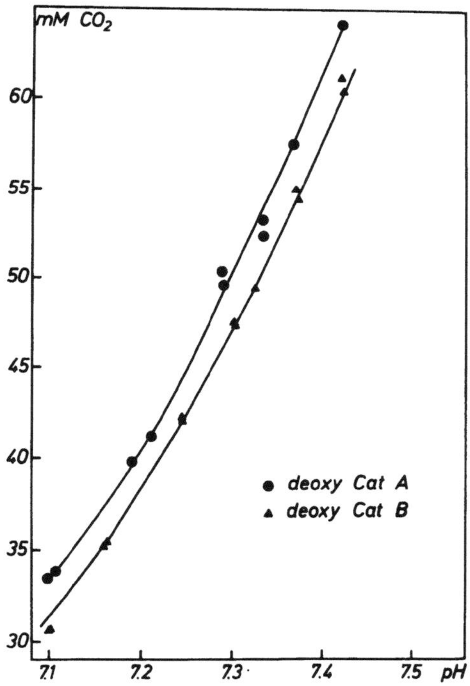

Fig. 4. Carbon dioxide binding curves of deoxygenated solutions of cat hemoglobin A and B in 160 mmol (Cl^- + HCO_3^-) 37°C, 200 g Hb/l, PCO_2 ~100 mm Hg (13.3 kPA)

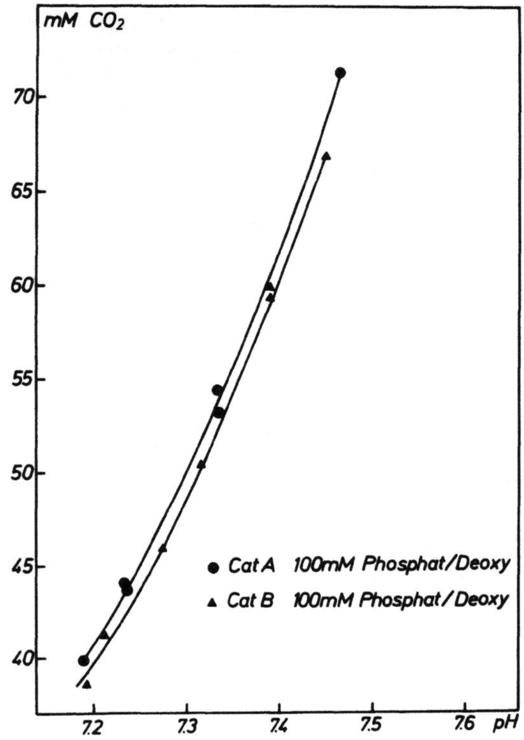

Fig. 5. Carbon dioxide binding curves of deoxygenated solutions of cat hemoglobin A and B in the presence of 100 mmol phosphate buffer. 37°C, 200 g Hb/l, PCO_2 ~100 mm Hg (13.3 kPa)

Table 2. Comparison of the fraction of carbamate bound to the β-chain (f_β) of cat hemoglobin A and human hemoglobin A and the amount of oxygen-linked carbamate bound to the α- and β-chain (Δf_β and Δf_α)

Cat hemoglobin A		Human hemoglobin A	
f_β deoxy	0.338 (0.161)	f_β deoxy	0.528
f_β oxy	0.219 (0.129)	f_β oxy	0.134
Δf_β	0.128 (0.03)	Δf_β	0.384
Δf_α	0.08 (0.03)	Δf_α	0.09

Values in brackets refer to the experiments done in the presence of 0.1 M inorganic phosphate.
In all other experiments total ionic strength was adjusted to 0.16 with NaCl.
T = 37°C, pH 7.2, hemoglobin concentration 200 g Hb/l, PCO_2 ~100 mm Hg (13.3 kPa).
Values for human hemoglobin were calculated with the data of Perrella et al. [18].

The main difference between both hemoglobins is the greatly dminished CO_2 affinity of the deoxy β-chain of cat hemoglobin A, whereas its oxy β-chain binds slightly more CO_2 than the human hemoglobin β-chain under physiological conditions.

The contribution of the α-chain of cat hemoglobin to CO_2 binding is small and not significantly different from the value reported for the human α-chain (Table 2). In human hemoglobin the contribution of the α-chain to oxygen-linked CO_2 binding is small, because of the involvement of its N-terminal α-amino group in the alkaline Bohr effect [14] and the same mechanism may operate in the cat hemoglobins.

Thus, the amount of oxygen-linked CO_2 binding seems to be primarily a function of the affinity of the deoxy β-chain towards CO_2, that is of K_z, the equilibrium constant of the N-terminal α-amino group and of K_c, the equilibrium constant for the carbamate reaction: $RNH_2 + CO_2 \overset{K_c}{\rightleftharpoons} RNH\,COO^- + H^+$.

Crystallographic analysis of human deoxyhemoglobin [1] shows that the large increase in CO_2 affinity upon deoxygenation of the β-chain reflects stabilizing electrostatic interactions between the carbamate anion and the cluster of positive charges surrounding the central cavity, notably lysine β 82, where the ϵ-amino group is in a position to form a salt bridge with the carbamate anion [1].

Structural changes which result in a different charge distribution in the central cavity or an outward movement of the β-chain N-terminal amino group relative to the charge cluster, severing in particular the possible interaction betwen carbamate and lysine β82, should result in a considerable reduction of CO_2 affinity of the deoxy β-chain. The few comparative data available seem to support this suggestion. Thus in human hemoglobin F_{II} [13] and in sheep hemoglobin A and B [4, 8] oxygen-linked CO_2 binding is greatly reduced. In these hemoglobins the β-chain α-amino groups are moved further apart from the central cavity [12, 19], compared to human deoxyhemoglobin, and they have substitutions or deletions affecting the charge pattern of the central cavity (His β143 → Ser in HbF_{II}; His β2 deleted in both sheep hemoglobins).

The replacement of His β2 by phenylalanine in the cat hemoglobins alters the charge distribution and is unlikely not to affect the orientation of the β-chain N-terminus with respect to neighboring residues especially lysine β82, and may thus be the main reason for the low CO_2 affinity of the deoxy β-chain.

The reduction of CO_2 binding to the deoxy α- and β-chains in the presence of inorganic phosphate confirms our previous results [7] and is in accordance with crystallographic data [2], which demonstrate inorganic anions bound to the α- and β-chain N-terminal regions of human deoxyhemoglobin. Since P_i also inhibits CO_2 binding to the oxy β-chain it seems that the β-chain binding site for P_i is not very different for oxy and deoxyhemoglobin.

Studies on human hemoglobin mutants [10, 16] have shown that lysine β82 is the important site for oxygen-linked chloride and P_i binding to the β-chain. However, only P_i and not chloride inhibits CO_2 binding in the cat or sheep hemoglobins. If one assumes that in these hemoglobins the carbamate anion is too distant from lysine β82 to align with its ϵ-amino group, then a noncompetitive binding of chloride and CO_2 seems possible, and the inhibitory action of P_i on carbamate formation must be attributed to its greater charge and bulk, rather than competition for the same site.

For human hemoglobin one should however expect competitive effects between CO_2 and chloride binding to the β-chain. At the α-chain a similar difference in the action of chloride and P_i is observed, since again only P_i inhibits carbamate formation. Nigen et al. [17] found on the other hand that the rate of reaction of the α-chain N-terminus with KCNO is greatly reduced by increasing chloride concentrations whereas P_i has only a weak inhibitory effect. They concluded that P_i and chloride are bound to different sites in the N-terminal region, which is also suggested by our results.

The apparent independence of oxygen-linked chloride and CO_2 binding in the cat hemoglobins is of advantage for the regulation of their oxygen affinity. Since these hemoglobins have a relatively low intrinsic oxygen affinity, they need a proportionally smaller decrease in oxygen affinity than, for example, human hemoglobin, to adjust to physiological conditions. This can be achieved through the combined action of weak effectors such as chloride or CO_2, and because their action is on low affinity hemoglobins, the resulting change in half saturation pressure (P_{50}) in absolute terms, can be quite substantial. In cat hemoglobin A the P_{50} at pH 7.2 and 37°C with 10 mmol chloride is 15.6 mm Hg (2.08 kPa) and increases to 29.2 mm Hg (3.89 kPa) with 100 mmol chloride and further to 36.7 mm Hg (4.89 kPa) when 40 mm Hg (5.3 kPa) CO_2 are present. The absolute change in P_{50} is 21 mm Hg (2.79 kPa), to be compared with the increase of P_{50} in human hemoglobin, which is 5.6 mm Hg with 10 mmol chloride at pH 7.2 and 37°C and increases to 27 mm Hg (3.6 kPa) with physiological concentrations of 2,3 DPG, chloride and CO_2.

References

1. Arnone A (1974) X-ray studies of the interaction of CO_2 with human deoxy hemoglobin. Nature (London) 247: 143–144
2. Arnone A, Benesch RE, Benesch P (1977) Structure of human deoxyhemoglobin modified specifically with pyridoxal compounds. J Mol Biol 115: 627–642
3. Bauer C (1970) Reduction of the carbon dioxide affinity of human hemoglobin solutions by 2,3 diphosphoglycerate. Respir Physiol 10: 10–19
4. Bauer C, Jung HJ (1975) A comparison of respiratory properties of sheep hemoglobin A and B. J Comp Physiol 102: 167–172
5. Bauer C, Schroeder E (1972) Carbamino compounds of hemoglobin in human adult and fetal blood. J Physiol (London) 227: 457–471
6. Bauer C, Baumann R, Engels U, Pacyna B (1975) The carbon dioxide affinity of various human hemoglobins. J Biol Chem 250: 2173–2176
7. Baumann R, Haller EA (1975) Cat hemoglobins A and B: Differences in their interaction with Cl^-, phosphate and CO_2. Biochem Biophys Res Commun 65: 220–227
8. Baumann R, Bauer C, Haller EA (1975) Oxygen-linked CO_2 transport in sheep blood. Am J Physiol 229: 334–339
9. Benesch R, Macduff G, Benesch RE (1965) Determination of oxygen equilibria with a versatile new tonometer. Anal Biochem 11: 81–87
10. Bonaventura J, Bonaventura C, Sullivan B, Feruzzi G, McCurdy PR, Fox J, Moo-Penn WF (1976) Hemoglobin providence. Functional consequences of two alterations of the 2,3 diphosphoglycerate binding site at position β-82. J Biol Chem 251: 7563–7571
11. Dayhoff MO (1975) Atlas or protein sequence and structure. National Biomedical Res Found, Washington DC

12. Frier JA, Perutz MF (1977) Structure of human fetal deoxy hemoglobin. J Mol Biol 112: 97–112
13. Gros G, Bauer C (1978) High pK value of the N-terminal amino group of the γ-chain causes the low CO_2-binding of human fetal hemoglobin. Biochem Biophys Res Commun 80: 56–62
14. Kilmartin JV, Rossi-Bernardi L (1969) Inhibition of CO_2-combination and reduction of the Bohr effect in hemoglobin chemically modified at its α-amino groups. Nature (London) 222: 1243–1246
15. Matthew JB, Morrow JS, Wittebort RJ, Gurd FRN (1977) Quantitative determination of carbamino adducts of α- und β-chains in human adult hemoglobin in presence of carbon monoxide and 2,3 diphosphoglycerate. J Biol Chem 252: 2234–2244
16. Mauk AG, Mauk MR, Taketa F (1976) The role of the β-82 Lysine in the binding of small anions to human hemoglobin Fed Proc 35: 1241
17. Nigen AU, Bass BD, Manning JM (1976) Reaction of cyanate with valine 1α of hemoglobin. J Biol Chem 251: 7648–7643
18. Perrella U, Guglielmo G, Mosca A (1977) Determination of the equilibrium constants for oxygen-linked CO_2-binding to human hemoglobin. FEBS Lett 78: 287–290
19. Perutz MF (1970) Stereochemistry of cooperative effects in hemoglobin. Nature (London) 228: 726–739
20. Taketa F (1974) Organic phosphates and hemoglobin structure function relationship in the feline. Ann NY Acad Sci 241: 524–537
21. Taketa F, Mauk AG, Mauk MR, Brimhall B (1977) The tryptic peptide composition of the beta chains of hemoglobins A and B of the domestic cat (felis catus). J Mol Evol 9: 261–272

The Influence of CO_2 on the First Binding Constant of Oxygen to Human Hemoglobin Ao

C. POYART, E. BURSAUX, and P. MATZKE [1]

1 Introduction

The oxylabile nature of CO_2 binding to mammalian hemoglobins (Hb) was recognized a long time ago [4, 13, 14]. Owing to its preferential binding to deoxygenated Hb, carbon dioxide decreases the affinity for oxygen of the tetramer and inhibits to a large extent the alkaline Bohr effect. This latter phenomenon was explained on the basis of the following equations describing the reactions involved in CO_2 binding to the α amino groups of the protein:

$$R\text{-}NH_3^+ \overset{K_z}{\rightleftharpoons} R\text{-}NH_2 + H^+ \tag{1}$$

$$R\text{-}NH_2 + CO_2 \overset{K_c}{\rightleftharpoons} R\text{-}NH\,COO^- + H^+ \tag{2}$$

It was later demonstrated that in the physiological pH range the α-amino group of the terminal valine of α and β chains are the sites for CO_2 binding in both oxy and deoxy Hb. The binding constants of CO_2 to these two pairs of sites were estimated from NMR studies [8, 9] and these indicate that the β chains binding sites have a higher affinity for CO_2 in deoxy Hb than the α chains binding sites. In the oxy configuration the affinity for CO_2 of the terminal valines in the β chains decreases largely and this explains the predominance of these binding sites in the amount of oxygen-linked CO_2 bound to Hb [10, 11]. However unsuitable for the estimation of the CO_2 binding constants, oxygen-binding studies performed on unmodified Hb solutions in the presence of CO_2 have been very useful for describing the overall effect of oxylabile CO_2 on the functional properties of various Hb [3]. They are still the only means to estimate the dynamics of this binding at the different steps of heme ligation. Up to now it is not known what is the effect of CO_2 on the four Adair constants describing in equilibrium conditions the O_2 binding to the four subunits of Hb [12]. The aim of this study was therefore to estimate the effect of CO_2 on K_1, the first association constant of O_2 to human adult Hb and an index of the affinity for O_2 of Hb in its T deoxy configuration. These effects were compared to those observed at the $P_{0.5}$ level which, according to the linked-functions theory [15], describe the overall influence of CO_2 on the tetrameric Hb.

1 Inserm, Unité 27, 42 rue Desbassayns de Richemont, 92150 Suresnes/France

2 Material and Methods

Oxygen-binding experiments were all performed on purified isolated Hb Ao solutions at a concentration of 6.2 to 9.3 x 10^{-4} mol x l^{-1} (10–15 mg/ml). Aliquots of the stock Hb solution (metHb content less than 2%) were diluted in $NaHCO_3$-NaCl buffers. In order to maintain a constant ionic strength ($\Gamma/2 = 0.1$) the amount of $NaHCO_3$ was varied inversely to that of NaCl depending on the desired pH and P_{CO_2} used. All experiments were performed at 25°C ± 0.1 with a CARY 219 spectrophotometer. K_1 and $P_{0.5}$ were calculated from Hill plots as described in detail previously [12]. Briefly, K_1 mm Hg^{-1} was computed from four to five points of the initial part of the oxygenation curve below 1.5% saturation which were measured in deoxy Hb solutions previously freed from traces of carbon monoxide. The Hb solutions were first equilibrated at atmospheric pressure with an argon-CO_2 gas mixture which corresponded to the desired P_{CO_2}. Small amounts of O_2 diluted in argon (Wösthoff pump) were added in the equilibration vessel with a Hamilton syringe through a rubber-capped side hole. $P_{0.5}$ were calculated from at least five points of the oxygenation curve between 30% and 70% saturation. Experiments in the absence of CO_2 were carried out in Hb solutions diluted in 0.05 M bis-Tris or Tris buffer containing 0.1 M Cl^- adjusted to the desired pH with HCl. The experimental protocol consisted of measurements of K_1 and $P_{0.5}$ at different pH values and for each pH value at various P_{CO_2} from 7 to 120 mm Hg (1 mm Hg = 0.133 kPa). In most of these experiments pH was measured in the deoxy and oxy Hb solutions (initial and final pH). No significant variation between the two sets of values has been observed. MetHb content at the end of the 90-min period of the experimental procedure was usually increased by not more than 2% from its starting value. To help a future comparison of the oxygenation parameters at various ionic strength, K_1 and $P_{0.5}$ values have been expressed as $(O_2)_1$ and $(O_2)_{50}$ where $(O_2)_1 = \alpha \times 1/K_1$ and $(O_2)_{50} = \alpha \times P_{0.5}$. α is the solubility coefficient of O_2 in aqueous solutions at $\Gamma/2 = 0.1$ in presence of NaCl and amounts to 13.45 μmol O_2 x l^{-1} x kPa^{-1} [2].

Calculations: The results were analyzed according to the linked-functions theory [1, 15]. The Bohr effect at constant pH and pCO_2 was calculated as the first derivative of the function relating log $(O_2)_1$ or log $(O_2)_{50}$ to pH variations at a given P_{CO_2}. Similarly the amount of oxygen-linked CO_2 at any given pH and P_{CO_2} was estimated as the first derivative of the function relating log $(O_2)_1$ or log $(O_2)_{50}$ to log (CO_2) mol x l^{-1} where (CO_2) is the concentration of molecular CO_2 dissolved in the aqueous solution. This was obtained as $\alpha \times P_{CO_2}$ where α_{CO_2} the solubility coefficient of CO_2 at $\Gamma/2 = 0.1$ and amounts to 3.3 x 10^{-4} mol x l^{-1} x kPA^{-1} at 25°C [5]. The difference in the Bohr effect with and without CO_2 was computed from the Bohr curves obtained in the presence and absence of CO_2. This quantity was divided by the amount of oxygen-linked CO_2 at any given pH and P_{CO_2} to calculate the amount of Bohr protons taken up per mol oxylabile CO_2 released. As discussed below, binding constants to the deoxy and monoliganded Hb have been tentatively calculated with the assumption that one single binding site was involved at the first step of oxygenation. This was done using an equation similar to that given by Szabo and Karplus (1976). From the pH dependence of the binding constants approximate values for the pK_z and pK_c of the amino-groups involved were estimated [13]. All calculations were carried out using standard least squares curve fitting procedures on a 370/138 IBM computer.

3 Results and Discussion

3.1 Influence of CO_2 on $(O_2)_{50}$ and the Alkaline Bohr Effect

Figure 1 depicts the variations of log $(O_2)_{50}$ with pH at various P_{CO_2}. It may be noted that at all pH values $(O_2)_{50}$ was increased upon addition of CO_2 compared to the values observed in its absence. The steepness of the curves was decreased at elevated pH and P_{CO_2} which indicates a reduction of the alkaline Bohr effect. Figure 2 shows the calculated Bohr curves. It is clear that addition of CO_2 induces a significant reduction of the alkaline Bohr effect which is reversed at high pH values as net proton uptake is observed upon oxygenation of Hb. One may note also that at neutral or acid pH the Bohr curves in the presence of CO_2 remains below the CO_2-free Bohr curve despite the discrete variations of $(O_2)_{50}$ with increasing P_{CO_2} in this pH range. This indicates an increase in the amount of Bohr protons consumed per mol oxygen-linked CO_2 released as expected at these pH values from Eq. (1) and the pK_z of the amino group involved in this reaction reported by different authors [6, 9, 10]. Variations of the amount of oxygen-linked CO_2 with varying concentrations of CO_2 are shown in Fig. 5 (upper part). Only three curves have been represented for clarity.

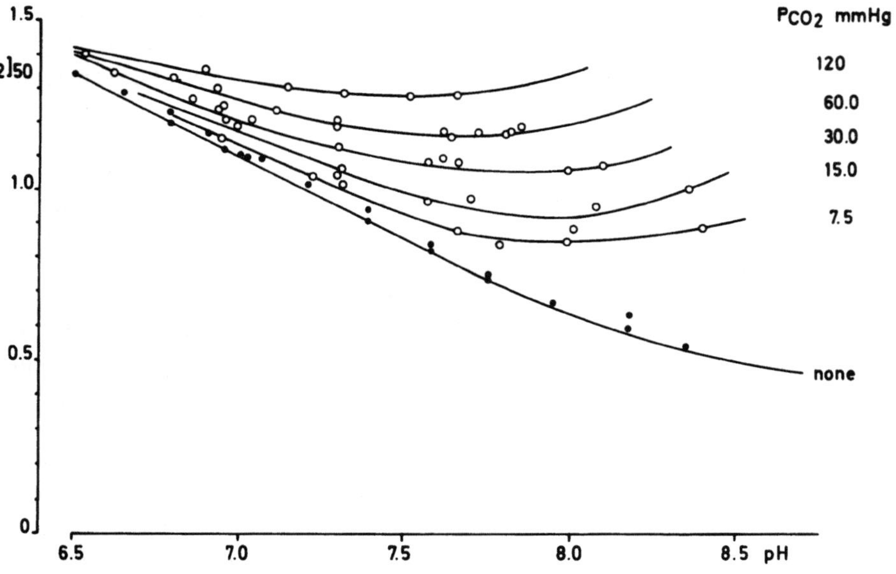

Fig. 1. Variations of log $(O_2)_{50}$ with pH at various P_{CO_2}. Conditions were: Hb = 6 – 9 x 10^{-4} mol heme, NaCl-$NaHCO_3$ buffer, $\Gamma/2$ = 0.1, 25°C. In the absence of CO_2 (*lower curve*) 0.05 M bis-Tris or Tris buffer plus 0.1 M Cl^- were used

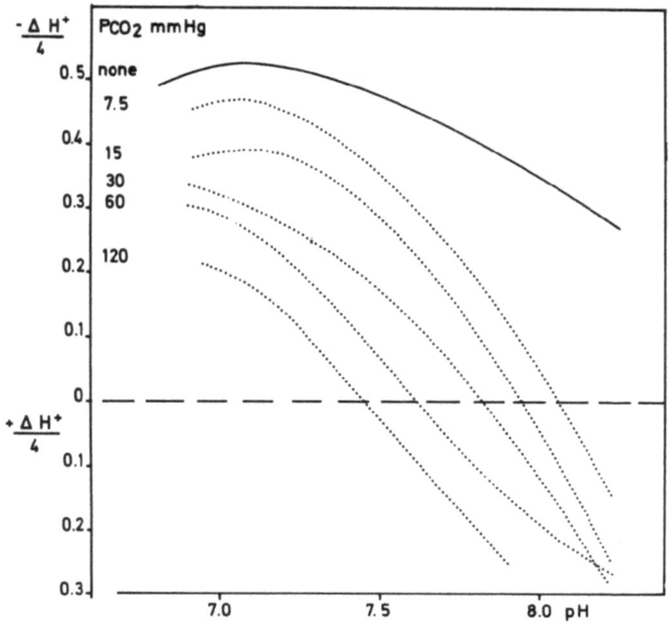

Fig. 2. Calculated Bohr curves at $(O_2)_{50}$ in the absence and presence of various amounts of CO_2. Conditions were those of Fig. 1

These CO_2-binding curves are progressively shifted to the left at increasing pH as expected from Eqs. (1) and (2). The present values are in quantitative agreement with those reported by others and measured with different techniques [3, 10]. For example 1.16 mol oxygen-linked CO_2 per mol tetramer was calculated at P_{CO_2} 40 mm Hg, pH 7.4 and at 25°C. At elevated CO_2 concentration the amount of oxygen-linked CO_2 decreased. This indicates the increasing binding of CO_2 to amino groups in oxy Hb with lower affinity for the effector than those in deoxy Hb. All these results are well in keeping with current knowledge concerning CO_2 binding to human Hb [3, 6, 7, 8, 9, 10].

3.2 Influence of CO_2 on $(O_2)_1$ and the Alkaline Bohr Effect

Figure 3 represents a typical experiment for the measurement of K_1 and $P_{0.5}$ at pH 7.75, in the presence and absence of CO_2. CO_2 influenced both parameters however differently as its effect on the O_2 affinity is much larger at $(O_2)_{50}$ than at $(O_2)_1$. We have performed such experiments at various pH and at three levels of P_{CO_2}. The results are shown in Fig. 4. Log $(O_2)_1$ was increased upon addition of CO_2, predominantly at alkaline pH. At pH values lower than 7.1 no significant difference could be evidenced in the values of $(O_2)_1$ with and without CO_2. One should note from Fig. 4 that CO_2 almost completely inhibited the alkaline Bohr effect at P_{CO_2} 30 or 60 mm Hg. Figure 5 compares the variations of the amount of oxygen-linked CO_2 with varying

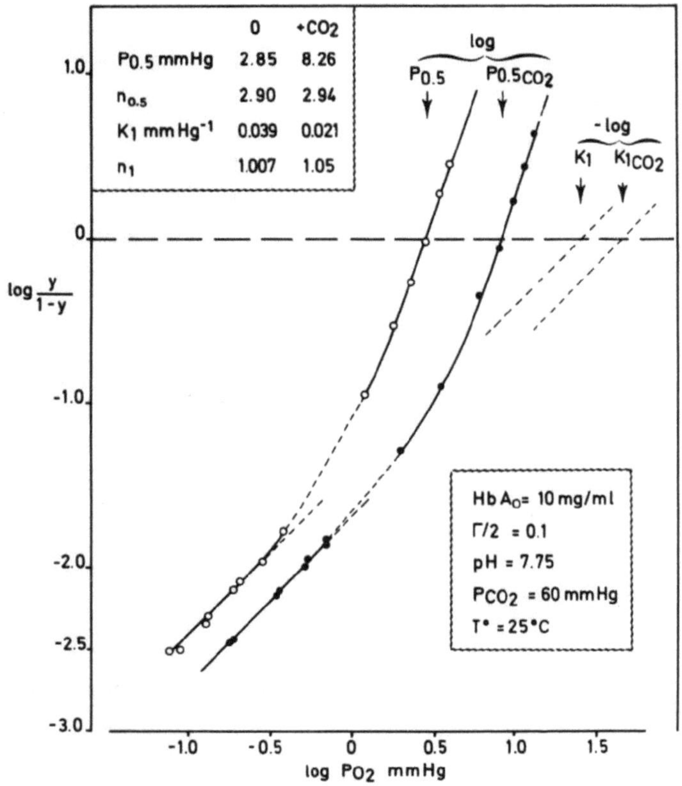

Fig. 3. Oxygen-binding experiment for the determination of K_1 and $P_{O \cdot 5}$ in human hemoglobin Ao. Conditions are given in the *inset* of the figure

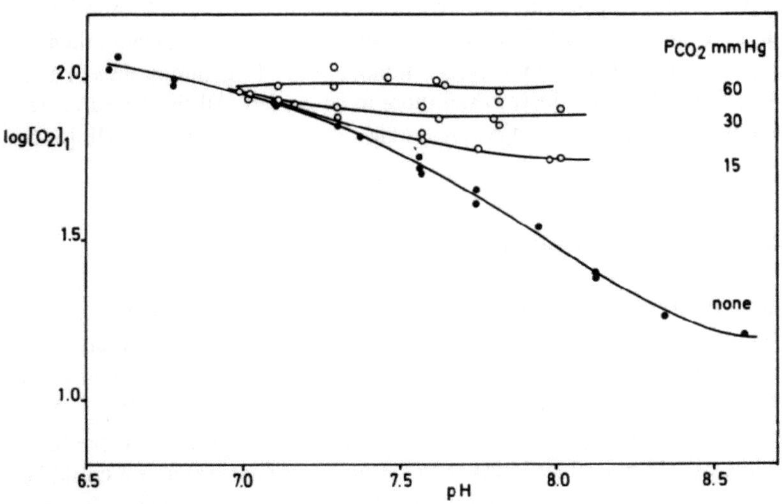

Fig. 4. Variations of log $(O_2)_1$ with pH in the presence or absence of CO_2. Conditions were as in Fig. 1

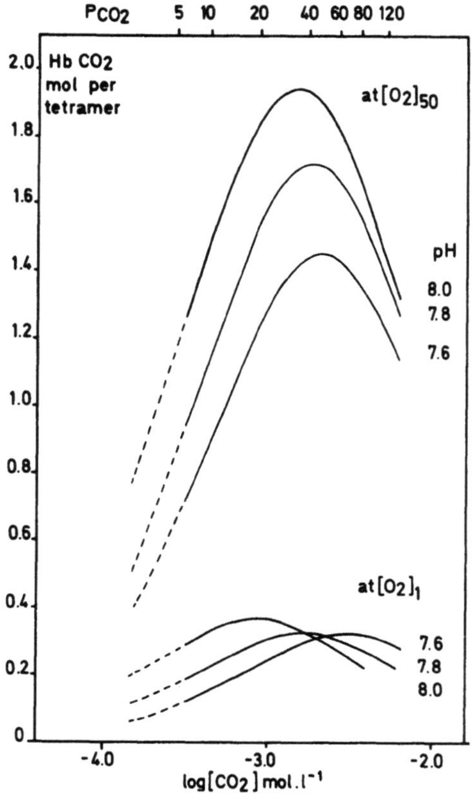

Fig. 5. Variations of the amount of oxygen-linked CO_2 to human hemoglobin Ao calculated at $(O_2)_{50}$ (*upper three curves*) and at $(O_2)_1$ (*lower curves*) as a function of CO_2 concentration

concentration of CO_2 calculated at $(O_2)_1$ and at $(O_2)_{50}$. The curves at $(O_2)_1$ were displaced to the left with increasing pH as those at $(O_2)_{50}$. These results demonstrated that part of the oxygen-linked CO_2 was released from Hb at the first oxygenation step, indicating an increase in the pK_c of a carbaminated amino group in the oxygenated subunit. The decreased amount of oxygen-linked CO_2 at high CO_2 concentration is similar to what was observed at the $(O_2)_{50}$ level. This indicates that CO_2 binds to monoliganded tetramer, and that this binding increased at high pH and P_{CO_2} values. From the curves shown in Fig. 5 one may calculate that the amount of oxygen-linked CO_2 released at $(O_2)_1$ represents approximately 20% (18%–21%) of the total oxygen-linked CO_2 in the pH range studied here. For example, it amounts to 0.23 mol compared to 1.16 mol per mol tetramer, pH 7.4, P_{CO_2} 40 mm Hg and 25°C, at $(O_2)_1$ and $(O_2)_{50}$ respectively. The large inhibition of the alkaline Bohr effect at $(O_2)_1$ upon addition of CO_2 together with the relatively small amount of oxygen-linked CO_2 released at this level of oxygenation raised the question of the binding site(s) involved. Table 1 shows that at pH values greater than 7.4 the amount of Bohr protons taken up on CO_2 release at $(O_2)_{50}$ is very close to unity, as expected from Eqs. (1) and (2) and the pK_z and pK_c reported by Morrow et al. [9]. By contrast, the ratio $\Delta H^+/\Delta HbCO_2$ at $(O_2)_1$ was much greater, with a mean value of 1.8 at the

Table 1. Amount of Bohr protons taken up per mol oxygen-linked CO_2 at various pCO_2 ($\Delta H^+/\Delta HbCO_2$)

pH	P_{CO_2} mm Hg	at $(O_2)_{50}$	at $(O_2)_1$
7.4	15	0.97	1.78
	30	1.02	1.96
	60	1.11	1.71
7.8	15	0.97	1.97
	30	1.0	1.80
	60	1.16	1.78

different pH and P_{CO_2} values studied. This was unexpected as, in this pH range, only Eq. (2) should play a significant role. The explanation for this difference may be sought in an inhibition of the intrinsic alkaline Bohr effect upon addition of CO_2. Indeed a decrease in the intrinsic Bohr effect of ca. 20%–25% would lower the ratio $\Delta H^+/\Delta HbCO_2$ at $(O_2)_1$ to a value close to 1. The inhibition of the intrinsic alkaline Bohr effect may be due either to a widespread decrease of the tertiary changes accompanying the first oxygenation, responsible in part for the pK shift of the alkaline Bohr groups, or to the blockade of one specific Bohr group by CO_2. In this regard it has been shown that Hb tetramers specifically carbamylated at the amino groups of the α chains evidenced a 25% reduction of the overall alkaline Bohr effect [7].

It is therefore postulated that CO_2 released at $(O_2)_1$ is linked to the oxygenation of the α subunits for the following reasons: (1) the relatively small amount of oxygen-linked CO_2 calculated at $(O_2)_1$, (2) the large inhibition of the alkaline Bohr effect which could best be explained by the blockade of one important Bohr group, (3) the CO_2 binding site in the α chains is known to behave as an alkaline Bohr group in the absence of CO_2. Support for this hypothesis was sought in the estimated equilibrium constants for the binding of CO_2 to deoxy (T_0) and monoliganded hemoglobin (T_1). It is known that the extraction of binding constants for CO_2 from CO_2 binding curves is difficult [10, 11]. We also failed to obtain meaningful values from the CO_2 binding curves calculated at $(O_2)_{50}$. From the oxygen-linked binding curves calculated at $(O_2)_1$ reasonable K_c and K_z values were found by fitting the curve with a simple equation for one site in T_0 and T_1. This was done on the assumption that either the α and β chains binding sites behave as one unique site or that one of the two pairs of binding sites predominated over the other. Table 2 gives the values for these binding

Table 2. Calculated binding constants for CO_2, pK_z and pK_c values for one amino group involved at the first oxygenation step

	Deoxy (T_0)	Oxy (T_1)
λ mol^{-1} pH 7.6	606 ± 4.7	160 ± 0.07
pK_z	7.81	7.59
pK_c	4.56	5.11

constants in T_0 and T_1 states at pH 7.6 and at 25°C. The pH dependence of these constants allowed the estimation of approximate values for the pK_z and pK_c of the group involved. It is shown in Table 2 that the pK_z of the amino group is 7.8 in T_0 and 7.6 in T_1. A slight increase in the pK_c of the carbaminated group was observed upon oxygenation of the first subunit. These values should be considered with caution as they are based on few experimental points and on the assumptions mentioned above. They give, however, some indication that according to the elevated pK_z value found, the α-amino groups in the α chains should be predominantly involved in the release of oxygen-linked CO_2 at the first oxygenation step. This would mean also that in the presence of CO_2 oxygenation of Hb starts at the α chains.

References

1. Baldwin J (1975) Structure and function of hemoglobin. Prog Biophys Mol Biol 29: 225–320
2. Bartels H, Bucherl E, Hertz CW, Rodewald G, Schwab M (1963) In: Methods in pulmonary physiology. Hafner, New York London, p 408
3. Bauer C, Baumann R, Engels U, Pacyna B (1975) The carbon dioxide affinity of various human hemoglobins. J Biol Chem 250: 2173–2176
4. Christiansen J, Douglas CC, Haldane JS (1914) The adsorption and dissociation of carbon dioxide by human blood. J Physiol (London) 48: 244–277
5. Edsall JT, Wyman J (1958) In: Biophysical chemistry. Academic Press, London New York, p 555
6. Garner MH, Bogardt RA, Gurd FRN (1975) Determination of the pK values for the α amino groups of human hemoglobin. J Biol Chem 250: 4398–4404
7. Kilmartin JV, Fogg J, Luzzana M, Rossi-Bernardi L (1973) Role of the α amino groups of the α and β chains of human hemoglobin in oxygen-linked binding of carbon dioxide. J Biol Chem 248: 7039–7043
8. Matthew JB, Morrow JS, Wittebort RJ, Gurd FRN (1977) Quantitative determination of carbamino adducts of α and β chains in human adult hemoglobin in presence and absence of carbon monoxide and 2,3 diphosphoglycerate. J Biol Chem 252: 2234–2244
9. Morrow JS, Matthew JB, Wittebort RJ, Gurd FRN (1976) Carbon 13 resonances of $^{13}CO_2$ carbamino adducts of α and β chains in human adult hemoglobin. J Biol Chem 251: 477–484
10. Perrella M, Bresciani D, Rossi-Bernardi L (1975) The binding of CO_2 to human hemoglobin. J Biol Chem 250: 5413–5418
11. Perrella M, Guglielmo G, Mosca A (1977) Determination of the equilibrium constants for oxygen-linked CO_2 binding to human hemoglobin. FEBS Lett 76: 287–290
12. Poyart C, Bursaux E, Bohn B (1978) An estimation of the first binding constant of oxygen to human hemoglobin A. Eur J Biochem 87: 75–83
13. Rossi-Bernardi L, Roughton FJW (1967) The specific influence of carbon dioxide and carbamate compounds on the buffer power and Bohr effect in human hemoglobin solutions. J Physiol (London) 189: 1–29
14. Roughton FJW (1949) The carbamino combination of CO_2 with reduced haemoglobin and oxy haemoglobin. Biochem J 44: 31P
15. Szabo A, Karplus M (1976) Analysis of the interactions of organic phosphates with hemoglobin. Biochemistry 15: 2869–2877
16. Wyman J (1964) Linked functions and reciprocal effects in hemoglobin: a second look. Adv Protein Chem 19: 223–286

Carbonic Anhydrase – Structure and Function

Chairmen: S. LINDSKOG and J. E. COLEMAN

Current Concepts of the Mechanism of Action of Carbonic Anhydrase

J.E. COLEMAN [1]

1 Introduction

The central role of CO_2 in life on this planet seems self-evident, even when one considers only two of its many roles; as the carbon source in the dark reactions of photosynthesis, and as the major buffer regulating extracellular pH in animals. Carbon dioxide plays the latter role despite the fact that it dissolves in water and hydrates to form a relatively strong acid, H_2CO_3 ($pK_a = 3.77$). H_2CO_3 is also in equilibrium with gaseous CO_2 as shown in Eq. (1),

$$CO_2 + H_2O \overset{K_1}{\rightleftharpoons} H^+ + HCO_3^- \quad \text{(with } CO_2 + H_2O \rightleftharpoons H_2CO_3 \rightleftharpoons H^+ + HCO_3^-) \tag{1}$$

and the apparent equilibrium constant which actually applies is given by Eq. (2),

$$K_1 = \frac{[H^+][HCO_3^-]}{[CO_2][H_2O]} = 10^{-6.35}, \tag{2}$$

so the effective buffering pK_a is near neutral pH.

The uncatalyzed hydration of CO_2 is a relatively rapid process, the reaction $CO_2 + H_2O \rightarrow H_2CO_3$ having a rate constant of $3.5 \times 10^{-2}\ s^{-1}$. A second hydroxide-dependent mechanism [Eq. (3)] is much faster, having a rate constant of $8.5 \times 10^3\ s^{-1}\ M^{-1}$.

$$CO_2 + {}^-OH \rightarrow HCO_3^- \tag{3}$$

This becomes the major pathway for the hydration reaction at pH 10 or above. Considering the rate constants typical of metabolic transformations at neutral pH, the fact that both plant and animal cells should evolve an enzyme which vastly accelerates this process $k_{cat} = \sim 10^6\ s^{-1}$, did not seem too surprising when Meldrum and Roughton (1933) and Keilin and Mann (1940) first described the enzyme (M.W. = ~30,000) from bovine erythrocytes. The precise understanding of its physiological role in erythrocytes has become somewhat clouded, however, with the discovery that two genes

1 The Department of Molecular Biophysics and Biochemistry, Yale University, New Haven, Connecticut 06510/USA

for this enzyme are expressed in the human erythrocyte, one coding for an enzyme with $k_{cat} \sim 10^3\ s^{-1}$ (HCAB), and expressed ~ ninefold over the other minor but more active isozyme (HCAC), $k_{cat} \sim 10^6\ s^{-1}$ (for review see Lindskog et al., 1971). To add further complexity, the low activity isozyme binds the Cl^- anion so tightly at the active site that it must be almost totally inhibited at the concentrations of Cl^- present in the red cell. Thus the minor more active enzyme, which binds Cl^- much less tightly, accounts for most of the activity (Maren et al., 1976). This fact, coupled with the observation that it is extremely difficult to demonstrate any major deleterious physiological consequences of complete inhibition of the red cell enzyme with sulfonamides (Maren, 1967), leaves some doubt about the exact physiological role of the two erythrocyte isozymes[2].

Carbonic anhydrase, however, is present in many different types of cells and its overall physiological role is perhaps better placed in context by noting from Eq. (1) that it catalyzes the production of protons and anions (HCO_3^-) from two neutral species, CO_2 and H_2O. Thus it provides a ready source of these ions for secretory and exchange processes across cell membranes. Effects of carbonic anhydrase inhibitors on these processes are well documented (Maren, 1967) and are discussed elsewhere in this symposium. Lastly, carbonic anhydrase enjoys the distiction of being the first Zn metalloenzyme discovered (Keilin and Mann, 1940), hence establishing the first molecular basis for the physiological function of the essential trace element zinc.

Kernohan (1964) was the first to carry out extensive kinetic studies of both the hydration and dehydration reactions and discovered that hydration followed a pH-rate profile sigmoid to high pH, while dehydration followed a pH-rate profile sigmoid to low pH (Fig. 1). Both were described by apparent pK_a's of ~ 7. Since the substrates do not have pK_a's in this region, these pH-rate profiles suggest that activity is controlled by a group at the active center of the enzyme with a pK_a of ~ 7, the deprotonated form being required for activity. This statement applies to both hydration and dehydration reactions. The apparent inverse pH-rate profiles are generated since, irrespective of the mechanism proposed, a proton [Eq. (1)] is a substrate in the dehydration reaction. Hence the pH dependence of dehydration is governed by the $[H^+]$ term in the kinetic expression (Lindskog and Coleman, 1973). Given these facts about the hydration of CO_2 and the properties of carbonic anhydrase it is not difficult to do a little paper chemistry and suggest that nature has simply capitalized on the chemical properties of the reactants assembled at the active center and insured that the very fast $CO_2 + {}^-OH \rightarrow HCO_3^-$ reaction, ordinarily predominating only at lethal pH values, is the predominant one at neutral pH. The Zn ion is the obvious agent for insuring the required concentration of hydroxide ion at the active site as the attacking species by forming a $Zn \cdot {}^-OH$ in a mixed complex with enzyme ligands at neutral pH. The activity-linked pK_a could then be assigned to the $Zn \cdot OH_2 \rightleftharpoons Zn \cdot {}^-OH + H^+$ dissociation. This is a rather straightforward proposal and has been suggested by numerous investigators.

2 This does not mean that alterations in the rate of attainment of CO_2 equilibria are not observed (see Chaps. by Silverman and Maren, this vol.). Such alterations could be deleterious under maximum metabolic stress

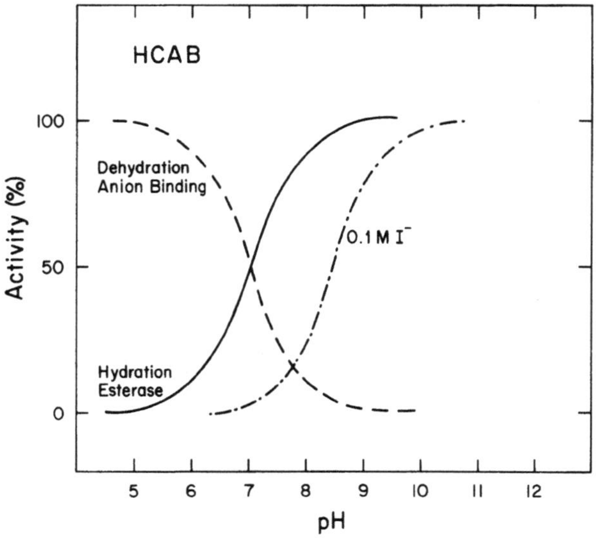

Fig. 1. Summary of the pH-dependent kinetic properties of human carbonic anhydrase B. (——) Hydration of CO_2 and hydrolysis of p-nitrophenyl acetate, pH-rate profiles. (– . – . –) Hydration of CO_2 in the presence of 0.1 M I^-, pH-rate profile. (- - -) Dehydration of HCO_3^- (H_2CO_3), pH-rate profile. The latter also correlates with the pH dependency of anion binding

Direct proof is another matter and the rest of this paper is devoted to a synopsis of the evidence. To anticipate the conclusion of this survey it appears that the Zn · $^-$OH mechanism remains the mechanism most compatible with all the data. Additonal subtleties such as participation of other groups or hydrogen-bonded networks in proton transfer or transient expansion of the Zn (II) coordination sphere to assist ligand exchange may play roles, but do not alter the basic features of the mechanism.

2 pH-Dependent Kinetic Properties of Carbonic Anhydrase

The most prominent pH-dependent function is, of course, the apparent existence of an acid and basic form of the enzyme, presumably reflecting a proton dissociation [Eq. (4)].

$$EH \rightleftharpoons E + H^+ \tag{4}$$

and responsible for the apparent pK_a of the pH-rate profiles (Fig. 1). It is well to introduce early the fact that this pK_a depends on the anion composition of the medium. Monovalent anions, e.g., Cl^-, Br^-, I^-, $\bar{S}CN$, $\bar{C}N$, inhibit the enzyme by shifting this pK_a to alkaline pH (shown in Fig. 1 for 0.1 M I^-). A variety of data including effects on the d-d absorption spectra of the cobalt enzyme (Lindskog, 1966), quadrupole broadening of the $^{35}Cl^-$ NMR resonance by the Zn (II) enzyme (Ward, (1969), X-ray diffraction studies of the I^- complex (Lindskog et al., 1971), and

X-ray absorption edge spectroscopy of the ^{-}Cl complex of the Zn enzyme (Brown et al., 1977), establish beyond any doubt that the halide ions and probably the other anions as well form a rather normal coordination bond with the Zn ion at the active site in the 1:1 complexes [3]. Three other coordination bonds are made with three histidyl ligands of the protein (see below). The fact that pH dependence of the activity profile is preserved, but shifted to high pH in the presence of anions, implies that anion binding is also pH-dependent. As shown most clearly by $^{35}Cl^{-}$ NMR (Ward, 1969), the anion binding is pH-dependent, anions being most tightly bound to the acid form of the enzyme and binding disappearing at alkaline pH with the generation of active enzyme. Since Zn-binding anions influence the activity-linked pK_a, the group involved is likely very close to the metal ion. While the proton dissociation in Eq. (4) need not be directly involved in the reaction path, it is natural to invoke the group as being involved in the proton transfer, since hydration involves a proton transfer.

All the facts mentioned thus far are compatible with the assignment of the activity-linked pK_a to that of a coordinated water molecule. Hence Eq. (4) becomes

$$E - Zn \cdot OH_2 \rightleftharpoons E - Zn \cdot {}^{-}OH + H^{+}, \quad (5)$$

and the hydration mechanism becomes

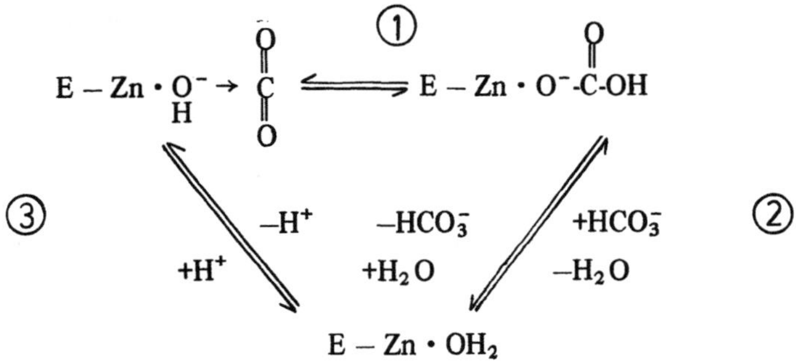

(6)

With such a mechanism the pH dependency of anion inhibition and anion binding for anions which clearly coordinate the metal ion receives a satisfactory explanation, i.e., they compete with the ^{-}OH anion.

In any mechanism, including Eq. (6), in which the activity-linked group is responsible for the transfer of the proton to solvent (in the hydration reaction) the rate of transfer of the proton from donor ($pK_a \sim 7$) to acceptor (H_2O, forming H_3O^{+}, $pK_a \sim 1.7$) cannot be greater than 10^3 s^{-1} (Lindskog and Coleman, 1973). This is immediately at variance with the observation that the most active carbonic anhydrases have $k_{cat} = 10^5 - 10^6$ s^{-1}. One could propose that the mechanism does not involve enzyme-mediated proton transfer, rather that the product is the neutral carbonic acid. The same dilemma arises here, however, in different form; namely problems with the rate of diffusion of H_2CO_3 to the active site or the rate or protonation of HCO_3^{-}, the predominant species. These problems have been discussed thorougly elsewhere (Lindskog and Coleman, 1973) and will not be covered here.

3 A 2:1 anion complex can form under some conditions. It has been well-documented for the CN^{-} ion which forms a highly "covalent" complex (Haffner and Coleman, 1973)

An adequate solution of the above dilemma was provided when it was pointed out that measurements of the initial velocity of carbonic anhydrase reactions are invariably done at high concentrations of buffer. Thus proton transfers prior to equilibrium involve a buffer sink [Eq. (7)]

$$EH + B^- \rightleftharpoons E + BH \qquad (7)$$

The proton transfer in Eq. (7) for two groups with pK_a's near neutral pH, can pro ceed at a rate near 10^7 s^{-1}, and hence proton transfer need not be rate-limiting.

One of the most significant advances in understanding carbonic anhydrase made in the last few years is the work of Silverman and his colleagues. Tu and Silverman (1975) have monitored the carbonic anhydrase-catalyzed interconversions of CO_2, HCO_3^- and H_2O, at equilibrium, by measuring the ^{18}O exchange between HCO_3^- and water and the ^{18}O exchange between ^{12}C- and ^{13}C-containing species of CO_2. They describe these data in terms of two constants, θ_{cat}, which is a measure of the first-oder rate constant describing the catalyzed dehydration of HCO_3^-, a process monitored by the exchange of ^{18}O between bicarbonate and water, and ϕ_{cat}, which describes the exchange rate of ^{18}O betwen the ^{12}C- and ^{13}C-containing species of CO_2. The results as a function of buffer concentration are shown in Fig. 2. The striking finding is that as the buffer is reduced below 10 mM, θ_{cat} decreases sharply, demonstrating that turnover is dependent upon a buffer-mediated process, most likely a proton transfer step. The constant, ϕ_{cat}, represents a step that retains ^{18}O in the CO_2 system and its behavior is consistent with a model in which ^{18}O labels the active site from bicarbonate to form ($E - Zn \cdot {}^{18}OH^-$). If the latter has a long lifetime, the probability is great that it will react with CO_2 and retain ^{18}O in the CO_2 system. This lifetime will be long if there is a slow rate of protonation of $E - Zn \cdot {}^{18}OH^-$ in the

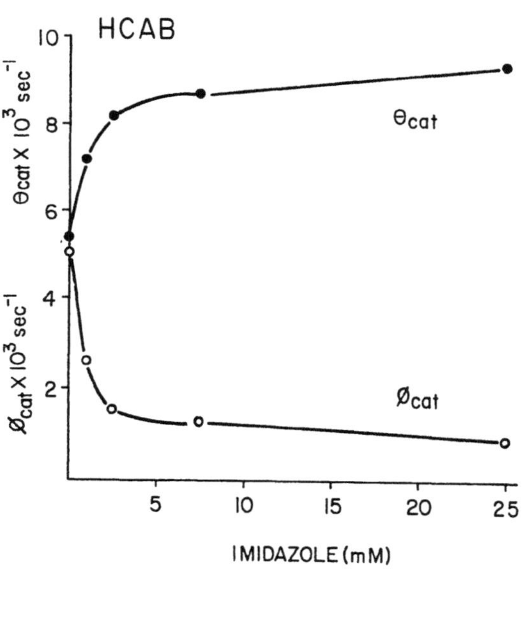

Fig. 2. The first-order rate constants at 25° and pH 7.0 for the catalyzed exchange of ^{18}O between bicarbonate and water, θ_{cat} (●), and for the catalyzed exchange of ^{18}O between ^{12}C- and ^{13}C-containing CO_2 species, ϕ_{cat} (○), as a function of imidazole concentration. The concentration of bovine carbonic anhydrase was 1.6 x 10^{-9} M and the total concentration of CO_2 species was 10 mM with ionic strength held constant at 0.2 using Na_2SO_4. The ^{13}C enrichment was 44%–46% in a final volume of 8 ml. (Tu and Silverman, 1975)

absence of buffers, and ϕ_{cat} will be large. On the other hand as proton transfer is facilitated, ϕ_{cat} will decrease as is observed. If proton transferring groups other than water could truly be reduced to zero, then θ_{cat} might be very close to the calculated $10^3\ s^{-1}$ for transfer to solvent. Further details are discussed in the paper of Silverman in this symposium.

Under physiological conditions, or at high protein concentrations, the buffer present is more than sufficient to facilitate the proton transfer and it is thus not rate-limiting. What is the rate-limiting step when proton transfer to the external acceptor is not? One might, for example, examine the steps collected in Eq. (6) as step 2, a ligand exchange, HCO_3^- for H_2O. The final part of this paper will address this question and present some ^{113}Cd NMR data bearing on it. First it is necessary to review some equilibrium data on the acid and basic forms of the enzyme.

3 pH-Dependent Equilibrium Properties of Carbonic Anhydrase

There are several equilibrium properties which distinguish the acid from the basic form of the enzyme. Among the most intersting are the changes in proton equilibria actually accompanying the binding of anions to the zinc ion. It has been possible to study this by using the $\bar{C}N$ complex, since the complex is so stable that at a ratio of 1 $\bar{C}N$ per enzyme molecule the complex is 100% formed between pH 5 and 11 (Coleman, 1967). HCN has a pK_a of 9.3, hence at the opposite ends of the pH scale shown in Fig. 3 there are two different reactions; HCN with the acid form of the enzyme, EH; and ^-CN with the alkaline form of the enzyme, E.

The results of carrying these reactions out at various pH values, using a pH-state to keep the pH constant, are plotted in Fig. 3. EH releases a proton on combination with HCN, while E releases an ^-OH ion (or takes up a proton) on the combination with $\bar{C}N$. Since Zn • $\bar{C}N$ is the active site species in both cases, one can assume that the H^+ in the first case comes from HCN and that there is no net change in the proton equilibrium of the enzyme. On the other hand there is indeed a net change of one in the enzyme-solvent proton equilibria on the combination of ^-CN with E.

To explain the biphasic equilibrium titration curve in Fig. 3 one has to invoke, in addition to the pK_a of HCN, an additional titrating group on the enzyme with a pK_a the same as that of the activity-linked group under these conditions. This titration is represented in Fig. 3 as the esterase activity pH-rate profile for the Zn enzyme and the transition between the d-d absorption spectrum of the acid and basic forms of the Co (II) enzyme (Lindskog, 1963; Coleman, 1967). These findings are completely compatible (if not uniquely so) with the Zn-OH_2 $\rightleftharpoons$ Zn-$\bar{O}H$ + H^+ equilibrium, since displacement of OH_2 from EH would not release a proton, while combination of ^-CN with E releases $\bar{O}H$. It is well to point out here that all such pH-dependent phenomena will be altered by metal-binding anions, and if an anion, e.g., Cl^-, is present, which undergoes a pH-dependent binding in the pH-region examined, then a pH-dependent change in properties will be observed. This does not, however, avoid (as proposed in the paper of Koenig et al. in this symposium) the necessity of introducing an enzyme ionization unless one proposes that a Zn-$\bar{O}H$ is a stable species down to pH 5 or below. Zn • ^-OH is still required to explain competition with anions

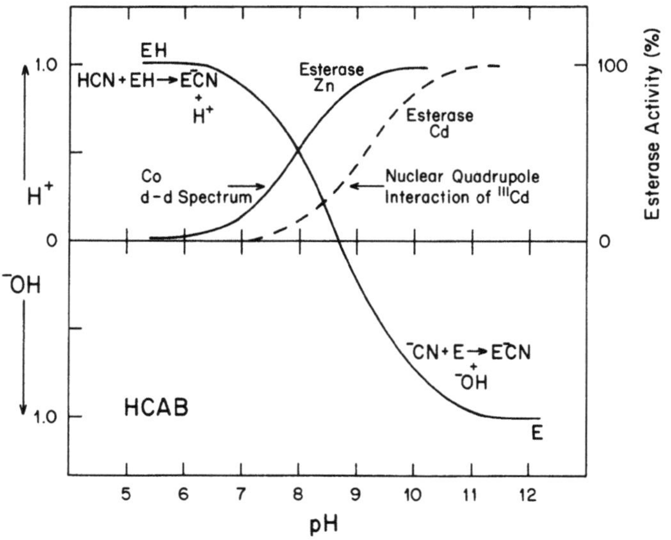

Fig. 3. Summary of pH-dependent equilibrium properties of human carbonic anhydrase B. (— biphasic) Change in proton equilibria (H^+ release, ^-OH release or H^+ uptake) at fixed pH values accompanying combination of HCN and $\bar{C}N$ with the active site of carbonic anhydrase. (—) pH dependence of esterase activity of the Zn (II) enzyme and of the change in the d-d transition absorption envelope of the Co (II) enzyme. (Data adapted from Coleman, 1967 and Lindskog, 1963). (- - -) pH dependence of esterase activity of Cd (II) carbonic anhydrase and of nuclear quadrupole parameters for ^{111}Cd (II) carbonic anhydrase (Data adapted from Bauer et al., (1976)

at high pH. Lack of an enzyme ionization also has the disadvantage of having to invoke another pathway for proton transfer.

4 Crystal Structure of the Active Center of Carbonic Anhydrase

The results of the X-ray diffraction studies of the human isozymes will be presented elsewhere in this symposium and are only briefly summarized here in Fig. 4. There are three histidyl nitrogen ligands from the enzyme in both the B and C isozymes from human erythrocytes, His 93, 95, and 118 in the C isozyme, 94, 96, and 119 in the B isozyme. The initial crystal structure of the C isozyme indicated that the fourth position in a distorted tetrahedral geometry is occupied by a solvent molecule (H_2O or ^-OH) (Lindskog et al., 1971). A distorted tetrahedral geometry is compatible with the d-d absorption and MCD spectra of the Co (II) derivative of the enzyme and its anion complexes (Coleman and Coleman, 1972), although a 5-ccordinate geometry around the high-spin Co (II) ion in the alkaline form of the enzyme cannot be ruled out. ESR spectra of the Co (II) enzyme show that in the presence of two cyanide ions a dicyanide complex of the Co (II) enzyme forms which forces a low spin Co (II) (Haffner and Coleman, 1975). While the dicyanide complex is not a catalytically competent complex, its formation indicates that there is room in the active center

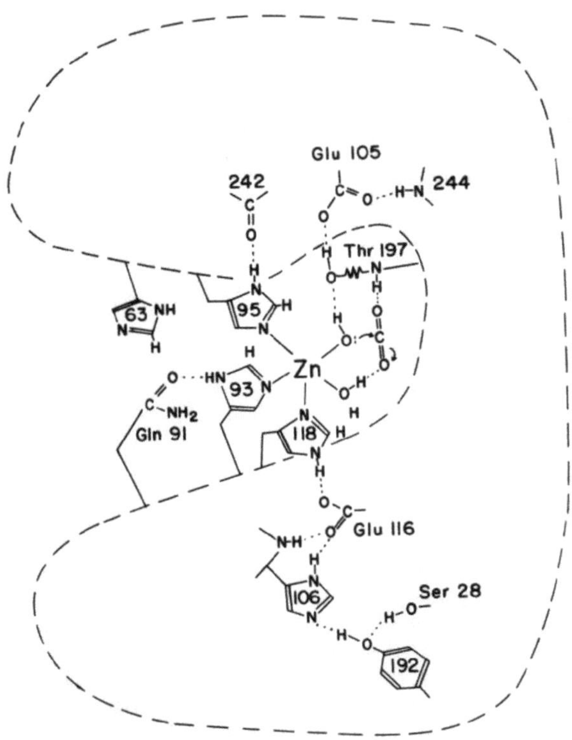

Fig. 4. Schematic representation of the molecular details at the active site of human carbonic anhydrase C. Based on the crystallographic findings (Lindskog et al., 1971) and drawn after suggestions of S. Lindskog (personal communication)

for two coordinating groups from solution. The drawing in Fig. 4 incorporates the speculation based on X-ray structures of imidazole (Kannan et al., 1977) and sulfonamide complexes which suggest that in addition to the primary ligand there may be an additional water molecule on the zinc, perhaps with a somewhat longer bond. These findings suggest the possibility that a five-coordinate intermediate might form. A 5-coordinate intermediate could have a mechanistic advantage by providing for a facile associative mechanism of ligand exchange with a transient expansion of the coordination sphere rather than a dissociative mechanism of ligand exchange (Haffner and Coleman, 1975). Figure 4 suggests a mechanism for Zn · $\bar{O}H$ attack on CO_2 which is aided by a hydrogen-bonded network involving the adjacent threonine 197, as proposed by Kannan et al. (1977).

Since the MCD of the acid form of the Co (II) enzyme is of the tetrahedral type, substantially different from the band structure of the high pH form (Coleman and Coleman, 1972), the possibility exists that the ionization at the active site is associated with a shift from 4 to 5-coordinate [Eq. (8)], the latter being the active species.

$$N_3\text{Zn}{-}OH_2 \rightarrow N_3\text{Zn}(\bar{O}H)(OH_2) \qquad (8)$$

This must remain speculative.

5 Paramagnetic Relaxation of Bulk Solvent Protons Induced by Co (II) Carbonic Anhydrase

If a water molecule is coordinated to the metal ion at the active site, and considering the rapid rate at which ligand exchanges are likely to occur, one might expect that the Co (II) enzyme should paramagnetically relax bulk solvent water due to exchange of proton-carrying species in and out of the active site. The process need not be ligand exchange, but any process that transfers a proton out of the active site cavity at the required rate.

In the high pH form, both low and high activity forms of Co (II) carbonic anhydrase relax solvent water (Fabry et al., 1970; Fabry, 1978). The most extensive and recent work on both types of enzyme, HCAB (Fig. 5) and bovine BCAB (Fig. 6) has been carried out by Fabry both in the presence and absence of buffers.

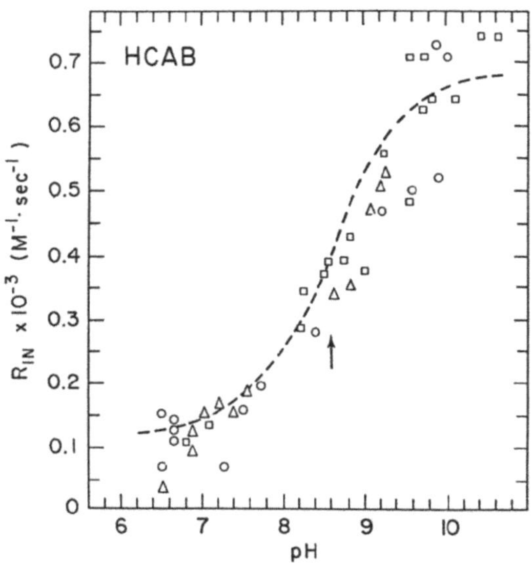

Fig. 5. Titration of 0.9 mM human-B cobalt carbonic anhydrase at 25 MHz and 25°C. △ R_{IN} in distilled water; ○ R_{IN} in 40 mM Tris/SO_4; □ R_{IN} in 32 mM sodium 2,2-diethyl malonate. The *smooth curve* is a calculated best fit to all the data points assuming a single ionization. Better fits to individual sets might be obtained by small shifts in the pK or end points or by assuming two closely spaced ionizations, but the data do not warrant such refinements. (Fabry, 1978)

The most striking qualitative result is that the relaxivity is pH-dependent along a sigmoid curve that is very close to that describing the pH dependence of enzymatic activity or the optical transition of the Co (II) enzyme between EH and E. Under some conditions the NMR pK_a is not the same as the optical transition pK_a (see below). For Co (II) HCAB the relaxivity completely disappears in the low pH form of the enzyme, both in the presence and absence of buffers (Fig. 5). In contrast, if no buffer is present, the relaxivity induced by Co (II) BCAB decreases along a similar sigmoid pH-curve, but remains substantial in the low pH form of the enzyme (Fig. 6). Ligand exchange reactions, i.e., Co-O bond breaking, are not necessarily expected to be pH-dependent in this manner. On the other hand the sharp pH dependency of the relaxivity may be for the most part explained by a change in the rate of transfer of

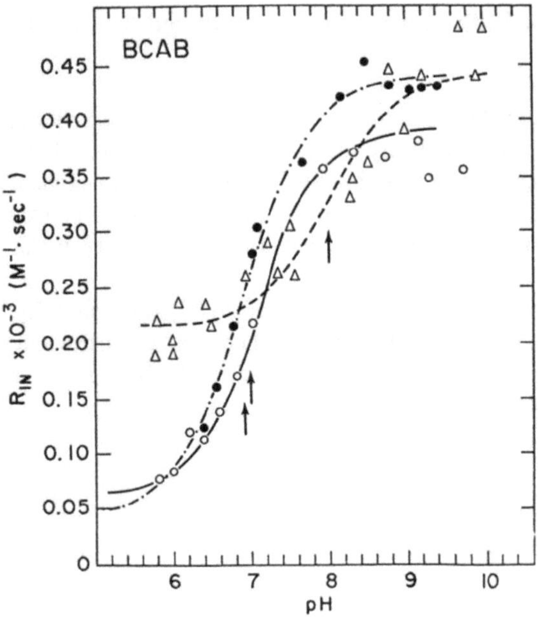

Fig. 6. Titration of millimolar bovine cobalt carbonic anhydrase at 25 MHz and 25°C. ○ R_{IN} in 32 mM sodium 2,2-diethyl malonate; △ R_{IN} in distilled water; ● R_{IN} in 40 mM Tris/SO_4. The *smooth curves* are calculated for a single proton ionization. (Fabry, 1978)

a proton in and out of the active center, a process expected to be pH-dependent and the explanation favored by Fabry (1978). In addition the magnitude and the pH profile for this relaxivity is altered by buffers (Fig. 6). In the presence of buffers the NMR pK_a is the same as the optical (d-d transition) or activity pK_a, but is not identical in the unbuffered case. Fabry (1978) suggests that the proton transfer may follow a path involving intramolecular transfers in the active center before the final acceptor (solvent or buffer). Similar intramolecular proton transfer has recently been suggested by Silverman (this vol.) and by Lindskog (this vol.) based on ^{18}O exchange studies and ^{13}C-NMR studies in H_2O and D_2O. Such steps may contribute to the rate-limiting step or steps when the basic proton transfer to acceptor is no longer rate-limiting in the presence of adequate buffer. The findings on the pH dependency of the paramagnetic relaxation are compatible with the Zn • $OH_2 \rightleftharpoons$ Zn • $\bar{O}H + H^+$ equilibrium, especially since conditions can be found for BCAB where significant relaxivity remains in the acid form (Fig. 6). The failure to observe significant relaxivity in the acid form in the past has been used to argue that no coordinated water molecule is present in EH (Koenig and Brown, 1972). As we suggest below, the intrinsic ligand exchange in the protonated E-Me • OH_2 may be too slow under most circumstances to result in significant solvent relaxation, the enhanced relaxivity at alkaline pH being due to facile proton transfer in and out of the cavity containing a Zn-$\bar{O}H$[4].

4 The decrease in relaxivity of bulk H_2O by the acid form of BCAB in the presence of buffers is not readily explained by the basic mechanisms described here. Fabry (1978) should be consulted for a discussion of this problem

6 Cd (II) Carbonic Anhydrase

With the exception of the Co (II) substitution, replacement of Zn (II) with other first transition and IIB metal ions, although giving rise to very stable complexes, apparently resulted in inactive enzymes. In connection with their studies on the perturbed angular correlation (PAC) of γ-rays by ^{111}Cd (II) carbonic anhydrase, Bauer et al. (1976) discovered that the Cd (II) enzyme, like the Zn (II) enzyme, was an esterase against p-nitrophenyl-acetate. The apparent pK_a of activity was shifted under their assay conditions from pH 7.3 to ~ 9.1 (see Fig. 3). From the PAC data nuclear quadrupole interaction parameters can be derived and the data show that low and high pH forms of the active site ^{111}Cd (II) complex exist. The transition between the two correlates with the pK_a of esterase activity. This change in quadrupole interaction was interpreted as compatible with the ionization of a coordinated water molecule, whose pK_a was shifted from 7.3 to 9.1 by the Cd (II) substitution. The interaction of ligands with Cd (II) when compared to Zn (II) is indeed marked by quantitative differences. This is illustrated by the enormous differences in the stability of Zn and Cd halides, e.g., the equilibrium constant for $ZnCl_4$ is 1 compared to 10^3 for $CdCl_4$. This differential affinity of Cl^- for Zn and Cd is reflected in the inhibition of the esterase activity of the two carbonic anhydrases, K_i (Cl^-)for Zn (II) HCAB = 11×10^{-3} M at pH 7.5, K_i (Cl^-) for Cd (II) HCAB = 0.28×10^{-3} M at pH 9.5 (Bauer et al., 1977). These findings are readily explained by the E-Me · $\bar{O}$H mechanism; are not so readily accomodated by models which assign the activity-linked ionization to an amino acid side chain.

^{113}Cd NMR of Carbonic Anhydrase. NMR of metal nuclei has generally been too insensitive to apply to metalloenzymes as a probe of the central metal ion structure. With an extensively modified 90 MHz Bruker NMR instrument we have obtained a sensitivity for ^{113}Cd comparable to that of ^{31}P which has allowed studies of the ^{113}Cd resonances from the active center ^{113}Cd (II) ions in the cadmium carbonic anhydrases (Armitage et al., 1978). One of the startling characteristics of ^{113}Cd NMR signals, in contrast to the nuclei whose NMR signals have frequently been employed in biochemical systems, is the relatively great range of chemical shift, over 600 ppm in the common compounds of ^{113}Cd (II). This relates in part to the large electron cloud, which amplifies the "paramagnetic contribution" to the shielding. In ^{113}Cd (II)-substituted metalloenzymes, where the protein ligands are nitrogen and/or oxygen, we have observed a chemical shift range of ~ 200 ppm (Armitage et al., 1978; Coleman and Chlebowski, 1979). Within a given Cd (II) enzyme a chemical shift range of ~ 100 ppm for the Cd nucleus in two coexisting states has been observed (Coleman et al., 1979; Coleman and Chlebowski, 1979).

Effects of Chemical Exchange on ^{113}Cd Resonances. Before presenting the spectra of the ^{113}Cd (II) carbonic anhydrases, one consequence of the large range of chemical shifts that may exist between exchanging states needs to be pointed out. If one calculates the exchange rates or lifetimes (τ values), which may be expected to modulate ^{113}Cd resonances, these chemical exchange rates turn out to be between 10^2 and 10^4 s^{-1}, rates which may be expected to characterize many of the conformational fluxes in protein molecules. These relatively slow rates are not immediately thought of as modulating NMR signals for NMR sensitive nuclei usually used as probes.

We have calculated the effect of variation of τ values on the ^{113}Cd NMR spectra for a system involving exchange between species of equal population differing in chemical shift by 100 ppm. The latter seems to be a reasonable chemical shift difference between states (see below). The exchange process may represent a fluctuation in the ligand geometry at the metal binding site, ligand exchange, or exchange of the central metal ion. This latter mechanism is perhaps the least favorable as a result of the very slow ($< 10^{-2}$ s^{-1}) dissociation rate constants and rapid association rate constants ($> 10^4$ s^{-1}) for metal ions in metalloenzymes. The calculations show that for τ values of 10^{-3} or 10^{-4} s complete broadening of the resonances of the two exchanging states occurs. The results have been published elsewhere (Coleman et al., 1979) and are not shown in complete form here. One need not assume equal populations of exchanging states, which is of particular relevance to the enzyme systems considered here. The equilibrium constants governing accessible conformers of protein macromolecules may in fact differ from unity significantly. The calculated line shapes for the ^{113}Cd case when the populations of exchanging species are 70:30 are shown in Fig. 7. This model is sufficient to explain why the minor resonance can elude detection with existing instrumental sensitivity. Even a 10% fractional population under these conditions can give rise to an increase in linewidth by a factor of ~ 7 for the dominant resonance when τ values are ~ 10^{-3} s. Increase in linewidth by a factor of 7 is frequently sufficient to make the observation of the major resonance ambiguous with present instrumentation. The unequal population model also predicts much smaller changes in the chemical shift of the dominant resonance than expected from the chemical shift difference between the states involved. Thus, one would predict that processes modulating the cadmium chemical shift will give rise to a clearly observable resonance only when the site lifetime is in the range $10^{-5} \geqslant \tau \geqslant 10^{-2}$ s.

The ^{113}Cd resonances of ^{113}Cd (II) HCAB and BCAB are shown in Fig. 8 from the data of Schoot Uiterkamp et al. (1979). The most striking feature is the great variation in linewidth depending most dramatically on the particular isozyme, but also on the presence of halide anions (Fig. 8 and 9). We believe that the differences in

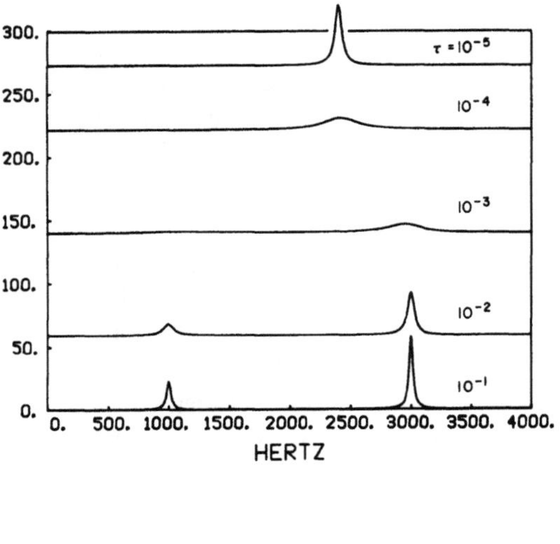

Fig. 7. Calculated line shapes for a two-site exchange with a separation in resonance frequency of 2000 Hz (100 ppm) and a natural linewidth of 35 Hz. τ values represent the lifetime of the dominant species. Populations are in the ratio 70:30 (Coleman et al., 1979)

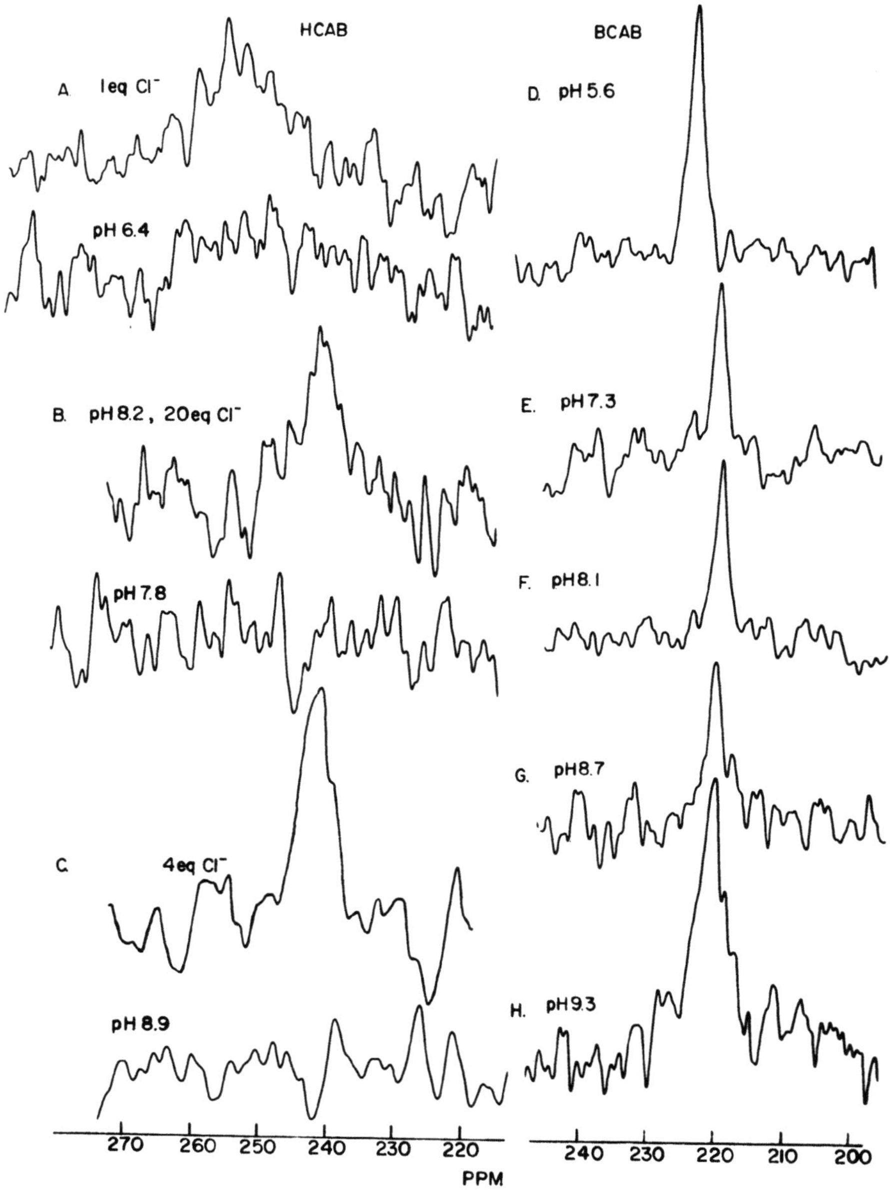

Fig. 8. ^{113}Cd NMR spectra of ^{113}Cd (II) human carbonic anhydrase B (HCAB) and bovine carbonic anhydrase B (BCAB) as a function of pH and chloride concentration. *A* ^{113}Cd (II) HCAB 4.5×10^{-3} M, pH 6.4; *insert* plus 4.5×10^{-3} M NaCl. *B* ^{113}Cd (II) HCAB 4.5×10^{-3} M, 0.025 M Tris-sulphate, pH 7.8; *insert* ^{113}Cd (II) HCAB 4.4×10^{-3} M, 8.8×10^{-2} M NaCl, pH 8.2. *C* ^{113}Cd (II) HCAB 6.1×10^{-3} M, 0.025 M Tris-sulphate, pH 8.9; *insert* plus 2.4 c 10^{-2} M NaCl. *D* through *H* ^{113}Cd (II) BCAB 6.0×10^{-3} M, pH as indicated. (Schoot Uiterkamp et al., 1979)

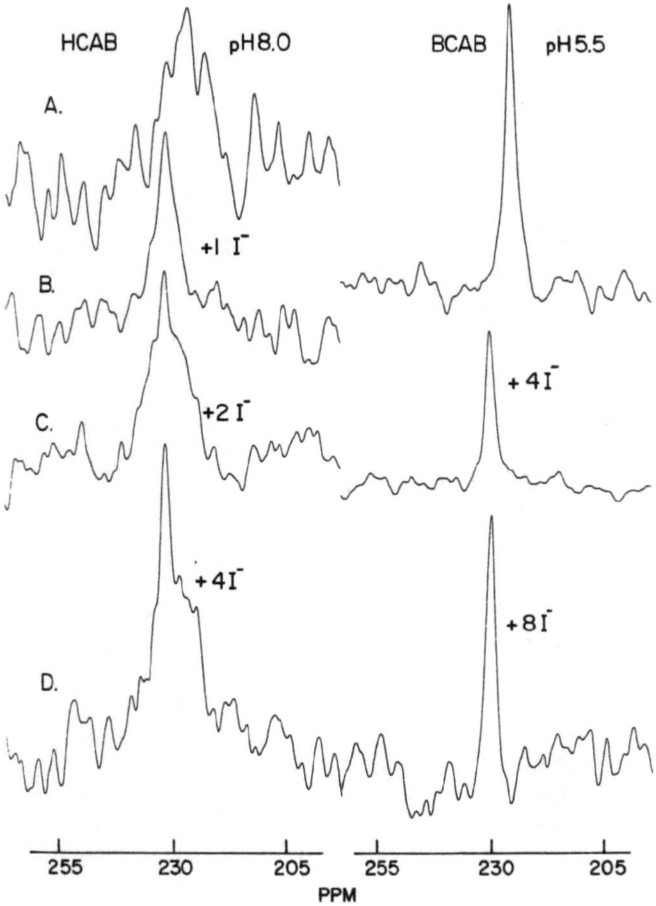

Fig. 9. ^{113}Cd NMR of ^{113}Cd (II) human carbonic anhydrase B (HCAB) at pH 8.0, 25°. (A); plus 1 eq. I^- (B), 2 eq. I^- (C), and 4 eq. I^- (D); and ^{113} Cd NMR of ^{113}Cd (II) bovine carbonic anhydrase B (E) at pH 5.5, 25°; plus 4 eq. I^- (F), and 8 eq. I^- (G). (Data from Schoot Uiterkamp et al., 1979)

linewidth represent varying degrees of chemical exchange broadening reflecting the rates of conformational flux between at least two states of each enzyme, each with a different chemical shift. As will be argued below the most likely process responsible for such flux is exchange of monodentate ligands from solution at the central metal ion. The arguments for exchange processes can be developed as follows. One of the few explanations for the complete disappearance of a ^{113}Cd (II) resonance as is observed for the anion-free form of HCAB at all pH values (Fig. 7) is the involvement of a chemical exchange process occurring at an intermediate rate. As outlined above, an appropriate rate for such an exchange for ^{113}Cd species is near 10^3 s^{-1}. Further support for the absence of ^{113}Cd resonance in HCAB as being due to an intermediate chemical rate is the appearance of a broad resonance as the temperature is lowered, suggesting that as exchange is slowed the enzyme conformational flux is moving from intermadiate to slow exchange.

In contrast, BCAB (HCAC is essentially the same), which shows a relatively narrow signal, could be considered to represent either fast exchange or slow exchange. That a fast chemical exchange process is influencing the ^{113}Cd signal of ^{113}Cd (II) BCAB is supported by the broadening of the resonance as the temperature is lowered from 24° to 4° (not shown).

^{113}Cd (II) NMR distinguishes the two forms of the bovine enzyme BCAB (the Arg 56 isozyme) and BCAA (the Glu 56 isozyme) (Gulian et al., 1977, 1978) by virtue of very different chemical shifts (80 ppm different) for the two ^{113}Cd (II) isozymes in the absence of halide (Fig. 10A). These two isozymes can be separated by DEAE Sephedex chromatography and show the typical different isoelectric points reported for these isozymes (Gulian et al., 1977, 1978). The ^{113}Cd chemical shift of one of these isozymes, BCAA, is much more sensitive to halide than the other, as shown by the ~ 100 ppm downfield chemical shift when Cl^- is added to BCAA (cf. Fig. 10A and B). Thus such a relatively small change as an anion ligand substitution can cause a 100 ppm change in the chemical shift of the ^{113}Cd coordination complex.

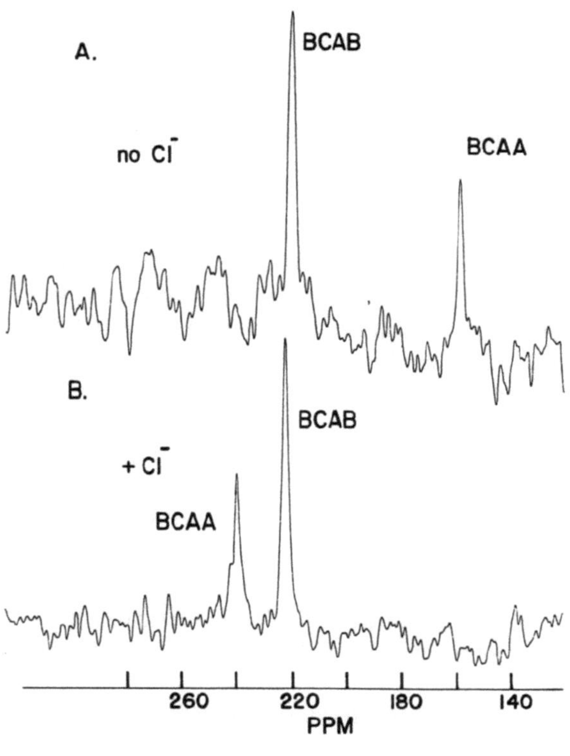

Fig. 10. ^{113}Cd NMR of ^{113}Cd (II) bovine carbonic anhydrase prepared from pooled bovine blood. *A* Spectra after the removal of all anions by ion exchange chromatography in H_2O. *B* After the addition of 4 eq; Cl^-. The species labeled BCAB and BCAA have been isolated separately by column chromatography and been shown to correspond in properties to the known isozymes of the bovine enzyme. (Data from Schoot Uiterkamp et al., 1979)

Summary of Exchange Processes at the Active Center. Do the chemical exchange processes that are detected by the ^{113}Cd resonances of the active site Cd (II) ion reflect any basic aspects of the carbonic anhydrase mechanism? In this section we present arguments that they do represent important rate-limiting steps in the mechanism, most likely ligand exchange reactions at the metal ion. As discussed in Section 5, the vanishing relaxivity of HCAB at low pH requires that the lifetime of a solvent molecule in the inner coordination sphere of the metal ion must be $> 5 \times 10^{-3}$ s, i.e., relatively slow exchange. This is much slower than frequently observed for a metal-coordinated solvent molecule, perhaps reflecting a complex hydrogen bond network at the active center. While the acid form of Co (II) BCAB show only ~ half the relaxivity of the alkaline form (Fig. 6), this value would still allow for a reasonable rapid exchange of coordinated solvent. It is possible that there are two contributions to relaxation, a facile pH-dependent proton exchange which may not involve Zn-O bond breaking, and a slower ligand exchange involving Zn-O bond breaking which is relatively pH-independent. This is suggested by relative invariance of the ^{113}Cd NMR signals as a function of pH for both HCAB and BCAB (Fig. 8).

A variant of this model, but one in many ways equivalent, is to suppose that the low pH form contains a slowly exchanging H_2O ligand in a 4-coordinate configuration, while the high pH form is 5-coordinate with a coordinated ^-OH and a distantly coordinated, rapidly exchanging H_2O, analogous to the distantly coordinated imidazole complex of HCAB which appears to be 5-coordinate. The ^{113}Cd NMR samples involve mM protein or greater, so regardless of whether buffer is present or not it is likely that proton transfer is no longer the rate-limiting step due to proton acceptors on the protein molecules themselves. Hemoglobin may serve such a function in the red cell.

When the rate of proton transfer is increased by the addition of buffer acceptors, another step becomes rate-limiting, having a rate of $\sim 10^3$ s^{-1} in HCAB and 10^5 – 10^6 s^{-1} in BCAB. This step may be the intrinsic ligand exchange rate (Zn-oxygen bond breaking) involving the exchange of HCO_3^- for H_2O or ^-OH. The relatively rapid ligand exchange rate in BCAB might explain the relaxivity remaining in the low pH form of the enzyme under some conditions (Fig. 6). Chemical exchange rates of 10^3 and $10^5 - 10^6$ s^{-1} for HCAB and BCAB respectively are also being sensed by the ^{113}Cd (II) nuclei in the two enzymes[5].

The correspondence of the magnitudes of their exchange rates and the rate-limiting steps in enzyme turnover may be coincidental, but it is perhaps not too speculative to suggest that the requisite chemical exchange processes may be conformational fluxes accompanying ligand exchange in the coordination sphere of Cd (II), much faster for the high activity isozymes than for HCAB. The variation in structure that slows the ligand exchange for HCAB may also relate to the longer lifetime of the anion ligands as well. The relation of intramolecular proton transfers and such con-

---

5 If the chemical exchanges modulating the ^{113}Cd (II) resonances reflect ligand exchange processes they may be relatively slower for the Cd (II) than for the Zn (II) enzyme, reflecting generally longer lifetimes for Cd (II)-coordinated ligands. While this difference may be relatively small for oxygen ligands, maximum CO_2 hydration activity of Zn (II)-HCAB in the presence of adequate buffer is over 10^4 s^{-1}, Cd (II)-HCAB may exchange ligands more slowly. This is clearly true for halides

formational fluxes cannot be discerned from this data. Comparative D_2O and H_2O studies are not complete as yet. The papers of Silverman and Lindskog in this symposium should be consulted for discussion.

At high pH, 8–9, the bicarbonate level of any solution exposed to the atmospheric CO_2 may reach 2 mM. It has been suggested therefore that HCAB might be in the form of the bicarbonate complex at alkaline pH unless the solution was degassed. Spectrophotometric titrations of Co (II) HCAB with HCO_3^- show, however, that considerably higher concentrations, 0.1 M, are needed before significant equilibrium concentrations of the bicarbonate complex are present. While a small percentage of the HCAB may normally be present as the HCO_3^- complex, adding HCO_3^- to both HCAB and BCAB does not have significant effect on the resonances. By the above model, such an addition would not be expected to alter the ligand exchange rate (Zn-O bond breaking) significantly.

This survey of the data relating to the catalytic and physicochemical properties of carbonic anhydrase can be summarized by the statement that all of the data are compatible with the postulate that the activity-linked ionization on the enzyme is the dissociation of a proton from a coordinated water molecule. The Zn-$\bar{O}$H species then becomes the attacking nucleophile in the hydration of CO_2.

Acknowledgments.Original work from the author's laboratory reported in this paper was supported by NIH Grants AM 09070-15, AM 18778-03, and NSF Grant PCM 76-82231

References

Armitage IM, Schoot Uiterkamp AJM, Chlebowski JF, Coleman JE (1978) ^{113}Cd NMR as a probe of the active sites of metalloenzymes. J Mag Res 29: 375–392

Bauer R, Limkilde P, Johansen JT (1976) Low and high pH form of cadmium carbonic anhydrase determined by nuclear quadrupole interaction. Biochemistry 15: 334–342

Bauer R, Limkilde P, Johansen JT (1977) Metal coordination geometry and mode of action of carbonic anhydrase. Effect of imidazole on the spectral properties of Co (II) and ^{111}Cd (II) human carbonic anhydrase B. Carlsberg Res Commun 42: 325–339

Brown GS, Navon G, Shulman RG (1977) X-ray absorption studies of halide binding to carbonic anhydrase. Proc Natl Acad Sci USA 74: 1794–1797

Coleman JE (1967) Mechanism of action of carbonic anhydrase. J Biol Chem 242: 5212–5219

Coleman JE, Chlebowski JF (1979) Molecular properties and mechanism of alkaline phosphatase. Adv Inorg Biochem 1: 1–66

Coleman JE, Coleman RV (1972) Magnetic circular dichroism of Co (II) carbonic anhydrase. J Biol Chem 247: 4718–4728

Coleman JE, Armitage IM, Chlebowski JF, Otvos JD, Schoot Uiterkamp AJM (1979) Multinuclear NMR approaches to the solution structure of alkaline phosphatase. In: Shulman RG (ed) Biological applications of magnetic resonance. Academic Press, London New York, p 345–395

Fabry ME (1975) The effect of buffers on nuclear magnetic relaxation of water protons by bovine and human-B cobalt carbonic anhydrase. J Biol Chem 253: 3568–3574

Fabry ME, Koenig SH, Schillinger WE (1970) Nuclear magnetic relaxation dispersion in protein solutions. J Biol Chem 245: 4256–4262

Gulian MM, Limozin N, Mallet B, DiCostanzo J, Charrel M (1977) Indépendance génétique de deux formes de l'anhydrase carbonique érythrocytaire bovine. Biochimie 59: 293–302

Gulian MM, Faure D, Buc J, Charrel M (1978) Anhydrases carboniques érythrocytaires bovines: étude comparée des activités anhydrasiques et estérasiques des différentes formes. Biochimie 60: 473–478

Haffner PH, Coleman JE (1973) Cu (II)-carbon bonding in cyanide complexes of copper enzymes. 13C splitting of the Cu (II) electron spin resonance. J Biol Chem 248: 6626–6629

Haffner PH, Coleman JE (1975) Structure of the active site of carbonic anhydrase as determined by ESR. J Biol Chem 250: 996–1005

Kannan KK, Petef M, Fridborg K, Cid-Dresdner H, Lövgren S (1977) Structure and function of carbonic anhydrases. FEBS Lett 73: 115

Keilin D, Mann T (1940) Carbonic anhydrase. Purification and nature of the enzyme. Biochem J 34: 1163–1176

Kernohan JC (1964) The activity of bovine carbonic anhydrase in imidazole buffers. Biochim Biophys Acta 81: 346–356

Koenig SH, Brown RD (1972) H_2CO_3 as substrate for carbonic anhydrase in the dehydration of HCO_3^-. Proc Natl Acad Sci USA 69: 2422–2425

Lindskog S (1963) Effects of pH and inhibitors on some properties related to metal binding in bovine carbonic anhydrase. J Biol Chem 238: 945–951

Lindskog S (1966) Interaction of cobalt (II)-carbonic anhydrase with anions. Biochemistry 5: 2641–2646

Lindskog S, Coleman JE (1973) The catalytic mechanism of carbonic anhydrase. Proc Natl Acad Sci USA 70: 25-5–2508

Lindskog S, Henderson LE, Kannan KK, Liljas A, Nyman PO, Strandberg B (1971) Carbonic anhydrase. Enzymes 5: 587–665

Maren TH (1967) Carbonic anhydrase: Chemistry, physiology and inhibition. Physiol Rev 47: 595–781

Maren TH, Rayburn CS, Liddell NE (1976) Inhibition by anions of human red cell carbonic anhydrase B: Physiological and biochemical implications. Science 191: 469–472

Meldrum NU, Roughton FJW (1933) Carbonic anhydrase: its preparation and properties. J Physiol (London) 80: 113–142

Schoot Uiterkamp AJM, Armitage IM, Coleman JE (1979) ^{113}Cd NMR of mammalian carbonic anhydrases. J Biol Chem (submitted for publication)

Tu, CK, Silverman DN (1975) The mechanism of carbonic anhydrase studied by ^{13}C and ^{18}O labeling of carbon dioxide. J Am Chem Soc 97: 5935–5936

Ward RL (1969) ^{35}Cl nuclear magnetic resonance studies of a zinc metalloenzyme carbonic anhydrase. Biochemistry 8: 1879–1883

Some Structural Aspects of Carbonic Anhydrase

I. BERTINI, C. LUCHINAT, and A. SCOZZAFAVA [1]

Any proposed model for the active site structure and the mechanism of action of carbonic anhydrase should take into account those experimental data which are more directly related to the metal ion environment. In particular, water proton relaxation measurements of solutions of carbonic anhydrase derivatives containing paramagnetic metal ions constitute one of the most reliable insights into the solvation of the metal itself and the possible hydrogen ion equilibria involved (Fabry et al., 1970).

Although the relevance of relaxation measurements is essentially qualitative, the systematic investigation on VO^{2+}, Mn^{2+}, Co^{2+}, Ni^{2+}, and Cu^{2+} carbonic anhydrase derivatives allowed us to establish some common features (Bertini et al., 1978a, b, 1979). In Fig. 1 the ^1H T_1^{-1} values of water solutions of various metallosubstituted carbonic anhydrases are reported as a function of pH. The arrows indicate the approximate pK_a relative to the acid-base equilibrium as determined through the electronic spectra and/or activity and inhibition measurements. All of the proton relaxation measurements indicate that (1) water is bound to all the above metal ions, since there is a large T_1^{-1} enhancement consistent with the expected electronic relaxation times; (2) the same number of exchangeable protons is bound over the entire pH range investigated, and therefore water does not appear to be responsible for the acid-base equilibrium of the enzyme.

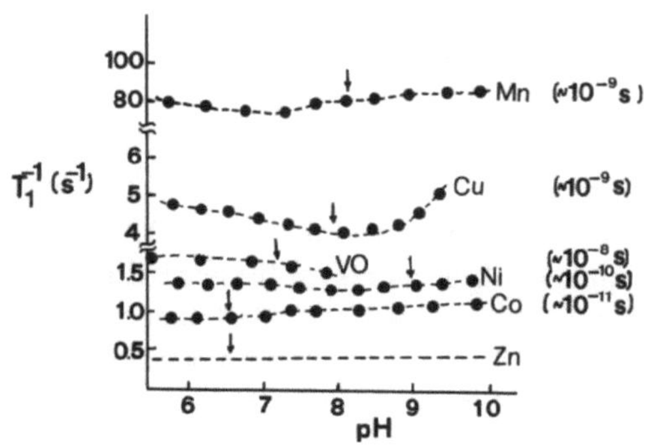

Fig. 1. T_1^{-1} values of the ^1H NMR signal of water solutions of metal substituted carbonic anhydrases 1×10^{-3} M, as a function of pH. The *arrows* indicate the pK_a of the enzymes; in *parentheses* the estimated values of the electronic relaxation times

1 Istituto di Chimica Generale ed Inorganica della Facoltà di Farmacia, Università di Firenze, Via Gino Capponi 7, 50121, Florence/Italy

Other mechanisms able to account for the acid-base equilibrium have been proposed (Pesando, 1975; Kannan et al., 1977). Furthermore, the hydrogen bond net and the charge within the active crevice supposedly play an important role which overlaps the "main" acid-base equilibrium. In particular, uncoordinated histidines which have pK_a around 6 and are present in the active site become statistically more and more protonated as the pH decreases from say 7, and therefore a complex variation of the chemical properties of the active site has to be expected, since there has to be a difference between protonated and unprotonated molecules. The electronic spectra of cobalt bovine carbonic anhydrase B as a function of pH, carefully recorded in unbuffered and HEPES buffered solutions (Barzi et al., 1979), are shown in Fig. 2.

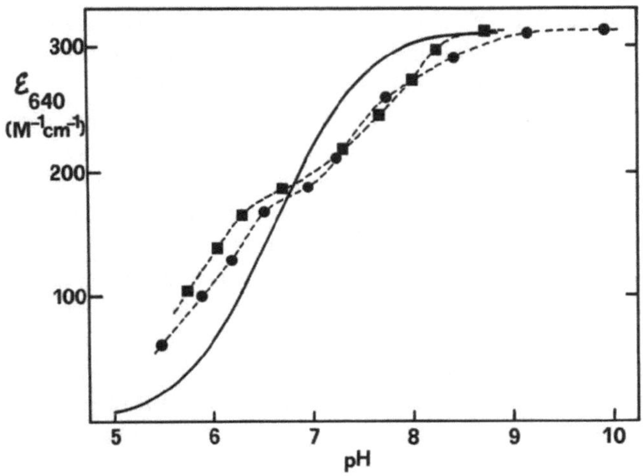

Fig. 2. Intensity of the 640 nm d-d transition of cobalt (II) bovine carbonic anhydrase B as a function of pH in unbuffered (■) and 0.1 M HEPES buffered (●) solutions. The *solid line* is calculated for a single $pK_a = 6.6$

It appears that they are not consistent with a single acid-base equilibrium. This has also been suggested from inhibition measurements (Whitney and Brandt, 1976), and was noted for the nickel derivative in the electronic spectra (Bertini et al., 1978b). Finally, the existence of acidic groups with low pK_a accounts for the anomalous increase in the affinity constants of negatively charged inhibitors (Bertini et al., 1978c). The suggestion (Kannan et al., 1977) of the Glu 106 residue as a candidate for a further dissociating group can also be framed in this scheme; the above residue, being connected to the metal-bound water through Thr 199 via hydrogen bond is expected to be sensitive to the binding of inhibitors, which are known to impose the "acidic" conformation to the active site.

References

Barzi D, Bertini I, Luchinat C, Scozzafava A (1979) The electronic spectra of cobalt (II) bovine carbonic anhydrase. Inorg Chim Acta 36: L431–432

Bertini I, Canti G, Luchinat C, Scozzafava A (1978a) Characterization of cobalt (II) bovine carbonic anhydrase and of its derivatives. J Am Chem Soc 100: 4873–4877

Bertini I, Borghi E, Luchinat C (1978b) Characterization of nickel (II) bovine carbonic anhydrase and its inhibitor derivatives. Bioinorg Chem 9: 495–504

Bertini I, Luchinat C, Scozzafava A (1978c) Binding affinity of bicarboxylate ions for cobalt (II) bovine carbonic anhydrase. Bioinorg Chem 9: 93–100

Bertini I, Canti G, Luchinat C, Scozzafava A (1979) Characterization of oxovanadium (IV) substituted bovine carbonic anhydrase B. Inorg Chim Acta 36: 9–12

Fabry (Riepe) ME, Koenig SH, Shillinger WE (1970) Nuclear magnetic relaxation dispersion in protein solutions. IV. Proton relaxation at the active site of carbonic anhydrase. J Biol Chem 245: 4256–4262

Kannan KK, Petef M, Fridborg K, Cid-Dresdner H, Lövgren S (1977) Structure and function of carbonic anhydrases. FEBS Lett 73: 115–119

Pesando JM (1975) Proton magnetic resonance studies of carbonic anhydrase. II. Group controlling catalytic activity. Biochemistry 14: 681–688

Whitney PL, Brandt H (1976) Effects of two ionizing groups on the active site of human carbonic anhydrase B. J Biol Chem 251: 3862–3867

Primary Structure of Carbonic Anhydrases[1]

L.E. HENDERSON[2]

In 1928, studies on the rate of escape of carbon dioxide from hemolyzed blood led to the recognition of a red cell substance catalyzing the reversible hydration of CO_2 [12]. By 1932, the substance was shown to be an enzyme and given the name carbonic anhydrase [26, 27]. In the following 10 years the enzyme was partially purified [28], shown to be a zinc-containing enzyme [15, 16] having a molecular weight of about 30,000 [33] and to contain one zinc ion per enzyme molecule [37]. World War II and its aftermath interrupted research on carbonic anhydrase for almost 20 years, but in the early 60's researchers in Sweden [23, 30] reported several homogenous preparations of the enzyme which differed in electrophoretic mobility and specific activity. These observations were quickly confirmed by others [18, 34, 35] with the resulting realization that human erythrocytes contained several different enzyme forms catalyzing the reversible hydration of carbon dioxide.

In the early investigations of human and bovine erythrocyte carbonic anhydrase polymorphism, the various forms of the enzymes were named in accordance to their electrophoretic mobilities [35, 32]. Human erythrocytes contained at least three electrophoretically distinct forms designated A, B, and C, whereas bovine erythrocytes contained two forms designated A and B. The human A and B enzyme had a relatively low specific activity when compared to the human C enzyme for catalysis of the reversible hydration of carbon dioxide. Both forms of the bovine enzyme have specific activities comparable to the human C enzyme in the same assay. It has since become apparent that erythrocytes of many species contain high and low specific activity forms of carbonic anhydrase. It is now conventional to classify carbonic anhydrases according to their specific activity using the human system as a reference. High specific activity carbonic anhydrases are classified as type C, and low specific activity enzymes are classified as type B [24].

The first indication that the various forms of human carbonic anhydrase might be proteins with distinctly different amino acid sequences rather than modified forms of some common sequence came from amino acid analysis data presented by Nyman and Lindskog in 1964 [32]. The amino acid compositions for the human A and B forms of carbonic anhydrase were indistinguishable from each other, as were the compositions for the bovine A and B forms, but the human C form had a distinctly unique amino acid composition. The amino acid compositions for the two distinctly different human enzymes (B and C) and the major bovine enzyme are given in Fig. 1.

---

1 By acceptance of this article, the publisher acknowledges the right of the U.S. Government to retain nonexclusive, royalty-free license in and to any copyright covering the article

2 Biological Carcinogenesis Program, Frederick Cancer Research Center, Frederick, Maryland 21701

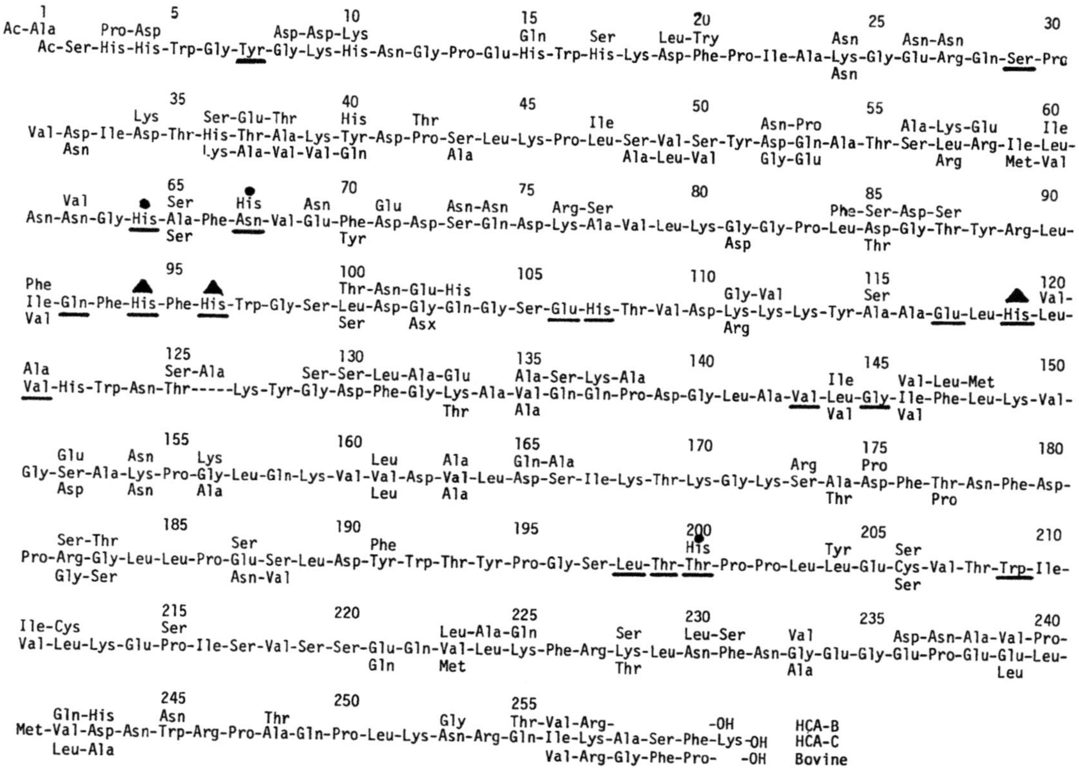

Fig. 1. The amino acid sequence of human carbonic anhydrase C (HCA-C, line 2) and residues that are different from HCA-C in the complete amino acid sequences of human carbonic anhydrase B (HCA-B, line 1) and bovine carbonic anhydrase CI (Bovine, line 3). Residues known to be within 8 Å of the active site Zn atom in the three-dimensional structure of HCA-C [29] are underlined, (—): histidine residues liganded to Zn, (▲): and histidine residues modifiable with active site directed reagents, (●)

The most notable differences between the amino acid content of the two human enzymes were in the values for serine and methionine. These differences were outside the range of experimental error and definitely indicated two distinctly different human enzymes. The bovine enzymes were found to be lacking cysteine. This ruled out the popular idea that the sulfur atom of cysteine was a metal ligand in the carbonic anhydrases.

In 1968, Nyman et al. [31] reported the isolation of a peptide representing the carboxyl terminal segment of the human B enzyme (residue 242–260) and corresponding peptides from the human C enzyme and the bovine enzyme. These peptides were derived by cleavage of methionyl bonds through the action of cyanogen bromide. The amino acid sequence of the peptides was determined and it was clear that the two human enzymes had different amino acid sequences, but showed considerable sequence homology when compared to each other and to the bovine enzyme. The authors proposed that the two human enzymes were the result of an ancient gene duplication and subsequent independent evolution. Moreover, it seemed that the two high activity forms (human C and bovine) were more closely related to each other than to the low activity human B enzyme. Thus, at an early stage in the structural analysis of the carbonic anhydrases, limited primary structural data proved adequate to establish a broad concept for evolutionary relationships between the enzymes that are widely accepted today.

Attempts to determine the amino terminal amino acid of the carbonic anhydrases had been unsuccessful until 1967, where Laurent et al. [19] reported the amino terminal end group of the human B enzyme to be N-acetyl-alanine and that of the human C enzyme to be N-acetyl-serine. They also isolated amino terminal peptides from the human B enzyme [17] but were unable to obtain a corresponding peptide from the human C enzyme. At that time Laurent correctly speculated that the N-acetyl-serine end group of the human C enzyme should be followed by a basic amino acid in the amino acid sequence of the protein.

Studies to determine the complete primary structure of the human B enzyme were undertaken in three laboratories. Nyman and Strid, working in Sweden, continued the work they had started with the isolation of the cyanogen bromide peptides. Laurent and coworkers, working in France, continued their work that began with peptides derived from amino terminal region of the B enzyme. There was considerable communication between the two European groups so that points of disagreement were resolved before publication of the results. By 1971, preliminary accounts published by the Swedish [24, 31, 1, 9] and French [19] groups gave almost 80% of the complete primary structure of the B enzyme. A third group working at the University of Wisconsin in the United States was also attempting the complete structure of the B enzyme.

In 1972 the Swedish group published the complete primary structure of the human B enzyme [2], and, in 1973, the Wisconsin group published their proposal for the complete primary structure of the B enzyme [21].

The two groups differed on 17 amino acid sequence assignments in their proposed structures for the B enzyme. At position 13–15 the Swedish group had proposed -Gln-Pro-Glu, while the Wisconsin group had proposed -Pro-Glu-Gln. The Swedish group had taken the data of the French group for residues 1 through 17 in their proposal and had made no effort to confirm the assignments. Subsequent reinvestigation of this region confirmed the proposal of the Wisconsin group for residues 13–15, while confirming the French assignments for the remainder of this region of the molecule [38]. All other discrepancies have been reinvestigated [38, 11], but there is still some disagreement between the Swedish and Wisconsin groups [2]. The amino acid sequence of the human B enzyme shown in Fig. 1 is that given by the Swedish group (Anderson, Nyman, and Strid [2]) with the Wisconsin group's (Lin and Deutsch [21]) corrections at positions 13–15. This revised structure for the human B enzyme has been confirmed by the French group (Laurent and coworkers [25, 7]).

In Sweden, work on the complete primary structure of the human C enzyme paralleled work on the B enzyme and by 1971 almost half of the structure had been reported [9]. The complete primary structure of the C enzyme was published by this group in 1973 [10]. Later in the same year the Wisconsin group also reported the complete primary structure of the C enzyme [22]. There had been no collaboration between the two groups other than that which appeared in the literature. The two proposed sequences differed in 11 positions. All points of difference were reinvestigated by the Swedish group and the results of the reinvestigation, together with all primary data relating to the structure of the C enzyme, have been published [11]. The primary structure of the C enzyme shown in Fig. 1 is the result of this reinvestigation and contains the following corrections as proposed by the Wisconsin group [22]: at positions 10 Asp to Asn, 61–62 Asx-Asx to Asn-Asn, 103 Glu to Gln; 106 Gln to Glu, and 232 Asp to Asn. Other discrepancies between the Swedish and Wisconsin proposals for the structure of the C enzyme were not resolved by the reinvestigation [3].

The complete primary structure of sheep carbonic anhydrase C was published by Tashian and coworkers in 1974 [39]. The French group published a preliminary account of their work on the bovine enzyme in the same year [6]. The complete primary structure of the bovine enzyme (now called bovine carbonic anhydrase C1) was reported by the French group in 1976 [5, 36] (Fig. 1).

The complete amino acid sequence of the human C enzyme and residues in the sequences of the human B enzyme and bovine enzyme that differ from the human C enzyme sequence are shown in Fig. 1. All three structures are aligned to give a maximum of number of identities and the numbering system is that of the human B enzyme since it is longer by one residue at the amino terminal end. It is necessary to introduce a one-residue gap at position 126 (Fig. 1) in the bovine and human C sequences to obtain the maximum number of identities with the human B sequence. The amino acid compositions of the human B and C enzymes and the bovine enzyme as determined from their amino acid sequences are in excellent agreement with their compositions as determined by direct amino acid analysis (Table 1).

The bovine and sheep enzymes are closely related type C carbonic anhydrases and show differences in amino acid sequence at only 18 positions out of 259 residues.

2 The proposal of Lin and Deutsch [21, 22] disagreed with the structure given in Fig. 1 in the following way: at positions 26–28 Asp-Gln-Asn; 165 -Glu

3 The proposal by Lin and Deutsch [22] for the structure of the C enzyme differs from that given in Table 2 as follows at positions: 74, Glu; 101, Asn; 137, Glu; 154–155, Pro-Lys and 205, Gln

Table 1. Amino acid compositions of carbonic anhydrases by direct analysis and amino acid sequence

	Human B		Human C		Bovine	
	Direct	sequence	direct	sequence	direct	sequence
Asx	31	31	29	29	32	32
Thr	14	14	13	12	15	14
Ser	30	30	19	18	16	16
Glx	22	22	25	24	24	23
Pro	18	17	17	17	20	19
Gly	16	16	22	22	20	20
Ala	19	19	13	13	17	17
Cys	1	1	1	1	0	0
Val	17	17	17	17	20	20
Met	2	2	1	1	3	3
Ile	10	10	9	9	5	5
Leu	20	20	25	26	26	26
Tyr	8	8	8	8	8	8
Phe	11	11	12	12	11	11
His	11	11	12	12	11	11
Lys	18	18	24	24	19	18
Arg	7	7	7	7	9	9
Trp	6	6	7	7	7	7
Total residues		260		259		259
Acetyl group		1		1		1
Protein M.W.		28,750		28,820		28,990

The amino acid compositions as determined by direct amino acid analysis of isolated proteins (data taken from [24]) and by amino acid sequence analysis of isolated proteins (data taken from Fig. 1)

The positions (numbering system used in Fig. 1) and nature of the amino acid sequence differences between the sheep and bovine enzymes are given in Table 2. The residues in the homologous positions in the human C enzyme are also listed in Table 2. From this comparison it can be seen that the human C enzyme is slightly more closely related to the bovine enzyme than to the sheep enzyme.

X-ray crystallographic studies on the human enzymes began in Upsalla, Sweden, about the same time as the work on their primary structures. An excellent three-dimensional model for the human C enzyme was constructed from high resolution X-ray crystallographic data about one year before the complete primary structure of the enzyme was known [13, 20]. The model was re-interpreted in the light of the primary structure when that data became available [29]. An equally excellent three-dimensional model for the human B enzyme was also constructed by the same group [14]. The X-ray crystallographic models for both enzymes are presented and discussed by Kannan, Chap. 51, this volume, so no effort will be made to treat this data here. However, to aid the discussions that follow, residues within 8Å of the zinc ion in the X-ray crystallographic model of the human C enzymes and the corresponding residues in the human B, bovine, and sheep enzymes are presented in Table 3 and indicated in Fig. 1.

Table 2. Amino acid sequence difference between sheep and bovine carbonic anhydrases

Position in Fig. 1	Residue		
	Sheep enzyme	Bovine enzyme	Human C
9	Glu	Lys	Lys
24	Asp	Asn	Lys
32	Asp	Asn	Asp
40	Pro	Gln	Tyr
50	Leu	Val	Ser
55	Ala	Thr	Thr
70	Phe	Tyr	Phe
163	Val	Ala	Val
174	Ala	Thr	Ala
182	Ser	Gly	Arg
186	Lys	Pro	Pro
187	Arg	Glu	Glu
188	Ala	Val	Ser
200	Asn	Thr	Thr
210	Val	Ile	Ile
216	Thr	Ile	Ile
228	Ser	Thr	Lys
258	Val	Gly	Ala

The residue and position in Fig. 1 where the amino acid sequence of sheep carbonic anhydrase differs from the amino acid sequence of bovine carbonic anhydrase and the corresponding residues in the human C structure. Data for the sheep and bovine structure from [39, 5, and 36]

The total number of conserved amino acid residues in the C enzyme structure compared to human B or bovine structures is shown in Table 4. In compiling the data in Table 4, Asp and Asn assignments were treated together as Asx and Glu and Gln assignments were treated together as Glx. This was done because of the difficulty and uncertainty sometimes encountered in making amide assignments. The human C enzyme shows 210 identities out of 259 residues (81%) when compared to the bovine enzyme, 206 identities (80%) when compared to the sheep enzyme, and 159 identities (61%) when compared to the human B enzyme. The sheep and bovine structures are more closely related to each other (93% identity) than they are to the human C enzyme, but all three type C enzymes are more closely related to each other than they are to human B enzyme. Among the type C structures, one of the most highly conserved regions of their amino acid sequences is from position 2 to 35 (Fig. 1 and Table 2) where there are only three amino acid substitutions. Another highly conserved region extends from position 61 to 132, where there are only eight amino acid substitutions in the known type C structures. Between these two conserved regions there is a variable region from position 36 to 60 containing 15 positions, where amino acid substitutions take place among the type C structures. In the three-dimensional model of the C enzyme, residues 36 to 46 appear to be on the surface of the molecule, whereas residues 47 to 60 are involved in β structure.

Table 3. Residues within 8 Å of the zinc ion in human carbonic anhydrase C and the residue in corresponding positions in human B, sheep and bovine carbonic anhydrases

Residue	Residue in human C	Residue in human B if different	Residue in sheep if different	Residue in bovine if different
7	Tyr			
29	Ser			
64	His			
65	Ala	Ser		
67	Asn	His		
92	Gln			
94	His			
96	His			
106	Glu			
107	His			
117	Glu			
119	His			
121	Val	Ala		
143	Val			
145	Gly			
198	Leu			
199	Thr			
200	Thr	His	Asn	Asn
209	Trp			

Residues in the three-dimensional model of human carbonic anhydrase C that are within 8 Å of the active Zn [29] and residues 1 in the corresponding position in the primary structure of the human B, sheep and bovine enzymes if different. The numbering system is that used in Fig. 1

The two longest consecutive stretches of variable residues among the type C structures are from position 36 through 40 and 256 through 260. Each variable stretch involves five residues and in the three-dimensional structure of the C enzyme these two parts of the peptide chain appear to be involved in interacting parallel β structures. In the human B enzyme, corresponding residues in positions 258 to 260 have been shown to contribute to the stability of the native three-dimensional structure. Removal of these residues by the action of carboxypeptidase A yields a degraded B enzyme which is much more readily denatured by guanidine hydrochloride than is the native enzyme [4]. These observations suggest that residues in positions 36 through 40 and 256 through 260 in the type C structures may be involved in maintaining a stable native conformation. The fact that both stretches appear to be variable in the type C structures may indicate that amino acid substitution in one stretch must be complemented by substitution in the other stretch in order to maintain a stable native state.

The conservative nature of each type of amino acid residue in the human C enzyme structure when compared to either the human B or bovine structure is shown in Table 4. The most highly conserved types of residue in all three structures are the tryptophans and the single methionine of the human C enzyme at position 241 (note that the human B enzyme has another methionine at position 148 and the bovine enzyme has additional methonines at positions 59 and 223). The least conservative amino

Table 4. Total number of residues in C enzyme conserved in the B enzyme and the bovine enzyme amino acid sequences

Residue	Total number conserved in sequence		
	C enzyme composition	B enzyme	Bovine enzyme
Asx	29	17	26
Thr	12	8	10
Ser	18	13	12
Glx	24	12	22
Pro	17	13	17
Gly	22	14	17
Ala	13	8	9
Cys	1	0	0
Val	17	10	11
Met	1	1	1
Ile	9	6	5
Leu	26	14	22
Tyr	8	6	7
Phe	12	7	10
His	12	6	11
Lys	24	14	17
Arg	7	4	6
Trp	7	6	7
Total	259	159	210

The total number of residues in the amino acid composition of the human C enzyme, and the total number of these residues that are identical in a comparison of the human C structure to the human B and to the bovine structures. The numbering system and alignment of structures is that given in Fig. 1

acid is the cysteine residue. The conservative nature of the serine and histidine residues deserves special comment.

As mentioned in the introduction, the most notable difference in the amino acid compositions of the types B and C carbonic anhydrases is in their serine content. The known type B enzymes have about 30 residues of serine, whereas the known type C enzymes have about 18 residues of serine [24]. On the basis of this observation, the serine content of newly isolated enzymes is sometimes used as one means of classifying an enzyme as type B or C. However, this classification is based on the relative specific activity of the enzyme. Accordingly, one might expect that amino acid substitutions, including numerous serine residues, would have some bearing on the specific activity of the enzyme. There is only one serine residue within 8Å of the active site Zn atom in the three-dimensional structure of the enzymes (Fig. 2; position 30 in all structures). As seen in Table 4, 13 of the 18 serine in the human C enzyme are conserved in the human B enzyme sequence, whereas only 12 are conserved in the bovine enzyme sequence. Serine would appear to be as variable among the type C enzymes as it is from type C to type B. Serine in the human B enzyme is replaced by another residue in the human C enzyme at 17 different positions in Fig. 1. Of these,

4 (at positions 65, 182, 206, and 228) are conserved in the bovine or sheep enzyme sequence (both type C); nine can be regarded as conservative replacements (serine for an alanine, threonine, aspartic acid, glutamine, or glycine) which probably do not affect conformation; the four nonconservative replacements (at His-17, His-36, Pro-215, and Phe-231) appear to be too far from the active site to influence specific activity. Considering all these points, it seems highly improbable to me that the serine content of a carbonic anhydrase has any direct relation to its specific activity. I would expect that as carbonic anhydrases are isolated from species more distantly related to man, type B or C characterization on the basis of serine content will become less distinct.

Histidine residues are seen to be among the most variable residues when comparing type B and C carbonic anhydrases, but amont the most conservative when comparing only type C carbonic anhydrases (Table 4). Human carbonic anhydrase C has 5 histidine residues within 8 Å of the active site zinc atom at positions 64, 94, 96, 107, and 119. All five of these histidine residues are conserved in the known carbonic anhydrase amino acid sequences. The histidine residues at positions 94, 96, and 119 are known to be ligands to the active site Zn atom and would be expected to be conserved in all carbonic anhydrases. The catalytic significance of histidines 64 and 107 is not known. The human B enzyme has two additional histidine residues within 8 Å of the Zn atom at positions 67 and 200. Both of these residues in the B enzyme have been modified by active site-directed reagents: His-67 by N-chloroacetylchlorothiazide [2, 41]; His-200 by bromoacetate and iodoacetate [2, 42, 40], iodoacetic acid [3], and bromopyruvic acid [8]. Of all these reagents, only bromopyruvic acid reacts with the human C enzyme and results in modifications of His-64 [8]. This modified form of the human C enzyme retains enzymic activity.

The two additional histidines at positions 67 and 200 in the active site of the human B enzyme would seem to be the most likely residues responsible for lower specific activity. It also seems reasonable that these two residues would contribute to the anion binding properties of the active site.

The type C carbonic anhydrases have five conserved histidines in the amino terminal 16 residues of their amino acid sequences at positions 3, 4, 10, 15, and 17 (Fig. 1). All of these histidines are replaced by other residues in the B enzyme sequence. This seems to be one of the most outstanding differences between the type B and C amino acid sequences. As more structures become known, this region of the amino acid sequence may become one of the most characteristic features for distinguishing these structures.

References

1. Anderson B, Nyman PO, Strid L (1969) Primary structure studies on human carbonic anhydrase B and its active site. In: Forster RE, Edsall JT, Otis AB, Roughton FJW (eds) CO_2: Chemical, biochemical, and physiological aspects. National Aeronautics and Space Administration, NASA Sp-188 Washington DC, pp 109–114
2. Anderson B, Nyman PO, Strid L (1972) Amino acid sequence of human erythrocyte carbonic anhydrase B. Biochem Biophys Res Commun 48: 670–677

3. Bradbury SL (1969) The carbomymethylation of human carbonic anhydrase B. I. The nature of the reaction. J Biol Chem 244: 2002–2009. II. The amino acid sequence around a reactive histidine residue. J Biol Chem 244: 2010–2016
4. Carlson U, Henderson LE, Nyman PO, Samuelsson T (1974) Studies on the influence of Carboxyl-terminal amino acid residues on the activity and stability of human erythrocyte carbonic anhydrase B. FEBS Lett 48: 167–171
5. Filippi-Foveau D, Sciaky M, Limozin N, Dalmasso C, Laurent-Tabusse G (1976) Structure primaire de l'anhydrase carbonique erythrocytaire bovine CI. I. Peptides trypsiques. Biochimie 58: 1057–1070
6. Foveau D, Sciaky M, Laurent G (1974) New data on the chemical relation between human B and C and bovine CI erythrocyte carbonic anhydrases. CR Acad Sci Paris Ser D 278: 959–962
7. Firaud N, Marriq G, Laurent-Tabusse G (1974) Structure primaire de l'anhydrase carbonique erythrocytaire B humaine: III. Sequence des fragments ICNBr et III CNBr (residus 149–26o) Biochimie 56: 1031–1044
8. Göthe PO, Nyman PO (1972) Inactivation of human erythrocyte carbonic anhydrase by bromopyruvate. FEBS Lett 21: 159–164
9. Henderson LE, Henricksson D, Nyman PO, Strid L (1972) On the primary structure of the B and C isoenzymes of human erythrocyte carbonic anhydrase. In: Rorth M, Astrup P (eds) Oxygen affinity of hemoglobin and red cell acid-base status. Alfred Benzon Symposium IV, Munksgard, Copenhagen and Academic Press, London New York, pp 341–352
10. Henderson LE, Henricksson D, Nyman PO (1973) Primary structure of human erythrocyte carbonic anhydrase C. Biochem Biophys Res Commun 52: 1388–1394
11. Henderson LE, Henricksson D, Nyman PO (1976) Primary structure of human carbonic anhydrase C. J Biol Chem 251: 5457–5463
12. Henriques OM (1928) Die Bindungsweise des Kohlendioxyds im Blute. Biochem Z 200: 1–24
13. Kannan KK, Liljas A, Waara I, Bergsten P-C, Lovgren S, Strandberg B, Bengtsson U, Carlbom U, Fridborg K, Järup L, Petef M (1971) Crystal structure of human erythrocyte carbonic anhydrase C. VI. The three-dimensional structure at high resolution in relation to other mammalian carbonic anhydrases. Cold Spring Harbor Symp Quant Biol 36: 221–231
14. Kannan KK, Notstrand B, Fridborg K, Lövgren S, Ohlsson A, Petef M (1975) Crystral structure of human erythrocyte carbonic anhydrase B. Three-dimensional structure at a nominal 2.2 Å resolution. Proc Natl Acad Sci USA 72: 51–55
15. Keilin D, Mann T (1939) Carbonic anhydrase. Nature (London) 144: 442–443
16. Keilin D, Mann T (1940) Carbonic anhydrase. Purification and nature of the enzyme. Biochem J 34: 1163–1176
17. Laurent B, Derrien Y (1969) Investigations of the structure of human and bovine carbonic anhydrases. In: Forster RE, Edsall JT, Otis AB, Roughton FJW (eds) CO_2: Chemical, biochemical, and physiological aspects. National Aeronautics and Space Administration, NASA SP-188, Washington DC, pp 115–120
18. Laurent G, Charrel M, Castay M, Nahom D, Marriq C, Derrien Y (1962) Identification of erythrocyte Y, X1 and X2 proteins with human carbonic anhydrase. C R Soc Biol 156: 1461–1464
19. Laurent G, Marriq C, Garcom D, Luccioni F, Derrien Y (1967) Sur les anhydrases carboniques erythrocytaires humaines: VI. Enchainement N-terminal de l'enzyme B. Bull Soc Chim Biol 49: 1035–1058
20. Liljas A, Kannan KK, Bergsten PC, Waara I, Fridborg SL, Strandberg B, Carlbom U, Järup L, Lovgren S, Petef M (1972) Crystal structure of human carbonic anhydrase C. Nature New Biol 235: 131–137
21. Lin KTD, Deutsch HF (1973) Human carbonic anhydrase. XI. The complete primary structure of carbonic anhydrase B. J Biol Chem 248: 1885–1893
22. Link KTD, Deutsch HF (1974) Human carbonic anhydrase. XII. The complete primary structure of the C isozyme. J Biol Chem 249: 2329–2337
23. Lindskog S (1960) Purification and properties of bovine erythrocyte carbonic anhydrase. Biochem Biophys Acta 39: 218–226

24. Lindskog S, Henderson LE, Kannan KK, Liljas A, Nyman PO, Strandberg B (1971) Carbonic anhydrase. In: Boyer PD (ed) The enzymes, 3rd edn, vol 5, pp 587–665
25. Marriq C, Sciaky M, Giraud N, Foveau D, Laurent-Tabusse G (1974) Structure primaire de l'anhydrase carbonique erythrocytaire B humaine: II. Clivage par le bromure de cyanogene et sequence des residue 1–148. Biochimie 55: 1361–1379
26. Meldrum NU, Roughton FJW (1932) Some properties of carbonic anhydrase, the carbon dioxide enzyme present in blood. J Physiol (London) 75: 15–16
27. Meldrum NU, Roughton FJW (1932) Discussion on recent advances in the study of enzymes and their action. Proc R Soc Bull 296–297
28. Meldrum NU, Roughton FJW (1933) Carbonic anhydrase; its preparation and properties. J Physiol (London) 80: 113–142
29. Notstrand B, Vaara I, Kannan KK (1975) Structural relation of human erythrocyte carbonic anhydrase isoenzymes B and C. In: Market C (ed) Isozymes, molecular structure. Proc 3rd Int Symp Isozymes. Adacemic Press, London New York, pp 575–599
30. Nyman PO (1961) Purification and properties of carbonic anhydrase from human erythrocytes. Biochim Biophys Acta 52: 1–12
31. Nyman PO, Strid L, Westermark G (1968) Carboxyl-terminal region of human and bovine erythrocyte carbonic anhydrases. 1. Amino acid sequences of terminal cynogen bromide fragments. Eur J Biochem 6: 172–189
32. Nyman PO, Lindskog S (1964) Amino acid composition of various forms of bovine and human erythrocyte carbonic anhydrase. Biochim Biophys Acta 85: 141–151
33. Petermann ML, Hakala NV (1942) Molecular kinetic and electrophoretic studies on carbonic anhydrase. J Biol Chem 145: 701–705
34. Rickli EE, Edsall JT (1962) Zinc binding and the sulfhydryl group of human carbonic anhydrase. J Biol Chem 237: pc 258
35. Rickli EE, Ghanzanfar SAS, Gibbons BH, Edsall JT (1964) Carbonic anhydrase from human erythrocytes: Preparation and properties of two enzymes. J Biol Chem 239: 1065–1078
36. Sciaky M, Limozin N, Filippi-Foveau D, Gulian JM, Laurent-Tabusse G (1976) Primary structure of bovine erythrocyte carbonic anhydrase CI: II. Complete amino acid sequence. Biochimie 58: 1071–1082
37. Scott DA, Fisher AM (1942) Carbonic anhydrase. J Biol Chem 144: 371–381
38. Strid L (1973) Personal communication
39. Tanis RJ, Ferrell RE, Tashian RE (1974) Amino acid sequence of sheep carbonic anhydrase C. Biochim Biophys Acta 371: 534–548
40. Whitney PL (1970) Inhibition and modification of human carbonic anhydrase B with bromoacetate and iodoacetate. Eur J Biochem 16: 126–136
41. Whitney PL, Fölsch G, Nyman PO, Malmström BG (1967) Inhibition of human erythrocyte carbonic anhydrase B by chloroacetyl sulfonamides with labeling of the active site. J Biol Chem 242: 4206–4211
42. Whitney PL, Nyman PO, Malmström BG (1967) Inhibition and chemical modification of human erythrocyte carbonic anhydrase B. J Biol Chem 242: 4212–4220

Evolution of Structure and Function in the Carbonic Anhydrase Isozymes of Mammals

R.E. TASHIAN [1], D. HEWETT-EMMETT [1], S.K. STROUP [1], M. GOODMAN [2], and Y.-S.L. YU [1]

1 Introduction

Comparative studies on the primary structures of the carbonic anhydrase isozymes, designated CA I and CA II, from a number of mammlian species have shown that nearly one half (46%) of the amino acid residues at homologous positions in these isozymes has remained invariant (cf. Tashian, 1977). Because of this high degree of sequence homology between the two carbonic anhydrase isozymes and the fact that the three-dimensional structures of human CA I and CA II are very similar (Kannan et al., 1975; Liljas et al., 1972; Notstrand et al., 1974), it is clear that the structural genes for these isozymes arose from the same ancestral gene. These isozymes are known to differ, sometimes markedly, in their stabilities to heat and denaturing reagents, specific activities, and inhibition by cyclic sulfonamides, halides, and anions (Cf. Lindskog et al., 1971; Maren, 1976; Tashian, 1977). Thus, some critical changes must have evolved in the structures (active site residues?) of these isozymes to account for the observed functional differences (Notstrand et al., 1974; Kannan et al., 1977). Recently, a third isozyme of carbonic anhydrase has been reported, which we have designated CA III, from mammalian skeletal muscle (Holmes, 1977; Koester et al., 1977; Tashian, 1977, 1978; Tashian et al., 1978) some of whose properties differ strikingly from those of CA I and II. For example, the CO_2 hydratase and esterase activities and sulfonamide binding affinity of the CA III isozymes are much lower than those of the CA I and II isozymes, especially the esterase activity and sulfonamide binding affinity (Holmes, 1977; Koester et al., 1977; Tashian et al., 1978; Carter et al., 1979). In addition, the conformational properties of CA III appear to be uniquely different from CA I and CA II (Koester and Noltmann, 1979). Because of these pronounced differences, we have examined here the complete primary structures of bovine CA I, CA II, and CA III in an attempt to determine those differences in their active site residues which might account for the observed dissimilarities in their activities and inhibition by sulfonamides. In addition, we will compare the partial and complete sequences of the homologous forms of these three isozymes from various mammalian species in an effort to trace their evolutionary origins and rates of evolution.

1 Department of Human Genetics, University of Michigan Medical School, Ann Arbor, Michigan 48109/USA
2 Department of Anatomy, Wayne State University School of Medicine, Detroit, Michigan 48201/USA

The CA I and CA II isozymes are also referred to as CA B and CA C, respectively, and Holmes (1976) has termed the muscle isozyme, CA A. We prefer the designations, CA III, for the muscle isozyme and CA I and CA II for the other isozymes for reasons which have been stated elsewhere (Tashian and Carter, 1976). The numbering of the residues are based on the human CA I sequence (see Table 2).

2 Materials and Methods

Carbonic anhydrase II was isolated from bovine hemolysates by affinity chromatography on a sulfonamide (Prontosil)-bound column as described by Osborne and Tashian (1975), and CA I was purified from bovine rumen by a modification of the procedure of Carter (1971) in which the soluble proteins from the rumen homogenate were passed through a sulfonamide-bound affinity column (Osborne and Tashian, 1975) rather than being separated by ion-exchange chromatography. Because bovine CA III bound onyl weakly to the sulfonamide column, it was purified from the 40% $(NH_4)_2SO_4$ fraction of bovine skeletal muscle homogenate on a DEAE-52 celulose column equilibrated and eluted with 0.01 M Tris, pH 8.7. The fractions containing CO_2 hydratase activity were then pooled and passed through a CM-32 cellulose column equilibrated with 0.01 M sodium phosphate buffer, pH 6.6, and eluted with a 0–0.2 M NaCl gradient as the final purification step. Human and gorilla CA III were initially prepared from skeletal muscle homogenates by either chloroform-ethanol extraction (Tashian et al., 1966) or from a 40% $(NH_4)_2SO_4$ fraction followed by chromatography on a sulfonamide-bound affinity column equilibrated with 0.01 M Tris, pH 8.7, and eluted with a 0.01 M Tris–0.05 Tris/0.4 M KI gradient, pH 8.7, followed by purification on CM-32 as with the bovine enzyme. All stages in the purification of the CA III isozymes contained 0.0005 M dithiothreitol.

The esterase activity toward p-nitrophenyl acetate and CO_2 hydratase activities were assayed essentially by the methods of Armstrong et al. (1966) and Nyman (1963), respectively.

The strategies that were used to sequence the bovine CA I and bovine, human, and gorilla CA III isozymes were similar to those employed for sequencing the sheep and rabbit CA II isozymes (Tanis et al., 1974; Ferrell et al., 1978). Detailed accounts of the bovine CA I and bovine, human and gorilla CA III sequences will be reported elsewhere.

A maximum parsimony computer algorithm (Moore, 1977; Goodman et al., 1979) was used to construct the phylogenetic tree (dendrogram) in Fig. 1 from which the rates of evolution of the different carbonic anhydrase isozymes were determined. The branching points for this tree were taken from the fossil record (McKenna, 1969).

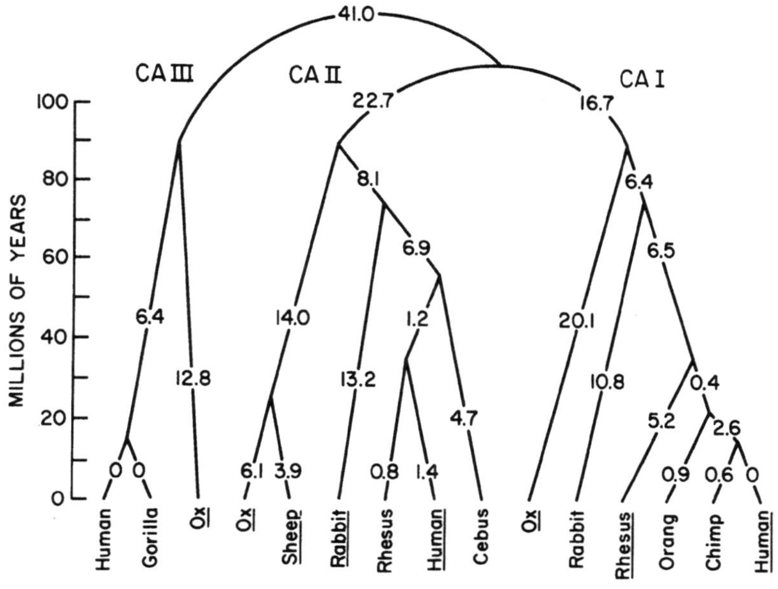

Fig. 1. Maximum parsimony phylogenetic tree (dendrogram) of the mammalian carbonic anhydrase isozymes, CA I, CA II, and CA III. The link lengths are the number of nucleotide substitutions per 100 codons. The sources of the sequences used to construct the dendrogram and the percentage of residues sequenced are as follows: Human CA I (100%), Andersson et al. (1972); Lin and Deutsch (1973); Giraud et al. (1974); chimpanzee CA I (63%), Tashian (1977) and E. Contel (unpublished); orangutan CA I (90%), Tashian (1977) and R.E. Ferrell (unpublished); rhesus CA I (100%), R.J. Tanis and D. Henriksson (unpublished) and Tashian (1977); rabbit CA I (60%), Tashian (1977) and R.E. Ferrell (unpublished); ox CA I (100%), Tashian (1977 and unpublished); cebus CA II (66%), Tashian (unpublished); human CA II (100%), Henderson et al. (1973, 1976); Lin and Deutsch (1974); rhesus CA II (55%), Tashian (1977); rabbit CA II (100%), Ferrell et al. (1978); sheep CA II (100%), Tanis et al. (1974); ox CA II (100%), Sciaky et al. (1976); Ox CA III (100%), Tashian (unpublished); human and gorilla CA III (30%), D. Hewett-Emmett (unpublished)

3 Results and Discussion

3.1 Comparative Activities and Sulfonamide Inhibition

The CO_2 hydratase and esterase activities are listed in Table 1. As can be seen, the esterase activity of the CA III isozyme is barely detectable in the assay system used, and its CO_2 hydratase activity is about one third that of the "low-activity" CA I isozyme. Greatly reduced hydratase and esterase activities have also been reported for the CA III isozymes of rabbit (Koester et al., 1977), human (Carter et al., 1979), and sheep (CO_2 hydratase) (Holmes, 1977).

The very low affinity for sulfonamide of bovine CA III (Table 1) also appears to be characteristic of the CA III isozymes from other mammals as a similar decrease in sulfonamide binding has also been reported for rabbit CA III (Koester et al., 1977) and human CA III (Carter et al., 1979).

Table 1. Comparative activities and sulfonamide inhibition for bovine carbonic anhydrases I, II and III

Isozyme	Source	Specific activity		Acetazolamide inhibition [c]
		Esterase [a]	CO_2 hydrase [b]	($I_{50} \times 10^7$)
CA II	Erythrocytes	62	38	1.1
CA I	Rumen	25	1.2	2.0
CA III	Muscle	<0.03	0.4	2400

a $v/[E_0]$ min^{-1} (p-nitrophyenyl acetate)
b ΔOD_{267} mg^{-1} s^{-1}
c T.H. Maren (personal communication)

3.2 Comparison of Active Site Residues

The primary amino acid sequences of bovine CA I, CA II, and CA III from which the following discussions are based are listed in Table 2.

The studies of Kannan et al. (1977) on the crystal structure of sulfonamides in the active sites of human CA I and CA II show that the sulfonamide group ($-SO_2NH_2$) of acetazolamide binds to the zinc ion at the bottom of the active site possibly through the nitrogen atom, and the ring structure and acetylamido group (CH_3CONH-) interact with residues 91, 121, and 131. More specifically, the sulfur atom of the heterocyclic ring 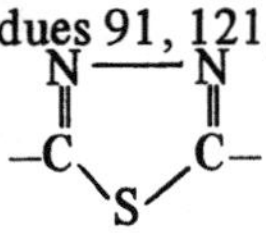 is in contact with residue 121, and the acetylamido group with residues 91 and 131. As seen in Table 3, all three bovine isozymes show Val at position 121; at position 131, the residues are Phe for CA I and CA II and Ile for CA III. Thus, the hydrophobicity of the residues at these two sites is preserved. At position 91, however, the hydrophobicity is retained for CA I (Phe) and CA II (Val), but not for CA III which is Arg at this position. If these residues are at similar locations in the active site of CA III, then the substitution of a basic residue for a non-polar one in the hydrophobic side of the active site cavity could produce hindrances which might account for the low affinity of CA III for acetazolamide. This same substitution could also be responsible for the negligible esterase activity of CA III. Of the eight residues on the hydrophobic side of the active site of human CA I which could be in contact with the ring structures of certain aromatic ester substrates such as p-nitrophenyl acetate (Argos et al., 1978), all are nonpolar in bovine CA III with the exception of Arg-91 (Table 3).

Concerning the low CO_2 hydratase activity of CA III, it would appear that in addition to Arg-91, the residues at positions 64, 67, and 69 on the hydrophilic side of the active site, which are Lys, Arg, and Val, respectively (Table 3), contribute to its low CO_2 hydratase activity. The presence of three basic residues (Lys-64, Arg-67, and Art-91) at the entrance to the active site along with a hydrophobic one (Val-69) in the hydrophilic side of the active site might be expected to alter considerably the charge interactions within the active site.

These speculations should be viewed with caution, however, as changes such as those in residues either adjacent to certain active site residues, or changes outside of the active site could also produce the observed activity differences. For example, the specific esterase activities of rhesus macaque CA I and orangutan CA I toward both p-nitrophenyl acetate and β-naphthyl acetate are considerably higher than the esterase activities of the CA I isozymes from all other mammalian species tested (cf. Tashian, 1977). Nevertheless, all of the active site residues (listed in Table 3) are the same for human and rhesus CA I and also for orangutan CA I with the exception of residues 106, 119, 207, 209, and 211 which have not been sequenced in orangutan (cf. Tashian, 1977).

3.3 Evolutionary Aspects

Comparative Rates of Evolution. Based on the most recent primary sequences of CA I, CA II and CA III from the species listed in Fig. 1, a phylogenetic tree or dendrogram was constructed to follow the evolution of the three isozymes during their descent in mammals. The computer program used to develop this dendrogram generates the ancestral order of branching and ancestral mRNA sequences which require the fewest number of nucleotide changes to account for the descent of the sequences seen in the nine contemporary species studied (Moore, 1977; Goodman et al., 1979).

The link lengths of the dendrogram (nucleotide replacements/100 codons) were used to calculate the rates at which mutations were fixed during the evolution of a particular isozyme from its origin to its present sequence in a particular species. The rates (nucleotide replacements/100 codons/10^8 years) for the different isozymes during different time periods are listed in Table 4. As can be seen, since the major radiation of placental mammals about 90 million years ago, the CA I and CA II isozymes seem to have been evolving at similar rates of about 20 nucleotide substitutions/100 codons/10^8 years. Although the data are limited, the CA III isozymes appear to have been evolving at about half the rates of the other two isozymes during the same period of time. Since the emergence of the higher primates, about 55 million years ago, the rates of the CA II isozymes appear to have been reduced almost fourfold, and during the past 35 million years the rates of the CA II isozymes have been reduced even further (about sevenfold). In contrast, the rates for the CA I isozyme have been reduced over the past 35 million years by about 50%. The picture that emerges is that, providing the fossil record dating is reasonably accurate, there is a general slowing down of the rates of evolution for the CA I and CA II isozymes in the higher primates, with the rates of the CA II isozymes being slower than those of the CA I isozymes. The CA III isozymes appear to be the most conservative of the isozymes, showing a much slower rate of evolution than the other two isozymes. Although only about one third of the human and gorilla CA III isozymes have been sequenced at the present time, nine of the active site residues have been sequenced (Hewett-Emmett, unpublished) and found to be identical with bovine CA III (Table 3).

In Table 5, 16 proteins or protein fragments which have been sequenced in both human and ox are listed along with the percentage of sites substituted since their origin from a common ancestor about 90 million years ago. Although only about one third of the homologous sites could be compared from the three human and

Table 2. Comparison of bovine CA III sequence with bovine and human CA I and CA II sequences

```
        1       10         20         30         40         50         60
HCAI    ASPDWGYDDK NGPEQWSKLY PIANGNNQSP VDIKTSETKH DTSLKPISVS YNPATAKEII
BCAI    ASPDWGYDGE NGPEHWGKLY PIANGNNQSP IDIKTSETKY DPSLKPRSVS YNPATAKEIV

BCAIII   AKEWGYADH NGPDHWHELF PNAKGENQSP IELNTKEISH DPSLKPWTAS YDPGSAKTIL

BCAII    SHHWGYGKH BGPZHWHKDF PIANGERQSP VNIDTKAVVQ DPALKPLALV YGEATSRRMV
HCAII    SHHWGYGKH NGPEHWHKDF PIAKGERQSP VDIDTHTAKY DPSLKPLSVS YDQATSLRIL
              b                        a

                70         80         90        100        110        120
HCAI    NVGHSFHVNF EDNDNRSVLK GGPFSDSYRL FQFHFHWGST NEHGSEHTVD GVKYSAELHV
BCAI    NVGHSFHVNF EDSDNRSVLK GGPLSESYRL FQFHFHWGIT DEDGSEHLVD GAKFSAELHL

BCAIII  NNGKTYRVVF DDTYDRAMLR GGPLAAPYRL RQFHLHWGSS DDHGSEHSVD GVKYAAELHL

BCAII   NNGHSFNVEY DDSQDKAVLK DGPLTGTYRL VQFHFHWGSS BBQGSEHTVD RKKYAAELHL
HCAII   NNGHAFNVEF DDSQDKAVLK GGPLDGTYRL IQFHFHWGSL DGQGSEHTVD KKKYAAELHL
        b  bb b b                b b b b          aa          a b
                                  a

               130        140        150        160        170        180
HCAI    AHWNSAKYSS LAEAASKADG LAVIGVLMKV GEANPKLQKV LDALQAIKTK GKRAPFTNFD
BCAI    VHWNSAKYPS FADAASKADG LALIGLLVKV GQANPNLQKV LDALKAVKNK NKKAPFTNFD

BCAIII  VHWNS-KFNS IATALKHADG IAVVGVFLKI GREKGEFQLL LDALDKIKTK GKEAPFNNFN

BCAII   VHWNT-KYGD FGTAAQQPDG LAVVGVFLKV GDANPALQKV LDALDSIKTK GKSTDFPNFD
HCAII   VHWNT-KYGD FGKAVQQPDG LAVLGIFLKV GSAKPGLQKV VDVLDSIKTK GKSADFTNFD
        b          b          b b b
```

```
                                190                 200                 210                 220                 230                 240
HCAI     P S T L L P S S L D F W T Y P G S L T H P P L Y E S V T W I I C K E S I S V S S E Q L A Q F R S L L S N V E G D N A V P
BCAI     P S T L L P S S L D F W T Y S G S L T H P P L L E S V T W I I F K E T I S A S S E Q L A Q F R C L L A N A E G D G E L L

BCAIII   P S C L L P A S R D Y W T Y H G S F T T P P D C E C I V W L L L K E P I T V S S D Q I A K L R T L Y S S A E N E P P V P

BCAII    P G S L L P Q V L D Y W T Y P G S L T T P P L L E S V T W I V L K E P I S V S S Q Q M L K F R T L N F N A E G E P E L L
HCAII    P R G L L P E S L D Y W T Y P G S L T T P P L L E C V T W I V L K E P I S V S S E Q V L K F R K L N F N G E G E P E E L
                                  a   a       b b b b           b   b   b
                                                a
```

```
                                250                 260
HCAI     M Q H N N R P T Q P L K G R T V R A S F
BCAI     I K Q N H R P P Q P L K G R T V K A S F

BCAIII   L V R N W R P P Q P I K G R I V K A S F K

BCAII    M L A N W R P A Q P L K N R Q V R G F P K
HCAII    M V D N W R P A Q P L K N R Q I K A S F K
```

a Residues hydrogen-bounded to His 94 and His 119, or to Zn-bound solvent molecule
b Residues in active site cavity
References: Human CA I and II (Andersson et al., 1972; Henderson et al., 1973, 1976; Lin and Deutsch, 1973, 1974; Giraud et al., 1974); ox CA II (Sciaky et al., 1976); ox CA I and III (Tashian, unpublished)
One letter code is that of Dayhoff (1972)

Table 3. Amino acid residues postulated to occur in the active sites of bovine carbonic anhydrases I, II, and III

Position Number	Polar residues CA I	CA II	CA III	Position Number	Nonpolar residues CA I	CA II	CA III
7	Tyr	Tyr	Tyr	91 [a, b]	Phe	Val (Ile)	*Arg* [c]
61	Asn	Asn	Asn [c]	121 [a, b]	Val (Ala)	Val	Val
64	His	His	*Lys* [c]	131 [a, b]	Phe(Leu)	Phe	Ile
65	Ser	Ser (Ala)	Thr [c]	141 [a, b]	Leu	Leu	Ile
67	His	Asn	*Arg* [c]	143 [b]	Leu (Val)	Val	Val
69	Asn	Glu	*Val* [c]	145	Gly	Gly	Gly
92 [a]	Gln	Gln	Gln	198 [a]	Leu	Leu	Phe
94 [a]	His	His	His [c]	201	Pro	Pro	Pro
96	His	His	His [c]	202	Pro	Pro	Pro
106	Glu	Glu	Glu [c]	207 [b]	Val	Val	Ile
119	His	His	His	209 [a, b]	Trp	Trp	Trp
199 [a]	Thr	Thr	Thr	211 [b]	Ile	Val	Leu
200 [a]	His	Thr	Thr				

a Within 5 Å of acetazolamide in human CA I and CA II (cf. Kannan et al., 1977)
b Within contact distance of aromatic structure of p-nitrophenyl acetate (cf. Argos et al., 1978)
c Only these residues have been sequenced in human and/or gorilla CA III, and all are identical to bovine CA III
Residues that are different in human CA I and II are in parentheses

Table 4. Comparative rates of evolution (nucleotide replacements /100 codons/ 10^8 years) for mammalian carbonic anhydrases I, II, and III

	CA I	CA II	CA III
	Eutherian ancestor (90 mya [a] to present)		
Human	17.7	19.6	7.1
Chimpanzee	18.4	–	–
Gorilla	–	–	7.1
Rhesus macaque	20.1	18.9	–
Cebus monkey	–	21.9	–
Rabbit	19.1	23.7	–
Ox	22.3	22.3	14.2
Sheep	–	19.9	–
	Platyrrhine/catarrhine ancestor (55 mya to present)		
Human	–	4.7	–
Rhesus macaque	–	3.6	–
Cebus monkey	–	8.5	–
	Catarrhine ancestor (35 mya to present)		
Human	8.6	4.0	–
Chimpanzee	10.3	–	–
Rhesus macaque	14.9	2.3	–
	Bovide ancestor (25 mya to present)		
Ox	–	24.4	–
Sheep	–	15.6	–
	African ape/human ancestor (15 mya to present)		
Human	0	–	0
Chimpanzee	4.0	–	–
Gorilla	–	–	0

a Million years ago

Table 5. Comparisons of the percentage of substituted sites in the primary sequences of human and bovine carbonic anhydrases with similar comparisons for other human and bovine proteins

Protein of protein fragment	Number of homologous sites compared	Percentage of sites substituted
Fibrinopeptides A + B	30	47
Growth hormone	186	39
Somatotropin	189	35
Insulin, C-peptide	30	30
Carbonic anhydrase I	84 [a]	24
Carbonic anhydrase II	84 [a]	24
Lactalbumin	123	23
Serum albumin	583	22
Hemoglobin β-chain	146	17
Myoglobin	152	17
Hemoglobin α-chain	141	12
Carbonic anhydrase III	84 [a]	12
Myelin basic protein	167	10
Cytochrome c	104	10
Insulin A + B chains	51	6
Corticotropin	39	3

a Eighty-four of the 261 possible homologous positions were compared; when the 261 positions of bovine and human CA I and II were compared, 21% of the sites were found to be substituted for both isozymes.
The sequences of the bovine and human proteins used for these conparisons were taken from Dayhoff (1972, 1976)

bovine carbonic anhydrases, an estimate of the comparative degrees of conservation could be determined for these isozymes. As shown, bovine CA III is about twice as conservative as CA I and II and is similar to that of the α-chain of hemoglobin, whereas the CA I and II isozymes are about as conservative as lactalbumin and serum albumin.

Functional Considerations. The CA III isozyme appears to be found primarily in red skeletal muscle of mammals and birds, and seems to be absent from smooth and cardiac muscle (Holmes, 1977; Moynihan, 1977; Carter et al., 1979). CA III also appears to be present in the red fibers, and not the white fibers, of skeletal muscle (Moynihan, 1977; Holmes, 1977).

While Van Goor (1940) claims to have detected CO_2 hydratase activity in muscle, it became generally accepted that carbonic anhydrase was absent in muscle. In reviewing carbon dioxide storage in muscle, Cherniack et al. (1974) stated that "calulations made from these models (which assumed carbonic anhydrase was absent in muscle) suggest that neither uncatalyzed rates of CO_2 hydration nor restrictions to CO_2 diffusion which seem possible physiologically, alone or together, account for the low CO_2 storage capacity of muscle in the first few minutes after a CO_2 change ...... it seems more likely that *some other factor* not yet considered in mathematical models *reduces immediate CO_2 storage capacity in muscle* (our italics)".

The findings that the primary structures of the CA III isozymes of skeletal muscle appear to be considerably conserved in mammals, and because of the relatively high levels of this enzyme in skeletal muscle (about 1 mg/g wet weight of muscle), it appears that CA III should play an important physiological role in muscle cells. An intriguing possibility is that CA III may not function simply as a CO_2 hydratase in muscle, but rather has some as yet undefined catalytic activity. The recent report by Koester (1979) that rabbit CA III has weak acid phosphatase activity suggests that CA III might function as an enzyme possessing catalytic activity toward phosphate groups (e.g., phosphatase, phosphorylase, phosphoryl transferase). The three basic residues, Lys-64, Arg-67 and Arg-91, at the entrance to the proposed active site of CA III lend some credence to this possibility, since the active site cavity of the phosphoryl transferase, adenylate kinase, includes one Lys, one His and six Arg residues which bind or interact with the four phosphate groups of ATP and AMP (Schulz et al., 1974). These authors suggest that the functions of these basic residues may be to neutralize the negatively charged phosphate groups. Also, arginyl residues at the active sites of several glycolytic enzymes have been shown to be involved in the binding of negatively charged, phosphorylated substrates and/or coenzymes (Riordan et al., 1977).

Origins of the Isozymes. We believe that the CA III isozyme is ancestral to the CA I and CA II isozymes for several reasons. First, the CA I and CA II isozymes are more similar to each other (i.e., they are separated by fewer nucleotide changes) than either is to CA III (Table 2, Fig. 1). Second, the CA I and CA II isozymes appear to be closely linked in mammals (Carter, 1972; DeSimone et al., 1973; Eicher et al., 1976) suggesting that CA I and CA II may have shared a common ancestor after the divergence of the lines leading to birds and mammals about 300 million years ago. Third, it appears that the CA III isozymes of chicken and mammalian muscle are homologous (Holmes, 1977) suggesting that the CA III isozymes of birds and mammals shared a common ancestor before the divergence of birds and mammals. The possibility that a low-activity carbonic anhydrase isolated from chicken intestine (Holmes, 1977) may be homologous to the low-activity CA I isozyme of mammals complicates the picture; because if it is true, it would mean that the CA I and II isozymes of mammals were formed by gene duplication before the divergence of avian and mammalian lines. Clearly, additional sequence data are needed from the carbonic anhydrase isozymes of nonmammalian species (especially from the active site regions) before further insights can be gained into the evolutionary origins of these isozymes.

Acknowledgment. This work was supported by U.S. Public Health Service Grant GM-24681.

References

Andersson B, Nyman PO, Strid L (1972) Amino acid sequence of human erythrocyte CA B. Biochem Biophys Res Commun 48: 670–677

Argos P, Garavito RM, Eventoff W, Rossmann MG, Branden CI (1978) Similarities in active center geometries of zinc-containing enzymes, proteases and dehydrogenases. J Mol Biol 126: 141–158

Armstrong J McD, Myers DV, Verpoorte JA, Edsall JT (1966) Purification and properties of human erythrocyte carbonic anhydrases. J Biol Chem 241: 5137–5149
Carter MJ (1971) The carbonic anhydrase of the rumen epithelial tissue of the ox. Biochim Biophys Acta 235: 222–236
Carter ND (1972) Carbonic anhydrase isozymes in *Cavia porcellus, Carvia aperea* and their hybrids. Comp Biochem Physiol 43 B: 743–747
Carter ND, Jeffery S, Shiels A, Edwards Y, Tipler T, Hopkinson DA (1979) Characterization of human carbonic anhydrase III from skeletal muscle. Biochem Genet 17: 837–854
Cherniack NS, Longobardo GS, Fishman AP (1974) The behavior of carbon dioxide stores of the body during unsteady states. In: Nahas G, Schaefer KE (eds) Topics in environmental physiology and medicine: Carbon dioxide and metabolic regulation. Springer, Berlin Heidelberg New York, pp 324–338
Dayhoff MO (1972) Atlas of protein sequence and structure, vol V. National Biomedical Research Foundation, Washington DC
Dayhoff MO (1976) Atlas of protein sequence and structure, vol V, Suppl 2. National Biomedical Research Foundation, Washington DC
De Simone J, Linde M, Tashian RE (1973) Evidence for linkage of carbonic anhydrase isozyme genes in the pig-tailed macaque, *Macaca nemestrina*. Nature (New Biol) 242: 55–56
Eicher EM, Stern RH, Womack JE, Davisson MT, Roderick TH, Reynolds SC (1976) Evolution of mammalian carbonic anhydrase loci by tandem duplication: Close linkage of *Car-1* and *Car-2* to the centromere region of chromosome 3 of the mouse. Biochem Genet 14: 651–660
Ferrell RE, Stroup SK, Tanis RJ, Tashian RE (1978) Amino acid sequence of rabbit carbonic anhydrase II. Biochim Biophys Acta 553: 1–11
Giraud N, Marriq C, Laurent-Tabusse G (1974) Structure primaire de l'anhydrase carbonique érythrocytaire B humaine. III. Séquence des fragments ICNBr et III CNBr (résidus 149–260) Biochimie 56: 1031–1043
Goodman M, Czelusniak J, Moore GW, Romero-Herrera AE, Matsuda G (1979) Fitting the gene lineage into its species lineage, a parsinomy strategy illustrated by cladograms constructed from globin sequences. Syst Zool 28: 132–163
Henderson LE, Henriksson D, Nyman PO (1973) Amino acid sequence of human erythrocyte carbonic anhydrase C. Biochem Biophys Res Commun 52: 1388–1393
Henderson LE, Henriksson D, Nyman PO (1976) Primary structure of human carbonic anhydrase C. J Biol Chem 251: 5457–5463
Holmes RS (1976) Mammalian carbonic anhydrase isozymes: evidence for a third locus. J Exp Zool 197: 289–295
Holmes RS (1977) Purification, molecular properties and ontogeny of carbonic anhydrase isozymes. Evidence for A, B and C isozymes in avian and mammalian tissues. Eur J Biochem 78: 511–520
Kannan KK, Notstrand B, Fridburg K, Lövgren S, Ohlsson A, Petef M (1975) Crystal structure of human erythrocyte carbonic anhydrase B. Three-dimensional structure at a nominal 2.2 Å resolution. Proc Natl Acad Sci USA 72: 51–55
Kannan KK, Vaara I, Notstrand B, Lovgren S, Borell A, Friedborg K, Petef M (1977) Structure and function of carbonic anhydrase: comparative studies of sulphonamide binding to human erythrocyte carbonic anhydrases B and C. In: Roberts GCK (ed) Drug action at the molecular level. University Park Press, Baltimore, pp 73–91
Koester MK (1979) Is carbonic anhydrase III also an acid phosphatase? Fed Proc 38: 727
Koester MK, Noltmann EA (1979) Unique conformational properties of muscle carbonic anhydrase III as demonstrated by circular dichroism spectrometry. Biochemistry 18: 343–348
Koester MK, Register AM, Noltmann EA (1977) Basic muscle protein, a third genetic locus isoenzyme of carbonic anhydrase? Biochem Biophys Res Commun 76: 196–204
Liljas A, Kannan KK, Bergstén P.-C, Waara I, Fridborg K, Strandberg B, Carlbom U, Järup L, Lövgren S, Petef M (1972) Crystal structure of human carbonic anhydrase C. Nature (London) 235: 131–137
Lin K-TD, Deutsch HF (1973) Human carbonic anhydrases. XI. The complete primary structure of carbonic anhydrase B. J Biol Chem 248: 1885–1893

Lin K-TD, Deutsch HF (1974) Human carbonic anhydrases. XII. The complete primary structure of the C isozyme. J Biol Chem 249: 2339–2337

Lindskog S, Henderson LE, Kannan KK, Liljas A, Nyman PO, Strandberg B (1971) Carbonic anhydrase. In: Boyer PD (ed) The enzymes, vol V. Academic Press, London New York, pp 587–665

Maren TH, Rayburn CS, Liddell ME (1976) Inhibition by anions of human red cell carbonic anhydrase B: physiological and biological implications. Science 191: 469–472

McKenna MC (1969) The origin and early differentiation of therian mammals. Ann NY Acad Sci 167: 217–240

Moore GW (1977) Proof of the populous path algorithm for missing mutations in parsimony trees. J Theoret Biol 66: 95–106

Moynihan JB (1977) Carbonic anhydrase activity in mammalian skeletal muscle and cardiac muscle. Biochem J 168: 567–569

Notstrand B, Vaara I, Kannan KK (1974) Structural relationship of human erythrocyte carbonic anhydrase isozymes B and C. In: Markert CL (ed) Isozymes, vol I. Academic Press, London New York, pp 575–599

Nyman PO (1963) A spectrophotometric method for the assay of carbonic anhydrase activity. Acta Chem Scand 17: 429–435

Osborne WRA, Tashian RE (1975) An improved method for the purification of carbonic anhydrase isozymes by affinity chromatogrphy. Anal Biochem 64; 297–303

Riordan JF, McElvany KD, Borders CL, Jr (1977) Arginyl residues: anion recognition sites in enzymes. Science 195: 884–886

Schulz GE, Elzinga M, Marx F, Schirmer RH (1974) Three-dimensional structure of adenyl kinase. Nature (London) 250: 120–123

Sciaky M, Limozin N, Filippi-Foveau D, Gullian J-M, Laurent-Tabusse G (1976) Structure primaire de l'anhydrase carbonique érythrocytaire bovine CI. Séquence complete. Biochimie 58: 1071–1082

Tanis RJ, Ferrell RE, Tashian RE (1974) Amino acid sequence of sheep carbonic anhydrase C. Biochim Biophys Acta 371: 534–548

Tashian RE (1977) Evolution and regulation of the carbonic anhydrase isozymes. In: Rattazzi MC, Scandalios JG, Whitt GS (eds) Isozymes: Current topics in biological and medical research, vol II. Alan R Liss Inc, New York, pp 21–62

Tashian RE (1978) Evolution of the carbonic anhydrases. In: Matsubara H, Yamanaka T (eds) Evolution of protein molecules. Japan Scientific Societies Press, Tokyo, pp 343–360

Tashian RE, Carter ND (1976) Biochemical genetics of carbonic anhydrase. In: Harris H, Hirschhorn K (eds) Advances in human genetics, vol VII. Plenum Publishing Corp, New York, pp 1–56

Tashian RE, Riggs SK, Yu Y-SL (1966) Characterization of a mutant human erythrocyte carbonic anhydrase: carbonic anhydrase Ic Guam. The amino acid substitution and carboxyl esterase and hydratase activities. Arch Biochem Biophys 117: 320–327

Tashian RE, Stroup SK, Yu Y-SL, Henriksson D (1978) Primary structure of an isozyme of carbonic anhydrase isolated from bovine skeletal muscle. Fed Proc 37: 1797

Van Goor H (1940) Die Verbreitung und Bedeutung der Carbonanhydrase. Enzymologia 8: 113–128

Carbonic Anhydrase Activity in Intact Red Cells as Measured by Means of ^{18}O Exchange Between CO_2 and Water[1]

R.E. FORSTER[2] and N. ITADA[3]

The enzymatic activity of carbonic anhydrase inside the red cell is of practical importance, because it is this activity which is of physiological interest and not the enzymatic activity of a dilute lysate, which is generally determined experimentally. We define this activity (A), in terms of the reaction

$$CO_2 + H_2O \rightleftharpoons H_2CO_3$$

as the specific rate of hydration or dehydration inside the cell divided by the same specific rate in the extracellular fluid or plasma.

It is difficult to measure A in intact red cells, because in the first place the rate of the chemical reaction is very fast and the process has to be measured with a rapid-reaction apparatus able to resolve changes in the ms range. In addition, transport of CO_2 into the cells and of H^+ and HCO_3^- out of the cell, will be impeded by the plasma membrane to some extent and gradients may even develop inside the cell itself. This will slow the enzyme catalyzed reactions and the observer will not be sure of the specific rate of the enzymatic process unless he knows the intracellular concentrations of the reactants and products.

Using the original "boat" technique of Roughton (1944) in which a manometer registers the change in pressure as CO_2 is absorbed or evolved in reactions taking place in an agitated closed vessel (boat), Booth (1938) found that a suspension of intact red cells produced less acceleration of CO_2 hydration than an equivalent concentration of lysate. Booth surmised, as we would today, that intracorpuscular CO_2 hydration proceeded at its catalyzed rate until end products H^+ and HCO_3^- accumulated in the cell, at which time the overall process became rate-limited by the uncatalyzed reaction in the suspending fluid and the exchange of ions across the cell membrane. He summarized these limitations on the effectiveness of cell carbonic anhydrase as an "availability factor".

In 1965, Constantine et al. measured the rate of CO_2 uptake by a human red cell suspension with a rapid mixing technique that could resolve changes in P_{CO_2} over several milliseconds. This was an improvement over the "boat" method and by using the initial slope of the P_{CO_2} disappearance curve they should have obtained a measure of the rate of CO_2 hydration before the intracellular concentration of end

1 Supported by a Grant HLB 19737 from the National Institutes of Health
2 Department of Physiology G4 School of Medicine, University of Pennsylvania, Philadelphia, Pennsylvania, 19104/USA
3 Present address: Kokuritsu Kyoto Hospital, Fushimi-ku, Kyoto, 612/Japan

products could have risen. They found that the intracellular CO_2 hydration velocity constant was only 60% of that expected from extrapolations of measurements of dilute hemolysate, raising a question that the uptake was rate-limited in part by something other than the carbonic anhydrase activity, or that the chemical activity of the enzyme at its concentration inside the cell was different from that in dilute solution, in spite of the support of Kernohan et al. (1963) for this assumption.

Kernohan and Roughton (1966) used a continuous-flow mixing apparatus as a rapid calorimeter measuring temperature changes produced by the reactions in the flowing stream with thermocouples. They found that the carbonic anhydrase inside the bovine red cell at 25°C had an activity of 25,000-fold and that there appeared to be no diffusion limitations on the enzyme function. There is no clear reason why the "availability factor" would be obviated by calorimetry, but these observations have not been followed up.

Itada and Forster (1977) developed a method for the determination of A by following the time course of disappearance of $C^{18}O^{16}O$ as the ^{18}O exchanges with the 55 M pool of unlabeled water as follows,

$$
\begin{array}{llll}
C^{18}O^{16}O + H_2{}^{16}O & \underset{2/3}{\overset{k_u}{\rightleftharpoons}} H_2C^{18}O^{16}O_2 & \rightleftharpoons H^+ + HC^{18}O^{16}O_2^- & \\
 & \swarrow k_v \quad 1/3 & & \\
C^{16}O^{16}O + H_2{}^{18}O & & \text{practically instantaneous} & (1)
\end{array}
$$

When $H_2C^{18}O^{16}O_2$ is dehydrated there is a 2/3 probability that labeled CO_2 and unlabeled water will be formed and a 1/3 probability that labeled water and unlabeled CO_2 will be formed. The concentration of HOH is so great that it effectively represents an infinite sink for ^{18}O. We follow the disappearance of $C^{18}O^{16}O$ and not the hydration-dehydration of CO_2-H_2CO_3 directly. Because the exchange rate is 1/3 as fast as the elemental reaction velocity, and because the labeled ^{18}O in CO_2 and HCO_3^- act as one pool, the half-time for $C^{18}O^{16}O$ disappearance is 60 times that of the elemental hydration of CO_2 or the dehydration of H_2CO_3. For example the half-time for the uncatalyzed elemental reaction of CO_2 with a buffer system at pH 7.4 and 37°C is 3.7 s, while the half-time for the disappearance of $C^{18}O^{16}O$ is over 4 min. This means the isotope exchange can be followed with a more slowly responding analytical instrument as compared to the elemental reaction, which requires a rapid mixing apparatus.

We carry out the isotopic exchange in an anaerobic stirred thermostated glass vessel of 10 ml whose floor is separated from the ion source of a mass spectrometer by only a thin Teflon membrane supported mechanically on a sintered glass disc. CO_2 diffuses through the Teflon into the mass spectrometer which determines the partial pressure of $C^{18}O^{16}O$ (mass 46) and $C^{16}O_2$ (mass 44). The half-time for the response of the mass spectrometer to a step change in the partial pressure of either isotope of CO_2 in the reaction mixture is 3 s, practically all of which delay represents the mixing time of the fluid in the reaction chamber.

During the disappearance of $C^{18}O^{16}O$ there are no changes in CO_2, H^+, or HCO_3^- concentrations inside the red blood cell as elemental chemical equilibrium occurs within the first few seconds of the reaction. The end product, $H_2{}^{18}O$, does not accumulate to any significant extent in the cell because firstly it is diluted in 55 M of unlabeled water and secondly because the labeled water can easily diffuse across the cell membrane and dilute in the even larger extracellular water pool. Therefore we believe that this method should provide an estimate of intracellular carbonic anhydrase activity relatively uninfluenced by the major limitations of the earlier methods referred to above.

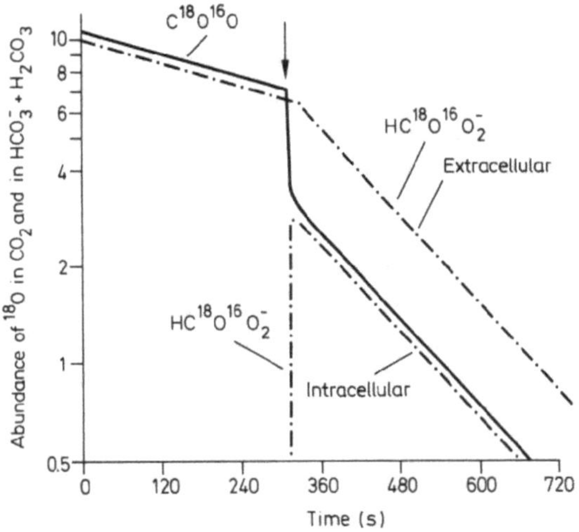

Fig. 1. Graph of $\log_{10}$ of abundance of ^{18}O in CO_2 (*solid line*) or in (HCO_3^- + H_2CO_3) (*dash-dot line*) plotted against time. 20 micro L cell suspension added at *arrow* to give v = 0.0006. The curves for ^{18}O abundance in CO_2 measured on the cells of HBJ at pH 7.4 and 37°C. Curves for ^{18}O abundance in HCO_3^- intracellular and extracellular calculated from the theoretical model. ^{18}O abundance in CO_2 = $[C^{18}O^{16}O](2[C^{18}O^{16}O] + 2[C^{16}O_2])^{-1}$, ^{18}O abundance in HCO_3^- = $[HC^{18}O^{16}O_2^-](3[HC^{18}O^{16}O_2^-] + 3[HC^{16}O_3^-])^{-1}$. Natural abundance was subtracted

The time course of $C^{18}O^{16}O$ during an experiment with red cells is shown in Fig. 1. When a cell suspension is added to the reacting mixture of labeled bicarbonate, the $C^{18}O^{16}O$ disappearance rate increases in a double exponential form

$$C^{18}O^{16}O = C_m (n-m)^{-1} e^{-mt} + C_n (n-m)^{-1} e^{-nt}, \qquad (2)$$

We have proposed a model for the time course of disappearance of $C^{18}O^{16}O$ in the presence of red cells based on the following assumptions: the system may be considered as two phases, intracellular and extracellular; chemical reactions of CO_2 proceed as dexcribed in Eq. (1) in both phases, except that the effective reaction velocity constant inside the cell is increased by the carbonic anhydrase and equals (Ak_u); CO_2 diffuses so rapidly across the red cell membrane that no significant P_{CO_2} gradient exists; HCO_3^- on the other hand has a lower and measurable permeability. The assumption that no P_{CO_2} gradient occurs across the red cell membrane has been disputed (Silverman et al., 1976). However, since we calculate later a value for intracellular carbonic anhydrase activity which is not greatly different from the value expected from measurements on dilute lysate, and since our calculations are based on the assumption that the gradient for CO_2 concentration across the red cell membrane is negligible, it appears unlikely that any transmembrane CO_2 concentration gradient is important.

In this model C_m, C_n, m, and n in Eq. (2) become algebraic expressions which are functions of: K_{HA}, the dissociation equilibrium constant for H_2CO_3 in M; K' which equals $[H^+] \times [HCO_3^-] / [CO_2]$; $[H^+]$; surface area of the red cell over the volume of the red cell; fractional red cell water volume (v); k_u and k_v the hydration and dehydration velocity constants respectively in s^{-1}; A the enzymatic activity of the carbonic anhydrase; and P, the permeability of the cell membrane to HCO_3^- in cm/s.

Mathematically there are only two unknowns, A and P, and if we obtain m and n experimentally we should be able to solve for the unknowns. Experimentally we cannot measure m because the half-time of the first rapid phase in $C^{18}O^{16}O$ disappearance is faster than the mixing time in the reaction vessel. Therefore we measure instead the relative magnitude of the first rapid phase which equals $C_n/(n - m)$, and which we call the reciprocal of the "step ratio". Unfortunately, even though we possess the two experimental knowns, we cannot extract an explicit mathematical statement for A and P from the theoretical equations. Originally we used a graphical approximation to solve for single values of A and P, combining the data from three or more experiments at different red cell concentrations, but it turned out that this technique underestimated A (Itada et al., 1978). Dr. Martin Pring has produced a computer program which solves for A and P given single values of the step ratio and the exponential constant for the second lowe phase, n. It is this technique which we use at present.

The uncatalyzed disappearance of $C^{18}O^{16}O$, that is prior to the addition of cells in Fig. 1, is linear on the logarithmic plot and is described by the following equation, first derived by Mills and Urey (1941):

$$k_u = \frac{-2 \times 0.693}{t_{1/2}} \times \left\{ 1 + \frac{[H^+]}{K'} \pm \left((1 + \frac{[H^+]}{K})^2 - \frac{4}{3} \frac{[H^+]}{K'} \right)^{1/2} \right\}^{-1} \quad (3)$$

In the presence of lysate or carbonic anhydrase in homogenous solution, the disappearance of $C^{18}O^{16}O$ is also linear and the process is described by Eq. (3) except that k_u should be replaced by $A \cdot k_u$ where A for a homogenous solution is defined as the specific rate of CO_2 hydration or H_2CO_3 dehydration divided by k_u or k_v respectively.

In order to determine the reproducibility of the technique, 17 experimental curves were obtained at relative red cell water volumes, indicated by v, of 0.00035, 0.00018 and 0.00175, pH of 7.4 and temperature of 37°C on the blood of HBJ. The average values of A and the standard error of the mean at each v are presented in Fig. 2. According to the theoretical model, A should not vary with v, since v is simply

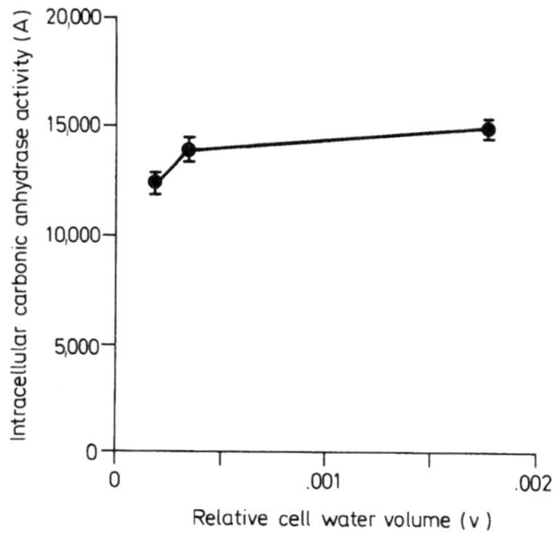

Fig. 2. A graph of the intracellular carbonic anhydrase activity, A, against the relative red cell water volume in the reacting mixture. The *vertical bars* indicate ± standard error of the mean. *Each point* is the average of 17 measurements on the blood of normal man (HBJ). Extracellular pH is 7.4, temperature is 37°C. $NaHCO_3$ concentration is 0.0244 M. NaCl concentration is 0.145 M

a measure of the number of red cells present. At v much less than 0.0005, the step ratio is so close to unity that the calculated values of A and P are sensitive to small variations in the experimental data. Therefore, we have tended to give less weight to the values at low v. The value of A for lysate from the same concentration of cells as in Fig. 2 at pH 7.4 and 37°C, was 20,340 as extrapolated to intracellular concentrations.

There are at least three additional significant factors which should be taken into account in estimating intracellular carbonic anhydrase activity from measurements on lysate; intracellular pH and chloride concentration, and a decrease in CO_2 concentration within the cell owing to the effects of simultaneous chemical reaction and diffusion.

Cl^- inhibits carbonic anhydrase and intracellular $[Cl^-]$ is 0.096 M while that in the extracellular fluid is 0.145 M, which would lead to a higher enzyme activity inside the cell. We measured A of dilute lysate in KCl and NaCl solutions up to 1 M and found that 0.096 M KCl reduced A only 8% while 0.145 M NaCl reduced A 44%. Therefore, A measured on lysate in 0.145 M NaCl should be increased by 1.64 to approximate intracellular conditions.

Decreasing pH lowers A, and the interior of the normal human red cell is 0.18 pH units less than the pH of 7.4 in the suspending fluid (Funder and Wieth, 1966), which should reduce A by the factor 0.82 according to the data in Fig. 3.

As CO_2 diffuses within the cell it is simultaneously consumed by chemical reactions so that P_{CO_2} gradients occur. Roughton developed theoretical expressions which can be used to estimate the average P_{CO_2} inside the cell in relation to that outside under conditions where there are no significant end products (Forster, 1964; Constantine et al., 1965).

The relation is

$$\frac{\text{Average intracellular } P_{CO_2}}{\text{Average extracellular } P_{CO_2}} = \frac{\tanh\,[b\,(Ak_u/D)^{1/2}]}{b\,(Ak_u/D)^{1/2}} \tag{4}$$

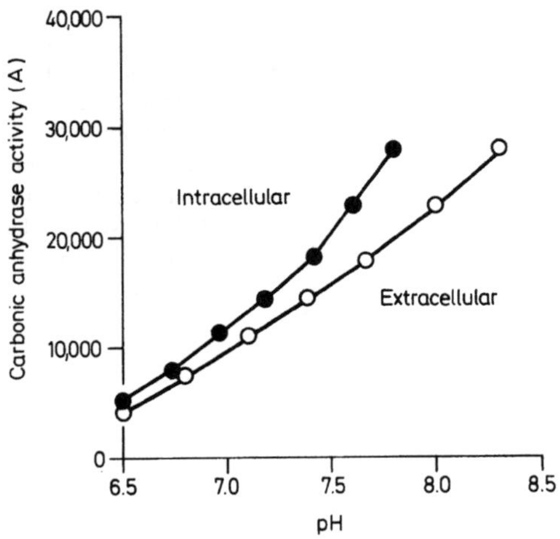

Fig. 3. A graph of intracellular carbonic anhydrase activity, A, for the blood of HBJ against extracellular pH (*open circles*) and calculated intracellular pH (*closed circles*). Temperature is 37°C. The data points at each pH are the average of 3–4 experimental runs at fractional red cell water volume equal to 0.00035–0.0015

b is the half-thickness of the red cell, which we will take to be 7×10^{-5} cm. A is 16,700 as discussed above, k_u is $0.18\ s^{-1}$, D is the diffusion coefficient for CO_2 in the interior of the red cell taken to be $6.6 \times 10^{-6}\ cm^2/s$ (Forster, 1969). If we substitute these values into Eq. (4) we obtain 0.60. This means that when CO_2 enters the cell, the average P_{CO_2} within the cell is only 60% of that outside, which reduces the chemical reaction rate in the cell in proportion. An analogous situation obtains when CO_2 is leaving the cell. This phenomenon is unrelated to diffusion through the membrane and occurs entirely within the cell substance. Therefore we would expect the average catalytic activity of carbonic anhydrase within the cell to appear to be only 60% of what we would find in a perfectly mixed homogenous solution of cell contents under the exact same chemical conditions.

When we now combine the effects of Cl^- inhibition, pH and simultaneous reaction + diffusion, we find that the predicted value of A inside the red cell becomes

$$1.64 \times 0.84 \times 0.60 \times 20{,}340 = 16{,}411$$

The average value of A at higher v in Fig. 2 is 14,200, 85% of this prediction. In view of possible errors in these calculations, we conclude that there is no real difference between the enzymatic activity of lysate in dilute solution outside the cell as compared with the highly concentrated conditions within the intact cell.

We have investigated the influence of pH on A by determining the time course of $C^{18}O^{16}O$ disappearance in normal fresh human red cells at 37°C and extracellular pH from 6.5 to 8.3. We computed A using Dr. Pring's program and the results are shown in Fig. 3. Intracellular pH can be calculated from extracellular pH according to the relationship of Funder and Wieth (1966) and a graph of A versus intracellular pH constructed. This has been done in the figure and the slope of the line is similar to data of Kernohan (1964) and of Roughton (1944). We conclude that the effect of acidity on the carbonic anhydrase enzymes is nearly the same within the cell as is in more dilute solutions.

In summary we have used the exchange of ^{18}O in CO_2 with the ^{16}O in water in red cell suspensions to estimate the enzymatic activity of carbonic anhydrase [A] and found that its value is close to that expected from measurements on dilute lysate and that A inside the cell increases with increasing pH to the same extent as reported by other authors in solution. We believe that this method is unique in permitting the study of the kinetics of an enzyme with a high turnover number in the intact cell.

Note added in proof. The authors have recently determined that the carbonic anhydrase inhibitor ethoxyzolamide reduces enzyme activity (A) in intact human red cells and their hemolysate to the same extent as measured by the ^{18}O technique. Comparable experiments with acetazolamide are difficult to carry out because the drug permeates the red cell membrane so slowly. This may explain the results of Silverman et al. (1976).

References

Booth VH (1938) Carbonic anhydrase activity inside corpuscles. Enzyme-substrate acessibility factors. J Physiol (London) 93: 117–128

Constantine HP, Craw MR, Forster RE (1965) Rate of the reaction of carbon dioxide with human red blood cells. Am J Physiol 208: 801–811

Forster RE (1964) Rate of gas uptake by red cells. In: Fenn WO, Rahn H (eds) Handbook of physiology. Respiration, vol I. American Physiological Society, Washington, pp 827–837

Forster RE (1969) The rate of CO_2 equilibration between red cell and plasma. In: Forster RE, Edsall JT, Otis AB, Roughton FJW (eds) CO_2: Chemical, biochemical, and physiological aspects. National Aeronautics and Space Administration, Washington, pp 275–284

Funder J, Wieth JO (1966) Chloride and hydrogen ion distribution between human red cells and plasma. Acta Physiol Scand 68: 234–245

Itada N, Forster RE (1977) Carbonic anhydrase activity in intact red blood cells measured with ^{18}O exchange. J Biol Chem 252: 3881–3890

Itada N, Peiffer L, Forster RE (1978) Intracellular enzyme activity. In: Frontiers of biological energetics, vol I. Academic Press, London New York, pp 715–724

Kernohan JC (1964) The activity of bovine carbonic anhydrase in imidazole buffers. Biochim Biophys Acta 81: 346–356

Kernohan JC, Roughton FJW (1966) Thermal measurements of the enzymatic activity of carbonic anhydrase in concentrated haemoglobin solutions and in red blood cells. J Physiol (London) 186: 138–139P

Kernohan JC, Forrest WW, Roughton FJW (1963) The activity of concentrated solutions of carbonic anhydrase. Biochim Biophys Acta 67: 31–41

Maren TH, Rayburn CS, Liddell NE (1976) Inhibition by anions of human red cell carbonic anhydrase B: Physiological and biochemical implications. Science 191: 469–472

Mills GA, Urey HC (1940) The kinetics of isotopic exchange between carbon dioxide, bicarbonate ion, carbonate ion and water. J Am Chem Soc 62: 1019–1026

Roughton FJW (1943–1944) Some recent work on the chemistry of carbon dioxide transport by the blood. Harvey Lect 39: 96–142

Silverman DN, Tu C, Wynns GC (1976) Depletion of ^{18}O from $C^{18}O_2$ in erythrocyte suspensions. The permeability of the erythrocyte membrane to CO_2. J Biol Chem 251: 4428–4435

Crystal Structure of Carbonic Anhydrase

K.K. KANNAN [1]

1 Introduction

Carbonic anhydrase (carbonate hydrolase EC 4.2.1.1) is one of the most powerful catalyst in nature for the simplest of all enzyme catalysed reactions $CO_2 + H_2O \rightleftharpoons HCO_3 + H^+$. This important enzyme is widely distributed in nature. The enzymes isolated from mammalian erythrocytes have been extensively studied by physico-chemical methods [36, 38] and a large volume of data accumulated over years since their first discovery in 1932 by Meldrum and Roughton [41] in red cells.

Human erythrocyte carbonic anhydrase form B (HCAB) and form C (HCAC) have been crystallized and their three-dimensional structures established by X-ray diffraction methods [22, 24, 33, 42]. The chemical sequences of these isoenzymes have also been well established in a number of laboratories [1, 16, 30, 39]. Carbonic anhydrase consists of about 260 amino acids and has a molecular weight of about 30,000. One firmly bound zinc ion per molecule is essential for the activity of the carbonic anhydrases. The zinc can be removed by dialysis with 1,10-phenanthraline [35] or 2.3-dimercaptopropanol (BAL) [23, 54] and replaced by divalent transition metal ions, Cd^{2+} and Hg^{2+}. Besides the zinc carbonic anhydrase Co^{2+} carbonic anhydrases exhibit high activity. Cd^{2+} and Mn^{2+} carbonic anhydrases also show some activity, while all other metal substitutions destroy enzymic activity.

The hydration of CO_2 and the dehydration of carbonic acid or bicarbonate is considered the in vivo catalyzed reaction of the carbonic anhydrases [36]. It has also been shown that the enzymes catalyse the hydration of certain aldehydes [45] and the hydrolysis of various esters like p- and o-nitro phenyl acetate [46, 53, 56]. The carbonic anhydrases are inhibited by monovalent anions and by aromatic or hetero-cyclic sulfonamides [36, 38]. Imidazole is a competitive inhibitor of the CO_2 hydration reaction in the case of HCAB [28]. The three-dimensional structures of a number of sulfonamide complexes of HCAB and HCAC have been reported [4, 26], as well as those of some anions [26, 57]. The high resolution structure of the HCAB-imidazole complex has also been established [25]. The crystal structure of the native carbonic anhydrases and the inhibited enzymes have given us a better understanding of the substrate binding sites as well as a probable mode of action of the enzymes.

---

1 Neutron Physics Division, Bhabha Atomic Research Centre Trombay, Bombay 400 085/India

2 Materials and Methods

Human erythrocytes obtained from the Uppsala University hospital blood bank were washed three times with an equal volume of cold 0.9% NaCl solution and centrifuged to remove the plasma and leucocytes. The haemolysate was prepared by adding an equal volume of cold distilled water, and was then dialysed against 0.02 M Tris-HCl at pH 8.2. The carbonic anhydrases were separated from other proteins by three stages of DEAE-Sephadex-A50 column chromatography [2, 22]. The HCAB and HCAC thus purified were dialysed against 0.15 M phosphate-borate buffer at pH 8.2 and purified further by zone electrophoresis using ethanolyzed cellulose powder as support medium. The HCAB was then crystallized after concentration as described by Kannan et al. [23]. The HCAC fraction was concentrated and dialysed against 0.05 M phosphate buffer, pH 6.0, for further purification on SE-Sephadex-A50 column to remove the last traces of HCAB. The enzyme was eluted from the column with 0.2 M Na_2HPO_4 buffer. The HCAC thus purified was concentrated and crystallized as a complex with methyl mercury thioglycolic acid (MMTGA) and the native HCAC crystals were prepared by dialysing these crystals against 2.5 M ammonium sulfate, 0.05 M Tris HCl solution, pH 8.5, containing 0.5 M cysteine to remove mercury [22]. The purity of the enzyme preparations was continuously checked by their esterase activity, the hydrolysis of ρ-nitrophenyl acetate [46, 53, 56].

The crystals of HCAB and HCAC diffract X-rays very well to better than 1.8 Å resolution. The heavy atom derivatives were prepared by soaking the crystals with the respective heavy atom salts in 2.5 M ammonium sulfate, pH 8.5 or by cocrystallization. The space group and the details of the heavy atom derivatives used are summarized in Table 1.

Precession cameras were used to collect the X-ray diffraction data on photographic films from the native enzyme crystals and the heavy atom bound enzyme crystals, using Ni-filtered (fine focus) cuKα radiation. These diffraction photographs were measured on an automated microdensitometer designed and constructed by Mr. V. Klimecki, Chemistry Department, Uppsala University, in collaboration with us at the Wallenberg Laboratory Uppsala. The measured data was processed using a programme system developed by us [17] for the CD 3600 and IBM 370/155 computers at the Uppsala University computing centre (UDAC).

The heavy atom positions were determined by difference Patterson synthesis with projection data, as well as three-dimensional data, and in some instances with the help of difference Fourier synthesis. The difference Fourier maps were also made use of to locate minor heavy atom sites. The heavy atom positions were refined by alternate cycles of phase calculation and least squares refinement. The "best Fouriers" [5] were calculated to about 2 Å resolution and interpreted independently in two optical comparators [40, 48] and the molecular models built for the two isoenzymes using Kendrew-Watson skeletal model parts. A detailed description of the interpretation of the HCAB and HCAC electron density maps has been given in other publications [21, 22, 24, 33, 42] and only the essential features of the structures will be discussed here.

Table 1. Crystallographic data

	HCAB	HCAC
Crystallization	Seeding of 5% HCAB in 2.3 M ammonium sulphate and 0.05 M Tris-HCl at pH 8.7	Dialysis of HCAC-MMTGA against 2.3 M ammonoium sulphate and 0.05 M Tris- HCl at pH 8.5 crystals of HCAC-MMTGA dialyzed against 0.01 M cystein in 2.3 M ammonium sulphate and 0.05 M Tris-HCl at pH 8.5
Space group unit cell dimension	Orthorhombic: a = 81.5, b = 74.6, c = 37.1 Å; $P\ 2_1\ 2_1\ 2_1$ 4 in unit cell, i.e., 1 per asymmetric unit	Monoclinic: a = 42.7, b = 41.7, c = 73.0 Å = 104.6°, $P2_1$ 2 in unit cell, i.e., 1 per asymmetric unit
Number of molecules		
Heavy atom	1. Bal + $HgCl_2$, Hg^{2+} replaces Zn^{2+} and also reacts with the single -SH group at CYS 212 2. $K_2Au(CN)_2$ gold cyanide reacts at the active site 3. Mercury acetate and benzene sulfonamide (?) AMSULF. Benzene sulfonamide is bound at the active site and mercury acetate reacts with the single -SH group, CYS 212 4. Diamox (acetazolamide) binds at the active site	Methyl mercury thioglycolic acid (MMTGA) reacts with the single -SH group at CYS 206 $K_2Au(CN)_2$ gold cyanide reacts at the active site Mercury acetate and benzene sulfonamide (?), AMSULF. Benzenesulfonamide is bound at the active site and one mercury acetate is bound to HIS 64 in the active site and one mercury to the single -SH group CYS 206 Iodine binds at the active site
Nominal resolution of crystallographic structure determination	2.2 Å 10,500 reflections	2.1 Å 11,200 reflections
Average figure of merit the accuracy of the structure determination: F.M.	0.73	0.74

3 Tertiary Structure

3.1 Secondary Structure

The amino acid sequences of HCAB and HCAC are highly homologous (Fig. 1) (see for example [42]) and, as could be expected, the three-dimensional structures of the isoenzymes also turned out to be highly homologous. The degree of homology was determined by a computer program, kindly provided to us by Drs. S.T. Rao and M.G. Rossmann [47]. In this computation the α carbon atom coordinates of the isoenzymes were compared by a nonlinear least squares procedure and the rotation and

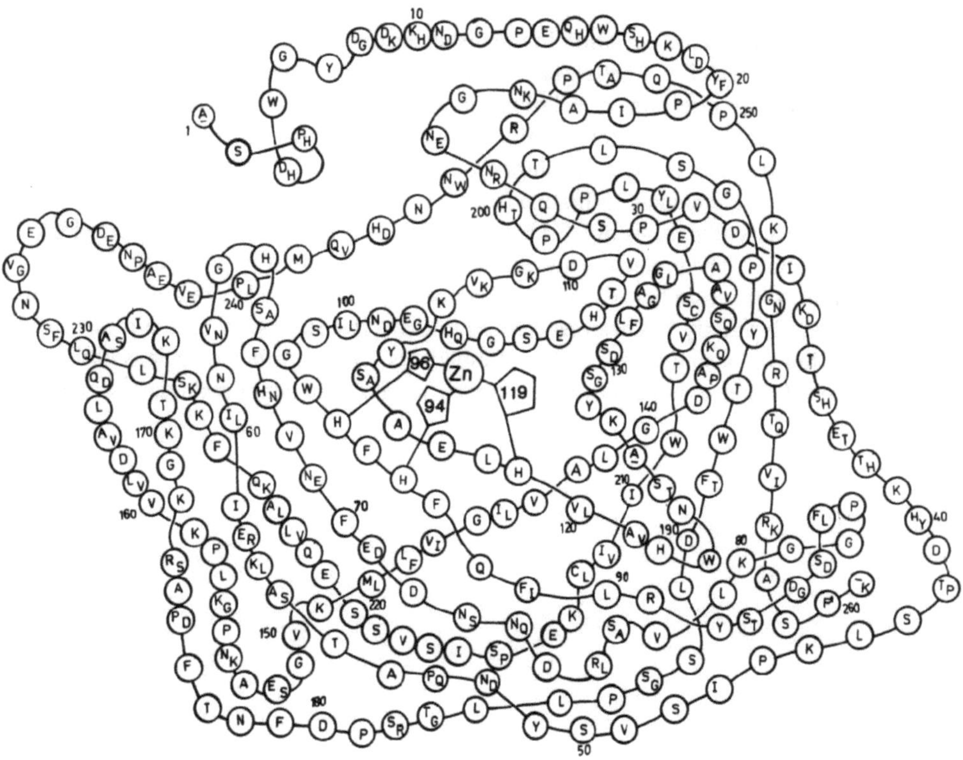

Fig. 1a. A schematic drawing of the tertiary structure of carbonic anhydrase showing the homology in the primary structure using the one letter code (G = gly, S = ser, A = ala, V = val, L = leu, I = ile, F = phe, P = pro, W = try, M = met, T = thr, C = cys, K = lys, R = arg, H = his, Y = tyr, D = asp, E = glu, N = asn, Q = gln)

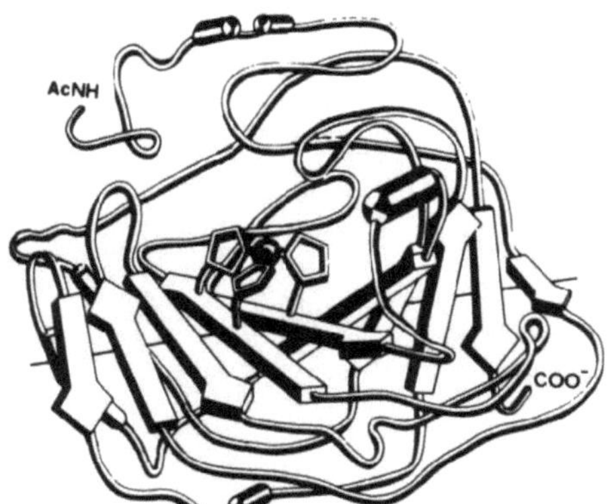

Fig. 1b. A schematic drawing of the tertiary and secondary structure of carbonic anhydrase

translation parameters were calculated for one molecule with respect to the other. The root mean squared deviation in the distances for the 255 equivalent atoms was determined to be 1.7 Å. The largest deviation of 5 Å occurred in the tertiary structure where there were insertions or deletions in one molecule or the other [42]. The rotation and translation parameters obtained from these calculations were used to produce the comparative stereoscopic drawings produced by the ORTEP program of Dr. Carol N. Johnson [18], included in this article.

The tertiary structures of HCAB and HCAC are given as the α-carbon drawings in Figs. 2 and 3. The molecules are ellipsoidal in shape, about 41 x 42 x 55 Å in dimension, with a number of cavities. The essential zinc ion is located in a deep conical cavity. This cavity has been recognized as the active site cavity by the presence of the zinc ion and also by the location of inhibitors. The enzyme molecule is itself folded up through a number of secondary structures. The major secondary structure

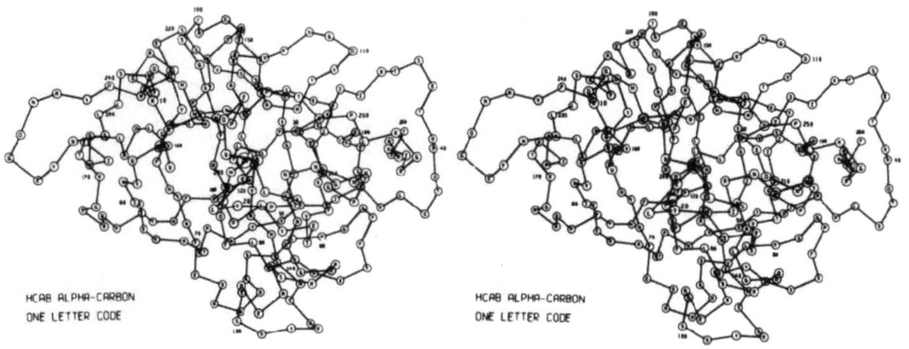

Fig. 2. Stereoscopic drawing of the α-carbon atoms of HCAB. With one letter code of amino acid sequence

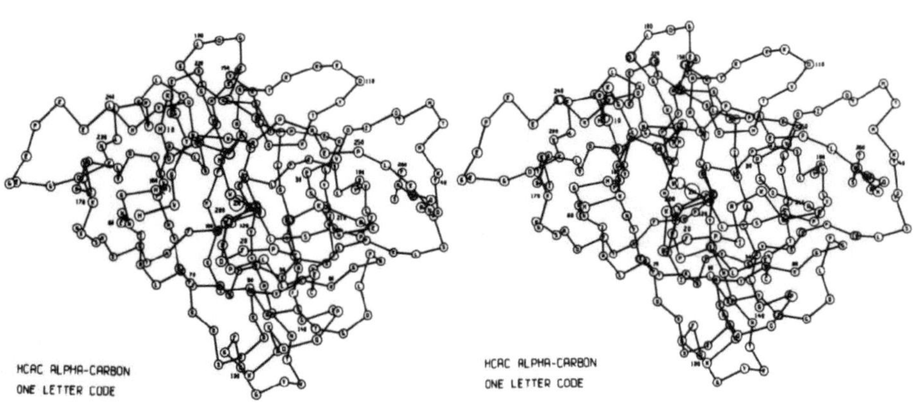

Fig. 3. Stereoscopic drawing of the α-carbon atoms of HCAC, in the same orientation as HCAB. With one letter code of amino acid sequence

in the carbonic anhydrases is a central twisted pleated sheet made up of 10 chain segments involving about 30% of the amino acid residues (see Figs. 4 and 5). The chain segments are arranged predominantly in antiparallel pleated sheet arrangement. There are however two pairs of chain segments which run as parallel pleated sheet segments (Table 2).

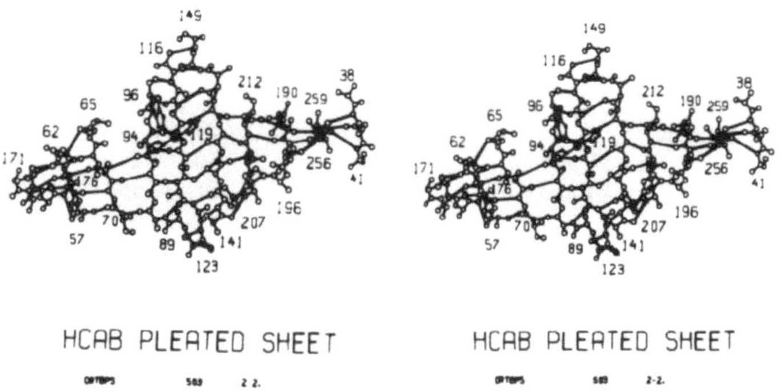

Fig. 4. Stereoscopic drawing of the central pleated sheet in HCAB

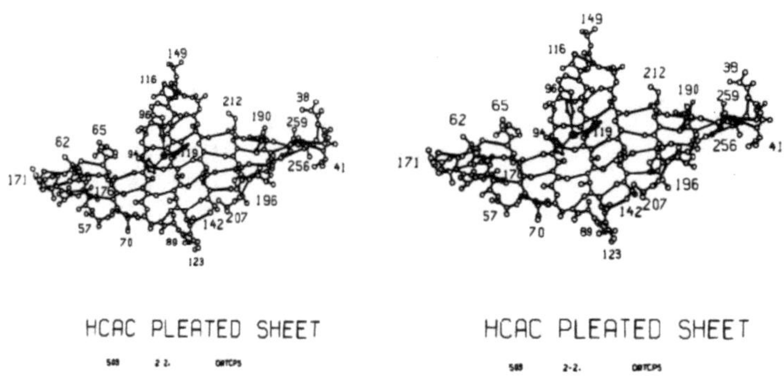

Fig. 5. Stereoscopic drawing of the central pleated sheet in HCAC

There are a number of minor pleated sheet structures located on the surface of the molecule often made up of two chain segments. Table 2 lists all the amino acids involved in β-structure formation in HCAB and HCAC along with other secondary structures. Recent Raman scattering studies have indicated 19% helix and 39% β-structure in HCAB [8] in close agreement with 17% helix and 40% β-structure observed in the carbonic hydrase structure. There are only two well-defined α-helices while the others are distorted 3_{10} helices (see Table 2). The folding together of these secondary structures produces a number of bends, 10 in number. Eight of these conform to the 3_{10} bend type [9, 32, 58].

Table 2. Secondary structure in carbonic anhydrase

Helix		Type of turn		β-structure chains involved	Chain segment in central pleated sheet and chain segment to which hydrogen bonded	Residues involved
HCAB	HCAC	HCAB	HCAC			
		Reverse	3_{10}-I			8–11
		3_{10}-I	3_{10}-I			13–16
A-α	A-α					16–20
B-3_{10}	B-3_{10}					20–25
		Open Reverse	Open Reverse			27–31
				a1 and a2		30 and 249
				b1 and b2		31–33 and 108–110
	Cα	Reverse				34–38
					10 parallel to 9	41–39
				c1 and c2		45–51 and 82–76
		3_{10}-I	Open			51–54
					2 antiparallel to 1 and 3	62–55
		Reverse	Reverse			62–65
					3 antiparallel to 2 and 4	65–71
		3_{10}-I	3_{10}-I			81–84
				d1 and d2		87–90 and 79–77
				e1 and e2		87–88 and 126–124
					4 antiparallel to 3 and 5	96–89
		3_{10}-I	Reverse			109–112
					5 antiparallel to 4 and 6	116–124
		3_{10}-I	3_{10}-I			124–128
Dα-3_{10}	Dα-3_{10}					130–137
		3_{10}-I	3_{10}-I			137–140
					6 antiparallel to 5 and parallel to 7	151–140
				f1 and f2		147–150 and 215–218
E_1-3_{10}	E_1-3_{10}					155–162
E_2-3_{10}	E_2-3_{10}					162–168
					1 antiparallel to 2	171–176
F-3_{10}	F-3_{10}					181–185
					8 antiparallel to 7 and 9	196–191
		Open Reverse	Open Reverse			199–203
					7 antiparallel to 8 and parallel to 6	206–212
G-α	G-α					219–229
				g1 and g2		229–231 and 241–239
		3_{10}-I	Open			233–236
		3_{10}-I	Open			251–254
					9 antiparallel to 8 and parallel to 10	259–256

As is evidenced from Table 2, the secondary structures are highly homologous in the tertiary structure, though the primary structures are not always that homologous (Fig. 1). Thus out of the 70 residues that make up the central pleated sheet structure (28 or 40%) are non-homologous residues. The changes are however mostly conservative in character. A number of changes are observed in one or the other isoenzyme, in the outermost pleated sheet segments 3–6 which form part of the active site cavity The eight amino acid replacements that occur in these pleated sheet segments that form the active site region tend to change the overall size of the active site cavity. All the deletions and insertions observed in one structure or the other occur on the surface in loop regions or in the amino or carboxyterminus of the molecules and not in the pleated sheet segments. This would be an important requirement for keeping the integrity of the pleated sheet structure of carbonic anhydrases throughout the evolutionary period. Any deletion or insertion in the middle of any chain segment forming the pleated sheet would disturb the active site geometry as well as the tertiary structure homology to a great extent.

3.2 Side Chain Packing

The packing of the side chains in the carbonic anhydrases follows the general rule of hydrophobic in and hydrophilic out, as in all other soluble globular proteins observed so far.

Three histidine residues are liganded to the metal ion and of the two buried histidines, one (HIS 122) is hydrogen-bonded to a tyrosine (TYR 51) while the other (HIS 107) is involved in a complex hydrogen bond net work [42] involving a glutamic acid, (GLU 117) and tyrosine (TYR 194). Six histidines in HCAB and seven histidines in HCAC are located on the surface of the molecule. Four tyrosyl residues are exposed in the two isoenzymes. Three tryptophenyl residues are buried in the interior of the molecule. Most of the phenylalanyl residues are tightly packed in the interior of the molecule and are involved in the formation of different aromatic clusters [42]. All the 17 prolyl residues occur on the surface of the molecule PRO 30 and PRO 202 occur in the cis conformation [24, 42] in two reverse open turns. No hydrogen bonds are formed in these two turns.

The location of certain side chains has been the subject of a number of solution studies. Thus the location of the histidine residues has been investigated using NMR methods in different laboratories [6, 29, 43, 44] and their pK values have been assigned in an effort to pinpoint the groups responsible for the catalytic activity of the enzymes. The titratability of tyrosines has been investigated by Riddiford [49]. He has also investigated the availability of histidyl, tyrosyl, carboxylic group and tryptophanyl residues by the acid difference spectra and found no evidence for buried carboxyl groups or lysyl residues. Recently Craig and Gaber [8] using laser Raman spectra found 3.7 buried and 4.7 exposed tyrosyl residues in HCAB, which agrees with that found in the structure [24, 42]. While the solution studies can reveal the nature of the side chains, the three-dimensional structure can give the time-averaged position of the side chains. The combination of these two reveals a wealth of information about the dynamics of the enzyme system [6, 8, 29, 43, 44].

4 Active Site of Carbonic Anhydrases

4.1 Native Zinc Enzyme

The active site cavity in carbonic anhydrase is easily located by the presence of the essential zinc ion. The placement of the inhibitors used in the crystal structure investigation defines the extent of the active site cavity and the side chains involved in forming this essential cavity. Of the 260 amino acids the carbonic anhydrases comprise only about 50 are actually lining the active site and of these 25 have their side chains located in the active sites (see Table 3). The critical cavity is about 12 Å deep and the opening of the cavity is about 10 Å in diameter. The zinc ion is located at the apex of the cone.

Table 3. Active site residues in HCAB and HCAC

Hydrophilic region		Hydrophobic region	
HCAB	if different in HCAC	HCAB	if different in HCAC
TYR 7		SER 65	ALA 65
ASN 61		PHE 91	ILE 91
HIS 64		ALA 121	VAL 121
HIS 67	ASN 67	LEU 131	PHE 131
ASN 69	GLU 69	LEU 141	
GLU 92		VAL 143	
HIS 94		GLY 145	
HIS 96		PRO 201	
HIS 119		CIS-PRO 202	
THR 199		THR 204	LEU 204
HIS 200	THR 200	SER 206	CYS 206 (shielded)
		VAL 207	
		ILE 211	VAL 211

Three histidyl residues HIS 94, HIS 96, and HIS 119 form the protein ligands to zinc, while a solvent molecule forms the fourth ligand, giving the zinc ion a distorted tetrahydral geometry. The histidyl side chains are hydrogen-bonded to various other side chains. Thus HIS 94 is hydrogen-bonded to GLN 92 and HIS 119 is hydrogen-bonded to GLU 117 which is hydrogen-bonded to HIS 105. GLU 117 and HIS 105 are both located inside the molecule (see Figs. 6 and 7). The zinc ligands may thus be expected to be neutral ligands which do not compensate the positive charges on the zinc ion. The charge on the zinc ion would therefore be expected to play an important role in the high catalytic efficiency of the carbonic anhydrases. This situation is in contrast to that found in the other metallo enzymes like carboxypeptidas A [37], liver alcohol dehydrogenase [10] and thermolysine [7] where one of the three zinc ligands is a negatively charged side chain.

The common feature observed in all enzyme active sites is the general division of the cavity or crevice into two halves, one predominantly composed of hydrophic side chains and the other of hydrophilic side chains. This feature is also observed in

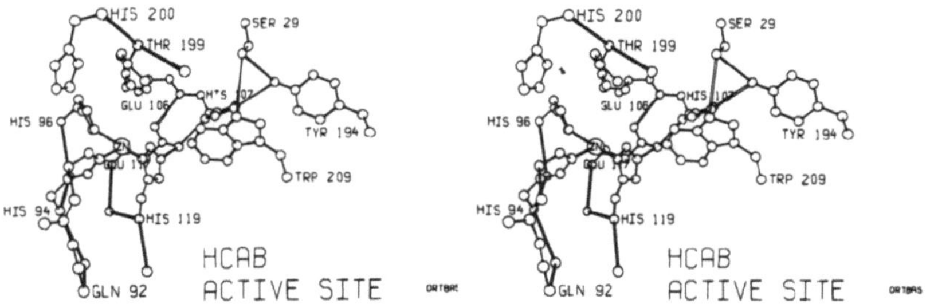

Fig. 6. Hydrogen bonding in HCAB active site region

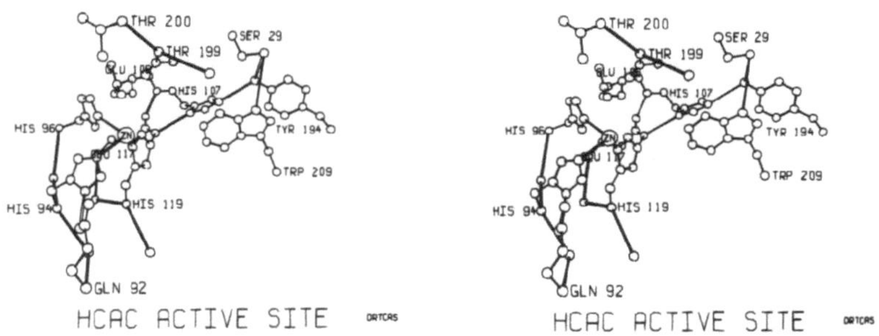

Fig. 7. Hydrogen bonding in HCAC active site region

the carbonic anhydrases (see Table 3). The residues located in the active site cavities of HCAB and HCAC are also depicted in the stereo ORTEP drawings of Figs. 8 and 9.

The important deviations in the active site residues have been discussed by Notstrand et al. [42] and are also seen in Figs. 8 and 9. Thus SER 206 in HCAB replaces CYS 206 in HCAC in the hydrophobic part of the cavity. LEU 204 in HCAC is located in the active site while the side chain of TYR 204 in HCAB is rotated out of the active site cavity. Due to this conformational difference in the two isoenzymes the side chain of SER 206 is inside the active site cavity of HCAB, while CYS 206 is shileded from the active site by LEU 204 in HCAC. This change makes the active site of HCAB a little more open in this region and also introduces a hydrophilic residue in the hydrophobic part of the active site. The other major changes in the hydrophobic half of the active site cavity are the replacement of ILE 91 and PHE 131 in HCAC by PHE 91 and LEU 131 respectively in the B isoenzyme. Also VAL 121 and VAL 135 in HCAC are replaced by ALA 121 and ALA 135 in HCAB. The changes in the hydrophilic half of the cavity involve the replacement of ASN 67 and THR 200 in HCAC by HIS 67 and HIS 200 respectively in HCAB. These two changes shield HIS 64 to some extent in HCAB while in HCAC this residue is easily accessible for active site directed modifications [20]. In HCAB, however, only HIS 67 and HIS 200 have been modified in a similar fashion [60]. All these modifications of the active site histidine residues reduce the activity of the enzymes considerably but do not abolish the activity.

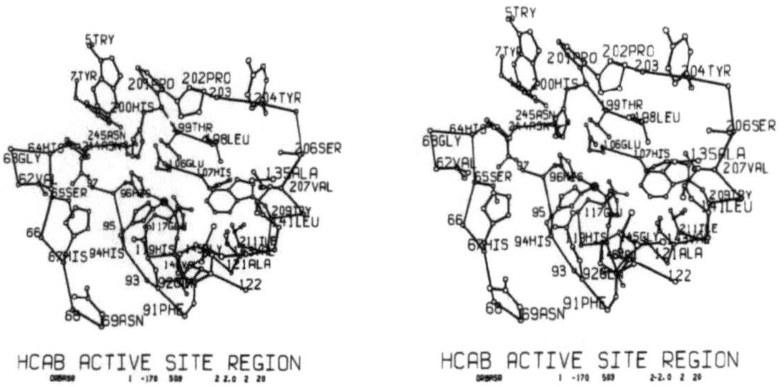

Fig. 8. Stereoscopic drawing of the active site residues in HCAB

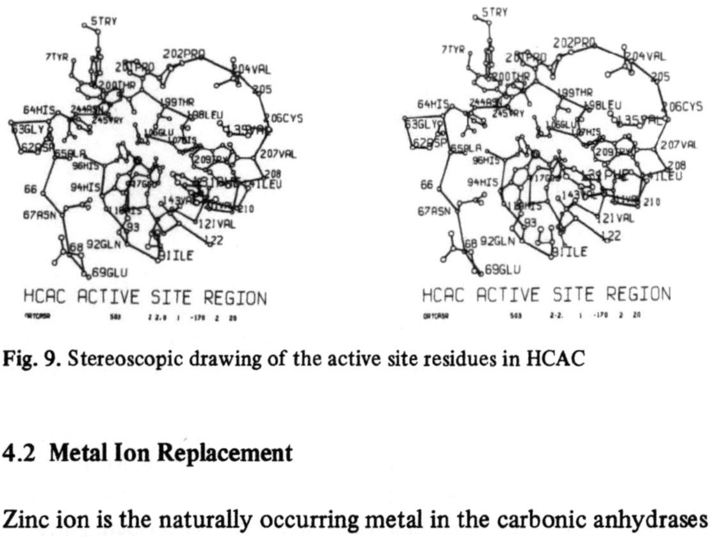

Fig. 9. Stereoscopic drawing of the active site residues in HCAC

4.2 Metal Ion Replacement

Zinc ion is the naturally occurring metal in the carbonic anhydrases. This metal also imparts the highest activity for these enzymes. The tightly bound metal ion is not removed even by prolonged dialysis against metal-free buffers at normal pH values. The metal ion can however be removed by dialysing at low pH values with 1,10-phenantraline [35, 36] or with 2,3-dimercaptopropanol (BLA) even at higher pH values [23, 54]. The apo enzyme thus prepared is devoid of any enzymatic activity and does not bind the anion or sulfonamide inhibitors [36]. The activity of the enzymes is wholly restored by the addition of Zn^{2+} and partly restored when Co^{2+}, Mn^{2+}, or Cd^{2+} is bound to the apo-enzyme. Co^{2+} carbonic anhydrase shows the highest carbonic anhydrase activity for the substituted metals. Co^{2+}, Cu^{2+}, Mn^{2+}, and Hg^{2+} carbonic anhydrases have been investigated by X-ray diffraction analysis [13, 21, 25]. The Hg^{2+} HCAB has been studied in detail to high resolution and reported by Kannan [21]. The Co^{2+}, Cu^{2+}, Mn^{2+}, and Hg^{2+} HCAC have been studied by difference electron density in two projections [25]. These investigations revealed that all the metal substitutions except Hg^{2+} bind to the apo-enzyme at the same position as the Zn^{2+}.

The Hg^{2+} is, however, displaced by about 0.8 Å from the zinc site in the two isoenzymes [13, 21, 25]. The three-dimensional high resolution investigation of HCAB revealed that the $HgCl_2$ which was substituted instead of Zn^{2+} has a different ligand distribution compared to the native zinc ion [21]. The active zinc enzyme study discussed above revealed a distorted tetrahedral ligand geometry around the zinc ion, which is drastically altered to a distorted octahedral geometry around the inactive Hg^{2+} ion. The mercury ion is displaced away from HIS 94 and towards THR 199 by about 0.8 Å from the zinc site. HIS 96 moves away from the metal but HIS 119 is in the same position as in the native enzyme. The two chlorides which were linearly coordinated to Hg^{2+} in the inorganic $HgCl_2$ molecule have rearranged in the enzyme such that one of them is axial to HIS 119 and is in hydrogen-bonding distance to HiS 200 and the other one occupies a position close to the solvent ligand site. HIS 94, HIS 96, THR 199 and a chloride are in almost one plane and at a greater distance to Hg^{2+} compared to HIS 119 and the other chloride which are tightly bound to the Hg^{2+}. These six ligands provide a distorted octahedral ligand geometry around the Hg^{2+}. The Hg^{2+} HCAB study has thus revealed that there is a large flexibility in the formation of the ligands in the carbonic anhydrases. Even though there is no high resolution three-dimensional investigation of Hg^{2+} HCAC, the HCAB study can be extended to HCAC also on the basis of the similar directional displacement observed for the Hg^{2+} with respect to Zn^{2+} in HCAC [13, 25]. The ligand pattern for Hg^{2+} in HCAC would thus be expected to be very similar to that in HCAB. It is also evident from this investigation that the ligand geometry around the metal is one other important contributing factor towards the high catalytic efficiency of the zinc carbonic anhydrases. The active metal ion has a distorted tetrahedral geometry, while the inactive metals deviate from this.

5 Inhibitor Interaction with Carbonic Anhydrase

5.1 Anionic Inhibitors

The inhibition of the carbonic anhydrases by most monovalent anions follows the so-called Hofmeister lyotropic series [14]. The order of these inhibitors according to their inhibitory power of carbonic anhydrase are

$$SH^- < CN^- < NCO^- < SCN^- < N_3^- < I^- < HCO_3^- < B^- < CH_3COO^- < Cl^- < F^-$$

The anions are competitive inhibitors of the dehydration reaction of HCO_3^- catalysed by the carbonic anhydrases. Thus an understanding of their binding site should throw some light on this substrate binding site.

It has been shown that $Au(CN)_2^-$ also inhibits the enzymes but the inhibition kinetics is complex and not well understood [15, 34]. The crystal structure of HCAB-$Au(CN)_2$ has been established to high resolution [21, 22]. The three-dimensional difference Fouriers revealed the general direction of stretching of the linear $Au(CN)_2$ molecules in the active sites of the respective isoenzymes. The solvent molecule observed at the fourth coordination site is present in these enzyme anion complexes of HCAB and HCAC. However, it has not been possible to locate unambiguously the position of the two (CN) groups in the crystal structures. The crystal structure refinement in progress by Kannan and coworkers may reveal the position of these important groups of $Au(CN)_2^-$. An interesting observation regarding gold cyanide inhibition is the acceleration of the bromopyrate modification reaction of HIS 64 in HCAC in the presence of this inhibitor [34]. The inhibition of the enzyme by $Au(CN)_2^-$ facilitates this modification as with the other active site directed modifications where the enzyme is inhibited first before the irreversible modification of a suitably located histidine residue in the active site.

Several other anion HCAC complexes have been investigated to high resolution by X-ray diffraction analysis. These investigations of the inhibitor complexes of HCAC with I^-, Br^- and Cl^- respectively have established that the halides bind by inner sphere coordination to the zinc ion and replace the solvent molecule at the fourth coordination site. This is in contrast to that observed for $Au(CN)_2$ described above. The bond distances of the halides to the zinc ion have been established to be 3.7 Å for I^-, 3.4 Å for Br^- and 3.3 Å for Cl^-. The distances are somewhat larger than the sum of the ionic radii of Zn^{2+} and the respective halides, but still too short for any other ligand to interpose between the halides and the metal ion. Inner sphere coordination proposed on the basis of numerous spectroscopic studies [36] is thus confirmed by the structure investigation. The HCAB-I^- complex has been investigated by difference Fourier and least squares analysis in two projections to 2Å resolution only and confirms the results of the HCAC-I^- studies, thus establishing a similarity in the I^- inhibition of the two isoenzymes.

5.2 Sulfonamide Inhibition

Aromatic and heterocyclic sulfonamides are very specific inhibitors of the carbonic anhydrases [38], when the sulfonamide group is unsubstituted. The inhibition of carbonic anhydrase by sulfonamides was first demonstrated by Keilin and Mann [27]. Sulfonamides have long been used for the treatment of glaucoma. Side effects such as acidosis, tiredness and kidney disorders have, however, been reported recently for patients on long-term treatment of glaucoma with sulfonamides [11].

Kinetic studies of sulfonamide inhibition have shown these to be competitive to the dehydration of H_2CO_3 reaction and the hydrolysis of esters, as well as the hydration reaction of CO_2 under certain circumstances [36].

The inhibitor potency of different sulfonamides towards the different isoenzymes varies considerably. Often ring substitutions have drastic effects on the inhibitory properties. These are similar to the variations observed in the esterase hydrolysis of p,0, and m-nitrophenyl acetates by HCAB and HCAC. The orientation of the heterocyclic or the aromatic ring of the inhibitors and their specific interactions with the side chains would thus throw light on the differences in the inhibitory properties as well as the esterase activity.

The detailed interactions of a number of sulfonamides with HCAB and HCAC have been recently reported by Kannan et al. [26]. The important features of the inhibitor binding is the five coordination observed for the zinc ion, replacement of the solvent ligand and the interaction of the aromatic or heterocyclic ring with the active site residues.

The inhibitor complexes, HCAB-acetazolamide, HCAB-benzene sulfonamide, have been investigated to 2 Å resolution by X-ray diffraction analysis. The molecular interactions of the sulfonamide inhibitor diamox with the enzymes are shown in Figs. 10 and 11.

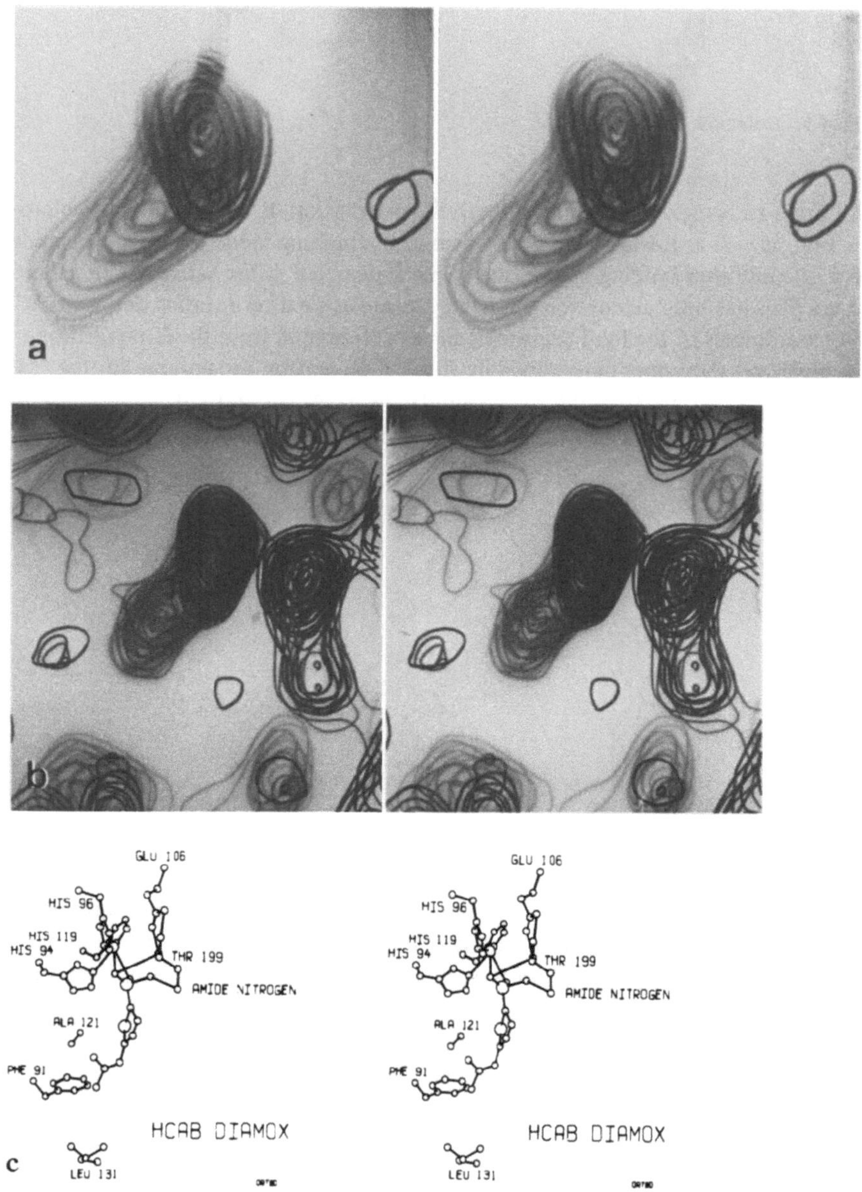

Fig. 10. a, b Difference electron density maps of HCAB acetozolamide inhibitor. **c** Acetazolamide binding to HCAB

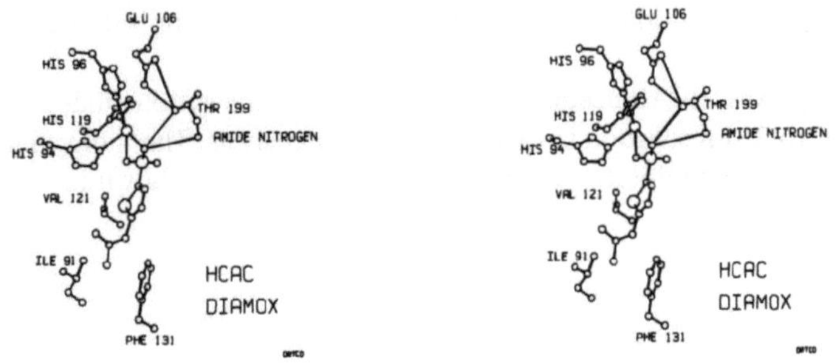

Fig. 11. Acetazolamide binding to HCAC

In order to bring out the similarities in the orientation of the aromatic or heterocyclic ring, as well as the intricate differences in the binding mode of the sulfonamide moiety, the inhibitor binding in one isoenzyme is depicted in the active site of other enzymes. This has been achieved by drawing the molecule after rotating and translating the coordinates of the inhibitor molecule, as determined from the X-ray diffraction analysis, by the values determined by the Rao-Rossmann programme for the α-carbon atom coordinates of the HCAB and HCAC (Figs. 12 and 13).

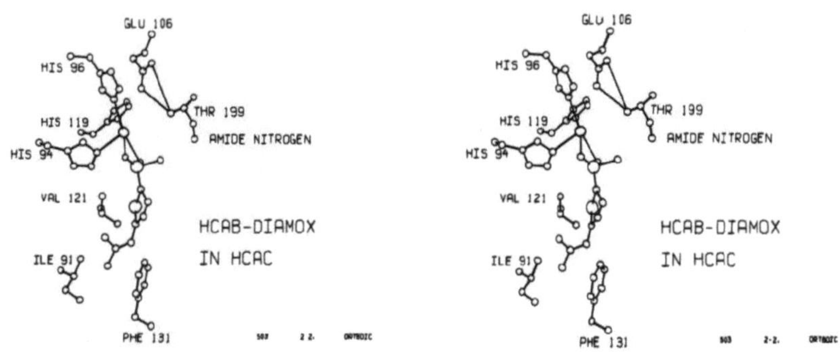

Fig. 12. Acetazolamide molecule bound to HCAC drawn in the active site of HCAB

The striking feature of the inhibitor binding is the replacement of the solvent molecule at the fourth coordination site of zinc by one of the atoms of the sulfonamide group and the presence of another atom of this group at a fifth coordination site near a hydrophobic pocket. The atoms of the sulfonamide group are also hydrogen-bonded to protein groups (Figs. 10 and 11). Thus THR 199 which is hydrogen-bonded to the zinc coordinated solvent molecule in the native enzyme is now hydrogen-bonded to one of the atoms of the sulfonamide group. THR 100 is also hydrogen-bonded to a buried glutamic acid residue GLU 106 both in the native and the inhibited isoenzymes.

The aromatic or the heterocyclic part of the inhibitor interacts with the side chains in the active site cavity. They are oriented just in the middle of the hydrophobic

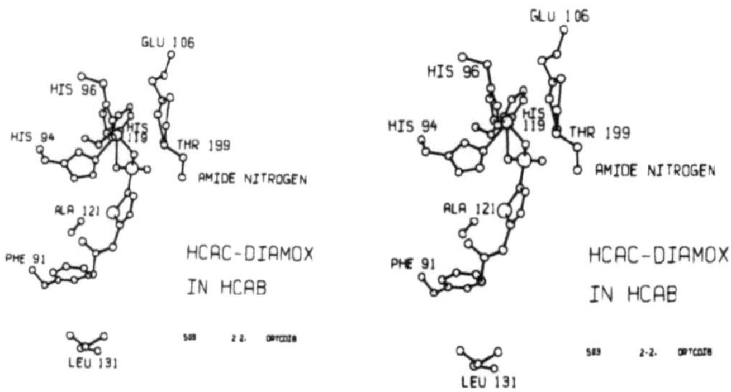

Fig. 13. Acetazolamide molecule bound to HCAB drawn in the active site of HCAC

and the hydrophilic parts of the cavity. The sulfur atom forming part of the heterocyclic ring of the acetazolamide inhibitor is in Van der Waals contact with the ALA 121 in HCAB and with VAL 121 in HCAC. This probably accounts for one important difference in the mode of interaction of this class of inhibitor. It may be expected that the ortho or meta substitution in benzene sulfonamides would interact with residue 121 in the isoenzymes and the difference in this side chain would give rise to differing strength of Van der Waals interaction. It may also be expected that the variations in the esterase activity for the ortho respective para nitrophenyl-acetate would also arise among other factors from the variation of the residue 121 in the different isoenzymes. Residues within 5 Å of the inhibitor molecules are listed in Table 4. Other residues which differ between the isoenzymes like the active site histidines HIS 200 and HIS 67 in HCAB in the place of THR 200 and ASN 67 would also be expected to influence the inhibitor binding.

Table 4. Amino acid residues in the active site of HCAB and HCAC within 5 Å of the sulfonamide inhibitors

Hydrophobic residues		Hydrophylic residues	
HCAB	HCAC if different	HCAB	HCAC if different
GLN 92		PHE 91	ILE 91
HIS 94		ALA 121	VAL 121
THR 199		LEU 131	PHE 131
HIS 200	THR 200	LEU 141	
		LEU 198	
		TRP 209	

5.3 Imidazole Inhibitor

Imidazole, a commonly used buffer substance in enzyme investigations was shown by Khalifah [28] to be a competitive inhibitor of the hydration of CO_2 by HCAB. The three-dimensional high resolution structure of HCAB-imidazole complex has been presented by Kannan et al. [25]. The salient feature of the HCAB-imidazole structure is the binding of imidazole as a fifth ligand to zinc, without replacing the solvent molecule at the fourth coordination site. The plane of the imidazole is well determined in the X-ray diffraction analysis (Fig. 14). The ring is oriented in an unusual way in contrast to normal coordination to metals by imidazoles. The plane of the imidazole is almost perpendicular to the plane of HIS 94. The solvent molecule at the fourth coordination site is located just above the middle of the imidazole ring (Fig. 14). The imidazole is located in the hydrophobic pocket where one of the atoms of the sulfonamide group also was identified. The distance of the imidazole nitrogen to zinc was determined to be about 2.8 Å, indicating weak binding to zinc. The solvent

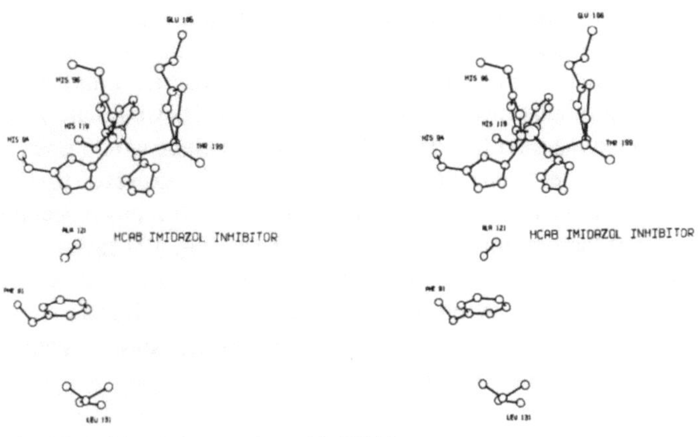

Fig. 14. Imidazole interaction with HCAB

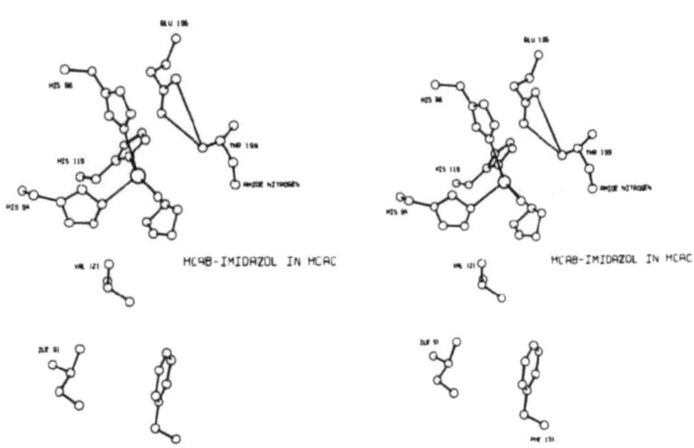

Fig. 15. Imidazole molecule bound to HCAB drawn in the active site of HCAC

molecule liganded to zinc is hydrogen-bonded to THR 199 which is hydrogen-bonded to GLU 106. This situation is identical to that observed in the native enzyme.

It is not clear why imidazole does not inhibit HCAC in the same way as HCAB. Model building experiments (Fig. 15) have not thrown sufficient light on this question. One reason for this might again be the replacement of ALA 121 in HCAB by VAL 121 in HCAC in the vicinity of the imidazole site. which might restrict the orientational freedom of the bulky imidazole group.

6 Discussion

It is now well established from the results above and from other physicochemical investigations [36, 42, 8] that the carbonic anhydrase isoenzymes are highly homologous both in their primary and tertiary structures. The X-ray diffraction results have revealed that the overall tertiary structure homology deviates at certain points, even though the primary structure homology is preserved at the corresponding points [21]. This might have been caused by the differences in the microscopic environment brought about by the differences in the side chains at these places in the two isoenzymes. The highly homologous active site also revealed local differences which might be very important for the differences in the activity as discussed above in connection with the sulfonamide inhibition. The pertinent question to discuss, however, is the role of the three-dimensional structure in the catalytic function of the isoenzymes.

The metal ion has been shown to be essential for the catalytic activity of the enzymes. The structure has revealed a distorted tetrahedral coordination for the zinc ion in the native structure and distorted octahedral coordination for the inactive mercury ion. The zinc ion has been shown to be 5-coordinated when inhibited by sulfonamides or imidazole. Spectral investigations have indicated that Co^{2+} in carbonic anhydrases might have four to five ligands [36, 3]. NMR investigations by several workers have shown that there is a fast exchanging water molecule associated with the metal at high pH, while there is a slow exchanging one at low pH [31, 12]. The X-ray diffraction analysis of the native enzymes at high pH revealed only one solvent molecule at the fourth coordination site. The inhibitor binding studies discussed above, however, revealed that the zinc is 5-coordinated. This discrepancy may be explained on the basis that two solvent molecules are coordinated to the metal at high pH together with the three protein ligands and that the solvent at the fifth ligands site is the fast exchanging water which is not located properly in the X-ray diffraction analysis. In any event the volume of evidence accumulated points to the fact that the active metals are capable of binding five ligands. This may be very important for the catalytic efficiency of the carbonic anhydrases. The uncompensated positive charge on the metal would be another important factor to be considered in this respect.

The anions and the sulfonamides are competitive inhibitors of the dehydration reaction and the esterase reaction, while imidazole is a competitive inhibitor of the hydration reaction catalysed by HCAB. The results discussed above show that the anions bind to the metal at the fourth coordination site displacing the solvent molecule. The sulfomanide inhibitors also bind by replacing the solvent molecule, as well as to the fifth coordination site. The imidazole is a unique inhibitor in that it binds to the fifth coordination site without displacing the solvent molecule at the fourth coordination site. Recently Walpert et al. [59] have shown that imidazole is indeed a unique inhibitor of HCAB and that this inhibitor binds without affecting the active site ionizations. Their results are in conformity with the X-ray diffraction results of Kannan et al. [25].

X-ray diffraction experiments have not been possible so far to investigate the binding of any of the substrates, CO_2 or HCO_3^- or esters. Stein et al. [50] have probed substrate binding site by ^{13}C NMR and shown that the carbon atom in CO_2 and HCO_3^- is located at about 3.5 Å from the metal. This confirms the interpretation made by Kannan et al. [25] from the imidazole binding studies that the substrates bind to zinc as weak fifth ligands.

It has been generally accepted [36, 19] that the main functional group of carbonic anhydrase is a hydroxyl ion bound to the metal at high pH and a water at the low pH. It has also been proposed [19] that the proton transfer rate in the mechanism of carbonic anhydrases is buffer-mediated. However, Tu and Silvermann [55] have shown by ^{18}O exchange studies that the buffer-mediated proton exchange alone cannot account for the catalytic efficiency of the carbonic anhydrases. Steiner et al. [51, 52] have interpreted the substantial hydrogen isotope effect on the kinetic parameters of HCAC to indicate an internal proton transfer process in the enzyme.

A number of mechanisms have been proposed for the catalytic properties of the carbonic anhydrases. Some of these have involved metal coordinated histidines and hydroxyls, active site histidines, and glutamic acid. The mechanisms involving histidines for facilitated proton transfer and water molecules interspersed between the metal and the substrates have become untenable in the face of current experimental evidence. The zinc-hydroxyl mechanism has generally been favoured as the most probable one for carbonic anhydrase catalysed reaction [36, 6]. However, as pointed out earlier, this mechanism does not account for all the existing experimental evidence and also it does not provide a satisfactory explanation for all the protein groups that make up even the active site cavity. This also involves an unusually low pK value for the hydroxyl ion. The glutamic acid mechanism has been criticized for invoking unusually high pK value for this amino acid residue. The main advantage of this mechanism is, however, that it associates protein groups together with the metal ion in the mechanism of action of the carbonic anhydrases.

Main features of any mechanistic proposal, however, should take into account the binding of the substrates at the fifth coordination site of the metal [25], the hydrogen bonding observed from the ligand at the fourth coordination site to THR 199 and its hydrogen bonding to GLU 106 and the charge on the zinc ion. It is intersting to note in this regard that the mechanism proposed by Kannan et al. [25] for the hydration of CO_2 and dehydration of HCO_3^- also fits well with the hydrolysis of esters. The carboxyl oxygen of the ester would bind as weak fifth ligand to the metal similar to imidazole. The aromatic part of the ester would be oriented in the same way as the aromatic or heterocyclic groups of the sulfonamide inhibitors. The constraints due to Van der Waals interaction discussed above for the ring substitutions of sulfonamides would apply effectively for the ester substrate and explain the differences in specificity for 0 and p nitrophenyl acetates. The crystal structure of p-nitrophenylacetate shows that the bond angle at the oxygen atom that is attached to the phenyl ring is 118.5°, the acetate group is in the trans-configuration to the ring and the plane of the acyl group is rotated from that of the phenyl ring by 64.2°. This will fit well with the proposals of Kannan et al. [25] for this substrate binding at the fifth coordination site and bring the carbon atom to be hydrated in a favourable position and orientation to the water or hydroxyl at the fourth coordination site in a very similar way to that proposed for CO_2 hydration. A tetrahedral intermediate would be formed and the hydrolysis of the ester bond would accompany the hydration.

Acknowledgments. I would like to thank Bror Strandberg, Kerstin Fridborg, Seved Lövgren, Marianne Petef, Gunnel Johansson, Birger Notstrand, Ivar Vaara, Hilda Cid, Agneta Borell, Anders Liljas, Jan Lundin, Per Christer Bergsten and Georgi Petef for the excellent collaboration and comradeship throughout this work at the Wallenberg Laboratory, Uppsala University, Uppsala, Sweden. I would like to thank my colleagues Dr. M. Ramanathan and Mr. Surender Sinh for help with the figures. I would like to thank Mr. K. Rajasekaran, Miss K Kalyani and Mrs. Anna-Greta Lundquist for typing the manuscript. I would also like to thank Dr. R. Chidambaram and Dr. P.K. Iyengar for the interest and encouragement.

References

1. Andersson B, Nyman PO, Strid L (1972) Amino acid sequence of human erythrocyte carbonic anhydrases. Biochem Biophys Res Commun 48: 670–677
2. Armstrong JMcD, Myers DV, Verpoorte JA, Edsall JT (1966) Purification and properties of human erythrocyte carbonic anhydrases. J Biol Chem 241: 5137–5149
3. Bauer R, Limkilde P, Johanssen JT (1977) Metal coordination geometry and mode of action of carbonic anhydrase. Effect of imidazole on the spectral properties of Co(II) and Cd (II) human carbonic anhydrase B. Carlsberg Res Commun 42: 325–339
4. Bergstén PC, Waara I, Lövgren S, Liljas A, Kannan KK, Bengtsson U (1972) Crystal structure of human erythrocyte carbonic anhydrase C. V. Complexes with some anion and sulfonamide inhibitors. In: Rorth M, Astrup P (eds) Proc of the Alfred Benzon Symp IV, vol IV. Munksgaard, Copenhagen, pp 363–383
5. Blow DM, Crick FHC (1959) The treatment of errors in the isomorphous replacement method. Acta Crystallogr 12: 794–802
6. Campbell ID, Lindskog S, White AI (1977) Proton magnetic resonance studies of histidines in human, Rhesus monkey and bovine carbonic anhydrases. Biochem Biophys Acta A84: 443–452
7. Colman PM, Jansonius JN, Matthews BW (1972) The structure of thermolysin. An electron density map at 2.3 Åresolution. J Mol Biol 70: 701–724
8. Craig WS, Graber BP (1977) Laser Raman scattering from an enzyme of well documented structure. Human carbonic anhydrase B. J Am Chem Soc 99: 4130–4134
9. Crawford J, Lipscomb WN, Shellman CG (1973) The reverse turns as a polypeptide conformation in globular proteins. Proc Natl Acad Sci USA 70: 538–542
10. Eklund H, Nordström B, Zeppezauer E, Söderlund G, Ohlsson I, Boiwe T, Söderberg B-O, Tapio O, Brändén C-I (1976) Three-dimensional structure of horse liver alcohol dehydrogenase at 2.4 Åresolution. J Mol Biol 102: 27–59
11. Epstein DL, Grant WM (1977) Carbonic anhydrase inhibitor. Side effects – serum chemical analysis. Arch Opthalmol 95: 1378–1382
12. Fahry ME, Koenig SH, Schillinger WE (1970) Nuclear magnetic relaxation dispersion in protein solution. IV. Protonrelaxation at the active site of carbonic anhydrase. J Biol Chem 245: 4256–4262
13. Fridborg K, Kannan KK, Liljas A, Lundin J, Strandberg B, Strandberg R, Tilander B, Wirén G (1967) Crystal structure of human erythrocyte carbonic anhydrase C. III. Molecular structure of the enzyme and an enzyme inhibitor complex at 5.5 Åresolution. J Mol Biol 25: 505–516

14. Fridovich I (1963) Inhibition of acetoacetic decarboxylase by anions. The Hofmeister Lyotropic Series. J Biol Chem 238: 592–598
15. Göthe PP, Nyman PO (1972) Inactivation of human carbonic anhydrases by bromopyruvate. FEBS Lett 21: 159–164
16. Henderson LE, Henriksson D, Nyman PO (1973) Amino acid sequence of human erythrocyte carbonic anhydrase C. Biochem Biophys Res Commun 52: 1388–1394
17. Järup L, Kannan KK, Liljas A, Strandberg B (1970) A program system for the evaluation of electron density maps from x-ray diffraction data, with special reference to biological macromolecules Comput Programs Biomed 1: 74–76
18. Johnson CK (1965) Thermal ellipsoid drawing. Oakridge National Laboratory Publication
19. Johnson BH, Steiner H, Lindskog S (1976) Participation of buffer in the catalytic mechanism of carbonic anhydrase. FEBS Lett 64: 310–314
20. Kandel M, Gornall AG, Wong S-CC, Kandel SI (1970) Some characteristics of human, bovine and horse carbonic anhydrases as revealed by inactivation studies. J Biol Chem 245: 2444–2450
21. Kannan KK (1978) Structure and function of carbonic anhydrases. In: Srinivasan R (ed) Proceedings on biomolecular structure, conformation, function and evolution. Pergamon Press, England
22. Kannan KK, Liljas A, Waara I, Bergstén PC, Fridborg K, Lövgren S, Strandberg B, Bengtsson U, Carlbom U, Järup L, Petef M (1971) Crystal structure of human erythrocyte carbonic nahydrase. VI. The three dimensional structure at high resolution in relation to other mammalian carbonic anhydrases. Cold Spring Harbor Symp 36: 221–231
23. Kannan KK, Gridborg K, Bergstén PC, Liljas A, Lövgren S, Petef M, Strandberg B, Waara I, Adler L, Falkbring SO, Göthe PO, Nyman PO (1972) Structure of human carbonic anhydrase B. I. Crystallization and heavy atom modifications. J Mol Biol 63: 601–604
24. Kannan KK, Notstrand B, Fridborg K, Lövgren S, Ohlsson A, Petef M (1975) Crystal structure of human erythrocyte carbonic anhydrase B. Three-dimensional structure at a nominal resolution of 2.2 Å resolution. Proc Natl Acad Sci USA 72: 51–55
25. Kannan KK, Petef M, Fridborg K, Cid-Dresdner H, Lövgren S (1977) Structure and function of carbonic anhydrases. Imidazole binding to human carbonic anhydrase B and the mechanism of action of carbonic anhydrases. FEBS Lett 73: 115–119
26. Kannan KK, Vaara I, Notstrand B, Lövgren S, Borell A, Fridborg K, Petef M (1977) Structure and function of carbonic anhydrase: Comparative studies of sulphonamide binding to human erythrocyte carbonic anhydrases B and C. In: Roberts CGK (ed) Proc on drug action at the molecular level. Macmillan Press, pp 73–93
27. Keilin D, Mann T (1939) Carbonic anhydrases. Nature (London) 144: 442–443
28. Khalifah RG (1971) The carbon dioxide hydration activity of carbonic anhydrase. I. Stop flow kinetic studies on the native human isoenzymes B and C. J Biol Chem 246: 2561–2573
29. Khalifah RG, Morlay PJ (1978) Personal communication
30. Kuang-Tzu DL, Deutsch HF (1973) Human carbonic anhydrases. XI. The complete primary structure of carbonic anhydrase B. J Biol Chem 248: 1885–1893
31. Lanir A, Gradstajn S, Havon G (1975) Temperature and frequency dependence of solvent proton relaxation rates in solution of manganese (II) carbonic anhydrase. Biochemistry 14: 242–248
32. Lewis PN, Momany FA, Scheraga HA (1973) Chain reversals in proteins. Biochem Biophys Acta 303: 211–229
33. Liljas A, Kannan KK, Bergstén PC, Waara I, Fridborg K, Strandberg B, Carlbom U, Järup L, Lövgren S, Petef M (1972) Crystal structure of human carbonic anhydrase C. Nature New Biol 235: 131–137
34. Lindskog S. Personal communication
35. Lindskog S, Nyman PO (1964) Metal binding properties of human erythrocyte carbonic anhydrases. Biochem Biophys Acta 85: 462–474
36. Lindskog S, Henderson HE, Kannan KK, Liljas A, Nyman PO, Strandberg B (1971) Carbonic anhydrase. In: Boyer PD (ed) The enzymes, 3rd ed, vol 5, pp 587–665

37. Lipscomb WN, Hartsuck JA, Quiocho FA, Reeke GN Jr (1969) The structure of carboxypeptidase A. IX. The x-ray diffraction results in the light of the chemical sequence. Proc Natl Acad Sci USA 64: 28–35
38. Maren TH (1967) Carbonic anhydrase chemistry, physiology and inhibition. Phys Rev 47: 595–838
39. Marriq C, Sciaky M, Giraud M, Foveau D, Laurent-Tabusse G (1973) Structure primaire de l'anhydrase carbonique érythrocytaire B humaine. II. Clivage par le bromure de cyanogéne et séquence des résidus 1–148. Biochemie 55: 1361–1379
40. Matthews BW, Jansonius JN, Colman PA, Schoenborn BP, Dupourque D (1972) Three-dimensional structure of thermolysin. Nature New Biol 238: 37–41
41. Meldrum NU, Roughton FJW (1933) Carbonic anhydrase: its preparation and properties. J Physiol (London) 80: 113–142
42. Notstrand B, Vaara I, Kannan KK (1975) Structural relationship of human erythrocyte carbonic anhydrase isozymes B and C. In: Markert CL (ed) Isozymes. Mol Struct I: 577–599
43. Pesando JM (1975) Proton magnetic resonance studies of carbonic anhydrases. I. Identification of histidine resonances. Biochemistry 14: 675–681
44. Pesando JM (1975) Proton magnetic resonance studies of carbonic anhydrase. II. Group controlling catalytic activity. Biochemistry 14: 681–688
45. Pocker Y, Meany JE (1967) The catalytic versatility of erythrocyte carbonic anhydrase. II. Kinetic studies of the enzyme catalysed hydration of pyridine aldehydes. Biochemistry 6: 239–246
46. Pocker Y, Stone JT (1967) The catalytic versatility of erythrocyte carbonic anhydrase. III. Kinetic studies of the enzyme catalysed hydrolysis of p-nitrophenyl acetate. Biochemistry 6: 668–678
47. Rao ST, Rossmann MG (1973) Comparison of super secondary structures in proteins. J Mol Biol 76: 241–256
48. Richards FM (1968) The matching of physical models to three-dimensional electron density maps. A simple optical device. J Mol Biol 37: 225–230
49. Riddiford LM (1965) Acid difference spectra of human carbonic anhydrases. J Biol Chem 240: 168–172
50. Stein PJ, Merril SP, Henkens RW (1977) Carbon-13 nuclear magnetic relaxation study of cobalt carbonic anhydrase. Evidence on the location of enzyme bound CO_2 and HCO_3^-. J Am Chem Soc 99: 3194–3195
51. Steiner H, Jonsson BH, Lindskog S (1975) The catalytic mechanism of carbonic anhydrase. Eur J Biochem 59: 253–259
52. Steiner H, Jonsson BH, Lindskog S (1976) The catalytic mechanism of human carbonic anhydrase C: Inhibition of CO_2 hydration and ester hydrolysis by HCO_3^-. FEBS Lett 62: 16–20
53. Tashian RE, Douglas DP, Yu YSL (1964) Esterase and hydrase activity of carbonic anhydrase. I. From primate erythrocytes. Biochem Biophys Res Commun 14: 256–261
54. Tilander B, Strandberg B, Fridborg K (1965) Crystal structure studies on human erythrocyte carbonic anhydrase C. II. J Mol Biol 12: 740–760
55. Tu C, Silvermann DN (1977) Comparison of ^{18}O exchange catalyzed by isoenzymes of carbonic anhydrase. J Biol Chem 252: 3332–3337
56. Verporte JA, Metha S, Edsall JT (1967) Esterase activities of human carbonic anhydrases B and C. J Biol Chem 242: 4221–4229
57. Vaara I, Lövgren S, Liljas A, Kannan KK, Bergstén PC (1972) Functional aspects of the three-dimensional structure of the active site of carbonic anhydrase. In: Brewer GJ (ed) Two proceedings on hemoglobin and red cell structure and function. Plenum Publishing Corporation, New York, pp 169–187
58. Venkatachalam C (1968) Stereochemical criteria for polypeptides and proteins. V. Conformation of a system of three linked peptide units. Biopolymers 6: 1425–1436
59. Walpert HR, Strader CD, Khalifah RG (1977) Interaction of the unique competitive inhibitor iminidazole with human carbonic anhydrase B. Biochemistry 16: 5717–5721
60. Whitney PL, Fölsch G, Nyman PO, Malmström B (1967) Inhibitor of human erythrocyte carbonic anhydrase B by chloroacetyl sulfonamides with labelling of the active site. J Biol Chem 242: 4206–4211

Kinetics and Mechanistic Implications of CO_2 Hydration Activity of Human Erythrocyte Carbonic Anhydrases

R.G. KHALIFAH [1]

1 Introduction

The reversible hydration of carbon dioxide is recognized as one of the simplest physiological reactions requiring biochemical catalysis. The mechanism of its catalysis by carbonic anhydrase has, nevertheless, been the subject of unabated discussion and controversy during the last decade. This account attempts to review the major features of the catalyzed reaction and their mechanistic implications. More detailed discussions of specific aspects are discussed elsewhere in these proceedings (cf. this vol. S. Lindskog, D.N. Silverman, and others). The present review is limited largely to data on the major human erythrocyte carbonic anhydrases whose crystal structure is now known to high resolution. These are the low specific activity isozyme HCAB (or CA I) and the high specific activity isozyme HCAC (or CA II). In addition, chemically modified derivatives of both of these isozymes have been studied and are of importance to understanding the mechanism of the enzyme. It is hoped that future work will soon provide the kinetic parameters of the newly discovered muscle isozyme (cf. Tashian, this vol.) and will increase the comparative kinetic data base.

2 Kinetic Parameters for Native HCAB and HCAC

2.1 Methods

The most extensive steady-state kinetic data at 25°C have been obtained with stop-flow methodology, A pH indicator technique is used to monitor proton concentration changes that result from the overall reaction:

$$CO_2 + H_2O \rightleftharpoons H^+ + HCO_3^- \qquad (1)$$

Since the apparent ("overall") pK_a for this process is about 6.3, it becomes extremely difficult to follow the dehydration reaction above pH 8 or the hydration reaction at low pH. Valid data are obtained if the pH excursion during kinetic runs is maintained below 0.05 pH units. Other technical difficulties should be recognized and have been discussed (Khalifah, 1971).

---

1 Department of Chemistry, University of Virginia, Charlottesville, Virginia 22901/USA

2.2 Hydration of CO_2

A complete study of the CO_2 hydration activity of HCAB and HCAC has been reported (Khalifah, 1971). Figures 1 and 2 show the variation of the Michaelis-Menten parameters K_m^h and k_{cat}^h, i.e., V_{max}^h/E_0, over the pH range of 5.8 to 8.8. Since K_m^h is obviously pH-independent, the pH variation of the ratio k_{cat}^h/K_m^h shown in Fig. 2 gives the variation of k_{cat}^h itself. The plot is chosen for accuracy, as the errors in k_{cat} and K_m are found to be positively correlated. The data for HCAC have been confirmed by the work of Steiner et al. (1975) who have extended the results to include measurements of the deuterium isotope effect. In addition, the CO_2-HCO_3^- exchange catalyzed by HCAB and HCAC has been followed by ^{13}C NMR linewidth measurements and has been found to slightly exceed the kinetic rates (Koenig et al., 1974; Simonsson et al., 1979), superceding earlier results (Koenig et al., 1973) that gave low rates for HCAB.

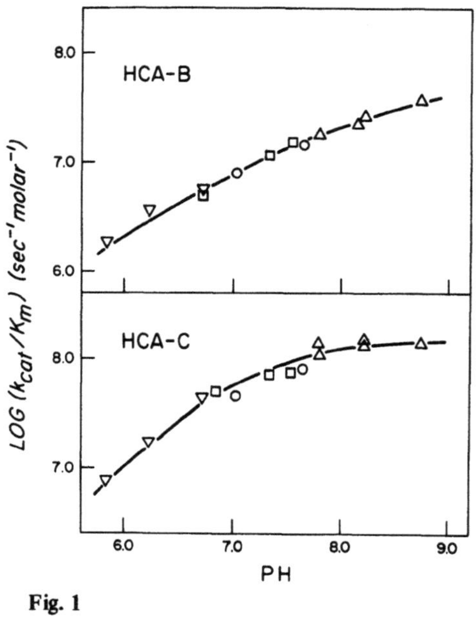

Fig. 1

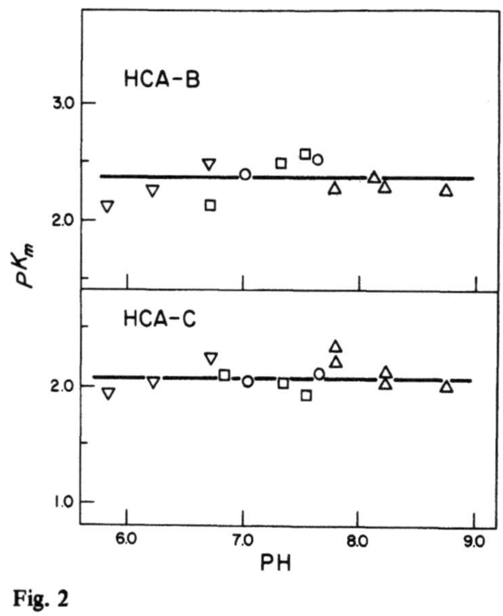

Fig. 2

Fig. 1. Effect of pH on the ratio k_{cat}^h/K_m^h for HCAB and HCAC at 25°C and 0.2 ionic strength. (Khalifah, 1971)

Fig. 2. Effect of pH on the Michaelis constant K_m for HCAB and HCAC at 25°C and 0.2 ionic strength. (Khalifah, 1971)

2.3 Dehydration of HCO_3

The dehydration kinetics of carbonic anhydrase are much less characterized and have yielded apparently conflicting behavior. Magid (1968) carried out initial studies on both isozymes at 1.6°C in phosphate buffers using a pH-stat method coupled to continuous purging of CO_2. HCAB was found to have a pH-independent K_m^d, while k_{cat}^d increased with decreasing pH. On the other hand, he found that for HCAC k_{cat}^d was pH independent, while K_m^d decreased with decreasing pH. The data on HCAC are at variance with the stop-flow results of Steiner et al. (1975) carried out at 25°C, who found K_m^d to be constant while k_{cat}^d increased and levelled off at low pH. These results have been confirmed by the observations of Pocker and Bjorquist (1977) on the bovine high activity enzyme. Stop-flow studies on HCAB thus remain to be carried out at the higher temperature, although the ratio k_{cat}^d/K_m^d can be calculated from the hydration data in conjunction with the Haldane relation (see below).

3 Kinetics of Modified HCAB and HCAC

3.1 Covalent Modifications

A number of derivatives have been studied where His_{200} of HCAB or His_{63} of HCAC were alkylated specifically at their N^τ positions. Both side-chains are known to reside in the active sites. Bromoacetate or iodoacetate is used to carboxymethylate His_{200} (Whitney et al., 1967; Bradbury, 1969), while bromopyruvate is used to carboxyketoethylate it (Göthe and Nyman, 1972). We have prepared further derivatives by reducing the carboxyketoethyl derivative with $NaBH_4$ to yield a carboxyhydroxyethyl derivative (CHE-HCAB) and by reacting HCAB with 3-bromopropionate. The latter reagent gave multiple products and is not discussed further here. For HCAC, a carboxyketoethyl derivative of His_{63} was prepared by the method of Göthe and Nyman (1972). This was further derivatized either by $NaBH_4$ reduction to give CHE-HCAC, or by reaction with phenylhydrazine and p-nitrophenylhydrazine to give two addition products. Table 1 summarizes the specific activities of these products using the Wilbur-Anderson assay.

Table 1. Covalently modified derivatives of human carbonic anhydrases B and C

Enzyme derivatives	Relative activity [a]
HCAB (sp. act. 14,000)	100
N^τ-Carboxymethylhistidine$_{200}$-HCAB (CmHCAB)	2.7
N^τ-Carboxyketoethylhistidine$_{200}$-HCAB (CKE-HCAB)	1.8
N^τ-Carboxyhydroxyethylhistidine$_{200}$-HCAB (CHE-HCAB)	1.0
HCAC (sp. act. 40,000)	100
N^τ-Carboxyketoethylhistidine$_{63}$-HCAC (CKE-HCAC)	30
N^τ-Carboxyhydroxyethylhistidine$_{63}$-HCAC (CHE-HCAC)	30
CKE-HCAC phenyl hydrazone	15
CKE-HCAC p-nitrophenylhadrazone	15

a Measured by the Wilbur-Anderson assay for CO_2 hyration activity. Note the difference in specific activity of HCAB and HCAC

3.2 CO_2 Hydration Parameters

Some of the above derivatives were chosen for a full stop-flow study of the pH dependence of their Michaelis-Menten parameters in CO_2 hydration. The data for the HCAB derivatives are shown in Figs. 3 and 4, while corresponding data are shown in Fig. 5 for CKE-HCAC, i.e., N^{τ}-carboxyketoethylhistidine$_{63}$ HCAC. All the chemically modified forms differ quantitatively and qualitatively from their parent enzyme forms. Particularly noteworthy is the low apparent pK_a for k^h_{cat} in CmHCAB (about 6.5). The esterase residual activity of this form shows a pK_a close to 9.2, as does the visible spectral titration of its Co^{2+} derivative (Whitney, 1970; Taylor et al., 1970).

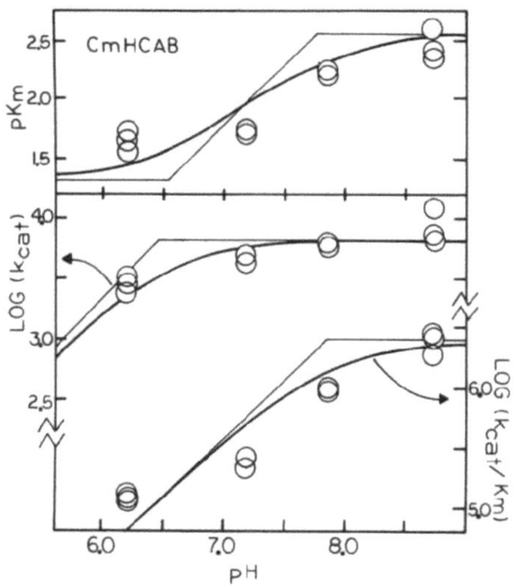

Fig. 3

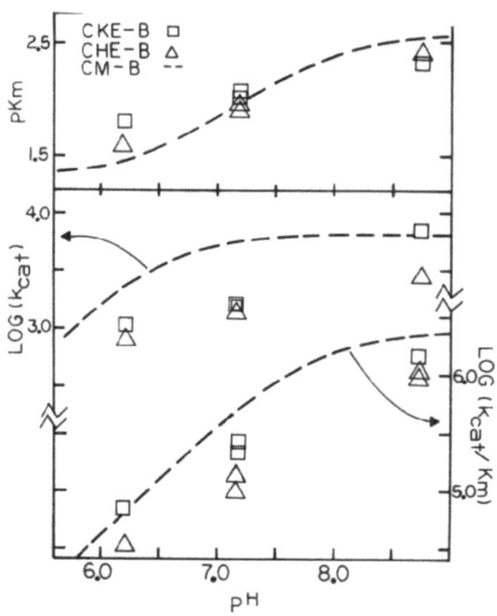

Fig. 4

Fig. 3. Effect of pH on k^h_{cat}, K^h_m, and k^h_{cat}/K^h_m for CmHCAB (N^{τ}-carboxymethylhistidine$_{200}$-HCAB) at 25°C and 0.2 ionic strength. The lines suggest that only 2 pK values influence all the parameters

Fig. 4. Effect of pH on k^h_{cat}, K^h_m, and k^h_{cat}/K^h_m for CmHCAB (*dashed curve*, taken from Fig. 3), CHE-HCAB, and CKE-HCAB at 25°C and 0.2 ionic strength

4. HCAB Inhibition by Imidazole

Imidazole has been previously discovered to inhibit the CO_2 hydration activity of HCAB in a competitive manner with respect to the CO_2 substrate (Khalifah, 1971). This makes it unique in this regard among all the known inhibitors of the enzyme. The crystal structure of the imidazole complex of HCAB has been recently solved by Kannan et al. (1977) who have found it to bind uniquely among inhibitors as a fifth ligand of the zinc without displacing the solvent ligand. Using UV difference spectroscopy, we have confirmed the unusual binding mode of this inhibitor when compared

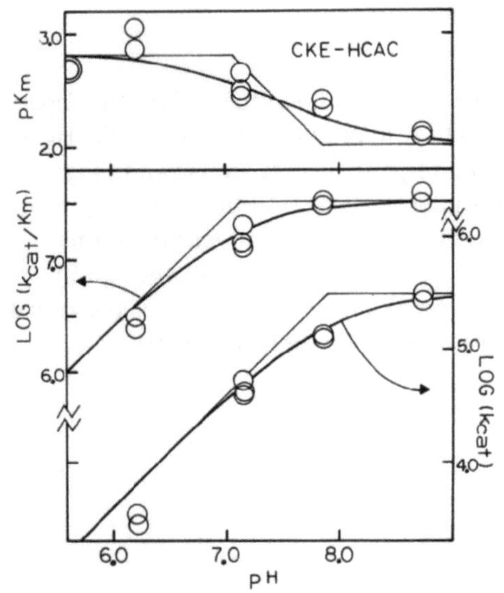

Fig. 5. Effect of pH on k^h_{cat}, K^h_m, and k^h_{cat}/K^h_m for CmHCAC (N^{τ}carboxyketo-ethylhistidine$_{63}$-HCAC) at 25°C and 0.2 ionic strength

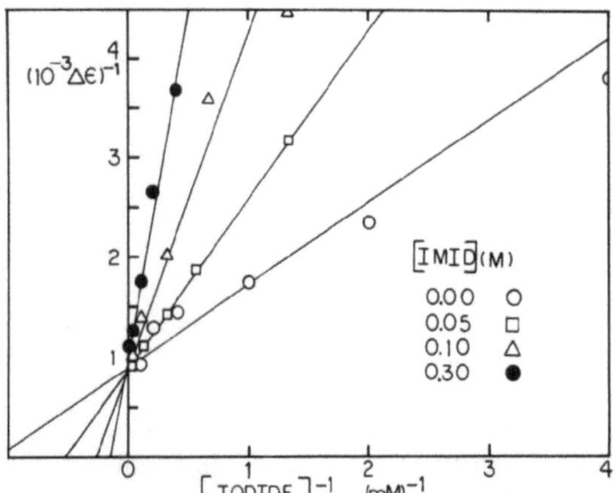

Fig. 6. Influence of imidazole on the iodide-induced UV difference spectral change at 289 nm showing the competitive nature of the binding of these two inhibitors to HCAB

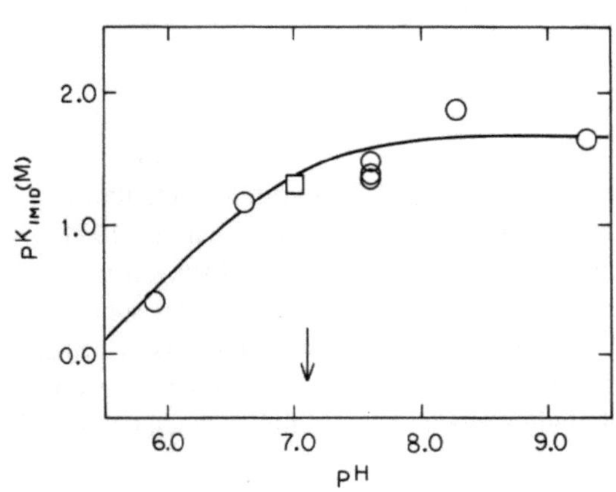

Fig. 7. Influence of pH on the dissociation constant K_i for the binding of imidazole to HCAB (taken from Wolpert et al., 1977)

to anionic or sulfonamide inhibitors (Wolpert et al., 1977). It is noteworthy that imidazole was also found to bind completely competitively with iodide, as shown by its inhibition of the iodide-induced UV difference spectrum (Fig. 6). This competition permitted a study of the pH dependence of imidazole binding. Figure 7 shows that the binding appears to be independent of any active site ionization and depends only on the state of protonation of imidazole itself.

5 Some Consequences of the Kinetics

5.1 Hydration Product: H_2CO_3 or HCO_3^- ?

The rapid ionization of H_2CO_3 and its low pK_a both serve to prevent a direct determination of whether H_2CO_3 or HCO_3^- is the product released from the enzyme in the catalytic cycle. However, the measured kinetics permit a distinction to be made. If H_2CO_3 is assumed as the product,

$$CO_2 + H_2O \rightleftharpoons H_2CO_3 \qquad K_h = 10^{-2.54}$$

then the ratio of product k_{cat}^d/K_m^d for this choice can be calculated using the Haldane relation. For HCAC we get:

$$k_{cat}^d/K_m^d = (k_{cat}^h/K_m^h)/K_h = 10^{8.13}/10^{-2.54} = 5 \times 10^{10}\ M^{-1}\ s^{-1}$$

Kinetic theory (Peller and Alberty, 1959) reveals that this ratio should be a lower bound for the rate constant for recombination of enzyme with this product. As first pointed out by DeVoe and Kistiakowsky (1961), this predicted rate is impossible for it exceeds the diffusion limit for this reaction, estimated as $1 \times 10^9\ M^{-1}\ s^{-1}$ (Khalifah, 1973). Most workers in the field have recognized the unlikely nature of this product (cf. Lindskog and Coleman, 1973), although others have attempted to rationalize such large rates in terms of rather implausible physical models (Koenig and Brown, 1972).

5.2 Active Site Ionization

The alternative choice of HCO_3^- as a product implies that the enzyme changes its state of protonation during the interconversion;

$$E \cdot CO_2 \overset{H_2O}{\rightleftharpoons} EH^+ \cdot HCO_3^- \rightleftharpoons EH^+ + HCO_3^-$$

To hydrate a second substrate molecule, the enzyme must thus deprotonate to the basic form,

$$EH^+ \underset{k_r}{\overset{k_d}{\rightleftharpoons}} E + H^+ \qquad K_a = k_d/k_r$$

Kinetic theory requires that this deprotonation rate k_d be not less than the overall turnover, i.e., k^h_{cat}. Around pH 7, k^h_{cat} approaches $4 \times 10^4\ s^{-1}$ for HCAB and $5 \times 10^5\ s^{-1}$ for HCAC. This is far greater than the $1 \times 10^3\ s^{-1}$ rate constant possible for an acid-base group with a pK_a near 7 with a diffusion-limited rate of protonation (reverse step), as pointed out by a number of authors (cf. Prince and Woolley, 1973; Lindskog and Coleman, 1973; Khalifah, 1973). Clearly, some mechanism is necessary to facilitate the deprotonation of the active site in order to account for the observed turnover rates. The resolution of this dilemma lay in the recognition that the deprotonation could be mediated by transfer to buffer acceptor species present in solution (cf. Lindskog and Coleman, 1973; Prince and Woolley, 1973; Khalifah, 1973; and earlier Alberty, 1962). This "Alberty mechanism" of deprotonation can be represented by:

$$EH^+ + B \rightleftharpoons E + BH^+$$

Moderate buffer concentrations could easily provide an effective rate of depronotation exceeding $1 \times 10^7\ s^{-1}$, in excess of observed turnover. The above analysis predicts decreases in turnover at lower buffer concentrations. Experimental evidence for this has been presented, although it is not clear whether the effects are as large as predicted (Jonsson et al., 1976; Silverman and Tu, 1977).

5.3 Kinetic Mechanism

The complex pH dependence of the Michaelis-Menten parameters for HCAB, HCAC, and, especially, their modified forms requires satisfactory explanation. A plausible mechanism for the hydration data has been proposed (Khalifah and Edsall, 1972):

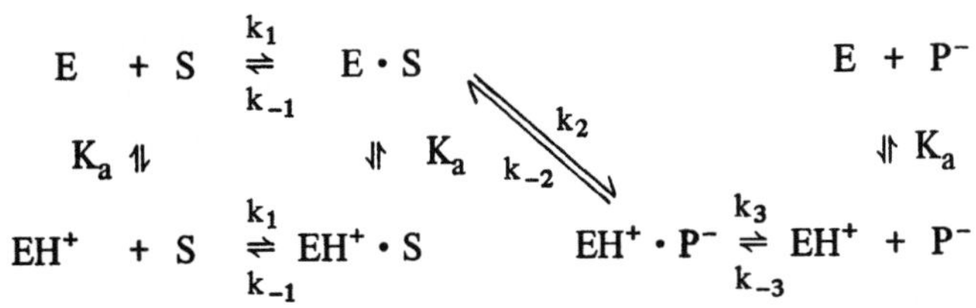

The main assumptions of the scheme are: (1) one critical ionization controls the activity. (2) Substrate CO_2 binds equally to both ionization states. (3) Product HCO_3^- binds only to the acid form, in analogy to the known binding of monovalent anionic inhibitors. The above mechanism has been particularly successful in accounting for the divergent pH variation of K_m. It also has provided a plausible explanation for the difference in apparent pK of k^h_{cat} of CmHCAB and the corresponding pK of the esterase activity and the cobalt spectral change (Whitney, 1970; Taylor et al., 1970), as shown earlier (Khalifah and Edsall, 1972). On the other hand, Lindskog and coworkers have found it inadequate to explain the observed ^2H effect on K^h_m (Steiner et al., 1975), as it predicts that K_m, especially at low pH, is equal to the CO_2 dissociation constant. Similarly, they find it inadequate to explain the pH-independent

K_m^d they measured, or to explain the product inhibition pattern for HCO_3^- in the hydration kinetics (Steiner et al., 1976). The alternative mechanism proposed by these workers on the basis of their extensive studies (cf. Simonsson et al., 1979) incorporates as a central feature the binding of substrate CO_2 only to the basic form of the enzyme. This assumption requires critical examination and experimental support. The detailed mechanism has not yet been applied to the kinetics of the modified enzymes, which should provide further tests.

5.4 CO_2 Binding Site and Imidazole Inhibition

Recent ^{13}C NMR studies suggest that CO_2 and HCO_3^- bind no further perhaps than the second coordination sphere of the zinc metal (Yeagle et al., 1975; Stein et al., 1977). The CO_2 competitive inhibitor has been found by crystallographic studies to bind at the zinc as a weak fifth ligand without displacing the water-hydroxide ligand (Kannan et al., 1977). Our spectral UV difference results fully support the proposed mode of binding of imidazole and, by analogy, CO_2 (see also Bauer et al., 1977). Consequently, the independence of imidazole affinity from active site ionizations suggests that CO_2 also should bind equally well to all forms of the active site, as in the kinetic scheme above. This is also reasonable in view of the neutral nature of CO_2. Due to the difference in size, it is not necessary to conclude that CO_2 must bind competitively with anionic inhibitors such as iodide.

5.5 Consequences of Covalent Modifications

The appreciable residual activity of CmHCAB and CKE-HCAC mitigates against a crucial role for His_{200} (in HCAB) and His_{63} (in HCAC) in the catalytic mechanism, although the latter may play a role in proton transfer (Steiner et al., 1975). In the case of CmHCAB, independent ^{13}C nuclear magnetic resonance evidence implicates a carboxymethyl carboxylate coordination to the zinc as the basis for the activity alterations (Khalifah, 1977; Khalifah et al., 1977; Jeffers et al., 1978). This provides clear support for a zinc-hydroxide mechanism for the enzymatic hydration.

5.6 Isozyme Activity Differences

The specific activity differences between HCAB and HCAC have received much attention since the discovery of the isozymes. The pH dependence of k_{cat}^h for HCAB suggests the influence of a secondary ionizing group in addition to the group controlling the activity. A strong interaction of His_{200} with zinc ligands in HCAB has been identified by ^{13}C NMR (Khalifah, 1977), suggesting that this residue perturbs, at the least, HCO_3^- binding. Consequently, the ionization of His_{200} may be reflected in the kinetics, and this residue may account also for the major differences between the isozymes. It is intriguing and suggestive, and perhaps relevant, that a simple change such as alkylation of His_{63} in HCAC can abolish the major differences in activity between the isozymes, as the superposition of Fig. 8 shows. The recently discovered

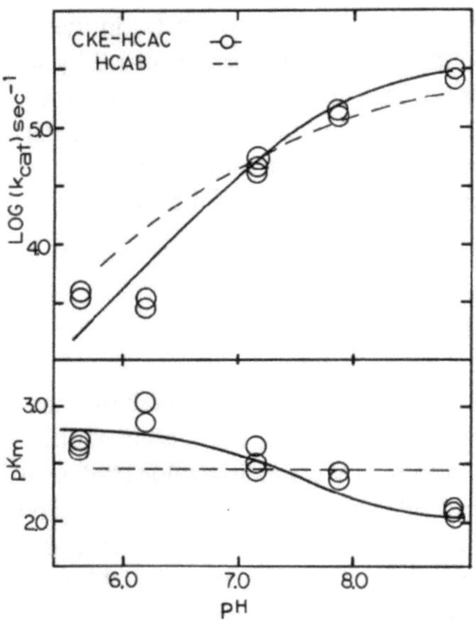

Fig. 8. Superposition of the CO_2 hydration catalysis parameters for CKE-HCAC (*experimental points* and *solid lines*) and native HCAB (*dashed lines*). The data are taken from Figs. 1, 2, and 5 above

third isozyme of carbonic anhydrase may well differ from HCAB in activity by virtue of a single substitution whose effect is analogous, for example, to reacting HCAB with iodoacetamide.

Acknowledgments. It is a great pleasure to express my indebtedness to Professor John T. Edsall who provided the impetus for this work, begun in his laboratory, and to whom this account is dedicated. Financial support for the research in my laboratory has been generously provided by a USPHS Grant from the National Institutes of Health (HL-17963).

References

Alberty RA (1962) The interpretation of steady state kinetic data on enzymatic reactions. Brookhaven Symp Biol 15: 18–31

Bauer R, Limkilde P, Johannsen JT (1977) Metal coordination geometry and mode of action of carbonic anhydrase. Effect of imidazole on the spectral properties of Co (II) and Cd (II) human carbonic anhydrase. Carlsberg Res Commun 42: 325–339

Bradbury S (1969) The carboxymethylation of human carbonic anhydrase B. The nature of the reaction. J Biol Chem 244: 2002–2009

DeVoe H, Kistiakowsky GB (1961) The enzyme kinetics of carbonic anhydrase from bovine and human erythrocytes. J Am Chem Soc 83: 274–280

Göthe PO, Nyman PO (1972) Inactivation of human erythrocyte carbonic anhydrases by bromopyruvate. FEBS Lett 21: 159–164

Jeffers PK, Sutherland WMcI, Khalifah RG (1978) Carbon-13 magnetic resonance of carboxymethylated human carbonic anhydrase B. Chemical shift and spin-lattice relaxation studies. Biochemistry 17: 1305–1312

Jonsson B-H, Steiner H, Lindskog S (1976) Participation of buffer in the catalytic mechanism of carbonic anhydrase. FEBS Lett 64: 310–314

Kannan KK, Petef M, Fridborg K, Cid-Dresdner H, Lövgren S (1977) Structure and function of carbonic anhydrases. FEBS Lett 73: 115–119

Khalifah RG (1971) The carbon dioxide hydration activity of carbonic anhydrase. I. Stop-flow kinetic studies on the native human isozymes B and C. J Biol Chem 246: 2561–2573

Khalifah RG (1973) Carbon dioxide hydration activity of carbonic anhydrase: Paradoxical consequences of the unusually rapid catalysis. Proc Natl Acad Sci USA 70: 1986–1989

Khalifah RG (1977) Histidine-200 alters inhibitor binding in human carbonic anhydrase B. A carbon-13 nuclear magnetic resonance identification. Biochemistry 16: 2236–2240

Khalifah RG, Edsall JT (1972) Carbon dioxide hydration activity of carbonic anhydrase: Kinetics of alkylated anhydrases B and C from humans. Proc Natl Acad Sci USA 69: 172–176

Khalifah RG, Strader DJ, Bryant SH, Gibson SM (1977) Carbon-13 nuclear magnetic resonance probe of active site ionizations in human carbonic anhydrase B. Biochemistry 16: 2241–2247

Koenig SH, Brown RD III (1972) H_2CO_3 as substrate for carbonic anhydrase in the dehydration of HCO_3^-. Proc Natl Acad Sci USA 69: 2422–2425

Koenig SH, Brown RD, Needham TE, Matwiyoff NA (1973) Kinetic parameters of human carbonic anhydrase B as determined from NMR linewidths of ^{13}C in CO_2 and HCO_3^-. Biochim Biophys Res Commun 53: 624–630

Koenig SH, Brown RD, London RE, Needham TE, Matwiyoff NA (1974) The kinetic parameters of carbonic anhydrase by ^{13}C NMR. Pure Appl Chem 40: 103–111

Lindskog S, Coleman JE (1973) The catalytic mechanism of carbonic anhydrase. Proc Natl Acad Sci USA 70: 2505–2508

Magid E (1968) The dehydration kinetics of human erythrocytic carbonic anhydrases B and C. Biochim Biophys Acta 151: 236–244

Peller L, Alberty RA (1959) Multiple intermediates in steady state kinetics. I. The mechanism involving a single substrate and product. J Am Chem Soc 81: 5907–5914

Pocker Y, Bjorquist DW (1977) Comparative studies of bovine carbonic anhydrase in H_2O and D_2O. Stop-flow studies of kinetics of interconversion of CO_2 and HCO_3^-. Biochemistry 16: 5698–5707

Prince RH, Wooley PR (1973) On the mechanism of carbonic anhydrases. Bioorg Chem 2: 337–344

Silverman D, Tu CK (1975) Buffer dependence of carbonic anhydrase catalyzed oxygen-18 exchange and equilibrium. J Am Chem Soc 97: 2263–2269

Simonsson I, Jonsson B-H, Lindskog S (1979) A ^{13}C nuclear magnetic resonance study of CO_2-HCO_3^- exchange catalyzed by human carbonic anhydrase at chemical equilibrium. Eur J Biochem 93: 409–417

Stein PJ, Merrill SP, Henkens RW (1977) Carbon-13 nuclear magnetic resonance relaxation study on cobalt carbonic anhydrase: Evidence on the location of enzyme bound CO_2 and HCO_3^-. J Am Chem Soc 99: 3194–3196

Steiner H, Jonsson B-H, Lindskog S (1975) The catalytic mechanism of carbonic anhydrase. Hydrogen-isotope effects on the kinetic parameters of the human C isoenzyme. Eur J Biochem 59: 253–259

Steiner H, Jonsson B-H, Lindskog S (1976) The catalytic mechanism of human carbonic anhydrase C: Inhibition of CO_2 hydration and ester hydrolysis by HCO_3^-. FEBS Lett 62: 16–20

Taylor PW, King RW, Burgen ASV (1970) Influence of pH on the kinetics of complex formation between aromatic sulfonamides and human carbonic anhydrase. Biochemistry 9: 3894–3902

Tu CK, Silverman DN (1977) Comparison of ^{18}O exchange catalyzed by isoenzymes of carbonic anhydrase. J Biol Chem 252: 3332–3337

Whitney PL (1970) Inhibition and modification of human carbonic anhydrase B with bromoacetate and iodoacetamide. Eur J Biochem 16: 126–135

Whitney PL, Nyman PO, Malmström BG (1967) Inhibition and chemical modification of human erythrocyte carbonic anhydrase B. J Biol Chem 242: 4212–4220

Wolpert HR, Strader CD, Khalifah RG (1977) Interaction of the unique competitive inhibitor imidazole with human carbonic anhydrase B. Biochemistry 16: 5717–5721

Yeagle PL, Lochmüller CH, Henkens RW (1975) ^{13}C nuclear magnetic resonance studies on the mechanism of action of carbonic anhydrase. Proc Natl Acad Sci USA 72: 454–458

Models for Metal Ion Function and Evolution of the Catalytic Step in Carbonic Anhydrase

P. WOOLLEY [1]

A plain-thinking chemist may show some surprise when confronted with carbonic anhydrase, which is a transition metal complex with catalytic activity dependent upon the presence of a metal ion, particularly zinc or cobalt [4, 21, 32], and with a catalytic efficiency so high that it has been spoken of as a "perfectly evolved enzyme" [8].

1 How Efficient is the Catalytic Step?

The degree of this "perfection" may be measured by examining the kinetic law usually written down for the enzyme [Eq. (1)] [2].

$$\begin{array}{ccccccc} E+P+B+H & \overset{k_1}{\rightleftharpoons} & EH+P+B & \overset{k_2}{\rightleftharpoons} & EHP+B & \overset{k_3}{\rightleftharpoons} ES+B \overset{k_4}{\rightleftharpoons} & E+S+B \\ k_6 \searrow\!\!\nwarrow & & \swarrow\!\!\nearrow k_5 & & & & \\ & E+P+BH & & & & & \end{array} \tag{1}$$

At given non-equilibrium concentrations of substrate CO_2, product HCO_3^- and hydrogen ion the concentration of enzyme necessary to provide a given rate of reaction can be calculated using steady-state kinetics, as long as all the rate constants are known. The evolving organism will try to increase the rates of each step if this can result in less enzyme being necessary to maintain the reaction rate. One can simulate evolution by repeating the calculation, increasing or decreasing the forward and backward rate constants of a chosen step. If this changes the amount of enzyme required, then there is selection pressure on the organism to change these rate constants in the direction corresponding to a lower enzyme requirement. For carbonic anhydrase the result of such a calculation, based on Eq. (1) and taking rate constants from the literature, is shown in Fig. 1. It is seen that there is still some selection pressure to accelerate the catalytic step, but not much – about 2/3 of the enzyme could be saved by increasing k_3 and k_{-3} tenfold; a further increase would be valueless.

By this criterion the enzyme is not perfectly evolved, but if, as has often been suggested [8, 16], the substrate and product binding steps are diffusion-controlled, then the enzyme is well on the way to evolutionary perfection, since it cannot very

---

1 Max-Planck-Institut für molekulare Genetik, Ihnestraße 63, D-1000 Berlin 33 (Dahlem)

2 At least two radically different mechanisms have recently been proposed [17, 26]. If either of these should prove to be right then application of the method of this section may lead to different conclusions

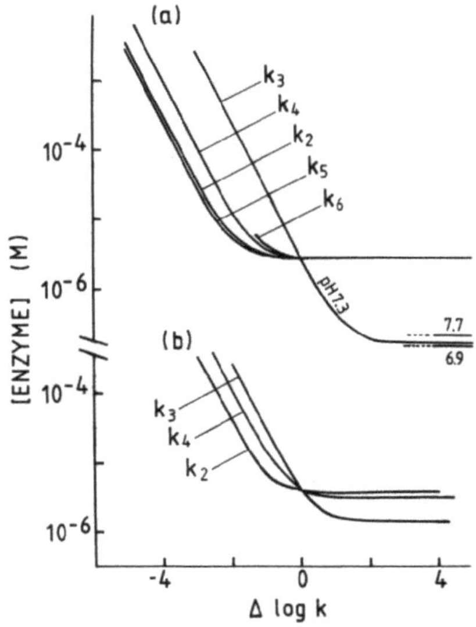

Fig. 1a, b. The amount of enzyme required to catalyze the turnover in Eq. (1), at a given deviation from equilibrium (1%) and turnover rate (10^{-3} M s^{-1}). N.B. (1) the enzyme requirement is directly proportional to these as long as these are small; (2) the $k_{\pm 1}$ step is very slow and may be neglected. Values used: k_{-2}/k_2, 0.032 M; k_3, 6×10^5 s^{-1}; k_{-3}, 10^6 s^{-1}; k_4/k_{-4}, 8.3×10^{-3} M; $k_5 = k_{-5}$; k_6, 800 s^{-1}; k_{-6}, 10^{10} M^{-1} s^{-1}; [HCO_3^-], 6×10^{-3} M; [CO_2], 6.06×10^{-4} M (equilibrium would require 6×10^{-4} M); pH follows from Haldane equation and is 7.3; [B] + [BH] = 0.01 M. **a** Typical diffusion-controlled values for rate constants involving diffusion to and from the enzyme: k_2, 1.5×10^9 M^{-1} s^{-1}; k_{-4}, 3×10^9 M^{-1} s^{-1}; k_5, 10^9 M^{-1} s^{-1}. **b** Typical literature values for k_2 (1.5×10^8 M^{-1} s^{-1}) and k_{-4} (3×10^8 M^{-1} s^{-1}), with k_5 also proportionately reduced (10^8 M^{-1} s^{-1})

well catalyse the diffusion of remote molecules up to itself. In fact the rate constants of the binding steps, typically 3.10^8 M^{-1} s^{-1}, are very low for diffusion-controlled reactions; normal values would be at least ten times higher [13, 7], and there is perhaps a need to reinvestigate these rates using the usual tests for diffusion control (the dependence upon temperature and viscosity).

If the rate is further influenced by an intramolecular proton transfer, as postulated recently [27, 31], then this may be included in the calculation: use of the rate constants for this which are given in ref. [3] leads to the same conclusion, i.e., that the catalytic step has evolved almost, but not quite, to perfection.

In order to illustrate the high efficiency of the enzyme, a phenomenological comparison between carbonic anhydrase and the hydroxide ion is made in Table 1. Hydroxide is chosen because all of the many reactions catalysed by carbonic anhydrase are also catalysed by OH^- acting formally as a nucleophile attacking a carbon atom which bears a partial positive charge. The third column of Table 1 shows the concentration of OH^- which cause a first-order decomposition of substrate as fast as in the enzyme-substrate complex (column 1). This concentration is fairly low for all the substrates except the "natural" substrate CO_2. So, as we might expect, the enzyme is most efficient in the reaction for which Nature needed it.

Table 1. Comparison of carbonic anhydrase with hydroxide ion in the hydration or hydrolysis of different substrates

Substrate	k (s^{-1})	k_{OH^-}(M^{-1} s^{-1})	Effective [OH^-] (M)
CO_2	10^6	8×10^3	125
Acetaldehyde	10^3	4×10^3	0.25
Pyridine carboxaldehydes	10^2	5×10^3	0.02
p-Nitrophenyl acetate	10	9.5	1.1

2 Metal Complexes as Models

The intriguing question of why the enzyme needs zinc or cobalt in order to function has been discussed at length [5, 24], with the conclusion that the only property common to these two metals alone is flexibility of co-ordination, expressed for example in the greater affinity of the metal for the ligand 2,2′, 2″-triamino-triethylamine ("tren"), which tends to form irregular complexes, than for triethylene tetramine ("trien") which tends to form regular ones. This gave rise to the prediction by A.G. Maddock (personal communication) that in spite of indications to the contrary [3] the cadmium ion must also be active in the enzyme, since it too satisfies this criterion of flexibility. Recent experiments have shown that cadmium is indeed active [2, 20].

In order to rationalise many kinetic data the simple mechanism shown in Eq. (2) was proposed [25].

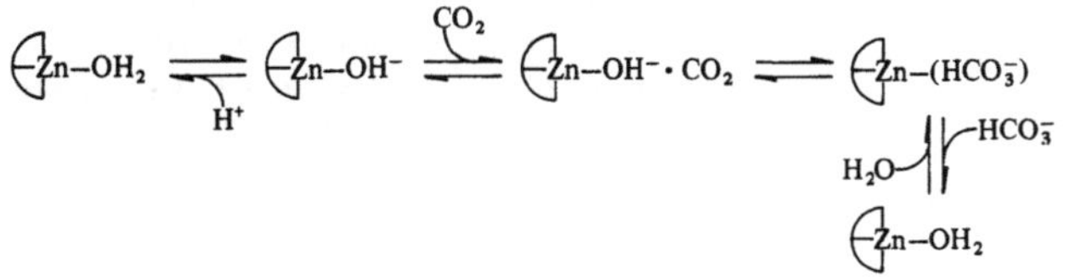

(2)

This suggestion was not well received, since the water co-ordinated to a free zinc ion does not lose a proton at neutral pH and since, even if a zinc-bound hydroxide ion could occur in the enzyme, the positively-charged, electron-withdrawing zinc ion should reduce the nucleophilic strength of OH^-, making it into a poor catalyst. To investigate this, use was made of chemical models for the enzyme. This term has been discussed elsewhere [28]; a model is not intended to mimic an enzyme, nor to show what the enzyme does; a model is a compound designed to show how chosen structural elements of the enzyme behave. Models do not yield information on the actual working of the enzyme, but they can verify or eliminate the possiblity of suggested mechanisms.

The obvious model for the ionisation of water bound to a metal ion in enzyme is water-bound to a metal in a simple complex, such as $(M.trien)^{2+}$. Figure 2 shows that the trien complex of zinc loses a proton more readily than those of cobalt, nickel and copper. This seems to be due to the adoption by the metal of a co-ordination number below six. The same result was obtained for tren complexes [30]. It has further been shown [1] that water bound to the tren complexes of Co^{2+}, Ni^{2+}, Cu^{2+}, and Zn^{2+} is also made more acidic if the environment of the metal is made less polar.

In order to eliminate uncertainty in these results due to the flexibility of the ligands, a range of macrocyclic ligands was made [28]. The structure of such a ligand is shown in Fig. 3a along with the conformation of its zinc complex in a crystal structure (Fig. 3b). Again, the complexes of the flexible metal ions Zn^{2+} and Co^{2+} showed five-co-ordination (four N, one H_2O), and the metal-bound water was ionisable, whereas water bound to the steroechemically rigid ions Cu^{2+} and Ni^{2+} ionised with a pK_a above 12. For the latter ions there was evidence of six-co-ordination (four N, two H_2O).

In the active form of carbonic anhydrase the co-ordination number appears to be four or five (three N, one or two H_2O) [14, 15]. It is therefore a priori to be expected

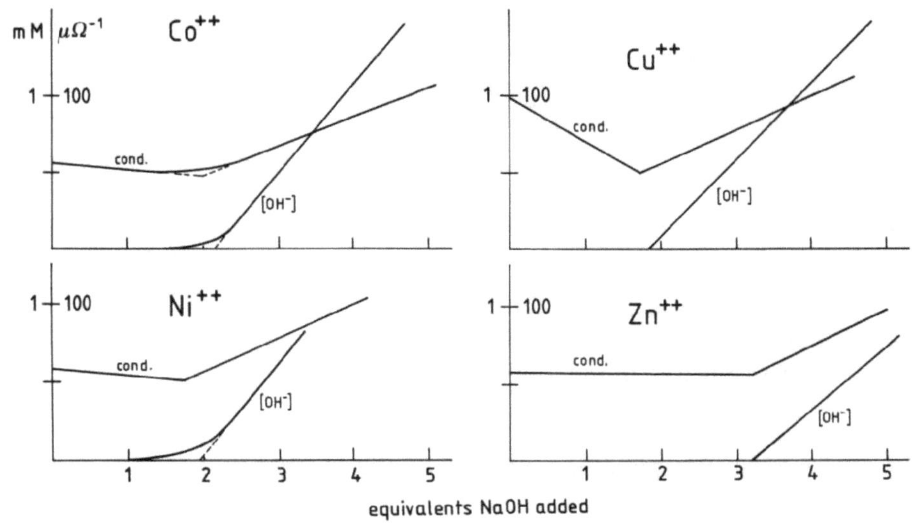

Fig. 2. Titration of equimolar mixtures of trien.2HCl and $CoCl_2$, $NiCl_2$, $CuCl_2$ or $ZnSO_4$ with sodium hydroxide [30] The zinc complex allows the ionisation of an additional proton. Method of Jonassen and Thielemann [11]

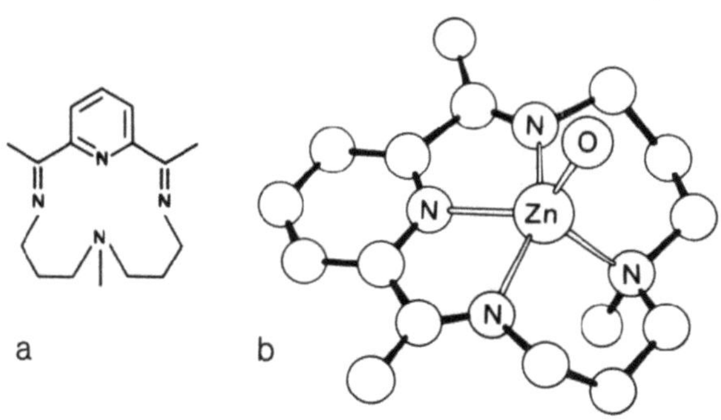

Fig. 3a, b. The macrocyclic complex used as a model for carbonic anhydrase. a The ligand. b Representation of the three-dimensional structure of the zinc complex. Details and literature in [28]

that (one of) the water molecule(s) bound to zinc in the enzyme should ionise at a fairly low pH. This expectation is based on the behaviour of the simple complexes, and means that, should the zinc-bound water in the enzyme be shown not to ionise, a further explanation will be needed.

On account of their stability the macrocyclic complexes have also proved suitable for studies of catalysis of the hydration of acetaldehyde and of carbon dioxide, an unnatural and a natural substrate for carbonic anhydrase. It was possible to show that the zinc macrocycle with an OH^- bound to the metal catalyses the hydration of both these substrates [28, 29], Acetaldehyde is attacked directly (nucleophilic catalysis) by the zinc-bound hydroxide ion [29]. and this mechanism is also expected to apply to carbon dioxide, since CO_2 hydration is as a rule not susceptible to the alternative form of catalysis available (general base catalysis) [5].

The study of model complexes thus leads to two empirical generalisations about the behaviour of metal-bound water.

1. The acidity of this water rises when the total co-ordination number of the metal is reduced.
2. The metal-bound hydroxide ion is a nucleophile (how strong, we shall see in a moment).

Naturally it remains a matter for enzymological investigation to determine whether carbonic anhydrase fits in with these generalisations or, on account of its more complex structure, provides an exception.

3 Comparison Between the Enzyme and the Models

The strength of a nucleophile is measured by the logarithm of its reaction rate with an electrophile, and this is related empirically to the strength with which the nucleophile binds a proton. Such a relationship can be demonstrated by the Brønsted plot. Examples of this are seen in Fig. 4, taken from ref. [29], for the attack of bound hydroxide on three substrates: propionic anhydride, acetaldehyde and carbon dioxide. In each case all the simple nucleophiles except water fall onto a straight line, confirming a common mechanism of reaction.

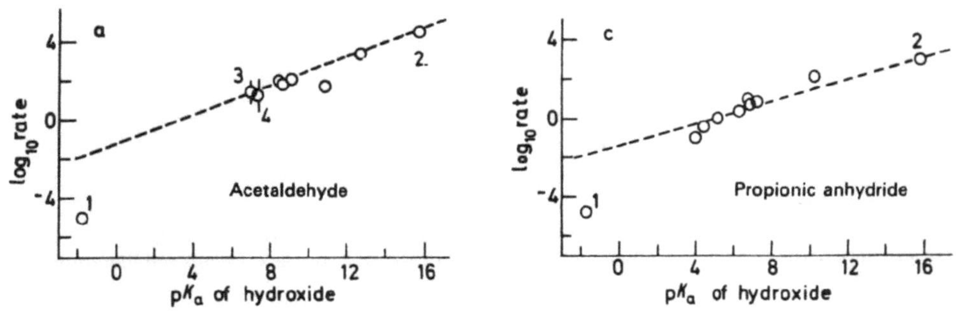

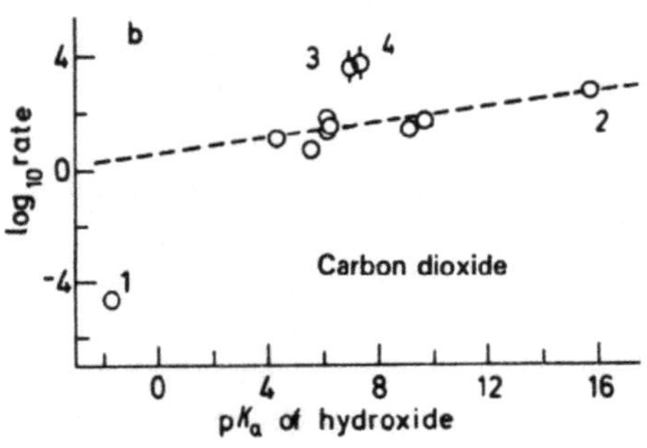

Fig. 4 a–c. Brønsted plots showing the correlation between the strength of a range of hydroxide-bearing nucleophiles. For each substrate the catalysts are: *1* H_2O; *2* OH^-_{aq}; *3* bovine carbonic anhydrase; *4* human carbonic anhydrase C; ***other points*** metal hydroxo complexes, from various authors' work. Data for 0°C. Details and literature in [29]

This comparison becomes more interesting when it is extended to include carbonic anhydrase. Taking the first-oder rate constant for decomposition of the (carbonic anhydrase.acetaldehyde) complex and assuming that the effective concentration of zinc-bound hydroxide in the active site is 55 M (give or take half an order of magnitude), we can calculate an effective second-order rate constant for this attack, which falls directly into line with the corresponding constants for the simple model compounds (Fig. 4a). Repeating the calculation for CO_2 (Fig. 4b) shows that the enzyme is a catalyst about 150-fold better than the apparent pK_a of its postulated hydroxide ion would suggest. This observation, in conjunction with the two above generalisations about metal-bound water, leads to a third conclusion:

3. Although the hydration by carbonic anhydrase of acetaldehyde may be due to simple nucleophilic attack by a metal-bound water, the hydration of carbon dioxide cannot be due to this alone, since it proceeds too rapidly.

This gives a more quantitative basis to our earlier observation that the enzyme is particularly efficient in catalysing the reaction of its natural substrate.

It has been suggested [12] that the function of carbonic anhydrase is solely to desolvate the reactants OH^- and CO_2, since the calculated activation energy for their reaction in vacuo is zero. This may be correct for the enzyme, but it is unlikely to be correct for the catalysts in Fig. 4b, since the bound hydroxide ions with the highest pK_a values should presumably be the most strongly solvated, and yet these are those that react the most rapidly with CO_2.

4 The Binding and Activation of Carbon Dioxide

How could the enzyme further accelerate the hydration of carbon dioxide? Kannan has suggested an internal proton transfer [14, 15] based on the structure of the active site, and various other schemes have been proposed such as desolvation, a charge relay system for protons, or binding of the CO_2 to the zinc. I would like to suggest a further factor, which is based on a guess about the nature of the substrate binding site for CO_2. First of all, it is difficult to see what properties a site must have in order to bind a CO_2 molecule. It cannot be simply apolar, since CO_2 is not particularly apolar (it is almost as soluble in water as in cyclohexane [19]); it cannot be composed of groups forming specific hydrogen bonds to the oxygens of CO_2, since these are *sp* hybridised with very weak lone pairs; it cannot provide simple co-ordination to the zinc ion, since CO_2 is a very weak ligand indeed, with no known co-ordination complexes involving 3d transition metals. CO_2 possesses no permanent dipole moment. It has however a quadrupole moment, with excess negative charge on each oxygen atom, and its enhanced solubility in dipolar, non-hydroxylic solvents [17, 19] and its strong interaction in vacuo with carbon disulphide (P. Madden, personal communication) which has an opposite quadrupole moment, suggests that the quadrupole interaction could provide the strength of binding to the enzyme. (Khalifah has pointed out that N_2O, a molecule very similar to CO_2, is not bound to the CO_2 site in carbonic anhydrase [17]. If the difference in electronegativity between the terminal and central atoms determines the quadrupole moment then on this basis CO_2 should indeed be preferentially bound.)

Thus we have a further factor which can accelerate the enzyme-catalysed hydration of CO_2 (and of CO_2 only, since no other known substrate will be able to fit properly into its binding site):

4. If the excess negative charge on the oxygen atoms is used to help bind the CO_2, then whatever electron-attracting group does this will also attract further charge onto the oxygens when the CO_2 is attacked by a nucleophile, lowering the energy of the transition state and thus accelerating the attack.

We may express this quantitatively using a crude calculation. If the CO_2 is in a field gradient then the energy of interaction, proportional to the excess charge (q electrons) on each oxygen, can be written as Aq. We assume that the orientation of the CO_2 at the active site is parallel to the field gradient within 15° in any direction and that in the absence of electrostatic effects the CO_2 competes on a molecule-for-molecule basis with H_2O to occupy the site. The generally accepted Michaelis constant for CO_2 is around 10 mM, making a binding constant (if it is one) of 100 M^{-1}. If further binding effects increase the binding by a factor S, then Eq. (3) follows.

$$100\ M^{-1} = \frac{1 - \cos 15^\circ \approx 0.03}{55\ M} \cdot S \cdot \exp \frac{Aq}{kT} \tag{3}$$

Now let the charge transferred to the CO_2 from the nucleophile in the transition state be x electrons, of which a fraction y is distributed between the oxygens and $(1 - y)$ is on the carbon. The new energy of interaction is proportional to the new charge on each oxygen, $(q + xy/2)$ electrons. The transition state is therefore stabilised with respect to the ground state by an energy $[A(q + xy/2) - Aq]$ or $Axy/2$, and the reaction accelerated by a factor $\exp(Axy/2\,kT)$. If $q = 0.2$ [23], and a value for S is assumed, then Eq. (3) leads to a value for A/kT; the gradient of Fig. 4b suggests a value around 0.15 for x. For various values of S and y the resulting acceleration is shown in Table 2. In a favourable case (low S, high y; the high electronegativity of oxygen will tend to keep y close to 1) this mechanism could clearly supply the two orders of magnitude of acceleration which the simple model catalysts lack.

Table 2. Acceleration of the hydration of CO_2 expected due to electrostatic interaction at the binding site, for various values of S, x and y

	S = 1	S = 10	S = 100
x = 0.15, y = 0.6	15	9	5
x = 0.15, y = 0.8	37	19	9
x = 0.15, y = 1.0	91	39	16
x = 0.20, y = 1.0	412	130	41

A more elaborate model, which dispenses with the arbitrary assumption that the CO_2 molecule is "bound" if its orientation is "correct" within 15° shows that such an acceleration has a minimum value of 60 for $q = 0.2$, $x = 0.15$ and $S = y = 1$.

5 Further Comments

The study of small molecules has thus shown that a mechanism such as suggested by Kannan [14, 15] is capable of giving the amount of catalysis that the enzyme shows – the amount expected from a nucleophilic hydroxide ion at neutral pH along with an extra polarisation of CO_2 due to its specific binding. If this should indeed be an approximation to what the enzyme really does, then three comments follow.

1. A pK_a of 7 for the nucleophile represents a further aspect of the enzyme which is "ideally evolved" for operation at a pH of 7; a higher or a lower pK_a would be disadvantageous [9].

2. An oligopeptide with zinc-binding properties could easily reproduce the activity of the model catalysts. So why is the large protein necessary? One reason may be to lower the co-ordination number of the metal (a co-ordination of 3N + $1H_2O$ has not yet been achieved in models) and thus to make the pK_a "ideal". Another is clearly to provide a site for the binding and perhaps the polarisation of CO_2. This is illustrated in a speculative way in Table 3, which indicates how much may be due to each of these factors in the development of k_{-3}/K_m – the kinetic parameter which is of importance to the evolving enzyme as long as the enzyme is not saturated [10]. A similar treatment of the reverse reaction (HCO_3^- dehydration) cannot be carried out at present since no information is available concerning the proton transfer step $M\text{-}OH\text{-}CO_2^- \rightleftharpoons M\text{-}O\text{=}CO_2H$.

Table 3. Expected catalysis of CO_2 hydration by peptides containing Zn^{2+}. The apparent second-order rate constant at pH 7 and 25°C is given

Catalyst	Rate constant ($M^{-1}\ s^{-1}$)
(Peptide-N_4 • $ZnOH_2$), pK_a 8.5	$200^{a} \times 10^{-(pK - pH)} = 6.3$
↓ reduced co-ordination no.	
(Peptide-N_3 • $ZnOH_2$), pK_a 7	$200 \times 10^{-0.15 \times \Delta pK}$ [b] $\times 1/2 = 60$
↓ CO_2 binding site	
(Peptide-N_3 • $ZnOH_2$), pK_a 7 $K_{CO_2} = 100\ M^{-1}$	$60 \times 100\ M^{-1} \times 55\ M = 3.3 \times 10^5$
↓ polarisation of bound CO_2	
Hypothetical enzyme, working at subsaturation [CO_2]	$3.3 \times 10^5 \times \sim 10^2 = \sim 3.3 \times 10^7$
Bovine carbonic anhydrase at subsaturation [CO_2]	$8.3 \times 10^7 \times 1/2 = 4.3 \times 10^7$ [c]

a Figure 4b implies 50 M^{-1} at 0°C, which together with the activation energy of carbonic anhydrase, 9 kcal/mol, gives 200 M^{-1} at 25°C. $10^{-(8.5-7)}$ of the enzyme is active at this pH

b The value 200 is reduced, when the pK_a is lowered, by a factor based on the Brønsted β (0.15, Fig. 4b). Half the enzyme is active at this pH

c From Kernohan [16]. Again, half the enzyme is active

3. It would be dangerous to infer from the foregoing that carbonic anhydrase may be more than heuristically compared to a hydroxide ion in a macrocyclic metal complex. The visible spectrum of cobalt-substituted carbonic anhydrase changes considerably when the pH is lowered [22]. The protonation of the model hydroxo-

cobaltous macrocyclic complex, on the other hand, produces only very small changes in its visible spectrum [30]. As observed above, a model cannot lead to positive deductions about an enzyme, but it can lead to negative ones. Enzymes are usually more complicated.

References

1. Coates JH, Gentle GJ, Lincoln SF (1974) Influence of local hydrophobic environment on acid dissociation constants of co-ordinated water molecules. Nature (London) 249: 773–775
2. Coleman JE (1979) Current concepts of the mechanism of action of carbonic anhydrase. This symposium
3. Coleman JE (1967) Metal ion dependent binding of sulphonamide to carbonic anhydrase. Nature (London) 214: 193–195
4. Coleman JE (1973) Metal ions in enzymic catalysis. In: Eichhorn E (ed) Inorganic biochemistry. Elsevier, Amsterdam, pp 488–548
5. Dennard AE, Williams RJP (1966) The catalysis of the reaction between carbon dioxide and water. J Chem Soc A: 812–816
6. Dennard AE, Williams RJP (1966) Transition metal ions in metalloenzymes. Trans Metal Chem 2: 115
7. Eigen M, Hammes GG (1963) Elementary steps in enzyme reactions. Adv Enzymol 25: 1–38
8. Fersht A (1977) Enzyme structure and mechanism, reading: Freeman WH, p 259
9. Jencks WP (1969) Catalysis in chemistry and enzymology. McGraw-Hill, New York, pp 83–85
10. Jencks WP (1969) Catalysis in chemistry and enzymology. McGraw-Hill, New York, pp 253–257
11. Jonassen HB, Thielemann H (1963) Polyamin-Hydroxo-Komplexe von Übergangsmetallen. Z Anorg Allg Chem 320: 274–282
12. Jönsson B, Karlström G, Wennerström H (1978) Ab initio molecular orbital calculations on the water-carbon dioxide system. J Amer Chem Soc 100: 1658–1661
13. Jönsson B, Wennerström H (1978) Diffusion control in reversible enzyme reactions. Applications to carbonic anhydrase. Biophys Chem 7: 285–292
14. Kannan KK (1979) Crystal structure of carbonic anhydrase. This symposium
15. Kannan KK, Petef M, Fridborg K, Cid-Dresdner H, Lövgren S (1977) Structure and function of carbonic anhydrases. FEBS Lett 73: 115–119
16. Kernohan JC (1965) The pH-activity curve of bovine carbonic anhydrase and its relationship to the inhibition of the enzyme by anions. Biochim Biophys Acta 96: 304–317
17. Khalifah RG (1971) The carbon dioxide hydration activity of carbonic anhydrase. Stopped-flow kinetic studies on the native human enzymes B and C. J Biol Chem 246: 2561–2573
18. Koenig SH, Brown RD, Jacob GS (1979) The pH-independence of carbonic anhydrase activity. Apparent pK_a due to inhibitiion by HSO_4^-. This symposium
19. Landolt-Börnstein (1962) Bartels J et al. (eds) Zahlenwerte und Funktionen, 6. Aufl, Bd. 2, 2. Teil, Bandteil 6 (Lösungsgleichgewichte I.) Springer, Berlin Heidelberg New York, pp 1-34–1-120
20. Lee WK (1979) to be published
21. Lindskog S (1971) Carbonic anhydrase. In: Boyer PD (ed) The enzymes. Academic Press, London New York, pp 587–665
22. Lindskog S (1963) Effects of pH and inhibitors on some properties related to metal binding in bovine carbonic anhydrase. J Biol Chem 238: 945–951
23. McLean AD (1963) Molecular orbitals of CO_2. J Phys Chem 38: 1347–1359
24. Prince RH, Woolley P (1972) Metal ion function in carbonic anhydrase. Angew Chem 84: 461–471; Angew Chem Int Edn 11: 408–417

25. Smith E (1949) The mode of action of the metal-peptidases. Proc Natl Acad Sci USA 35: 80–90
26. Steiner H, Jonsson B-H, Lindskog S (1975) The catalytic mechanism of carbonic anhydrase. Hydrogen-isotope effects on the kinetic parameters of the human C isozyme. Eur J Biochem 59: 253–259
27. Steiner H, Jonsson B-H, Lindskog S (1976) The catalytic mechanism of human carbonic anhydrase C: Inhibition of CO_2 hydration and ester hydrolysis by HCO_3^-. FEBS Lett 62: 16–20
28. Woolley P (1975) Models for metal ion function in carbonic anhydrase. Nature (London) 258: 677–682
29. Woolley P (1977) Metal-ion-assisted catalysis of nucleophilic attack. Part 4. Co-operation between metal ion and hydroxide. JCS (Perkin II) 318–324
30. Woolley P (1973) Ph D Thesis, University of Cambridge
31. Wyeth P (1976) Ph D Thesis, University of Cambridge
32. Wyeth P, Prince RH (1977) Carbonic anhydrase. Inorg Perspect Biol Med 1: 37–71

Consequences of Metal Ion Substitutions in Human Carbonic Anhydrase B. A Carbon-13 Nuclear Magnetic Resonance Comperative Study

R.G. KHALIFAH and P.J. MORLEY [1]

1 Introduction

The replacement of the intrinsic zinc cofactor of carbonic anhydrase by spectroscopically and magnetically useful metal ions has provided powerful insights into its active site structure and its mechanism of action (Lindskog et al., 1971). However, although only Co^{2+} reactivates the CO_2 hydration activity of the enzyme, many other replacement studies have been interpreted under the implicit assumption that the new metals do not significantly perturb the active site as compared to the zinc. Our recent successful introduction of a ^{13}C magnetic resonance "probe" into the active site of human carbonic anhydrase B via carboxymethylation with [1-^{13}C] bromoacetate permits comparative studies on the effects of diverse metal replacements. The ^{13}C approach has permitted the study of the introduced carboxylate environment and the determination of the ring pK_a of the modified $CmHis_{200}$ in the active site (cf. Strader and Khalifah, 1976; Khalifah, 1977; Khalifah et al., 1977; Jeffers et al., 1978; Khalifah and Jeffers, 1978).

2 ^{13}C NMR Studies

2.1 Diamagnetic Spectra

Figure 1 shows typical ^{13}C NMR spectra of N^{τ}-carboxymethylhistidine$_{200}$ HCAB (or CmHCAB) prepared with 90% [1-^{13}C] bromoacetate alkylation. The downfield carbonyl carbon region is shown, centered about 173 ppm from Me_4Si. The natural abundance background is seen as the broad hump (cf. Fig. 1C), while the tall resonance represents the single enriched carboxyl carbon covalently attached to His_{200}. The resonance is easily detected in CmHCAB (Fig. 1B), in metal-free apoCmHCAB (Fig. 1A), and in diamagnetic derivatives of the enzyme, e.g., with Cd^{2+} and Hg^{2+} replacing the Zn^{2+}. The Co^{2+} derivatives are discussed below.

---

1 Department of Chemistry, University of Virginia, Charlottesville, Virginia 22901/USA

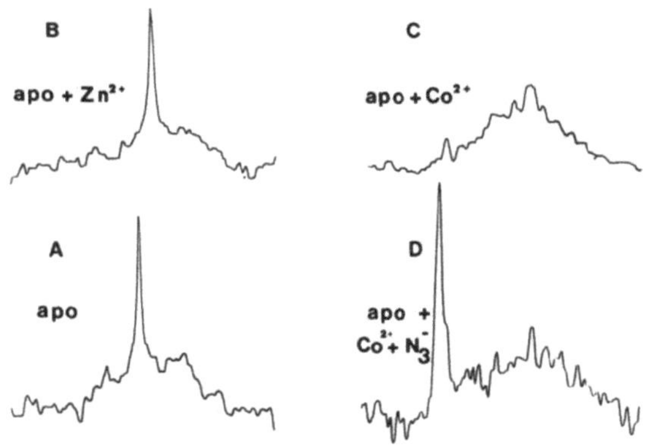

Fig. 1 A–D. ^{13}C NMR spectra (carbonyl region) at 25.15 MHz of derivatives of enriched CmHCAB. **A** ApoCMHCAB, **B** Reconstituted ZnCmHCAB, **C** Co (II) CmHCAB, **D** Co (II) CmHCAB + azide

2.2 Titration Behavior and Inhibition

The chemical shift of the enriched carboxylate is sensitive to pH and reflects the ionization of the imidazole ring to which it is attached (protonation at N^{π}). In the case of only the zinc CmHCAB, a further perturbation is seen from an ionizing group with a pK_a of 9.2, as shown by the dashed curve of Fig. 2. The chemical shift plateau around pH 8 in CmHCAB is abnormal in being about 1 ppm upfield from its expected position from model compound studies (Khalifah et al., 1977). These two anomalies have been attributed to the coordination of the enriched carboxylate to the zinc at pH 8, with its disruption (normalization of the chemical shift) at pH 10 due to displacement by hydroxide and at pH 5 due to ring protonation and rotation. These anomalous features are eliminated when: (a) inhibitors are added (Fig. 2); (b) Zn^{2+} is removed; (c) Zn^{2+} is replaced by Cd^{2+} or Hg^{2+}. Table 1 summarizes the available $CmHis_{200}$ pK_a information. The diamox-induced increase in pK_a is unusually large and implicates a strong interaction.

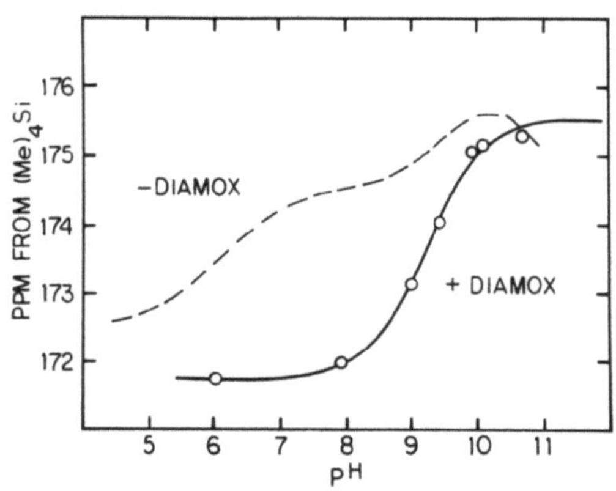

Fig. 2. pH dependence of ^{13}C-NMR signal of enriched ZnCmHCAB in presence (*experimental points*) and absence (*dashed curve*) of diamox at 25°C

Table 1. $CmHis_{200}$ pK_a in CmHCAB and derivatives

Enzyme	Free	I^-	CN^-	Diamox	N_3^-
Zn^{2+}	6.0 [a]	7.45 [b]	7.38	9.23	7.34 [b]
Apo	≅4 [c]	–	–	–	–
Hg^{2+}	5.8	6.0	–	–	–
Cd^{2+}	≅5	≪5	–	–	–
Co^{2+}	–	–	–	–	–

a Taken from Khalifah et al. (1977)
b Taken from Khalifah et al. (1977)
c Taken from Jeffers et al. (1978)

2.3 Coupling Constants to ^{113}Cd and ^{199}Hg

The reconstitution of the Cd^{2+} and Hg^{2+} derivatives of CmHCAB with 95% enriched spin 1/2 isotopes leads to no observable splittings in the enriched carboxylate resonance (Fig. 3). Carboxylate-metal coordination in model complexes (EDTA) leads to large coupling constants (13–35 Hz).

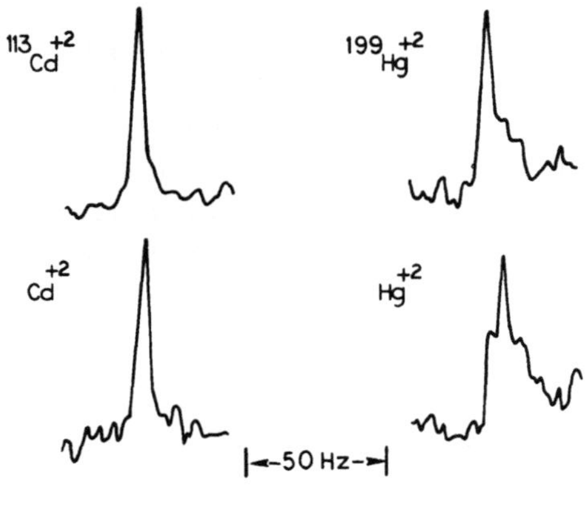

Fig. 3. ^{13}C NMR of enriched carboxymethyl carboxylate in Hg^{2+} and Cd^{2+} derivatives of CmHCAB. Metals were either at natural abundance or were >90% spin 1/2 isotopes. Metals present in 1.1 molar excess except HgCmHCAB where less than one equivalent resulted in appearance of apo shoulder downfield

2.4 Co^{2+} Paramagnetic Effects

The addition of Co^{2+} to enriched apoCmHCAB leads to the disappearance of the carboxylate resonance (Fig. 1C), suggestive of a close approach of the carboxylate to the metal with resulting contact shifts. However, inhibition of the Co^{2+} CmHCAB leads to the reappearance of the resonance (Fig. 1D), usually with a pronounced downfield pseudo-contact shift. Spin-lattice relaxation measurements define in such instances the metal-carboxyl carbon distances and prove the absence of direct coordination (Table 2).

Table 2. Approximate metal-carboxylate distances in inhibitor complexes of Co^{2+} CmHCAB [a]

Complex	T_{1p} (s)	r (Å) [b]
None	? .	Corrdinated?
N_3^-	0.43 –0.51	8.5–8.7
I^-	0.12 –0.14	6.8–7.0
Diamox	0.10 –0.12	6.7–6.8
CN^-	0.037–0.041	5.6–5.7

a Determination of T_1 by 180°-τ-90° pulse sequence

b Calculations of r (Å) incorporate a C value of 460 Å $s^{-1/3}$

3 Conclusions

Metalloderivatives of CmHCAB show great differences in two respects. (1) Coordination of the carboxylate occurs only in zinc CmHCAB (pH 8) and probably Co^{2+} CmHCAB. No coordination occurs in Cd^{2+}- and Hg^{2+}-derivatives, as evidenced by the normal titration behavior and the absence of coupling to their spin 1/2 isotopes. (2) The pK_a of the ring of $CmHis_{200}$ is very sensitive to the presence of a metal in the active site and the particular nature of the metal. It is also sensitive, of course, to the inhibition of the different metalloderivatives. The CmHis side-chain is probably at least 6 Å from the metal (Kannan et al., 1975), making the above results quite surprising. Metal replacement studies should be interpreted with great caution when being generalized to the native zinc enzyme.

Acknowledgments. This research was generously supported by a grant from the USPHS (HL-17963).

References

Jeffers PK, Sutherland WMcI, Khalifah RG (1978) Carbon-13 magnetic resonance of carboxymethylated human carbonic anhydrase B. Chemical shift and spin-lattice relaxation studies. Biochemistry 17: 1305–1312

Kannan KK, Notstrand B, Fridborg K, Lövgren S, Ohlsson A, Petef M (1975) Crystal structure of human erythrocyte carbonic anhydrase B. Three dimensional structure at a nominal 2.2-Å resolution. Proc Natl Acad Sci USA 72: 51–55

Khalifah RG (1977) Histidine-200 alters inhibitor binding in human carbonic anhydrase B. A carbon-13 nuclear magnetic resonance identification. Biochemistry 16: 2236–2240

Khalifah RG, Jeffers PK (1978) Dynamic C-13 NMR study of human carbonic anhydrase B. In: Agris PF (ed) Biomolecular structure and function. Academic Press, London New York, pp 363–368

Khalifah RG, Strader DJ, Bryant SH, Gibson SM (1977) Carbon-13 nuclear magnetic resonance probe of active site ionizations in human carbonic anhydrase B. Biochemistry 16: 2241–2247

Lindskog S, Henderson LE, Kannan KK, Liljas A, Nyman PO, Strandberg B (1971) Enzymes 5: 587–665

Strader DJ, Khalifah RG (1976) ^{13}C nuclear magnetic resonance probe of the active site of carbonic anhydrase. J Am Chem Soc 98: 5043–5044

Rate-Limiting Steps in the Catalytic Action of Carbonic Anhydrase

S. LINDSKOG [1]

With a maximum turnover number of 10^6 s^{-1} at 25°C, human carbonic anhydrase C is one of the most efficient of all known enzymes (Khalifah, 1971; Steiner et al., 1975). The catalyzed reaction is readily reversible:

$$CO_2 + H_2O \rightleftharpoons HCO_3^- + H^+ \tag{1}$$

In the usual assay in vitro as well as in vivo the interconversion of CO_2 and HCO_3^- is coupled to the titration of buffering systems (B/BH^+):

$$B + H^+ \rightleftharpoons BH^+ \tag{2}$$

To achieve this efficient catalysis the enzyme must have a mechanism which allows each one of the necessary elementary reactions to proceed at a high rate. Six chemical events that must take place during the catalyzed hydration of CO_2 are listed in Table 1. In the enzymic mechanism these events may occur as discrete reaction steps, or some of them may be concerted in a single reaction step.

Table 1. Elementary reactions in carbonic anhydrase catalyzed CO_2 hydration (and HCO_3^- dehydration)

1. Binding (dissociation) of CO_2
2. Binding (dissociation) of H_2O
3. Breaking (formation) of O-H bond in H_2O
4. CO_2-HCO_3^- transformation: formation (breaking) of O-C bond
5. Dissociaton (binding) of HCO_3^-
6. Dissociation (binding) of H^+

The purpose of this paper is to summarize the available evidence concerning the time course of these chemical events during the catalytic action of carbonic anhydrase and to discuss which steps limit the overall rate of catalysis under different conditions.

1 Avdelningen för biokemi, Umeå Universitet, S-901 87 Umeå/Sweden

1 Dissociation and Binding of H^+

For each turnover cycle of the catalyzed reaction (1) the equivalent of one H^+ ion must be transferred between the active site and components of the medium. It is often assumed that such H^+ transfer processes are so rapid that they can be considered at quasiequilibrium during catalysis. However, because of the rapid turnover rate of carbonic anhydrase this is not necessarily the case. In fact, this reaction step has been the subject of much debate and some confusion during the last few years (Caplow, 1971; Khalifah, 1971; Koenig and Brown, 1972; Khalifah, 1973; Prince and Woolley, 1973; Lindskog and Coleman, 1973).

A multitude of studies have shown that the active site of carbonic anhydrase contains an ionizing group with an apparent pK_a near 7 at physiological values of ionic strength (Lindskog et al., 1971; Pocker and Sarkanen, 1978). If the compulsory H^+ transfer could only take place between this active site group and H_2O or OH^- according to Eq. (3) and (4), then this reaction would limit the rate of turnover to about 10^3 s^{-1} (Eigen and Hammes, 1963).

$$EH^+ + H_2O \rightleftharpoons E + H_3O^+ \qquad (3)$$

$$EH^+ + OH^- \rightleftharpoons E + H_2O \qquad (4)$$

However, this is $10^2 - 10^3$ times less than the observed turnover rates, which means that the compulsory H^+ transfer must occur by some other mechanism. Thus, it is possible that H^+ is rapidly transferred directly between the active site group and a buffer component according to Eq. (5):

$$EH^+ + B \rightleftharpoons E + BH^+ \qquad (5)$$

If such a "buffer-mediated" H^+ transfer is actually required for rapid catalysis, the turnover rate is expected to depend on the buffer concentration when the buffer concentration is sufficiently low to make reaction (5) slow compared to other steps in the catalytic cycle.

We have studied the CO_2 hydration reaction under steady-state conditions in different buffers (Jonsson et al., 1976). When the buffer concentration is less than about 20 mM a buffer dependence is, indeed, observed. Both the apparent k_{cat} (V_{max}/E_{tot}) and the apparent K_m depend on the buffer concentration, while k_{cat}/K_m appears buffer-independent. This pattern is analogous to that of a "ping-pong" mechanism with CO_2 and buffer as the substrates. Table 2 shows the data obtained for some buffers often used in kinetic studies of carbonic anhydrase. Rate constants for the transfer of H^+ between EH^+ and the buffer base, B, were estimated to about 10^8 M^{-1} s^{-1}, which is not very far from the value expected for a diffusion-controlled reaction.

Table 2. Parameters for the buffer activation of the CO_2 hydration activity of human carbonic anhydrase C (Jonsson et al., 1976)

Buffer system	pK_a	pH	$(k_{cat})_{max}$ s^{-1}	K_m^{buf} mM	k_B $M^{-1}\ s^{-1}$
3,5-lutidine	6.2	6.65	2.0×10^5	6	0.5×10^8
2,2-diethylmalonate	7.3	7.8	7.4×10^5	11	0.9×10^8
N-methylimidazole	7.2	7.8	6.9×10^5	5.5	1.6×10^8
1,2-dimethylimidazole	8.2	8.9	1.0×10^6	6	2.0×10^8

The buffer K_m values, K_m^{buf}, were estimated on the basis of the total buffer concentration at CO_2 saturation. The rate constants, k_B, represent minimal values for the reaction of EH^+ with buffer base, B

2 The Splitting of H_2O

At buffer concentrations greater than about 20 mM the rate of CO_2 hydration does not depend on buffer. Apparently buffer-mediated H^+ transfer is not rate-limiting under such conditions. Steady-state kinetic studies in 50 mM buffers in H_2O and in D_2O have shown that k_{cat} as well as K_m have hydrogen isotope effects of 3–4 (Steiner et al., 1975). These results suggest that the rate-limiting step at substrate and buffer saturation involves a transfer of H^+ between two sites in the enzyme. Since k_{cat}/K_m has an isotope effect near 1, this H^+ transfer appears to be uncoupled from the CO_2-HCO_3^- transformation step. Further evidence on this point is presented in the next section. Our interpretation of these results is that a "proton transfer" group in the active site accepts a H^+ from the "catalytic" group and, subsequently, delivers it to the buffer. Possibly, this proton transfer group is His 63 which is a titratable active site group with a pK_a of 7.1 (Campbell et al., 1975). The most reasonable candidate for the catalytic group is a zinc-bound H_2O molecule (Campbell et al., 1977). If this assignment is correct, the rate-limiting, intramolecular H^+ transfer would involve the splitting of the zinc-bound H_2O molecule according to Eq. (6):

$$\text{E-Zn-OH}_2 \rightleftharpoons \text{H}^+\text{-E-Zn-OH}^- \tag{6}$$

When the observed pattern of isotope effects and the observed pH dependencies of the kinetic parameters were fitted to a kinetic model it was necessary to assume that only the proton transfer group, and not the catalytic group, can participate effectively in the buffer-mediated H^+ transport according to Eq. (5). Since the metal ion vicinity is known to interact with a variety of ligands, this assumption seems somewhat artificial. An alternative possibility, which is intuitively more attractive, but which so far lacks good experimental support, is that the intramolecular H^+ transfer takes place between the zinc-bound H_2O and an enzyme-bound buffer molecule according to Eq. (7) (Jonsson et al., 1976):

$$\text{E-Zn-OH}_2 + \text{B} \rightleftharpoons \text{E-Zn-OH}_2\text{-B} \rightleftharpoons \text{E-Zn-OH}^-\text{-BH}^+ \rightleftharpoons \text{E-Zn-OH}^- + \text{BH}^+ \tag{7}$$

The central step of this reaction sequence contains the rate-limiting H^+ transfer and involves the splitting of H_2O. The whole sequence represents the transport of H^+ from catalytic group to buffer.

3 The CO_2-HCO_3^- Transformation and Product Dissociation

The ^{13}C NMR resonance lines of $^{13}CO_2$ and $H^{13}CO_3^-$ in chemical equilibrium broaden in the presence of low concentrations of carbonic anhydrase (Koenig et al., 1973, 1974). This broadening is related to the catalyzed rate of exchange between CO_2 and HCO_3^-. The exchange rates, v_{exch}, were found to depend on the substrate concentration according to Eq. (8):

$$\frac{v_{exch}}{[E_{tot}]} = \frac{k_{cat}^{exch} \cdot [S]}{K_{eff} + [S]} \tag{8}$$

For the human C enzyme, the maximal exchange rate constant, k_{cat}^{exch}, is only weakly pH-dependent (Simonsson et al., 1979). The apparent dissociation constant for HCO_3^-, $K_{eff}^{HCO_3^-}$, is strongly pH-dependent and the binding becomes very weak at high pH values (Fig. 1). The maximal exchange rates are at least eightfold faster than

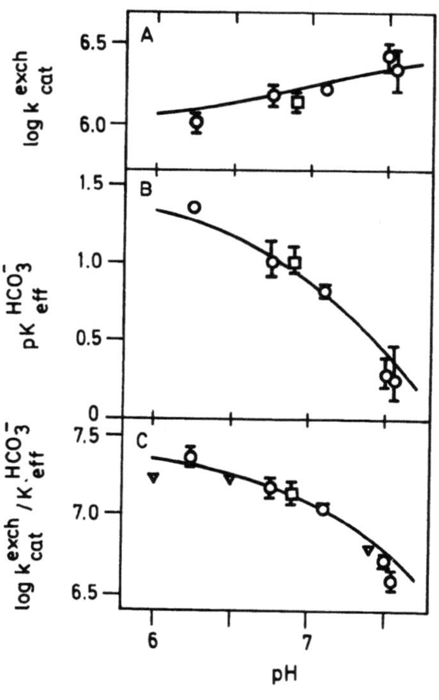

Fig. 1 A–C. pH dependence of exchange kinetic parameters for the human C enzyme. **A** log k_{cat}^{exch} (s^{-1}); **B** p$K_{eff}^{HCO_3^-}$ (M); **C** log $k_{cat}^{exch}/K_{eff}^{HCO_3^-}$ ($M^{-1}\ s^{-1}$). *Circles* represent values in 94% H_2O and 6% D_2O; *squares* values in 99.5 D_2O; ***triangles*** are values of $k_{cat}^{HCO_3^-}/k_m^{HCO_3^-}$ from steady-state measurements in 50 mM buffers, Values of pH are uncorrected. Temperature 25°C. (Simonsson et al., 1979)

Table 3. Observed values of the maximal exchange rate constant, k_{cat}^{exch}, for human carbonic anhydrase C, and values calculated from steady-state kinetic results assuming that exchange and turnover are limited by the same reaction steps (Simonsson et al., 1979)

pH	k_{cat}^{exch}, observed s^{-1}	k_{cat}^{exch}, calculated s^{-1}
6.25	1.0×10^6	1.0×10^5
6.75	1.5×10^6	1.8×10^5
7.55	2.2×10^6	1.3×10^5

expected if the rate of exchange were limited by reaction steps also limiting the rates of the turnover reactions (Table 3). The exchange proceeds rapidly in the absence of added buffers, and the addition of buffers has only small effects on the exchange parameters. Furtermore, the exchange is just as rapid in D_2O as in H_2O. From these observations it can be concluded that the chemical events that must be involved in the exchange reaction (CO_2 binding, CO_2-HCO_3^- transformation, HCO_3^- dissociation, and the reverse of these processes) are not concerted with the buffer-dependent step and the intramolecular H^+ transfer step which limit the rate of turnover at low and high buffer concentrations, respectively. The exchange results can be described by the following reaction sequence:

$$E\text{-}OH^- + CO_2 \rightleftharpoons E\text{-}OH^-\text{-}CO_2 \rightleftharpoons E\text{-}HCO_3^- \overset{H_2O}{\rightleftharpoons} E\text{-}OH_2 + HCO_3^- \quad (9)$$

The chemical transformation between CO_2 and HCO_3^- is, of course, of central importance in the catalytic mechanism of carbonic anhydrase. Apparently, evolution has resulted in a very efficient mechanism for this transformation which does not limit the turnover rates of the high specific activity C enzyme. It is interesting to note that all of the proposed mechanisms for this transformation, except the Zn-OH^- mechanism indicated in Eq. (9), are associated with H^+ transfer (Pocker and Sarkanen, 1978). Unfortunately, our exchange kinetic results are no definite proof for the Zn-OH^- mechanism. The possibility remains that the rate of exchange is limited by product dissociation rather than by the transformation step. This would require that the transformation is associated with rate constants approaching 10^7 s^{-1}, however.

The various features of the catalytic mechanism of carbonic anhydrase C which have been discussed above are integrated into the kinetic scheme shown in Fig. 2. The scheme is in agreement with the pH dependencies of all the observed steady-state and equilibrium exchange kinetic parameters. The "proton transfer" group is assumed to operate more or less independently of the state of the "catalytic" group.

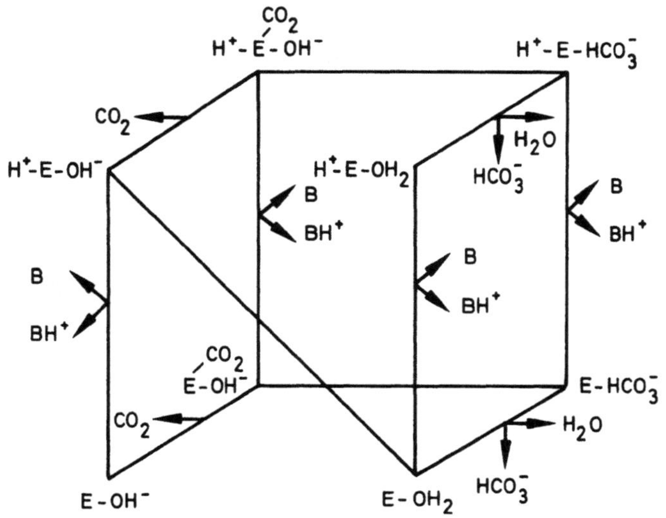

Fig. 2. Scheme of the catalytic mechanism of carbonic anhydrase C. Protonated "proton transfer" group is indicated by H^+ to the left of E. ***Vertical lines*** represent "buffer-mediated" H^+ transfer. The ***diagonal line*** represents the intramolecular H^+ transfer limiting the turnover rates at saturating concentrations of substrate and buffer

4 What Reaction Steps Limit the Rate of Carbonic Anhydrase Action in Vivo?

The previous discussion has shown that the turnover rate achieved by the human C enzyme at saturating substrate concentrations is limited by steps involved in the transfer of H^+ between zinc-bound H_2O and the medium, while CO_2-HCO_3^- transformation and product dissociation are rapid processes. However, the physiological concentrations of CO_2 and HCO_3^- are usually smaller than the K_m values measured at high buffer concentrations in vitro. Thus, it seems likely that the rate in vivo is limited either by the buffer-mediated H^+ transport or by substrate binding. The intracellular concentrations of low-molecular weight substances buffering around pH 7 might sometimes be suboptimal for the carbonic anhydrase reaction. However, recent observations by Silverman et al. (1978) suggest that hemoglobin can provide the required buffer function. Furthermore, there is evidence suggesting that carbonic anhydrase binds to hemoglobin so that H^+ transfer can take place in a complex between these two proteins (Silverman et al., 1979). These findings suggest that H^+ transport would not limit the rate of carbonic anhydrase action in vivo, at least not in the red cell.

When H^+ transport is rapid and substrate concentrations are smaller than K_m, the rate of CO_2 hydration, v_h, can be approximated by Eq. (10)

$$\frac{v_h}{[E_{tot}]} = \frac{k_{cat}^{CO_2}}{K_m^{CO_2}} \left([CO_2] - \frac{[H^+] \cdot [HCO_3^-]}{K_{CO_2}} \right) \tag{10}$$

where K_{CO_2} is the apparent acid dissociation constant of CO_2. The apparent second order rate constant, $k_{cat}^{CO_2}/K_m^{CO_2}$, is proportional to the rate constant for CO_2 binding and the probability that CO_2 binding will result in product formation. At pH 7, this rate constant is about 6×10^7 M^{-1} s^{-1}, which approaches the value for a diffusion-controlled reaction. Thus, it would seem that human carbonic anhydrase C has evolved into an almost optimally efficient catalyst. Further evolution could only have marginal effects on the catalytic efficiency of the enzyme. In this connection it is interesting to note that the concentration of carbonic anhydrase in the red cell and in some other tissues is quite high. A selective pressure on the organism to achieve a faster response to perturbations of the CO_2-HCO_3^- equilibrium could most simply be relieved by the production of more enzyme.

On the other hand, human carbonic anhydrase B appears to be far from optimally efficient, particularly since this isoenzyme is strongly inhibited by physiological concentrations of Cl^- (Maren et al., 1976). The question as to why this relatively inefficient isoenzyme [and the even less active muscle enzyme (Register et al., 1978)] has survived through evolution in the presence of the very efficient C enzyme has not yet been answered.

References

Campbell ID, Lindskog S, White AI (1975) A study of the histidine residues of human carbonic anhydrase C using 270 MHz proton magnetic resonance. J Mol Biol 98: 597–614

Campbell ID, Lindskog S, White AI (1977) Proton magnetic resonance studies of histidines in human, rhesus monkey, and bovine carbonic anhydrases. Biochim Biophys Acta 484: 443–452

Caplow M (1971) Bromine catalysis for carbon dioxide hydration and dehydration and some observations concerning the mechanism of carbonic anhydrase. J Am Chem Soc 93: 230–235

Eigen M, Hammes GG (1963) Elementary steps in enzyme reactions. Adv Enzymol 25: 1–38

Jonsson B-H, Steiner H, Lindskog S (1976) Participation of buffer in the catalytic mechanism of carbonic anhydrase. FEBS Lett 64: 310–314

Khalifah RG (1971) The carbon dioxide hydration activity of carbonic anhydrase. I. Stop-flow kinetic studies on the native human isoenzymes B and C. J Biol Chem 246: 2561–2573

Khalifah RG (1973) Carbon dioxide hydration activity of carbonic anhydrase: Paradoxical consequences of the unusually rapid catalysis. Proc Natl Acad Sci USA 70: 1986–1989

Koenig SH, Brown RD (1972) H_2CO_3 as substrate for carbonic anhydrase in the dehydration of HCO_3^-. Proc Natl Acad Sci USA 69: 2422–2425

Koenig SH, Brown RD, Needham TE, Matwiyoff NA (1973) Kinetic parameters of human carbonic anhydrase B as determined from NMR linewidths of ^{13}C in CO_2 and HCO_3^-. Biochem Biophys Res Commun 53: 624–630

Koenig SH, Brown RD, London RE, Needham TE, Matwiyoff NA (1974) The kinetic parameters of carbonic anhydrase by ^{13}C NMR. Pure Appl Chem 40: 103–113

Lindskog S, Coleman JE (1973) The catalytic mechanism of carbonic anhydrase. Proc Natl Acad Sci USA 70: 2505–2508

Lindskog S, Henderson LE, Kannan KK, Liljas A, Nyman PO, Strandberg B (1971) Carbonic anhydrase. In: Boyer PD (ed) The enzymes, vol 5, 3rd edn. Academic Press, London New York, pp 587–665

Maren TH, Rayburn CS, Liddell NE (1976) Inhibition by anions of human red cell carbonic anhydrase B: Physiological and biochemical implications. Science 191: 469–472

Pocker Y, Sarkanen S (1978) Carbonic anhydrase: Structure, catalytic versatility, and inhibition. Adv Enzymol 47: 149–274

Prince RH, Woolley PR (1973) On the mechanism of action of carbonic anhydrase. Bioorg Chem 2: 337–344

Register AM, Koester MK, Noltmann EA (1978) Discovery of carbonic anhydrase in rabbit skeletal muscle and evidence for its identity with "basic muscle protein". J Biol Chem 253: 4143–4152

Silverman DN, Tu CK, Wynns GC (1978) Proton transfer between hemoglobin and the carbonic anhydrase active site. J Biol Chem 253: 2563–2567

Silverman DN, Backman L, Tu CK (1979) Role of hemoglobin in proton transfer to the active site of carbonic anhydrase. Biol Chem 254: 2588–2591

Simonsson I, Jonsson B-H, Lindskog S (1979) A ^{13}C nuclear-magnetic-resonance study of CO_2-HCO_3^- exchange catalyzed by human carbonic anhydrase C at chemical equilibrium. Eur J Biochem 93: 409–417

Steiner H, Jonsson B-H, Lindskog S (1975) The catalytic mechanism of carbonic anhydrase. Hydrogen-isotope effects on the kinetic parameters of the human C isoenzyme. Eur J Biochem 59: 253–259

The pH-Independence of Carbonic Anhydrase Activity: Apparent pK_a Due to Inhibition by HSO_4^-

S.H. KOENIG, R.D. BROWN, III, and G.S. JACOB [1]

1 Introduction

The carbonic anhydrases (EC 4.2.1.1), zinc metallo-enzymes found, for example, in mammalian erythrocytes, catalyze the reversible hydration-dehydration reaction $CO_2 + H_2O \rightleftharpoons H^+ + HCO_3^-$ (cf. [8, 30, 40] for reviews). The various isozymes share several properties: (a) the intrinsic catalytic rates are among the fastest of any enzymatic system ($k_{cat} \gtrsim 10^5\ s^{-1}$); (b) the enzymatic activity has been found consistently to be pH-dependent, the reported pK_a ranging between 6 and 8, with the hydration activity greatest at high pH; (c) the enzymatic activity is inhibited by small monovalent anions, e.g., F^-, Cl^-, SH^-, SCN^-, NO_3^-, etc., as well as by a variety of sulphonamides, all of which appear to bind (with differing affinities) more strongly at lower values of pH; and (d) Co^{2+} ions can substitute for the essential Zn^{2+} ions (one per enzyme molecule of 29,000 daltons) with little alteration of the enzymatic properties, but with the introduction of optical absorption bands in the visible and, because of the paramagnetism of the Co^{2+} ions, an influence on the magnetic relaxation rates of solvent protons. It is now quite well established that the observed pH-dependence of the optical spectra, the proton relaxation rates, and the enzymatic activity (including the more recently discovered hydrolysis of esters and the hydration of aldehydes) all are correlated, certainly above pH 6, so that the pK_a observed for any property is characteristic of the pH-dependence of all these properties [8, 11, 16].

A wide variety of pH-dependent effects, including enzymatic activity [20, 21, 23, 37, 41], relaxation of solvent protons [11, 16] and of several halogen anions [48, 49, 50, 51], optical absorption [7, 29, 32], compleximetric titration [7], and inhibitor association constants [29, 33, 41] have been interpreted as evidence for two forms of carbonic anhydrase: a high-pH form, active in CO_2 hydration and ester hydrolysis, with a characteristic optical spectrum and a high relaxation rate for solvent protons for the Co^{2+}-enzyme; and a low-pH form [2] with a high affinity for

---

1 IBM Thomas J. Watson Research Center, Yorktown Heights, New York 10598/USA

2 There have been incorrect statements in the literature to the effect that the low-pH form of the enzyme catalyzes the dehydration of HCO_3^-. Because of the nature of the reaction $CO_2 + H_2O \rightleftharpoons H^+ + HCO_3^-$, the dehydration of HCO_3^- requires a proton as a co-substrate; thus lowering the pH produces an increase in dehydration activity. However, by microscopic reversibility, dehydration must be the reverse of hydration: a proton and a bicarbonate ion must be on an active enzyme molecule (high-pH form) concurrently to be turned over to $CO_2 + H_2O$, precisely as CO_2 and H_2O must similarly find an active enzyme molecule to be converted to $H^+ + HCO_3^-$. Thus the active form (whatever it is) must be active for both directions of the catalytic process

monovalent inhibitors. The two human isozymes, carbonic anhydrases B and C, and the bovine enzyme (which is very much like the human C) have been the most extensively studied, and their physical and chemical structures are now known quite well [8, 30]. Despite the apparent simplicity of the catalyzed reaction, and the considerable knowledge of the structures of carbonic anhydrases, the presumed ionization of the enzyme that distinguishes the high- and low-pH forms has not been identified.

It was earlier proposed that the ionization of a water ligand of the Zn^{2+} (or Co^{2+}) ion, at a pH near neutrality, allowed a nucleophilic attack by the metal-bound hydroxide on the bound CO_2, yielding HCO_3^- [10]. Although this view is still widely held, it is not in accord with early proton relaxation rate data which indicated that at high pH there is an exchanging water ligand [26], the resident lifetimes of OH^- ligands being too long to explain the large relaxation effects [11]. In addition, the results of experiments on the hydrolysis of esters and the hydration of aldehydes have also been used to argue that the ligand of the metal in the active enzyme is a water molecule [37]. Nonetheless the state of ionization of the water-ligand of the metal is still the subject of controversy.

The influence of various anions on carbonic anhydrase activity was studied early [43], and the shift of pK_a upon anion-binding has since been established for a wide variety of monovalent anions. The effect may be observed by measuring catalytic activity [29] or equivalently, for the Co^{2+}-substituted enzyme, by nuclear relaxation [11], and optical absorption [29]. Thus, the reported pK_a for some of the early measurements of the pH-dependence of carbonic anhydrase activity which were done in the presence of 80 mM Cl^- [20, 21] were artificially high due to the increasing affinity of monovalent anions for carbonic anhydrase at lower pH-values. A value of 6.4 could be calculated by extrapolating to zero chloride concentration.

Because of this interaction with monovalent anions, and despite the early report of sulfate and phosphate effects [43], recent investigations have used cationic buffers with SO_4^{2-} as the buffer anion and Na_2SO_4 to maintain ionic strength [23, 30, 38]. Even under these conditions, however, bovine anhydrase exhibits the usual pH-dependent activity, but with a pK_a that varies somewhat from one report to another [29, 38]. Also, the values reported for the pK_a have been decreasing with time due to the discovery that sufficient Cl^- ions from the measuring electrode may leak into a sample during a typical pH determination and substantially increase the apparent pK_a of the enzyme [4]. By using electrodes with SO_4^{2-} rather than Cl^- as the salt-contact, a pK_a as low as 6.1 was observed for the proton relaxation profile of the Co^{2+}-bovine enzyme [52], and most recently, using a special low leakage HCl electrode (see below), a pK_a of about 5.5 was obtained after exhaustic dialysis [16]. Thus the value of the pK_a of the postulated ionization on the enzyme responsible for activity is becoming progressively more elusive.

Recently another group has reported that (divalent) SO_4^{2-} anions interact with carbonic anhydrase [2, 3], though we have argued that their observations could be atrributed to an ionic-strength effect [16]. We now confirm that the presence of SO_4^{2-} ions does indeed influence the apparent pK_a of carbonic anhydrase; however, as we show, the effect is due to interactions with (monovalent) HSO_4^- ions. We present measurements of the pH-dependence of the optical absorption of Co^{2+}-substituted bovine carbonic anhydrase for several concentrations of sulfate, phosphate, and chloride ions, and show that in each case the data can be explained in terms of a model in which only monovalent anions bind to a single enzyme form. The pH-dependence of the monovalent anion concentration is significantly different for each of the three kinds of anion, and thus the dependence of the observed pK_a on total ion concentration is expected to be different for each ion, as is found to be the case.

The model is quite straightforward: The fourth ligand of the metal in the active enzyme is a water molecule that can be replaced by a "protonated" monovalent anion, by which we mean an anion in close association with a proton brought in to maintain the same total charge near the metal. The proton may well reside at a nearby proton acceptor. The proton-anion pair is regarded as competing with the entire water molecule; or alternatively, the monovalent anion may be regarded as competing with the OH^- moiety of this water ligand. There is no need to postulate more than one form of the enzyme in the pH range above ~ 5.5 in the absence of anions, and therefore, in our view, the long-sought ionization of carbonic anhydrase need never be considered; the expressions "high pH form of the enzyme" or "low pH form of the enzyme" become meaningless as regards enzymatic activity.

This model, as will be seen, can not only reconcile arguments that arise when one or another ionization on the enzyme is proposed but, when its implications are logically pursued, can explain a variety of phenomena, many of which have not been previously addressed. These include the temperature dependence of the enzymatic activity, the wide variation in reported values for the pK_a, the ability of some buffers (notably phosphate) both to enhance and to inhibit enzymatic activity and HCO_3^- binding, the existence of two pK_a's for the relaxation of solvent Cl^- ions, and the apparent existence of two HCO_3^- binding sites. In addition, it is the first model to reconcile the decade-old relaxation data [11] with enzymatic activity; all other authors have either ignored these data, or considered them artifactual in some sense.

2 Materials and Methods

Native bovine carbonic anhydrase, purchased from Sigma and purified by chromatography on DEAE-cellulose [18], was demetallized in the usual way [31] and lyophilized after extensive dialysis against distilled water. Co^{2+}-substituted enzyme in deionized water was prepared by dialysis of apoprotein against 1 mM $CoCl_2$ for one day, and then against deionized water, with frequent changes, for six days at 5°C. The final pH, 5.6, is near the isoelectric point for the enzyme. Sample pH was measured with a Radiometer PHM64 meter and combination electrode from Microelectrodes, Inc. [3]. Ultrahigh purity sodium chloride and sodium sulfate were obtained from J.T. Baker and EM Laboratories, respectively. All other chemicals were reagent grade.

Optical absorption spectra were measured at room temperature in cuvettes of 1 cm path length with a Cary Model 14 recording spectrophotometer equipped with a 0 to 0.1 optical density slidewire. Protein concentration was measured using

3 The KCl flow rate from the reference electrode is guaranteed no greater than 0.01 microliter per 24 hours thus obviating the possibility of chloride contamination of the sample

$A_{1\,cm}^{1\%} = 18$ at 280 nm. All glassware used in the titration experiments was immersed in a 0.1 M EDTA solution for at least one day, and rinsed with deionized water before use.

The pH-dependence of the optical properties of the exhaustively dialyzed enzyme in the presence of chloride, phosphate, and sulfate was measured in the following way: the particular anion was added as a concentrated salt solution to the dialyzed sample to give the desired anionic concentration. The pH was then adjusted with the appropriate acid (HCl, H_2SO_4, or H_3PO_4 for their respective anions), and a series of optical measurements taken as the pH was raised by addition of NaOH. The net dilution of the sample at the end of the titration was always less than 1%. The fraction of active enzyme in each sample was calculated from the amplitude of its 640 nm peak relative to the value observed at high pH. For the sulfate experiments, in the limit of low pH, there was no net absorbance at 640 nm. However, for the chloride and phosphate titrations a correction was necessary for absorbance at 640 nm attributable to the chloride-enzyme and phosphate-enzyme complexes, respectively.

3 Experimental Results

Figure 1 shows the fraction of active enzyme as a function of pH for a sample of the extensively dialyzed enzyme with no added anions, and for three other aliquots containing 0.006, 0.060, and 0.60 M NaCl. The solid lines through the data points of the chloride-containing samples result from a least-squares comparison of the chloride data with an expression derived from the model discussed below. The solid line through the data points for the extensively dialyzed enzyme sample was also the result of a least-squares comparison of data with an equation presented below.

Figure 2 shows the fraction of active enzyme as a function of pH for the extensively dialyzed enzyme sample and for three aliquots of this enzyme preparation with 0.005, 0.040, and 1.0 M total sodium phosphate added. A characteristic spectrum, substantially different from those reported for other anions [29], was observed at low pH. The lines through the phosphate data points again result from a least-squares comparison of the data with the model, with the value 7.2 used for the pK_a of the reaction $HPO_4^{2-} + H^+ \rightleftharpoons H_2PO_4^-$. The essence of the model in this case is that only the $H_2PO_4^-$ (monovalent) anion interacts with carbonic anhydrase. The qualitative difference between the results in Figs. 1 and 2, namely, the less drastic shift of the apparent pK_a with anionic concentration, arises because the concentration of monovalent anion varies significantly with pH in the case of phosphate, but is constant in the case of chloride.

Figure 3 shows the fraction of active enzyme as a function of pH for the extensively dialyzed enzyme sample and for three other aliquots of this enzyme preparation containing 0.006, 0.015, and 1.0 M Na_2SO_4. This is the first quantitative study of the influence of sulfate on the properties of the enzyme, though the existence of such an interaction has been reported [2, 3, 43]. Once again, the lines through the sulfate data result from a least-squares comparison of the data with the model discussed below. In this case the model involves the equilibrium $SO_4^{2-} + H^+ \rightleftharpoons HSO_4^-$, which has a pK_a of 1.94 [9], and the assumption that the enzyme interacts only with the (monovalent) anion HSO_4^-.

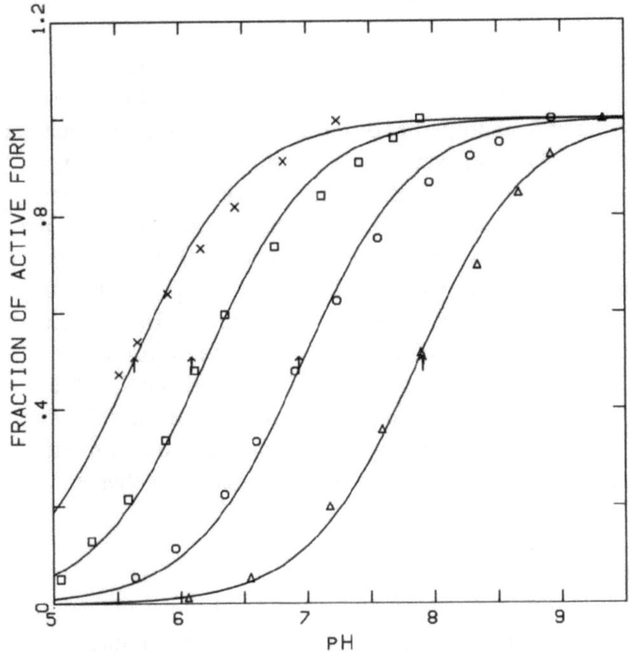

Fig. 1. The fraction of active Co^{2+}-bovine carbonic anhydrase as a function of pH for various total additions of NaCl: (×) no added salt; (□) 0.006 M; (○) 0.06 M; (△) 0.6 M. The *lines* represent the results of a least-squares comparison of the data with the model presented in the text

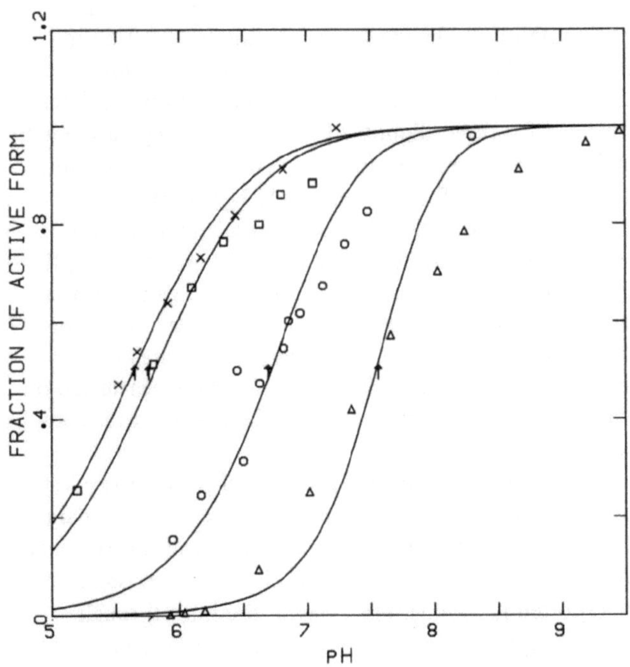

Fig. 2. The fraction of active Co^{2+}-bovine carbonic anhydrase as a function of pH for various total additions of sodium phosphate: (×) no added salt; (□) 0.005 M; (○) 0.04 M; (△) 1.0 M. The *lines* represent the results of a least-aquares comparison of the data with the model presented in the text

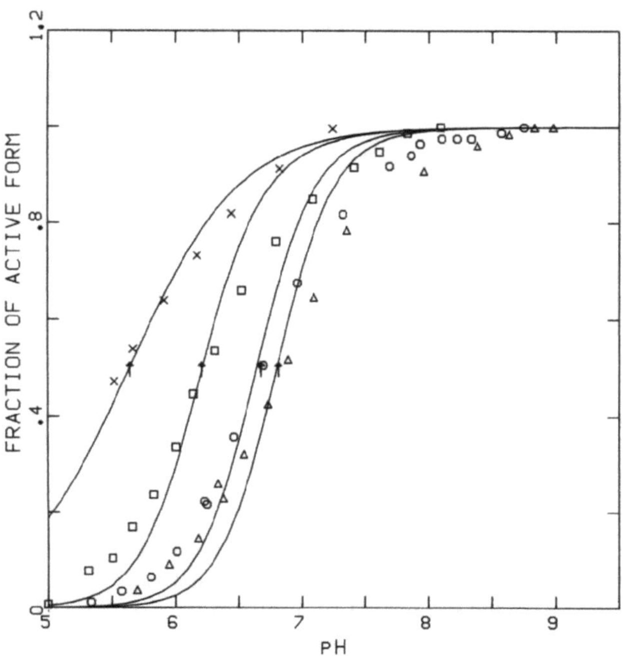

Fig. 3. The fraction of active Co^{2+}-bovine carbonic anhydrase as a function of pH for various total additions of Na_2SO_4: (×) no added salt; (□) 0.005 M; (○) 0.15 M; (△) 1.0 M. The *lines* represent the results of a least-squares comparison of the data with the model presented in the text

4 A Model for the Carbonic Anhydrases

We propose a model for the activity of carbonic anhydrase, motivated in part by the observation that the greater the effort and care taken in removing anions from solutions of carbonic anhydrase, the lower the observed pK_a; in part by the realization that HSO_4^-, as best one can estimate its binding affinity compared with other monovalent anions (see below), should be a good inhibitor of carbonic anhydrase activity; and in part by the desire to find a model that explains both activity and relaxivity data [11, 26, 52].

4.1 The Active Enzyme

We assume that, in the absence of inhibitors, there is only one form of the enzyme at all values of pH (above, say pH 5): the active form. Figure 4a gives a schematic representation of the catalytic site of the active form. M represents either a Zn^{2+} or Co^{2+} ion, known to be liganded to the nitrogens of three histidyl residues. The fourth ligand is a water molecule. PA represents a "proton acceptor", the nature of which will be considered below. The existence of a rapidly exchanging water molecule on the active Co^{2+}-enzyme was indicated by early results on the solvent proton relaxation rate [11, 26], which varies with pH as do the activity and the optical spectra

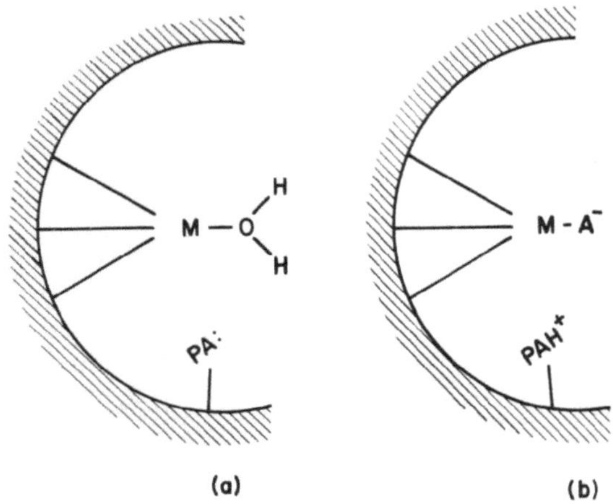

Fig. 4. Schematic drawings of **a** the active and **b** anion-inhibited forms of bovine carbonic anhydrase. *M* Zn^{2+} or Co^{2+}; *PA* proton acceptor

of the enzyme [16, 29]. In brief, the relaxation results set a lower limit to the off-rate of a proton on the metal, $\sim 10^5\ s^{-1}$, which is faster than an OH^- ion can exchange its proton with solvent. Unless either the water is a fifth ligand of the Co^{2+}, or OH^- exchanges by some unusual mechanism [42], the relaxation results require that the ligand of the metal of the active enzyme be a water that exchanges with solvent as an intact molecule.

4.2 Monovalent Anionic Inhibitors

There is a large class of monovalent anions that inhibit carbonic anhydrase activity; this includes the halogen ions, acetate, HCO_3^-, N_3^-, NO_3^-, ClO_4^-, HSO_3^-, SH^-, CN^-, OCN^-, and SCN^-. On the basis of the results in Figs. 2 and 3, we add $H_2PO_4^-$ and HSO_4^- to the list. Though the affinities of these anions for carbonic anhydrase vary widely (F^- has an affinity about 10^{-6} that of CN^-), they otherwise behave similarly. All bind progressively better as the pH is lowered and all produce optical spectra for the Co^{2+}-enzyme which differ in a characteristic manner from that of the active enzyme, but also differ somewhat from each other [29].

Figure 4b is a representation of the anion-inhibited form of the enzyme, where A^- represents any of the small anions but may represent equally well sulfonamide anions, as discussed below. The charge of the anion is balanced by that of a nearby proton on PA. The essence of our model is that the net charge in the region of the metal must be maintained upon ligand exchange. Either there is a water molecule on the metal or else a "protonated" monovalent anion with its associated proton residing on the proton acceptor. The proton-anion pairs may be regarded as competing with the water molecule ligand of the metal in the active enzyme, or thermodynamically, the anion may be regarded as competing with the OH^- moiety of the water molecule. In either view, the binding affinity of anions will be greater at lower values of pH, and the pH-dependence of the fraction, f, of active enzyme can be readily derived. The total concentration of enzyme $[E_T]$ may be written as $[E_T] = [EHOH]$

+ $[EH^+ A^-]$, (the sum of the concentrations of active and inactive enzyme). Regardless of whether we formally consider the competition as between the pair $H^+ A^-$ and HOH, or between A^- and OH^-, we can write:

$$EHOH + H^+ + A^- \rightleftharpoons EH^+ A^- + H_2O \tag{1}$$

$$\frac{[EHOH][H^+][A^-]}{[EH^+ A^-]} \equiv K_A \tag{2}$$

The ratio of inactive (inhibited) enzyme to active enzyme is then given by:

$$\frac{[EH^+ A^-]}{[EHOH]} = \frac{[H^+][A^-]}{K_A} = \frac{[H^+]}{K_{Aeff}} \tag{3}$$

where

$$K_{Aeff} = \frac{K_A}{[A^-]} \tag{4}$$

K_{Aeff} characterizes the pH-dependence of the activity of the enzyme in the presence of the anionic inhibitor A^-. The fraction of total enzyme that is active is then:

$$f = \frac{[EHOH]}{[E_T]} = \frac{1}{1 + \frac{[EH^+ A^-]}{[EHOH]}} = \frac{1}{1 + \frac{[H^+]}{K_{Aeff}}} \tag{5}$$

For the case of Cl^-, Fig. 1, $[A^-]$ is pH-independent so long as the fraction of bound Cl^- is small compared to the total, and f, using our model, is given by a simple titration curve with pK_{Aeff} proportional to log $[Cl^-]$. In the cases of phosphate and sulfate, Figs. 2, 3 $[H_2PO_4^-]$ and $[HSO_4^-]$ both depend on pH, and therefore K_{Aeff} is no longer constant; the functional form of f becomes somewhat more complex, though straightforward to calculate. We also note than in the absence of any deliberately added anion the titration exhibits a pK_a of 5.64 (Fig. 1) as determined by a fit of the data to Eq. (5). We tentatively ascribe this to a residual concentration of an unknown anion (alternative possibilities are considered in the Discussion Section). It is straightforward to generalize the expression for f, Eq. (5), to include competition between two anions; the result appropriate for the present is:

$$f = \frac{1}{1 + \frac{[H^+]}{K_{Aeff}} + \frac{[H^+]}{K_0}} \tag{6}$$

where $pK_0 = 5.64$. The solid lines for chloride, phosphate, and sulfate solutions in Figs. 1–3 were obtained by a least squares comparison of the data with Eq. (6)

taking into account the pH dependence of K_{Aeff} for phosphate and sulfate, as indicated below. In the case of chloride, Fig. 1, fitting the data for three concentrations of added chloride (0.006, 0.06, and 0.6 M) yielded K_A's of 5.2, 6.7, and 8.1 x 10^{-9} M^2, respectively. In the case of phosphate, Fig. 2, the concentration of $H_2PO_4^-$ was calculated using a second dissociation constant of 6.23 x 10^{-8} M; the first and third dissociation constants could be ignored over the pH range of the measurements. The curves in Fig. 2 are fits to the data for the three concentrations of added phosphate (0.005, 0.04, and 1.0 M) and gave K_A values of 23.4, 6.3, and 9.2 x 10^{-9} M^2, respectively. For sulfate, Fig. 3, the concentration of HSO_4^- was calculated using a K_d of 0.012 M. The curves in Fig. 3 are fits to the data for the three concentrations of added sulfate (0.005, 0.15, and 1.0 M) and gave K_A values of 2.2, 7.3, and 22.7 x 10^{-13} M^2. respectively. Variations in the apparent pK_a's due to the spreads in K_A's for the three anions are approximately ± 0.1 for Cl^- and $H_2PO_4^-$ and ± 0.3 for HSO_4^-.

4.3 Optical Spectra and Neutral Inhibitors

We propose that the optical spectra of the active and anion-inhibited forms of the CO^{2+}-substituted enzymes, Figs. 4a,b, differ mainly because of the internal ionization caused by a shift of the proton to PA. For the active enzyme the fourth ligand is neutral, and for the inhibited enzyme it is charged; this change of ligand field would cause the major differences observed in the spectra. The differences among anions would produce the smaller but observable anion-specific variations in the spectra of the different anionic forms.

Alcohols are known to be inhibitors of the esterase activity of carbonic anhydrase and appear competitive with anions [1, 46, 53]. Furthermore, it has been reported [39] that carbonic anhydrase catalyzes the decarboxylation of methyl carbonate to produce methanol and CO_2; the reaction is analogous to the enzyme-catalyzed dehydration (decarboxylation) of bicarbonate. If so, an enzyme-methanol form must exist in which methanol is directly bound to the metal. Methanol or ethanol, because of their high pK_a's for dissociation, would be expected to remain undissociated (like water) when bound to the metal, and thus the spectra of the Co^{2+}-enzyme in the presence of these neutral species should be that characteristic of the active enzyme, as observed [53][4].

4.4 Kinetic Considerations and Sulfonamide Inhibitors

The equilibrium model discussed above has a water molecule as a metal-ligand on the active enzyme, and an anion on the inactive enzyme with an associated proton so that the net charge remains the same in all cases. The optical spectra of the inhibited Co^{2+}-enzyme depends on whether or not the inhibitor is ionized and its proton shifted to PA. The question now arises as to the kinetics of the exchange of anions, protons, and water molecules between solvent and the region of the active site.

We generalize the static requirement of the model, that the net exchangeable charge in the active site region remain neutral, to the dynamic requirement that the exchange of anions and their associated protons be correlated. Thus for anions that bind weakly to the metal, such as F^- and Cl^-, we propose that the on-rates and

4 *Note added in proof:* We have since shown that the majority of the effects observed upon addition of alcohol relate to the increased solubility of ester substrates, and the concomitant increase in K_M. The binding of alcohols is far weaker than previously thought

off-rates of the ions are determined in the main by the rate of exchange of the proton on PA; coulomb forces would hold the ions near the metal until the proton left, and not permit ion-binding until the proton became bound. For anions with a higher affinity, such as SH^- or OCN^-, the situation would be reversed; the on- and off-rates of the proton would be rate-limited by the exchange of the ion. Aside from the logic that the dynamics should mimic the statics, there is ample evidence in the binding of various sulfonamides to support this view.

It has been known for some time that the various sulphonamides, which have pK_a's for ionization from 7 to 11, are potent inhibitors of the carbonic anhydrases [34, 35]. Kinetic measurements of sulfonamide-enzyme-complex formation as a function of pH indicate that it is the neutral forms of sulfonamides that interact with the active enzyme [25, 45]. The recent work [25] indicates that binding of sulfonamides is a two-step process, the first step is independent of the ionization state of the sulfonamide and of the presence of metal in the enzyme. The second step, which accounts for the observed pH dependent on-rate, is proportional to the concentration of neutral inhibitor. On the other hand, ultraviolet-visible [24], resonance Raman [28], and fluorescence [5] data indicate that the usual sulfonamide inhibitors exist as the ionized RSO_2NH^- species when fully bound. The presence of a charged rather than a neutral ligand on the Co^{2+} ion should alter the optical spectrum of the enzyme, as indeed it does [32].

The binding of CN^- to carbonic anhydrase may be analogous, in part, to the binding of sulfonamides. Kinetic data suggest that the main interaction is that of HCN with the active enzyme [44] (the solution pK_a for dissociation of HCN is near 9), whereas ^{13}C magnetic resonance data indicate that HCN is ionized once on the enzyme [12]. Thus the kinetics of ligand binding are not necessarily correlated with the state of ionization of the ligand once on the enzyme. Water, having a very high pK_a for dissociation exists undissociated on the protein, and exchanges as the neutral molecule. The sulfonamides and HCN, with somewhat lower pK_a's, exchange from the inner regions of the active site as neutral molecules, but most (see below) ionize once on the protein. Finally, our model proposes that the small highly acidic anion inhibitors exchange in a correlated fashion as anion and proton, and are always ionized on the enzyme.

5 Implications of the Model

A major implication of the proposed model is that the numerical value of the dissociation or inhibition constant of an anionic inhibitor is in some sense a meaningless quantity: it depends critically on the apparent pK_a of the enzyme solution which in turn depends on the composition of the buffer. As an example, the conclusion that two binding sites for bicarbonate have been identified for the bovine Co^{2+}-substituted enzyme [54] depends crucially on the value of the inhibition constant for I^-, measured by a different investigator [29] under different conditions. The constant was computed from data on a sample that contained 0.1 M Cl^-, and sufficient sulfate to produce a pK_a of 6.4. Thus the conclusion that there are two bicarbonate binding sites can be readily seen to be invalid, as therefore are the proposed enzymatic mechanisms based on this conclusion.

There has been a report of resonance Raman measurements of an unusual sulfonamide (Neoprontosil) both in solution and bound to carbonic anhydrase [36]. It is found that the neutral and anionic forms in solution can be characterized by their respective Raman spectra, and it is concluded that this inhibitor, when bound, is the neutral RSO_2NH_2. Our model predicts unequivocally that this complex would have an optical spectrum similar to that of the active enzyme.

Equations (1–6) may be readily generalized to show that the dissociation constant of a neutral inhibitor should have a pH dependence with a pK_a equal to the observed pK_a of the enzyme preparation. This has recently been found to be so for methanol [1].

There is a set of ostensibly unrelated observations that are explained by the kinetic assumption of the model; i.e., that the off-rate of a weakly bound anion such as Cl^- or HCO_3^- is determined by the off-rate of its associated proton. One example is the width of the magnetic resonance line of Cl^- ions in solutions of carbonic anhydrase. In the first of a series of papers [48], the width was shown to titrate with a pK_a equal to the pK_a of the enzyme, with a maximum width at low values of pH. The line-broadening was originally ascribed to a nuclear quadrupolar interaction of the Cl^- nucleus with the Zn^{2+} or Co^{2+} ions of the enzyme, and subsequent exchange of the Cl^- ligand into the solution. Later work, however, showed that the width is primarily exchange-limited, i.e., that the line width is proportional to the product of the off-rate of a liganded Cl^- ion and the probability that a Cl^- ion is bound [50]. Moreover, measurements were subsequently extended to lower values of pH, and a second titration was found. These data are readily explained. The Cl^- off-rate is, in our model, limited by the off-rate of the associated proton which is, in turn, proportional to the concentration of OH^- ions (in the absence of buffer); this explains the lower titration. This effect saturates when the relaxation is no longer limited by exchange. At higher values of pH, the concentration of $EH^+ Cl^-$ is reduced, due to the decrease in $[H^+]$; this explains the second titration. We would predict that if the proton off-rate were enhanced, for example by addition of buffer [17], the line-widths at low pH would increase. The maximum line width should equal (roughly) the off-rate divided by π. The maximum off-rate is given (approximately) by the product of $[OH^-]$ at the pH of the peak (~ 6) and the diffusion-limited rate constant, or about $10^{10}\ (M\ s)^{-1} \times 10^{-8}\ M = 10^2\ s^{-1}$; dividing by π gives a value of $30\ s^{-1}$, to be compared with the experimental value of about $20\ s^{-1}$.

The literature contains contradictory reports of the effect of phosphate buffer on carbonic anhydrase activity. Although phosphate was earlier found to be an inhibitor of enzymatic activity [43], recent reports have shown phosphate to be an activator of carbonic anhydrase activity as well [6, 37]. This anomalous behavior can readily be explained. At low concentrations of phosphate, its buffering properties enhance the proton exchange rate [17] which, in our model, normally limits the HCO_3^- off-rate and thus the catalytic activity. At higher concentrations of phosphate, inhibition by $H_2PO_4^-$ dominates, thereby reducing carbonic anhydrase activity. The net effect is thus a sensitive function of phosphate concentration and pH.

Implicit in the above is the idea that the rate limiting step in catalysis is the exchange rate of HCO_3^-, which in turn is controlled by the exchange of an associated proton. For given sample conditions, this rate should increase with increasing temperature, with an activation energy about that of the diffusion of ions in water; this also is consistent with the few data on the temperature dependence of the activity of carbonic anhydrase [22, 27].

Finally, compleximetric titration studies [7] show that when monovalent anions bind, at any pH, stoichiometric OH^- is liberated, totally consistent with our proposed model. Indeed, titration data (like relaxation data) often set very fundamental limitations on any proposed model, and it was our à priori requirement that the model agree with something so fundamental. Since the compleximetric titration data are complicated by a substrate ionization (HS and HCN), we demonstrate explicitly the consequences of our model for the case of HCN, and the agreement with the data [7]. Referring to Fig. 4, consider the limit of high pH, where HCN is ionized and the enzyme has an H_2O ligand. Inhibition then involves CN^- replacing the OH^- moiety of the water ligand, with a correlated shift of the proton to PA. Thus an OH^- is released to solution, as observed. In the limit of low pH, (~ 6), where HCN is the dominant form in solution and the enzyme has an A^- ligand with a proton on PA, binding of CN^- to enzyme is through an exchange of CN^- (from an HCN in solution) with a bound A^- to give an H^+ A^- pair in solution and the CN^--inhibited complex. A proton has been released this time, again totally consistent with the data [7].

6 Discussion

Many previous investigators have assumed that sulfate anions either do not interact with the active-site metal of carbonic anhydrase, or that the interaction is sufficiently weak to be disregarded. However a search of the literature shows that there is a systematic dependence of apparent pK_a on the concentration of SO_4^{2-} ions in solutions of bovine carbonic anhydrase [23, 29, 38]. Our data suggest that HSO_4^- binds to carbonic anhydrase and that the pK_a reported in many experiments is an artifact related to buffer composition. It is not surprising that this has not been noted earlier, since the ionization constant for HSO_4^- is ~ 10^{-2} and therefore its concentration generally low. We previously argued that the shift of pK_a in the presence of sulfate was due to ionic strength effects and that divalent sulfate is a noninteracting anion [16]. However, the results of these more extensive investigations of the shift of the pK_a in the presence of chloride, sulfate, or phosphate are most consistent with our model for the following reasons: (1) The shift of the apparent pK_a in the presence of chloride is porportional to log $[Cl^-]$ over a wide range of concentration. (2) If the divalent sulfate were a weak inhibitor, shifts in the pK_a with sulfate concentration would be expected to be similar to those observed for chloride. This is clearly not the case. Increasing the concentration of sulfate above 0.15 M shifts the pK_a very little compared to the large shift observed at lower concentrations, where the pK_a is correspondingly lower. The same diminution in pK_a shift at higher concentrations of phosphate is also observed. (3) The similarity between the titration data for sulfate and phosphate suggests that, like phosphate, a pH-dependent partitioning of sulfate between two forms, only one of which is inhibitory, is responsible for the decreasing shift in pK_a with increasing concentration of sulfate.

The affinity of monovalent anions follows the lyotropic or Hofmeister series [13, 47], an empirical ordering that correlates with ionic radius and the related interaction of the anions with water. Thus ClO_4^- binds to carbonic anhydrase about 10-fold better than Cl^-. The binding of HSO_3^- is also known [40], and assuming that HSO_4^- binds about 10-fold better than HSO_3^- (also using the various compilations of water-anion interactions [14]), one readily predicts that HSO_4^- should have approximately the relative affinity that we observe. For the case of chloride K_A is found to be essentially independent of Cl^- concentration. For sulfate and phosphate the variation of K_A with concentration may well be due to a concentration or ionic strength dependence of the dissociation constants for the acids.

Some discussion is in order regarding the source of the titration represented by the term K_0 in Eq. (6). We attributed it, tentatively, to residual anionic contamination, although the mathematical form of Eq. (6) is identical to that which arises by assuming that there are indeed two forms of the enzyme, high- and low-pH, and that anions bind only to the low-pH form [29]. However this latter view not only makes it very difficult to understand the source of this ionization, now having a pK_a no greater than 5.64, but it cannot correlate the wide range of data that our model does.

The strength of our model, the essence of which is that the net charge in the inner region of the active site never changes, neither statically nor dynamically, as the ligands of the metal change, is that it can explain and correlate a wide variety of phenomena, including the reasons for the variation of the optical spectra with the type of ligand, the wide range of values reported for the dissociation and inhibition constants of various anions, and a wide range of kinetic data, as discussed above. It is a particularly simple model, and obviates the need to invoke the existence of two forms of the enzyme above about pH 5. There are essentially no data at lower values of pH. We do not preclude another form at some lower value of pH, but the metals become labile near pH 5.

Our model requires a proton acceptor near the active site; this acceptor not only is protonated when small anionic inhibitors bind, but is almost certainly involved as an acceptor in the fundamental hydration reaction: $CO_2 + H_2O \rightarrow H^+ + HCO_3^-$, thereby allowing a concerted attack of OH^- on the CO_2. We make no commitment at this point as to the identity of the acceptor, but note that it could either be a histidine [15] with an abnormally low pK_a or an invariant threonine coupled with a glutamate proton-shuttle [19], somewhat akin to the mechanism of the several serine proteases. This would have a pK_a well below the range considered here.

Finally, there are two points to note. Though the shift of the apparent pK_a is reasonably well given by the equations used to fit the sulfate and phosphate data, Figs. 2, 3, the functional form could be improved. It must be emphasized again that our model is only first order. It ignores, for example, the likelihood that the various values for K_A are pH-dependent since there are ionizations that occur in the cleft that contains the active site. However, refinements at this level must await additional data. Second, none of the results in the present paper require assumptions that cannot be readily argued from extrapolations of published data. Rather, our model reduces the assumptions needed to account for the enzymatic properties of carbonic anhydrase while enlarging the range of phenomena explained. It also predicts many new ones that are being pursued.

Acknowledgments. We would like to acknowledge early and frank correspondence with Professor Bertini; attempts on our part [16] to interpret his recent results [2] were instrumental in the development of this report. Also Professor Henkens, in a recent discussion, kindly helped clarify the experimental procedures of his early work [54].

References

1. Appleton DW, Sarker B (1975) Carbonic anhydrase: studies of uncharged metal directed inhibitors. Bioinorg Chem 4: 309–321
2. Bertini I, Canti G, Luchinat C, Scozzafava A (1977) Evidence of exchangeable protons in the donor groups of the acidic form of cobalt bovine carbonic anhydrase B. Biochem Biophys Res Commun 78: 158–160
3. Bertini I, Canti G, Luchinat C, Scozzafava A (1978) Characterization of cobalt (II) bovine carbonic anhydrase and its derivatives. J Am Chem Soc 100: 4873–4877
4. Campbell ID, Lindskog S, White AI (1974) A study of the histidine residues of human carbonic anhydrase B using 270 MHz proton magnetic resonance. J Mol Biol 90: 469–489
5. Chen RF, Kernohan KC (1967) Combination of bovine carbonic anhydrase with a fluorescent sulfonamide. J Biol Chem 242: 5813–5823
6. Christiansen E, Magid E (1970) Effects of phosphate, HEPES, N_2O and CO on the kinetics of human erythrocyte carbonic anhydrases B and C. Biochim Biophys Acta 220: 630–632
7. Coleman JE (1967) Mechanism of action of carbonic anhydrase. Substrate, sulfonamide and anion binding. J Biol Chem 242: 5212–5219
8. Coleman JE (1973) Carbonic anhydrase. In: Eichhorn GL (ed) Inorganic biochemistry. Elsevier, Amsterdam, pp 488–548
9. Covington AK, Dobson JV, Wynne-Jones L (1965) Dissociation constant of the bisulphate ion at 25°C. Faraday Soc Trans 61: 2057–2062
10. Davis RP (1959) The kinetics of the reaction of human erythrocyte carbonic anhydrase. II. The effect of sulfanilamide, sodium sulfide and various chelating agents. J Am Chem Soc 81: 5674–5678
11. Fabry ME, Koenig SH, Schillinger WE (1970) Nuclear magnetic relaxation dispersion in protein solutions. IV. Proton relaxation at the site of carbonic anhydrase. J Biol Chem 245: 4256–4262
12. Feeney J, Burgen ASV (1973) Cyanide binding to carbonic anhydrase. A ^{13}C-nuclear-magnetic-resonance study. Eur J Biochem 34: 107–111
13. Fridovich I (1963) Inhibition of acetoacetic decarboxylase by anions. J Biol Chem 238: 592–598
14. Friedman HL, Krishnan CV (1973) Thermodynamics of ionic hydration. In: Franks F (ed) Water – a comprehensive treatise, vol III. Plenum Press, New York, pp 1–118
15. Gupta RK, Pesando JM (1974) Magnetic resonance study of exchangeable protons in human carbonic anhydrases. J Biol Chem 250: 2630–2634
16. Jacob GS, Brown RD, Koenig SH (1978) Relaxation of solvent protons by cobalt bovine carbonic anhydrase. Biochem Biophys Res Commun 82: 203–209
17. Jonsson BH, Steiner H, Lindskog S (1976) Participation of buffer in the catalytic mechanism of carbonic anhydrase. FEBS Lett 64: 310–314
18. Kandel M, Gornall AG, Wong SC, Kandel SI (1970) Some characteristics of human, bovine, and horse carbonic anhydrases as revealed by inactivation studies. J Biol Chem 245: 2444–2450
19. Kannan KK, Petef M, Fridborg K, Cid-Dresdner H, Lövgren S (1977) Structure and function of carbonic anhydrase. FEBS Lett 73: 115–119
20. Kernohan JC (1964) The activity of bovine carbonic anhydrase in imidazole buffers. Biochim Biophys Acta 81: 346–356
21. Kernohan JC (1965) The pH-activity curve of bovine carbonic anhydrase and its relationship to the enzyme by anions. Biochim Biophys Acta 96: 304–317

22. Kernohan JC (1972) The effect of temperature on the catalytic activity of bovine carbonic anhydrase. In: Brewer GJ (ed) Hemoglobin and red cell structure and function. Plenum Press, New York, pp 189–196
23. Khalifah RG (1971) The carbon dioxide hydration activity of carbonic anhydrase. I. Stop-flow kinetic studies of the native human isoenzymes B and C. J Biol Chem 246: 2561–2573
24. King RW, Burgen ASV (1970) Sulphonamide complexes of human carbonic anhydrases. Ultraviolet difference spectroscopy. Biochim Biophys Acta 207: 278–285
25. King RW, Burgen ASV (1976) Kinetic aspects of structure-activity relations: The binding of sulphonamides by carbonic anhydrase. Proc R Soc London Ser B 193: 107–125
26. Koenig SH, Brown RD (1972) H_2CO_3 as substrate for carbonic anhydrase in the dehydration of HCO_3^-. Proc Natl Acad Sci USA 69: 2422–2425
27. Koenig SH, Brown RD, London RE, Needham TE, Matwiyoff NA (1974) The kinetic parameters of carbonic anhydrase by ^{13}C NMR. Pure Appl Chem 40: 103–113
28. Kumar K, King RW, Carey PR (1976) Resonance Raman studies on some carbonic anhydrase-aromatic sulfonamide complexes. Biochemistry 15: 2195–2202
29. Lindskog S (1966) Interaction of cobalt (II) – carbonic anhydrase with anions. Biochemistry 5: 2641–2646
30. Lindskog S, Henderson LE, Kannan KK, Liljas A, Nyman PO, Strandberg B (1971) Carbonic anhydrase. In: Boyer PD (ed) The enzymes, vol V. Academic Press, London New York, pp 587–665
31. Lindskog S, Malmström B (1962) Metal binding and catalytic activity in bovine carbonic anhydrase. Biol Chem 237: 1129–1137
32. Lindskog S, Nyman PO (1964) Metal-binding properties of human erythrocyte carbonic anhydrase. Biochim Biophys Acta 85: 462–474
33. Lindskog S, Thorslund A (1968) On the interaction of bovine cobalt carbonic anhydrase with sulfonamides. Eur J Biochem 3: 453–460
34. Mann T, Keilin D (1940) Sulphanilamide as a specific inhibitor of carbonic anhydrase. Nature 146: 164–165
35. Miller WH, Dessert AM, Roblin RO (1950) Heterocyclic sulfonamides as carbonic anhydrase inhibitors. J Am Chem Soc 72: 4893–4896
36. Petersen RL, Li T, McFarland JT, Walters KL (1977) Determination of ionzation state by resonance Raman spectroscopy. Sulfonamide binding to carbonic anhydrase. Biochemistry 16: 726–731
37. Pocker Y, Bjorkquist DW (1977) Comparative studies of bovine carbonic anhydrase in H_2O and D_2O. Stopped-flow studies of the kinetics of interconversion of CO_2 and HCO_3^-. Biochemistry 16: 5698–5707
38. Pocker Y, Bjorkquist L, Bjorkquist DW (1977) Zinc (II) and cobalt (II) bovine carbonic anhydrases. Comparative studies and esterase activity. Biochemistry 16: 3967–3973
39. Pocker Y, Guilbert LJ (1974) Carbonic anhydrase catalyzed hydrolysis and decarboxylation. Kinetic studies of enzyme-catalyzed decomposition of mono- and disubstituted derivatives of carbonic acid. Biochemistry 13: 71–78
40. Pocker Y, Sarkanen S (1978) Carbonic anhydrase: Structure, catalytic versatility and inhibition. In: Meister A (ed) Advances in enzymology. John Wiley, New York, pp 149–274
41. Pocker Y, Stone JT (1968) The catalytic versatility of erythrocyte carbonic anhydrase. VI. Kinetic studies of noncompetitive inhibition of enzyme-catalyzed hydrolysis of p-nitrophenyl acetate. Biochemistry 7: 2936–2945
42. Prince RH, Woolley PR (1973) On the mechanism of action of carbonic anhydrase. Bioorg Chem 2: 337–344
43. Roughton FJW, Booth VH (1946) The effect of substrate concentration, pH and other factors upon the activity of carbonic anhydrase. Biochem J 40: 319–330
44. Taylor PW, Burgen ASV (1971) Kinetics of carbonic anhydrase-inhibitor complex formation. A comparison of anion- and sulfonamide-binding mechanisms. Biochemistry 10: 3859–3865
45. Taylor PW, King RW, Burgen ASV (1970) Influence of pH on the kinetics of complex formation between aromatic sulfonamides and human carbonic anhydrase. Biochemistry 9: 3894–3902

46. Verpoorte JA, Mehta S, Edsall JT (1967) Esterase activity of human carbonic anhydrases B and C. J Biol Chem 242: 4421–4229
47. Von Hippel PH, Schleich T (1969) The effects of neutral salts on the structure and conformational stability of macromolecules in solutions. In: Timasheff SN, Fasman GD (eds) Structure and stability of biological macromolecules, vol II. Marcel Dekker, New York, pp 417–574
48. Ward RL (1969) ^{35}Cl nuclear magnetic resonance studies of a zinc metalloenzyme carbonic anhydrase. Biochemistry 8: 1879–1883
49. Ward RL (1970) Zinc environmental differences in carbonic anhydrase isozymes. Biochemistry 9: 2447–2454
50. Ward RL, Cull MD (1972) Temperature and competitive anion binding studies of carbonic anhydrase. Arch Biochem Biophys 150: 436–439
51. Ward RL, Fritz KJ (1970) ^{35}Cl-NMR studies of Co^{2+} carbonic anhydrases. Biochem Biophys Res Commun 39: 707–711
52. Wells JW, Kandel SI, Koenig SH (1979) pH dependence of solvent proton relaxation in carbonic anhydrase solutions: Paramagnetic and diamagnetic effects. Biochemistry 18: 1989–1995
53. Westerik JOC, Sarkar B, Deslauriers R, Smith ICP (1978) The loci of binding of the specific inhibitors methanol and aniline to bovine carbonic anhydrase. J Biol Chem 253: 8499–8506
54. Yeagle PL, Lochmüller CH, Henkens RW (1975) ^{13}C nuclear magnetic resonance studies on the mechanism of action of carbonic anhydrase. Proc Natl Acad Sci USA 72: 454–458

Role of Buffer in Catalysis of the Hydration of CO_2 by Carbonic Anhydrase

D.N. SILVERMAN, C.K. TU, and G.C. WYNNS [1]

Measurements of the rates of ^{18}O exchange processes catalyzed by carbonic anhydrase have been a useful complement to the study by other kinetic techniques of the catalytic mechanism of CO_2 hydration. It was the observation of ^{18}O exchange between CO_2 and water that first confirmed the participation of buffer in the catalysis (Silverman and Tu, 1975) and gave an estimate of the residence time of oxygen at the active site (Silverman and Tu, 1976). In this report ^{18}O depletion from CO_2 catalyzed by carbonic anhydrase is used to describe the rate of exchange of water at the enzyme's active site and to elucidate further the role of buffer in the catalysis of CO_2 hydration.

1 Kinetic Method

The exchange of ^{18}O between CO_2 and water, first reported by Mills and Urey (1940), occurs because of the hydration-dehydration cycle.

$$CO^{18}O + H_2O \rightleftharpoons HCOO^{18}O^- + H^+ \rightleftharpoons COO + H_2{}^{18}O \tag{1}$$

This exchange is observed by mass spectrometric measurement of the atom fraction of ^{18}O in CO_2

$$\alpha = \frac{1/2\,(CO^{18}O) + (C^{18}O^{18}O)}{(COO) + (CO^{18}O) + (C^{18}O^{18}O)}$$

The exchange in Eq. (1) is first-order with a rate constant θ: $\alpha - \alpha_\infty = a_0 e^{-\theta t}$. Since the exchange is catalyzed by carbonic anhydrase, θ consists of a catalyzed and uncatalyzed component: $\theta = \theta_{uncat} + \theta_{cat}$.

A second type of ^{18}O exchange involving CO_2 was reported by Gerster et al. (1973): the exchange of ^{18}O between ^{12}C- and ^{13}C-containing species of CO_2. This exchange is observed by placing in solution ^{18}O-labeled bicarbonate not enriched in ^{13}C, and ^{13}C-labeled bicarbonate not enriched in ^{18}O. The appearance of ^{18}O in ^{13}C-contaning species is evidence of this exchange and can be measured by observing the ^{18}O exchange involving only ^{12}C- or ^{13}C-containing species of CO_2.

---

1 Department of Pharmacology, University of Florida College of Medicine, Gainesville, Florida 32610/USA

$$^{(12)}\alpha = \frac{1/2\,(^{12}CO^{18}O) + (^{12}C^{18}O^{18}O)}{(^{12}COO) + (^{12}CO^{18}O) + (^{12}C^{18}O^{18}O)}$$

It has been shown that the decrease in $^{(12)}\alpha - \alpha_\infty$ is biphasic (Gerster et al., 1973; Silverman and Tu, 1976).

$$^{(12)}\alpha - \alpha_\infty = a_1 e^{-\theta t} + a_2 e^{-(\theta + \phi)t}$$

where ϕ is a rate constant describing the exchange of ^{18}O between ^{12}C- and ^{13}C-containing species of CO_2. This exchange is also catalyzed by carbonic anhydrase: $\phi = \phi_{uncat} + \phi_{cat}$.

These ^{18}O exchange processes are observed by mass spectrometry using a CO_2 inlet vessel that allows continuous measurement of isotopic content of CO_2 in solution. The inlet, devised originally by Hoch and Kok (1963), comprises a cylindrical glass container the bottom of which is a membrane permeable to CO_2 and supported by a porous, stainless steel disk. CO_2 passing across the membrane enters a mass spectrometer (Finnigan 3000) where masses of the 44 through 49 peaks are followed by a Finnigan Programmable Ion Monitor. A volume (8 ml) of solution containing ^{18}O-labeled $KHCO_3$ (15 mM) is placed in the vessel, the pH is adjusted with NaOH or H_2SO_4, and after several minutes to allow approach to chemical equilibrium, the uncatalyzed rate of ^{18}O exchange is measured. After this procedure, carbonic anhydrase is added in a volume small enough ($<$ 0.1 ml) to make a negligible dilution to the contents of the vessel. All measurements were made at 25°C and in each experiment the total ionic strength of solution was maintained at 0.2 by addition of the appropriate amount of Na_2SO_4.

Human carbonic anhydrase C and bovine carbonic anhydrase were both purified from erythrocytes by affinity chromatography (Khalifah et al., 1977).

2 Results

The rate constants for ^{18}O exchange θ_{cat} and ϕ_{cat} were measured in the range of pH 6 to 8 in the presence of human carbonic anhydrase C at 1.6×10^{-9} M with a total substrate concentration, $[CO_2] + [HCO_3^-]$, of 15 mM. These data, shown in Fig. 1, are bell-shaped and are independent of the ^{13}C content of the substrate. There is a buffer dependence in θ_{cat} and ϕ_{cat} especially near neutral pH, as demonstrated in Fig. 1 for the buffer imidazole. This buffer dependence has been reported earlier and interpreted as support for the hypothesis that buffers participate as proton transfer agents in the catalytic hydration of CO_2 (Silverman and Tu, 1975).

In order to interpret the data in Fig. 1 and extract from it the exchange rate of water at the active site, we have devised the following simple, two-step model for ^{18}O exchange:

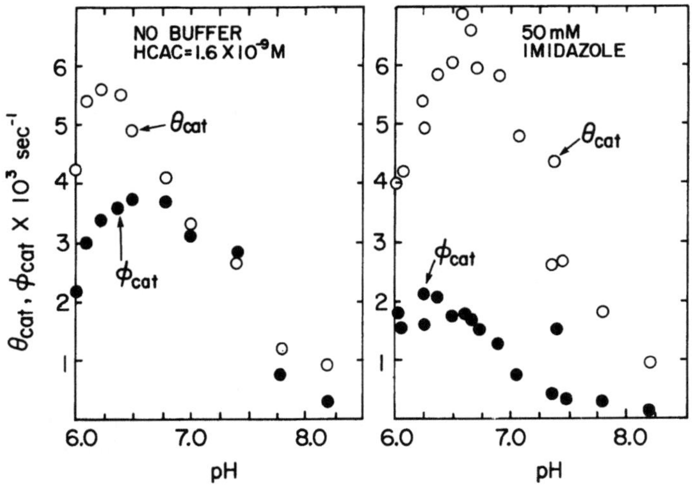

Fig. 1. *Left* The rate constant θ_{cat} (○) for the catalyzed exchange of ^{18}O between CO_2 and water, and ϕ_{cat} (●) for the catalyzed exchange of ^{18}O between ^{12}C- and ^{13}C-containing species of CO_2 at 25°C. This experiment was performed in the absence of buffers. The total concentration of all species of CO_2, (S_{tot}), was 15 mM and human carbonic anhydrase C was 1.6×10^{-9} M. *Right* Same except 50 mM of the buffer imidazole was present

$$HCOO^{18}O^- + E(OH_2) \xrightarrow{\text{Step 1}} E(^{18}OH) + CO_2 + H_2O \underset{+H^+}{\xrightarrow{\text{Step } H_2O}} E(OH_2) + CO_2 + H_2{}^{18}O$$

$$E(^{18}OH) \xrightarrow{^{13}CO_2,\ H_2O} E(OH_2) + H^{13}COO^{18}O^-$$

Using this model we have set up the differential equations to describe the decrease in the ^{18}O atom fraction of CO_2, HCO_3^-, and E (OH) in the same manner as Mills and Urey (1940). These equations describe the change in ^{18}O content with time as a function of the rates of the two steps, Step 1 and Step H_2O, which are responsible for the isotope exchange. The solutions to these kinetic equations involve only one very minor approximation to achieve a more workable solution and yield R_1 and R_{H_2O} in terms of θ_{cat} and ϕ_{cat} (Silverman et al., 1979):

$$R_1 \doteq 3[(CO_2) + (HCO_3^-)] \cdot (\theta_{cat} + \phi_{cat}) \tag{2}$$

$$R_{H_2O} = \frac{R_1 \theta_{cat}}{\phi_{cat}} \left(1 - \frac{(CO_2)}{3[(CO_2) + (HCO_3^-)]}\right) \tag{3}$$

The applicability of this model to the ^{18}O exchange processes catalyzed by the human C enzyme is tested in the following ways. (1) The Eq. (2) and (3) do not contain a factor representing ^{13}C content of CO_2, which must hold for a correct solution since we neglect the relatively small kinetic isotope effects. (2) The values of R_1 obtained applying this solution to the data in Fig. 1 agree with values of R_1 obtained

by other equilibrium kinetic methods: the measurement of R_1 by ^{13}C NMR (Simonsson et al., 1979) and the estimation of the binding of substrate to enzyme by inhibition of the esterase function of carbonic anhydrase (Steiner et al., 1976). Figure 2 compares values of R_1 in the range of pH 6 to 8 obtained in these various ways. The agreement in values of R_1 in this figure is taken as a verification of the technique.
(3) As shown in Fig. 2, R_1 obtained by the ^{18}O technique is independent of buffer concentration. This is in agreement with the results obtained by ^{13}C NMR measurements of R_1 which show no effect of buffer on the catalytic interconversion of CO_2 and HCO_3^- (Simonsson et al., 1979). The ^{18}O-exchange results do show a small buffer effect in R_1 at pH < 6.5. This is reproduceable and probably represents an inadequacy in the model.

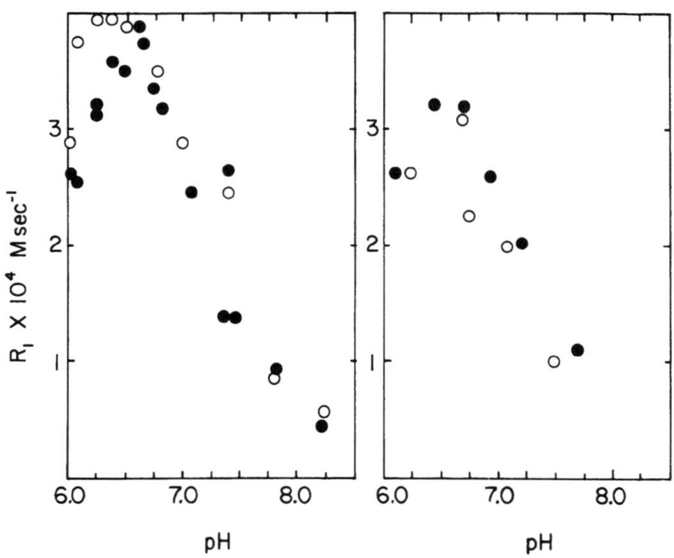

Fig. 2. *Left* Values of R_1 determined using Eq. (2) from measurements of θ_{cat} and ϕ_{cat} at 25°C (○) in the absence of buffer and (●) in the presence of 50 mM imidazole. (S_{tot}) was 15 mM and human carbonic anhydrase C was present at 1.6×10^{-9} M. *Right, Open circles* (○) are values of R_1 calculated from K_{eff} and k_{cat}^{exch} determined by effects of the C enzyme on the ^{13}C NMR spectra of equilibrium mixtures of $^{13}CO_2$ and $NaH^{13}CO_3$ reported by Simonsson et al. (1979). *Filled circles* (●) are values of R_1 calculated using k_{cat}^{exch} from ^{13}C NMR and K_{eff} from studies of inhibition of the esterase activity of the human C enzyme by equilibrium mixtures of CO_2 and HCO_3^- reported by Steiner et al. (1976). In each case (S_{tot}) = 15 mM and (E_{tot}) = 1.6×10^{-9} M

The rate of exchange from the active site of water containing the oxygen abstracted from bicarbonate, R_{H_2O} [calculated from the data of Fig. 1 using Eq. (3)], is given in Fig. 3. These data show that R_{H_2O} is quite buffer-dependent, especially near neutral pH. We can conclude that the buffer dependence of the ^{18}O exchange process lies in the steps described by R_{H_2O}, and not R_1.

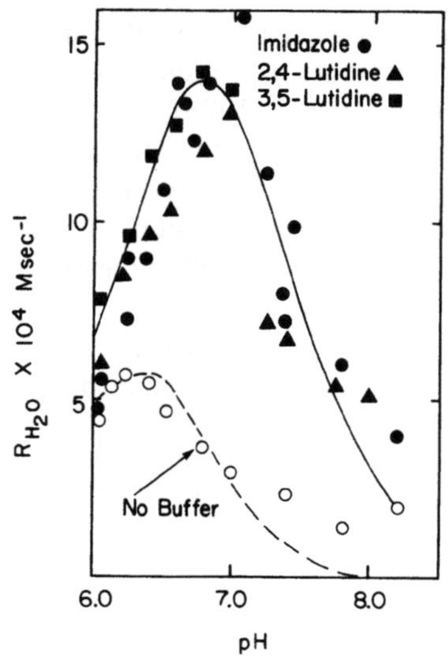

Fig. 3. R_{H_2O} calculated, using Eq. (3), from θ_{cat} and ϕ_{cat} measured (○) in the absence of buffer, and (●) in the presence of 50 mM imidazole, (▲) 50 mM 2,4-lutidine, or (■) 50 mM 3,5-lutidine. In each case (S_{tot}) = 15 mM and human carbonic anhydrase C was 1.6 x 10^{-9} M. The *solid line* is R_{H_2O} calculated from Eq. (4) with $K_1 = K_2 = 10^{-6.8}$ and $k_{-3} = 3.5 \times 10^6\ s^{-1}$. The *dashed line* is calculated from Eq. (5) for R_{H_2O} (no buffer) using a value of 6.4 for the pK_a of the internal proton transfer group

3 Discussion

The role of buffer in the catalytic mechanism of carbonic anhydrase was proposed by Khalifah (1973), Lindskog and Coleman (1973), and Prince and Woolley (1973) to explain the necessarily rapid proton transfer during the hydration of CO_2. The active site of carbonic anhydrase exists in two forms, a high pH form active in the hydration of CO_2 and a low pH form active in the dehydration of HCO_3^-. During catalysis of CO_2 hydration, the high pH form must be regenerated with a rate constant equal to or exceeding the turnover number, which is as large as $10^6\ s^{-1}$. The role of buffer is to provide a concentration of proton acceptor and proton donor which is large relative to $[OH^-]$ and $[H_3O^+]$. It is then a buffer-facilitated proton transfer that enhances the catalysis by permitting rapid proton transfer between enzyme and solution. These hypotheses have been confirmed by initial velocity (Jonsson et al., 1976) and ^{18}O exchange (Silverman and Tu, 1975) techniques which have shown the enhancement of catalytic activity of carbonic anhydrase by buffers. This effect reaches a plateau at a buffer concentration near 5 to 10 mM for buffers such as imidazole and 2,4-lutidine. For higher concentrations the buffer-facilitated proton transfer is not rate-determining.

Hence, when measured in the presence of 50 mM buffer as shown in Fig. 3, the values of R_{H_2O} are not dependent on proton transfer between external buffer and enzyme. Yet the values of R_{H_2O} in Fig. 3 indicate that the exchange of water from the active site depends on the ionization of two groups, one in the acidic form and one in the basic form. We conclude that both of these groups are on the enzyme; that is, the rate-limiting proton transfer is intramolecular. This interpretation is supported by an observed deuterium solvent isotope effect, $R_{H_2O}/R_{D_2O} \sim 5$ to 7 (Silverman et al., 1979). Moreover, this interpretation is consistent with the measurements of the maximal turnover numbers for hydration and dehydration in H_2O and

D_2O which also indicate rate-limiting intramolecular proton transfer steps in the catalytic pathway (Steiner et al., 1975; Pocker and Bjorkquist, 1977).

Carbon-13 NMR results indicate further that this rate-limiting proton transfer is not involved in the interconversion of CO_2 and HCO_3^-; rather, it is in the steps by which the low and high pH forms are interconverted (Simonsson et al., 1979).

The following scheme is a modification of the two-step model to include the internal buffer B, which could possibly be His 64 or Thr 199 hydrogen bonded to Glu 106 as proposed by Kannan et al. (1977).

$$HCOO^{18}O^- + E_{BH}^{OH_2} \overset{\text{Step 1}}{\rightarrow} E_{BH}^{^{18}OH} + CO_2 + H_2O \overset{\text{Step } H_2O}{\rightarrow} E_B^{OH_2} + H_2{}^{18}O + CO_2$$

$$E_{BH}^{^{18}OH} \xrightarrow{^{13}CO_2,\ H_2O} E_{BH}^{OH_2} + H^{13}COO^{18}O^-$$

In this scheme $EOH \rightleftharpoons EOH_2$ is considered to involve the catalytic group, possibly zinc-bound water, and is assumed not to be mediated by external buffers in agreement with the proposal of Steiner et al. (1975). R_1, the rate of interconversion of CO_2 and HCO_3^-, is assumed to be independent of the ionization state of the internal buffer B. The limiting step for R_{H_2O} is

$$E_{BH}^{^{18}OH} \overset{k_{-3}}{\rightarrow} E_B^{^{18}OH_2}$$

with a rate given by

$$k_{-3}\ [E\,(OH)\,BH] = k_{-3}\ (E_{tot}) \cdot \frac{(H^+)}{(H^+) + K_2} \cdot \frac{K_1}{K_1 + (H^+)} \tag{4}$$

The solid line in Fig. 3 is calculated from this expression with $k_{-3} = 3.5 \times 10^6\ s^{-1}$ and $K_1 = K_2 = 10^{-6.8}$. By fitting the steady-state rate equation to stopped-flow kinetic data Steiner et al. (1976) have obtained $k_{-3} = 7 \times 10^5\ s^{-1}$, $K_1 \doteq 10^{-6.9}$, $K_2 = 10^{-7.5}$.

In the absence of buffer there is no pathway for rapid interconversion of E_B^{OH} and E_{BH}^{OH}. Hence, we anticipate that with external buffer absent the fraction of the total enzyme with B unprotonated will be unable to exchange ^{18}O with water at a rate comparable to the rate R_1.

$$HCOO^{18}O^- + E_B^{OH_2} \rightarrow E_B^{^{18}OH} + CO_2 + H_2O$$

$$E_B^{^{18}OH} \xrightarrow{^{13}CO_2,\ H_2O} E_B^{OH_2} + H^{13}COO^{18}O^-$$

The following relationship approximates this situation:

$$R_{H_2O\ (\text{no buffer})} = R_{H_2O\ (\text{excess buffer})} \left[\frac{[E\,(OH)\,BH]}{[E\,(OH)\,BH] + [E\,(OH)\,B]} \right] \tag{5}$$

Using this relationship gives the dashed line in Fig. 3 when $R_{H_2O\ (excess\ buffer)}$ is taken as the solid line of Fig. 3 and the ionization constant of the internal transfer group is $10^{-6.4}$, a value considered experimentally indistinguishable from the value $10^{-6.8}$ obtained earlier.

The agreement between the data and the calculated curves in Fig. 3 is poor in places, especially for experiments in the absence of buffer at $pH > 7$, indicating the failure of the model to describe adequately the catalysis under all of the conditions used in this study. One explanation is the possibility of the binding of buffers to the enzyme, which could result in effects more complicated than those predicted by the simple two-step model. Binding by buffers has already been suggested in studies of carbonic anhydrase (Jonsson et al., 1976; Silverman et al., 1978) and could possibly explain the lower value of R_1 obtained in the presence of imidazole at values of $pH < 6.5$.

Finally, it is pertinent to compare the data for R_{H_2O} with the water proton relaxation data taken in the presence of cobalt bovine carbonic anhydrase (Fabry et al., 1970). Both techniques assign a value greater than $10^5\ s^{-1}$ for the rate constant of water exchange at the active site. Other results from these two techniques differ, however. The ^{18}O exchange measurements detect a buffer dependence in the rate of exchange with water and further predict a low rate of exchange of water at high pH. Both of these results differ from the proton relaxation studies.

Acknowledgment. This work was supported by Research Grant GM 25154 from the National Institutes of Health, U.S. Public Health Service.

References

Fabry ME, Koenig SH, Schillinger WE (1970) Nuclear magnetic relaxation dispersion in protein solutions. J Biol Chem 245: 4256–4262

Gerster R, Maren TH, Silverman DN (1973) A new approach to the study of aqueous solutions of CO_2. Proc 1st Int Conf Stable Isotopes Chem Biol Med, pp 219–228

Hoch G, Kok B (1963) A mass spectrometer inlet system for sampling gases dissolved in liquid phases. Arch Biochem Biophys 101, 160–170

Jonsson BH, Steiner H, Lindskog S (1976) Participation of buffer in the catalytic mechanism of carbonic anhydrase. FEBS Lett 64: 310–314

Kannan KK, Petef M, Fridborg K, Cid-Dresdner H, Lövgren S (1977) Structure and function of carbonic anhydrases. FEBS Lett 73: 115–119

Khalifah RG (1973) Carbon dioxide hydration activity of carbonic anhydrase: paradoxical consequences of the unusually rapid catalysis. Proc Natl Acad Sci USA 70: 1986–1989

Khalifah RG, Strader DJ, Bryant SH, Gibson SM (1977) Carbon-13 nuclear magnetic resonance probe of active-site ionizations in human carbonic anhydrase B. Biochemistry 16: 2241–2247

Lindskog S, Coleman JE (1973) The catalytic mechanism of carbonic anhydrase. Proc Natl Acad Sci USA 70: 2505–2508

Mills GA, Urey HC (1940) The kinetics of isotope exchange between carbon dioxide, bicarbonate ion, carbonate ion and water. J Am Chem Soc 62: 1019–1026

Pocker Y, Bjorkquist DW (1977) Comparative studies of bovine carbonic anhydrase in H_2O and D_2O. Biochemistry 16: 5698–5707

Prince RH, Woolley PR (1973) On the mechanism of action of carbonic anhydrase. Bioorg Chem 2: 337–344

Silverman DN, Tu CK (1975) Buffer dependence of carbonic anhydrase catalyzed oxygen-18 exchange at equilibrium. J Am Chem Soc 97: 2263–2269

Silverman DN, Tu CK (1976) Carbonic anhydrase catalyzed hydration studied by ^{13}C and ^{18}O labeling of CO_2. J Am Chem Soc 98: 978–984

Silverman DN, Tu CK, Wynns GC (1978) Proton transfer between hemoglobin and the carbonic anhydrase active site. J Biol Chem 253: 2563–2567

Silverman DN, Tu CK, Lindskog S, Wynns GC (1979) Rate of exchange of water from the active site of human carbonic anhydrase C. J Am Chem Soc 101: 6734–6740

Simonsson I, Jonsson B-H, Lindskog S (1979) A ^{13}C NMR study of CO_2-HCO_3^- exchange catalyzed by human carbonic anhydrase C at chemical equilibrium. Eur J Biochem 93: 409–417

Steiner H, Jonsson BH, Lindskog S (1975) The catalytic mechanism of carbonic anhydrase. Eur J Biochem 59: 253–259

Steiner H, Jonsson BH, Lindskog S (1976) The catalytic mechanism of human carbonic anhydrase C: inhibition of CO_2 hydration and ester hydrolysis by HCO_3^-. FEBS Lett 62: 16–20

^{35}Cl and ^{81}Br Nuclear Magnetic Resonance Studies of Carbonic Anhydrase

R.L. WARD [1]

1 Introduction

Mammalian erythrocytes often contain more than one type of carbonic anhydrase and human erythrocytes in particular yield two prinicpal forms of the enzyme, which are distinguished foremost by their relative specific activities in regards to the catalysis of the important reversible carbon dioxide hydration reaction. Although these isozymes exhibit many other differences, they have in common the stringent requirement of one zinc ion per molecular weight unit of about 30,000.

The binding of a single monovalent anion by carbonic anhydrase is sufficient to abolish the hydrolysis of CO_2 by the enzyme [6]. A model often proposed for this inhibition of carbonic anhydrase is competitive binding between the monovalent anion and water molecules for a zinc coordination site. ^{35}Cl NMR studies have substantiated the binding of Cl^- to the Zn (II) of carbonic anhydrase and have proved useful in studies of the properties of the active site Zn (II) of a number of the mammalian enzymes.

The nuclear relaxation properties of Cl^- (nuclear spin of 3/2) are determined by the interaction of the nuclear quadrupole moment with time varying electric fields experienced by the nucleus. Aqueous chloride ions are symmetrically solvated by water molecules and because of vanishingly small electric fields at the nucleus exhibit moderately narrow NMR absorption line widths of ~ 10 Hz. Distortion of the electric field at the nucleus, for example, by covalent bonding, yields broad NMR absorption line widths in the kHz range.

If chloride ions are capable of being in more than one type of environment with no exchange between these environments, the NMR spectrum will consist of a superposition of peaks each indicative of the particular environment. If the chloride ions exchange among these sites rapidly compared to the relaxation time in a particular site, then an averaged NMR absorption will be observed. In the particular case of interest here, carbonic anhydrase is present in small quantities, $10^{-4}-10^{-5}$ M, compared to aqueous chloride ions (~ 0.5 M). The NMR experiment consists in observing the effect of the carbonic anhydrase-chloride interaction on the more readily detectable 0.5 M Cl^- absorption.

In detail, the line width of the ^{35}Cl resonance, as measured by the full width at half-maximum amplitude, is related to the relaxation time of the nucleus by the expression $\pi\Delta\nu = 1/T_2$. When a moderately fast chemical exchange of the chloride

---

1 Lawrence Livermore Laboratory, Livermore, California 94550/USA

ion occurs between a free solvated environment and another environment in which the nucleus has an intrinsically different relaxation time, the observed relaxation time will be the weighted average of the relaxation time characteristic of the two environments [8]. Thus, if $1/T_2$ (0) represents the relaxation rate of the free chloride ion and $1/T_2$ the observed relaxation rate, then the relaxation due to the bound chloride ion is given by:

$$1/T_{2P} = 1/T_2 - 1/T_2\,(0) \tag{1}$$

When no chemical shift results from the different environment experiences by the chloride ion on binding

$$\frac{1}{T_{2P}} = f/(T_{2M} + \tau_M) \tag{2}$$

where T_{2M} is the transverse relaxation time of the protein bound chloride ion, τ_M is the exchange lifetime, and f is the fraction of the total chloride ion which is bound to the protein.

The relaxation mechanism for the ^{35}Cl nucleus in the absence of paramagnetic species is the interaction between the nuclear quadrupole moment of the nucleus (eQ) and the fluctuating electrostatic field gradient (eq) determined by the environment and sensed at the nucleus. The full NMR line width for a particular environment is given by

$$\Delta\nu_i = \frac{2}{5}\,\tau\,(e^2 q_i\,Q)\,g\,(t_c) \tag{3}$$

where

$$g\,(t_c) = 0.6\,t_c + \frac{t_c}{1 + \omega^2 t_c^2} + \frac{0.4\,t_c}{1 + 4\omega^2 t_c^2} \tag{4}$$

where t_c is the characteristic correlation time of the nucleus in that environment and ω is the nuclear precessional frequency.

It is important to distinguish between exchange-limited and rapid exchange cases and this can be done by examining the temperature dependence of the line width [11] or by comparing the ratio of the ^{35}Cl and ^{37}Cl line widths. For a nucleus which relaxes by a quadrupole related process, the rate is proportional to the correlation time for molecular tumbling. This rate decreases with increasing temperature and thus the line width would be expected to decrease with increasing temperature for the rapid exchange situation. By contrast, τ_M invariably decreases with increasing temperature. In the fast exchange situation Eq. (2) reduces to

$$\frac{1}{T_{2P}} = \frac{f}{T_{2m}} \tag{5}$$

while in the exchange limited case the equation reduces to

$$\frac{1}{T_{2P}} = \frac{f}{\tau_M} \tag{6}$$

For carbonic anhydrase, as will be shown, the exchange limited case pertains.

2 Results

The NMR line width of aqueous solutions of Cl^- up to about 1 molar is of the order of 10 Hz. Addition of Zn (II) in the mM concentration range to aqueous Cl^-produces an increase in the Cl^- line width. This increase varies linearly with the Zn (II) concentration and can be calculated from the expression

$$\Delta\nu_{obs} - \Delta\nu_{Cl^-} = 2 \times 10^3 \ [\text{Zn (II)}] \quad (7)$$

Although the relaxation mechanism for Cl^- bound to aqueous Zn (II) is quadrupolar in nature, i.e., it depends on t_c, and exchange limited for carbonic anhydrase, it is useful to compare their effectiveness in broadening chloride.

The first isozyme of carbonic anhydrase studied by the ^{35}Cl NMR technique was the bovine enzyme (BCA) [9]. In fact, the first experiment was carried out with such a large amount of the enzyme, 25 mg/ml, that the Cl^- resonance was broadened beyond detection. Subsequent studies revealed that carbonic anhydrase is approximately a thousand times more effective in broadening Cl^- than aqueous Zn (II). Furthermore, while the broadening produced by aqueous Zn (II) is pH-independent up to the point where Zn (II) begins to precipitate, the Cl^- broadening produced by CA exhibits a strong pH dependence. A pH titration of the Cl^- line width for a solution of horse CA is illustrated in Fig. 1. This titration curve illustrates a common

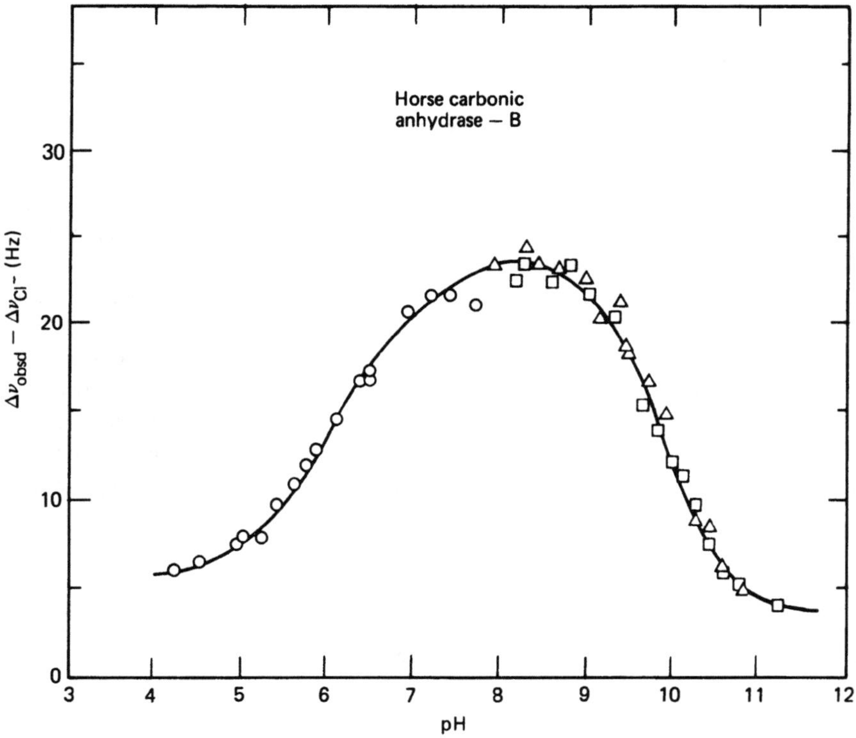

Fig. 1. $\Delta\nu_{obsd} - \Delta\nu_{Cl^-}$ (Hz) vs. pH for a solution containing 1.28 mg/ml of horse carbonic anhydrase B in 0.5 M NaCl. The pK value is 9.85

feature of all CA isozymes where the Cl^- broadening is large at pH 6–7 and decreases in a sigmoidal manner as the pH is increased. In addition, the midpoint of the pH curve in 0.5 M Cl^- can be correlated with the type of isozyme, i.e., a high or low activity form of the enzyme. All high activity forms, i.e., BCA, human carbonic anhydrase C (HCAC) and horse CA-C have a lower pH midpoint than the low activity forms (HCA-B, horse CA-B). The magnitude of the Cl^- broadening is a function of the individual enzyme and is pretty much independent of preparation (assuming a reasonably purified enzyme). There is some correlation of maximum broadening with activity classification (Table 1), but this correlation has not been extensively examined [10].

Table 1. Measured pK for isozymes of carbonic anhydrase in 0.5 M NaCl [a]

		pK	Molar relaxivity (Hz/M)
I.	Horse CA-B	9.85	5.7×10^5
	Human CA-B	9.22	6×10^5
II.	Human CA-C	7.42	1×10^6
	Horse CA-C	7.23	5×10^5
	Bovine CA-B	7.3	2×10^6

a The molar relaxivity is the maximum observed value in 0.5 M NaCl

Lindskog [2] and Lindskog and Thorslund [3] studied the effect of anions on the pKa of the group controlling the activity of carbonic anhydrase. Their data showed that anions that bind more tightly increase the pKa more than less inhibitory species. We have observed that the pKa values obtained from the pH variation of Cl^- line width also varies with the Cl^- concentration. An analysis of the Cl^- dependence in conjunction with kinetically determined inhibitor constants [14] (K_i of 2 mM for HCA-B and 200 mM at pH 6.5 for HCA-C) yields on extrapolation to zero chloride a pKa value of 7.0 for HCAB and 6.7 for HCAC. These values are in close agreement whereas the difference observed in 0.5 M Cl^- is 1.8 units. Differences in pKa values are magnified as the Cl^- concentration increases and the values divide into two groups which correspond to the high activity form of the enzyme (low pKa) and the low activity form of the enzyme (high pKa).

The Cl^- broadening produced by CA has been attributed to the binding of Cl^- to the single Zn (II) ion located at the bottom of the active site crevice described in the crystallographic studies [1, 4]. For both the human enzymes the Zn (II) ion is bound to three histidine residues, thereby leaving a formal +2 zinc ion with one coordination site available for a water molecule or Cl^-. The actual Zn (II) environment is one of a slightly distorted tetrahedral symmetry.

In order to substantiate the claim of a Zn (II)-Cl^- interaction, we have prepared Zn (II) free carbonic anhydrase and observed its effect on the Cl^- line width. ApoCA exhibits essentially no effect on the Cl^- line width when compared to the native enzyme. When the apoenzyme is titrated with aqueous Zn (II) ions and monitored by the Cl^- line width, the line width increases linearly until a maximum is reached at a mole ratio of one Zn (II) to one CA. The line width levels off after a 1:1 ratio has been reached and remains essentially constant as the Zn (II) concentration increases.

Cyanide inhibits the enzymatic activity of all carbonic anhydrases. A CN^- titration of native BCA is shown in Fig. 2 and demonstrates competition of Cl^- for the CN^- binding site. One concludes from this titration that there is only one available

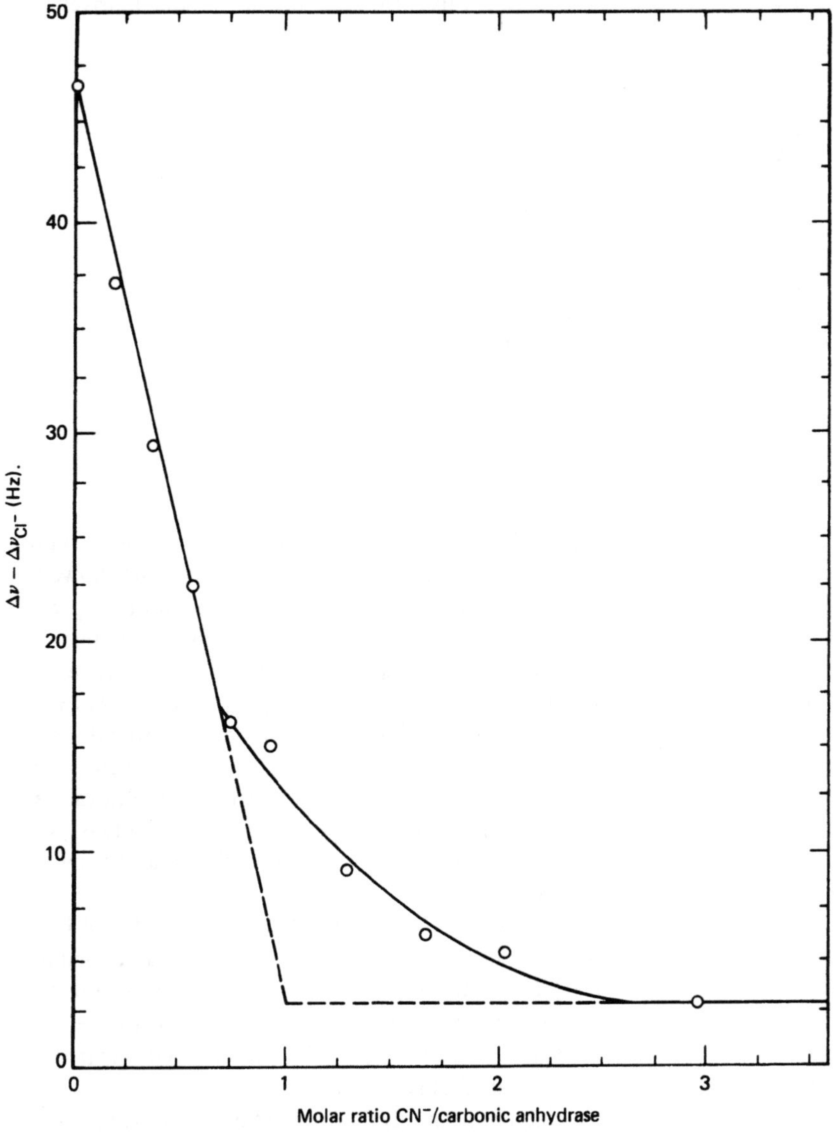

Fig. 2. $\Delta\nu_{obsd} - \Delta\nu_{Cl}$ (Hz) vs. molar ratio of CN^-/BCA for a solution containing 0.5 M NaCl, 3.8×10^{-5} M bovine carbonic anhydrase, and 0.05 M Tris buffer (pH 8.2)

Zn (II)-Cl^- binding site. Acetazolamide, a potent sulfonamide inhibitor of CA, also affects the Cl^- line width and completely removes the Cl^- broadning at a mole ratio of 1:1. Crystallographic studies reveal that it is the anion form of the sulfonamide which binds to the enzyme within the coordination site of the Zn (II) ion. Co (II) substituted carbonic anhydrase possesses properties quite similar to the native Zn (II) enzyme [12]. Co (II) broadens the Cl^- NMR line width in a manner similar to that of the native Zn (II) enzyme. Figure 3 is a titration of apobovine CA with μl amounts of $CoCl_2$. The Cl^- broadening reaches a maximum at a 1:1 stoichiometry of Co (II) to BCA which agrees with titrations of apocarbonic anhydrase with Zn (II). Similarly, Co (II) in excess of 1:1 has little effect on the Cl^- line width in agreement with the Zn (II) results. A pH titration of Co (II) BCA and Co (II) HCA-B appears in Fig. 4. Like the native enzyme, the high and low activity forms of the Co (II) enzyme exhibit a difference in the pKa's measured in 0.5 M Cl with the low activity form possessing the higher pKa.

Smith and Bryant [7] have examined the effect of Hg (II) and Cd (II) substituted carbonic anhydrase on the Cl^- line width and observed that the line width was not broadened. This is quite interesting since crystallographic studies reveal an octahedral Hg (II) coordinated to four amino acids and two Cl^- ions. As indicated later, in the discussion of ^{81}Br NMR studies, this is undoubtedly a result of slow exchange of bound Cl^-.

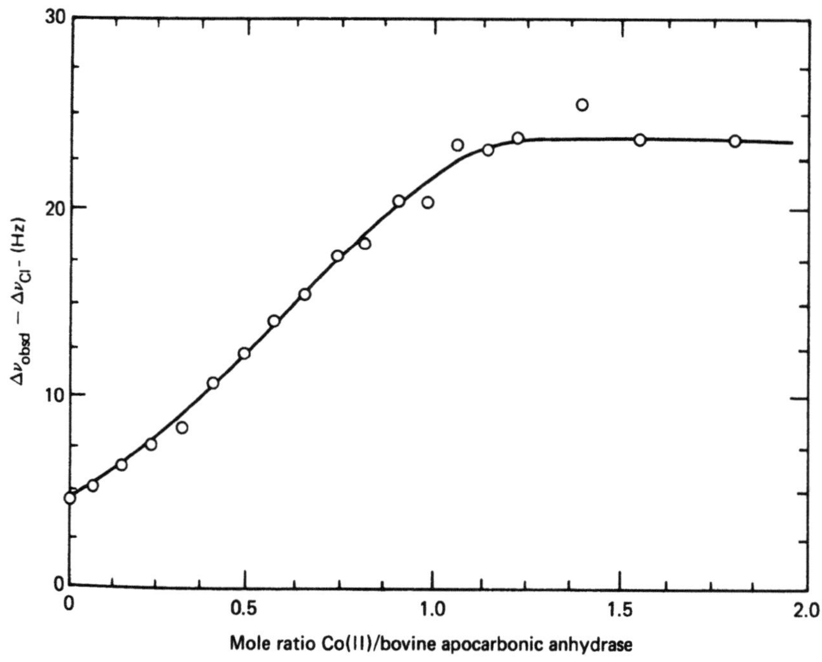

Fig. 3. Line width, $\Delta\nu_{obsd} - \Delta\nu_{Cl^-}$ (Hz), vs. mole ratio Co^{2+}/apo bovine carbonic anhydrase. The solution was 0.5 M NaCl, 0.54 mg per ml of apo bovine carbonic anhydrase and 0.05 M Bis-Tris, pH 6.15

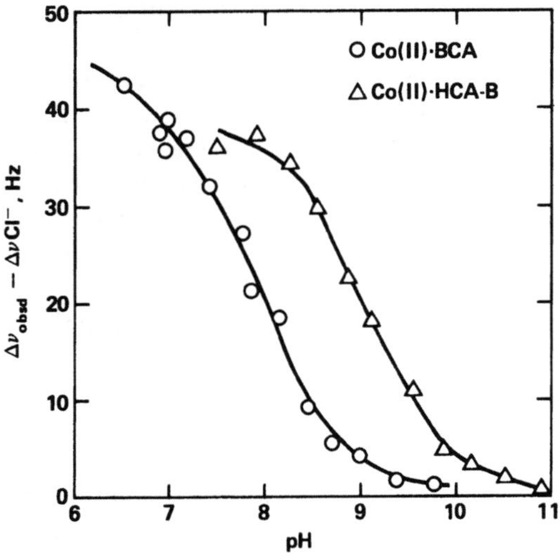

Fig. 4. Line width, $\Delta\nu_{obsd} - \Delta\nu_{Cl^-}$ (Hz) vs. pH for a solution of apo HCAB, 0.9 mg/ml, and 3.1×10^{-5} M Co (II) and a solution of apo BCA, 0.86 mg/ml, and 2.86×10^{-5} M Co (II) both in 0.5 MCl^-

3 Mechanism of Chloride Ion Relaxation

Equation (2) relates the increase in line width of Cl^- in the presence of CA to the inverse sum of two terms, T_{2M} and τ_M. T_{2M} is a measure of the effect of the electric field gradient experienced by Zn (II) bound Cl^- and the molecular tumbling of the molecule. τ_M is the lifetime of Cl^- bound to the Zn (II). As indicated, a temperature study can sort out these effects. Figure 5 displays the effect of temperature on the net line width of HCAB at pH 6.5 and 8.6. Since the net Cl^- line width increases with temperature, the lifetime of Zn (II) bound Cl^-, τ_M, is the dominating factor in the Cl^- relaxation. The value of k_{off} for the reaction Zn (II)-Cl $\underset{k_{on}}{\overset{k_{off}}{\rightleftharpoons}}$ Zn (II) + Cl^- at 25° C is 1×10^6 s^{-1} and 7.3×10^5 s^{-1} at pH 6.5 and 8.6 respectively [11]. Additional evidence that τ_M determines the relaxation time was derived from comparative $^{35}Cl/^{37}Cl$ line width measurements. If the relaxation is controlled by T_{2M} of Eq. (2), then the ratio of the $^{35}Cl/^{37}Cl$ line widths would be in the ratio of the quadrupole moments squared or 1.61. If τ_M, i.e., the exchange rate, determines the relaxation rate, then the ratio of the $^{35}Cl/^{37}Cl$ line widths should be 1.0. We measured this ratio at 14.09 kG and found it to be 1.0, substantiating that the lifetime of Cl^- bound to the Zn (II) of carbonic anhydrase dominates the relaxation behavior.

^{81}Br NMR behavior is quite similar to ^{35}Cl in that its relaxation is dominated by quadrupolar interactions. ^{81}Br line widths are ~ 45 times larger than ^{35}Cl, primarily because of the larger quadrupole moment. We have examined high (BCA) and low activity (HCAB) forms of carbonic anhydrase using ^{81}Br NMR and observed a striking difference in the effect of the two forms on the ^{81}Br line width.

BCA produces an effect on the ^{81}Br line width that is in all aspects quite similar to the results reported for the ^{35}Cl studies. HCAB, however, produces a nonspecific broadening of the ^{81} Br line width that is not responsive to either CN^-, acetazolamide,

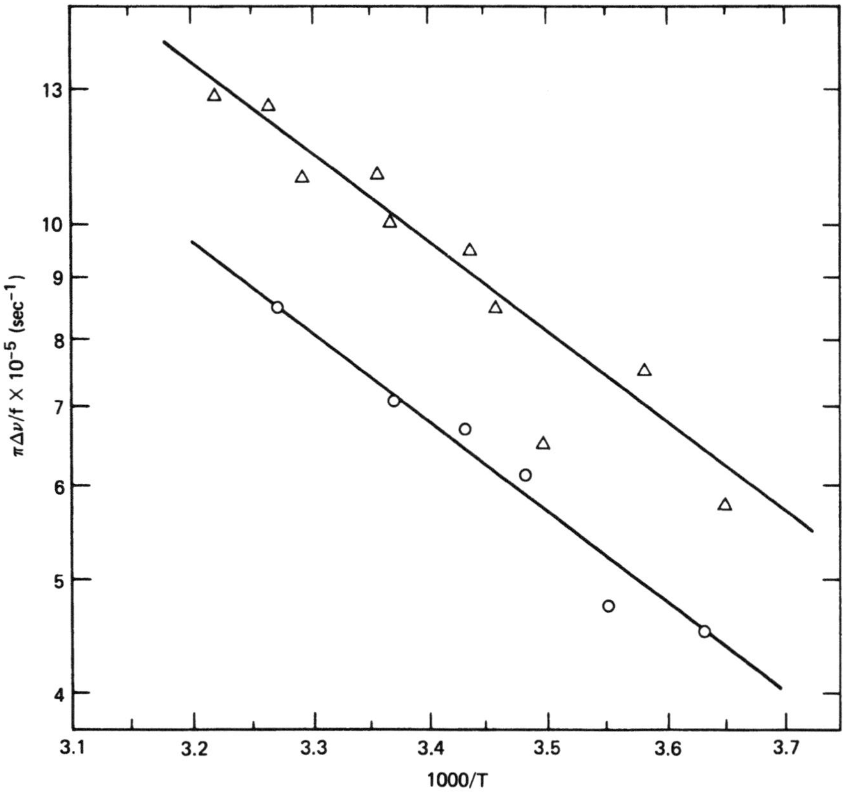

Fig. 5. Arrhenius plot of net chloride line broadening for human carbonic anhydrase B (*circles*) pH 6.4 (1.5 mg/ml) and (*triangles*) pH 8.59 (3 mg/ml) in 0.5 M NaCl. The f factor was calculated using a value of K_i of 0.2 M

or pH, Fig. 6. Competitive Br^-/Cl^- experiments using ^{35}Cl NMR indicate that Br^- competes with Cl^- for the Zn (II) binding site. We, therefore, conclude that the lifetime of Br^- bound to be Zn (II) ion of HCAB is much longer than for BCA and the slowness of the exchange prevents an observable increase in the aqueous Br^- line width. Interestingly, a chemically modified form of HCAB, which has an uncharged carboxyamidomethyl group attached to His-200 located near the active site, does exhibit ^{81}Br NMR results similar to the ^{35}Cl studies, Fig. 7.

4 Acid Tritration Results

Some of the pH titration curves, Fig. 1, exhibit a decrease in line width as the pH is decreased below neutrality. A comparison of this effect with UV difference spectral measurements has been made for HCAC and HCAB dissolved in 0.15 M Cl^- solutions [10]. The Cl^- line width measurements in the acid region are different for the two enzymes. The Cl^- measurements for the C enzyme closely parallel those of Riddiford [5].

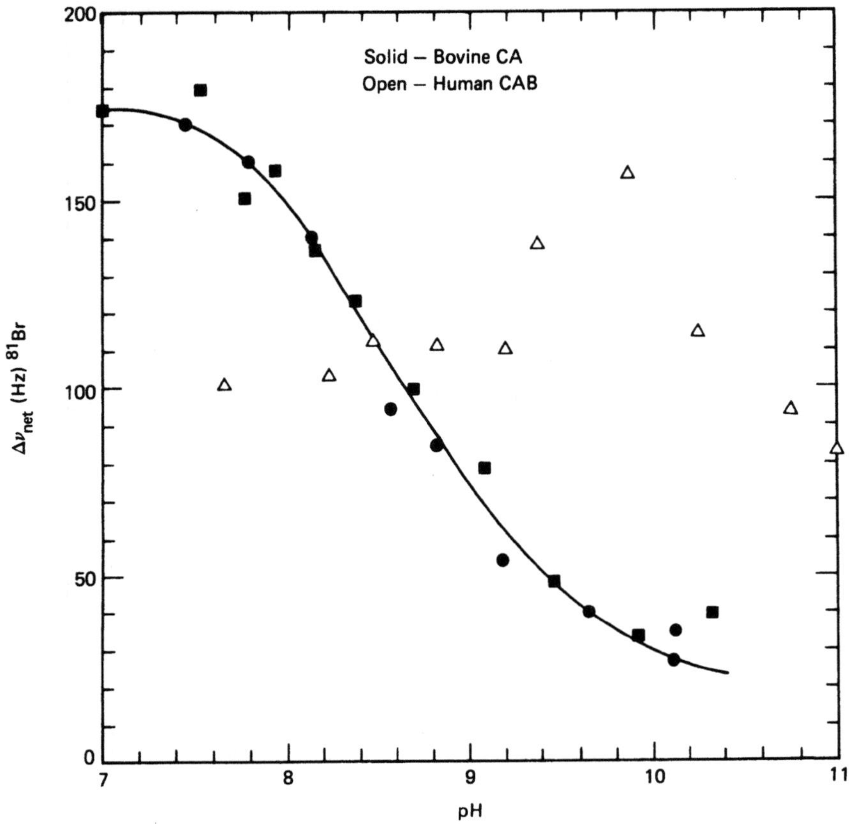

Fig. 6. Net line width, $\Delta\nu_{net}$, i.e., $\Delta\nu_{obsd} - \Delta\nu_{Br^-}$, vs. pH solutions of bovine carbonic anhydrase (*solid symbols*), 0.5 mg/ml, and human carbonic anhydrase B (*open symbols*), 4 mg/ml, in 0.5 M NaBr. The pKa for the bovine enzyme is 8.6

A small displacement in the curves may be due to differences in temperature, 32.9°C vs. room temperature. A sharp change in the Cl^- line width at pH 5 for the C enzyme is irresversible. If the pH of one solution is changed from 6 to 4.9 the line width decreases to a minimum value and remains narrow as the pH is raised again to ~ 6, where the protein precipitates. The Cl^- line widths as a function of pH for the B enzyme, in the acid region, do not parallel Riddiford's curve. The decrease in Cl^- broadening with decreasing pH is much more gradual and is reversible. At pH 4.1, however, the protein begins to come out of solution. It is in this pH region that HCAB undergoes an irreversible conformational change [5]. Although the reason for the decrease in Cl^- line width as the pH is decreased is not known, it is known that in the pH region, 5–7, the Zn (II) remains firmly bound to the protein and therefore the decrease in Cl^- line width cannot be attributed to removal of Zn (II).

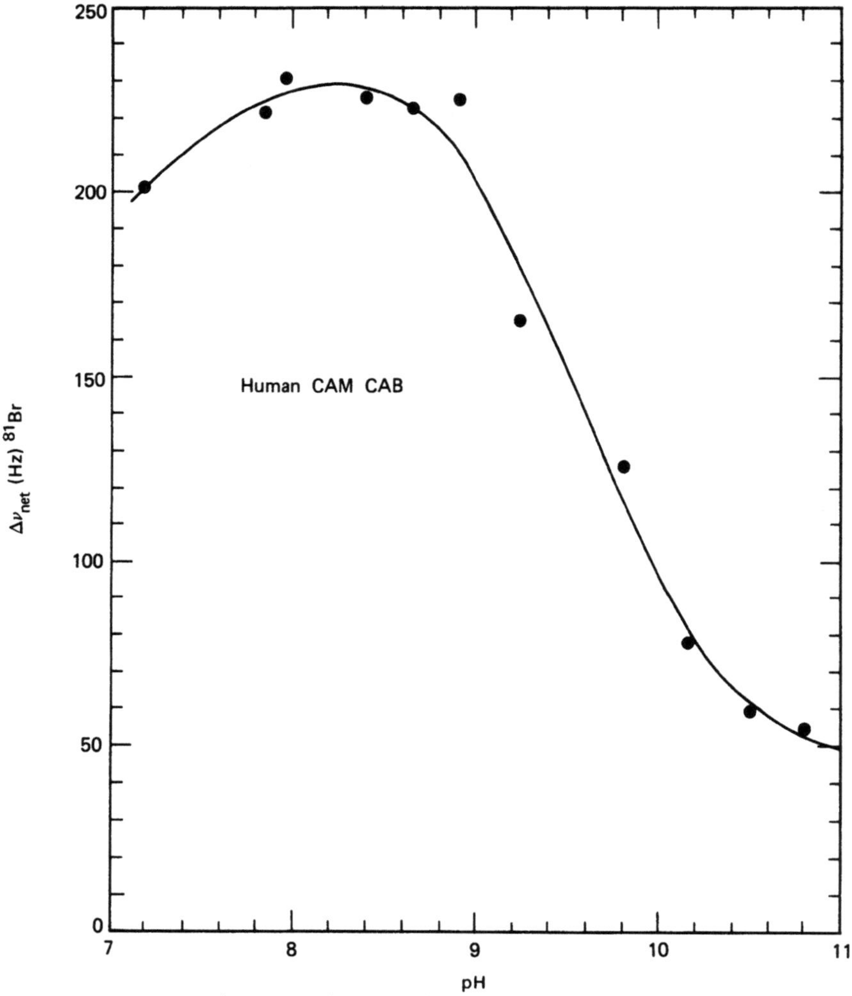

Fig. 7. Net line width, $\Delta\nu_{net}$, i.e., $\Delta\nu_{obsd} - \Delta\nu_{Br^-}$, vs. pH for a solution of modified (carboxyamindomethyl) human carbonic anhydrase B, 3.4 mg/ml, in 0.5 M NaBr. The pKa is 9.6

Acknowledgments. I would like to thank my associates and coworkers who have made much of this work possible and, in particular, Jim Happe, Al Maddux, and Bert Holder. I would also particularly like to express my thanks to Professor John T. Edsall whose encouragement and supplies of the human and horse enzymes made this work both possible and exciting.

References

1. Kannan KK, Notstrang B, Fridborg K, Lövgren S, Ohlsson A, Petef M (1975) Crystal structure of human erythrocyte carbonic anhydrase B. Three-dimensional structure at a nominal 2.2-Å resolution. Proc Natl Acad Sci USA 72: 51–55
2. Lindskog S (1966) Interaction of cobalt (II)-carbonic anhydrase with anions. Biochemistry 5: 2641–2646

3. Lindskog S, Thorslund A (1966) On the interaction of bovine cobalt carbonic anhydrase with sulfonamides. Eur J Biochem 3: 453–460
4. Notstrand B, Vaara I, Kannan KK (1975) Structural relationship of human erythrocyte carbonic anhydrase isozymes B and C. In: Markert CL (ed) Proc Int Conf Isozymes, vol I. Molecular structure. Academic Press, London New York, p 575
5. Riddiford LM (1965) Acid difference spectra of human carbonic anhydrases. J Biol Chem 240: 168–172
6. Roughton FJW, Booth VH (1946) Manometric determination of the activity of carbonic anhydrase under varied conditions. Biochem J 40: 309–319
7. Smith RJ, Bryant RG (1975) Metal substitutions in carbonic anhydrase: A halide ion probe study. Biochem Biophys Res Commun 66: 1281–1286
8. Swift TJ, Connick RE (1962) NMR-relaxation mechanisms of O^{17} in aqueous solutions of paramagnetic cations and the lifetime of water molecules in the first coordination sphere. J Chem Phys 37: 307–320
9. Ward RL (1969) ^{35}Cl nuclear magnetic resonance studies of a zinc metalloenzyme carbonic anhydrase. Biochemistry 8: 1879–1883
10. Ward RL (1970) Zinc environmental differences in carbonic anhydrase isozymes. Biochemistry 9: 2447–2453
11. Ward RL, Cull MD (1972) Temperature and competitive anion-binding studies of carbonic anhydrase. Arch Biochem Biophys 150: 436–439
12. Ward RL, Fritz KJ (1970) ^{35}Cl-NMR studies of Co^{2+} carbonic anhydrases. Biochem Biophys Res Commun 39: 707–711
13. Ward RL, Whitney PL (1973) ^{81}Br-NMR studies of carbonic anhydrase. Biochem Biophys Res Commun 51: 343–348
14. Whitney PL, Brandt H (1976) Effects of two ionizing groups on the active site of human carbonic anhydrase B. J Biol Chem 251: 3862–3867

NMR Studies of Carbonic Anhydrase

J.M. PESANDO [1] and R.K. GUPTA [2]

1 Spatial Arrangement of p-Nitrophenol and Acetate at the Active Sites of Mn (II) and Co (II)-Substituted Human Carbonic Anhydrase C as Observed by NMR Spectroscopy

1.1 Introduction

Hydrolysis of p-nitrophenylacetate to p-nitrophenol and acetate and the reversible hydration of CO_2 by the zinc metalloenzyme carbonic anhydrase appear to involve similar catalytic mechanisms (Pocker and Sarkanen, 1978). The essential Zn^{2+} ion at the active site can be replaced by several divalent cations, but only the Zn^{2+}, Co^{2+}, Mn^{2+}, and Cd^{2+} enzymes retain catalytic activity. The activity of Zn^{2+} and Co^{2+}-substituted enzmyes is comparable, while that of the Mn^{2+} and Cd^{2+} forms is ~ 10% and ~ 30% that of the native enzyme, respectively (Pocker and Sarkanen, 1978).

Since the catalytically active Co^{2+} and Mn^{2+}-substituted enzymes have paramagnetic metal ions at the active sites, distances of the magnetic nuclei of bound molecules from the paramagnetic reference point can be determined. This information can be used to derive the configuration of substrates or products at the active site and might provide new insights into the mechanism of enzyme action. This manuscript reports our studies of the arrangement of p-nitrophenol and acetate, products of the essentially irreversible hydrolysis of p-nitrophenylacetate, at the active site of human carbonic anhydrase C. Products of catalysis were studied to avoid the complications of a time-dependent system which would be unavoidable in studies of the substrate itself. Since the products contain multiple nuclei (^1H and ^{13}C), several metal-nuclear distances can be determined.

1.2 Materials and Methods

Metal-free carbonic anhydrase from human erythrocytes was purified as previously described (Pesando, 1975a). All measurements were made on the catalytically most active C isoenzyme. Full activity could be restored upon the addition of 1 equivalent of zinc sulfate to the apoenzyme. Co^{2+}- and Mn^{2+}-carbonic anhydrases were prepared by direct addition of 0.9 equivalent or less of the respective sulfates to solutions of the apoenzyme at pH ~ 8.

---

1 Sidney Farber Cancer Institute, Harvard Medical School, Boston, Massachusetts 02115/USA
2 The Institute for Cancer Research, The Fox Chase Cancer Center, Philadelphia, Pennsylvania 19111/USA

The magnitude of paramagnetic effects of a metal-enzyme complex on nuclear relaxation rates can yield information on metal-nuclear distances. The paramagnetic effects on the longitudinal ($1/T_{1p}$) and transverse ($1/T_{2p}$) nuclear relaxation rates are given by (Mildvan and Gupta, 1978):

$$\frac{1}{fqT_{1p}} = \frac{1}{T_{1M} + \tau_M} \tag{1}$$

$$\frac{1}{fqT_{2p}} = \frac{1}{T_{2M} + \tau_M} \tag{2}$$

where T_{1M} and T_{2M} are the longitudinal and transverse relaxation times of the bound ligand and τ_M is the residence time of the ligand in the paramagnetic environment. When the dipolar interaction predominates, the distance dependencies of $1/T_{1M}$ and $1/T_{2M}$ are described by the following equations:

$$\frac{1}{T_{1M}} = \frac{2}{15}\,\frac{S(S+1)\,g^2\beta^2\gamma_I^2}{r^6}\left\{\frac{3\tau_c}{1+\omega_I^2\tau_c^2} + \frac{7\tau_c}{1+\omega_S^2\tau_c^2}\right\} \tag{3}$$

$$\frac{1}{T_{2M}} = \frac{2}{15}\,\frac{S(S+1)\,g^2\beta^2\gamma_I^2}{r^6}\left\{2\tau_c + \frac{1.5\,\tau_c}{1+\omega_I^2\tau_c^2} + \frac{6.5\,\tau_c}{1+\omega_S^2\tau_c^2}\right\} \tag{4}$$

where τ_c is correlation time for the metal-nuclear dipolar interaction and other symbols have their usual meaning (Mildvan and Gupta, 1978).

1.3 Results and Discussion

Addition of Mn^{2+} or Co^{2+} to a solution containing metal-free HCAC and p-nitrophenol with or without acetate causes a sizable enhancement in the $1/T_1$ of the ortho (o-) and meta (m-) protons of p-nitrophenol and of the methyl protons and carboxyl and methyl carbon atoms of acetate. The magnitude of these effects indicates that they arise from paramagnetic interactions of nuclear spins with unpaired electrons of Mn^{2+} or Co^{2+}. The effects of Zn^{2+}, a diamagnetic metal, are much smaller (see below). Further, the addition of Zn^{2+} to a solution containing Mn^{2+} or Co^{2+} reduces the observed relaxation effects due to displacement of the enzyme-bound paramagnetic metal by Zn^{2+}. The displacement experiments indicate that the observed paramagnetic effects arise from the enzyme-bound paramagnetic metal and are not from traces of unbound metal invariably present. In nonenzymatic control experiments with binary metal-ligand complexes, the observed effects are an order of magnitude smaller.

Zn^{2+}-HCAC exerts only a small but measurable effect on the longitudinal nuclear relaxation rates of p-nitrophenol and acetate. The normalized magnitudes of these effects ($1/fT_{1d}$) measured at four different temperatures are shown in Table 1. These diamagnetic effects of Zn^{2+}-HCAC were subtracted from the observed total effects to calculate the paramagnetic effects of Mn^{2+} and Co^{2+}. In the absence of acetate, the normalized paramagnetic effects of Mn^{2+}-HCAC on the o- and m-protons of

Table 1. Diamagnetic effects of Zn (II)-HCAC on the protons of p-nitrophenol and acetate ions in the E·p-nitrophenol·acetate complex [a]

Proton	$(fT_{1d})^{-1}$, s^{-1}			
T (°C) =	10	15	20	30
p-Nitrophenol (m-)	98 ± 14	82 ± 12	65 ± 9	35 ± 5
p-Nitrophenol (o-)	87 ± 8	74 ± 7	61 ± 6	37 ± 4
Acetate ($-CH_3$)	31 ± 5	26 ± 4	20 ± 3	10 ± 2

a [p-nitrophenol] = 121 mM; [acetate] = 128 mM. Measurements were made at a proton resonance frequency of 100 MHz

Table 2. Paramagnetic effects of Mn (II)-HCAC on the $1/T_1$ of p-nitrophenol protons [a]

Proton	[p-Nitrophenol] mM	$(fT_{1p})^{-1}$ s^{-1}
m-	33	10,400
o-	33	52,600
m-	66	11,900
o-	66	59,500
m-	117	11,300
o-	117	57,500
m-	131	12,200
o-	131	58,500
m-	141	11,600
o-	141	57,500

a In the absence of acetate. Measurements were made at a proton resonance frequency of 100 MHz (T = 15°C)

p-nitrophenol at several concentrations of the latter are shown in Table 2. Except perhaps at the lowest concentration (33 mM) where the observed effect may be slightly smaller, the paramagnetic effects of Mn^{2+}-HCAC on the protons of p-nitrophenol are independent of concentration, indicating saturation of the binding site under our experimental conditions. Since the paramagnetic effect of Mn^{2+} is larger on ortho protons, the data indicate that the phenol group of p-nitrophenol points towards the enzyme-bound Mn^{2+}.

The paramagnetic effect of Mn^{2+} on $1/T_2$ of the ring protons of p-nitrophenol was ~ 16-fold larger than that on $1/T_1$ at 100 MHz. The over-an-order-of-magnitude difference between the observed T_{1p} and T_{2p} ($T_{2p} \ll T_{1p}$) directly indicates that $1/T_{1p}$ is not exchange-limited. The absence of exchange-limitation on $1/T_{2p}$ was established by a study of the temperature-dependence of the magnitude of this paramagnetic effect of Mn^{2+}-HCAC on a saturating level of p-nitrophenol. A ~20% decrease in $1/fT_{2p}$ with a 20°C increase in sample temperature was observed. Exchange-limited paramagnetic relaxation would show a stronger temperature-dependence in the opposite direction. Although the magnitude and nature of the observed paramagnetic effects themselves indicate that they arise within the Mn^{2+}-HCAC-p-nitrophenol complex (Mildvan and Gupta, 1978) and are not of outersphere origin, this was confirmed by displacement experiments where a saturating level of acetazolamide reduced the observed effects close to the values found with the Zn^{2+}-enzyme.

The presence of acetate decreases the magnitude of the paramagnetic effect of Mn^{2+}-HCAC on the protons of p-nitrophenol. This effect of acetate on the observed paramagnetic effects experienced by p-nitrophenol protons saturated at $\leqslant$ 120 mM acetate since a further increase in the concentration of acetate to 200 mM produced no further decrease in the observed effects. Higher concentrations of p-nitrophenol did not reverse the effect of acetate. Since acetate binding at the active site is weak with $K_I = 0.8 \pm 0.2$ M (Lanir and Navon, 1974), these results indicate the binding of acetate to an ancillary site which alters the arrangement and conformation of enzyme-bound p-nitrophenol relative to the paramagnetic reference point. As in the absence of acetate, the paramagnetic effect on $1/T_2$ of p-nitrophenol protons is much larger (18 ± 2-fold) than that on $1/T_1$ in the presence of acetate as well, indicating that $1/T_{1p}$ is not exchange-limited. The temperature-dependence of $1/T_{2p}$ indicates that this too is not exchange-limited. The magnitudes of paramagnetic effects of Mn^{2+}-HCAC on the $1/T_1$ of protons of p-nitrophenol with both acetate and p-nitrophenol simultaneously present on the enzyme are shown in Table 3. Even though the $1/T_1$ is in the fast exchange region since $T_{1p} \gg T_{2p}$, the magnitude of $1/fT_{1p}$ increases with increasing temperature in a manner opposite to that found for $1/fT_{2p}$. Such a behavior indicates that $\omega_I^2 \tau_c^2 \gg 1$. In contrast to the effects on p-nitrophenol, for acetate the $1/T_{1p}$ and $1/T_{2p}$ effects are of similar magnitude and both increase with increasing temperature. Further, the paramagnetic effects on acetate protons show a much stronger temperature dependence (Table 3) compared to that found for $1/T_{1p}$ of the protons of p-nitrophenol. The observed magnitudes and temperature dependencies of $1/T_{1p}$ and $1/T_{2p}$ indicate that the relaxation of acetate methyl group is exchange-limited (Lanir and Navon, 1974; Bertini et al., 1977).

Table 3. Temparature dependence of the paramagnetic effects of Mn (II)-HCAC on the protons of p-nitrophenol and acetate [a]

Proton	$(fT_{1p})^{-1}$, s^{-1}		
T (°C) =	10	20	30
p-Nitrophenol (m-)	5,900 ± 300	6,300 ± 300	6,700 ± 300
p-Nitrophenol (o-)	16,000 ± 1,000	20,000 ± 1,000	25,000 ± 1,000
Acetate ($-CH_3$)	3,500 ± 200	7,300 ± 400	11,100 ± 600

a [p-nitrophenol] = 107 mM; [acetate] = 121 mM. Measurements were made at a proton resonance frequency of 100 MHz

To determine the orientation of the acetate group at the active site of Mn^{2+}-HCAC in the presence of p-nitrophenol, paramagnetic effects of enzyme-bound Mn^{2+} on the carbon atoms of acetate enriched with ^{13}C at both carbon positions were measured. Sizable paramagnetic effects of Mn^{2+} (Table 4) were observed on both

Table 4. Paramagnetic effects of Mn (II)-HCAC on the ^{13}C relaxation rates of the carboxyl and methyl carbons of acetate [a]

Nucleus	$(fT_{1p})^{-1}$ s^{-1}	$(fT_{2p})^{-1}$ s^{-1}	T_{1p}/T_{2p} [b]
$^{13}COO^-$	70,900	100,000	1.41
$^{13}CH_3$	52,800	77,600	1.47

a Saturating [p-nitrophenol]; [acetate] = 103 mM. Measurements were made at a ^{13}C resonance frequency of 25.1 MHz (T = 15 °C)

b $T_{1M}/T_{2M} = 2.23$ ($\tau_c = 8 \times 10^{-9}$ s).

$1/T_1$ and $1/T_2$ of the two carbon atoms, the magnitude of the effect being somewhat greater for the carboxyl carbon atom. As with methyl protons, the paramagnetic effects of Mn^{2+} on the two ^{13}C atoms were also found to be exchange-limited as revealed by their temperature dependencies.

The paramagnetic effects on both p-nitrophenol and acetate are simultaneously removed by the addition of stoichiometric amounts of acetazolamide, a potent inhibitor of the enzyme. The latter observation indicates that the observed paramagnetic effects arise mostly at the active site of Mn^{2+}-HCAC and that the paramagnetic effects of the ancillary binding site for acetate are negligible under our conditions, i.e., in the presence of p-nitrophenol.

To investigate the possibility that differences in the activity of Mn^{2+}- and Co^{2+}-enzymes arise from small variations in the arrangement of substrates at the active site, we also examined the paramagnetic effects of Co^{2+}-HCAC on the protons of p-nitrophenol and acetate, both simultaneously present. The paramagnetic effects of cobaltous ion are approximately two orders of magnitude smaller than those of Mn^{2+} but still are well above the diamagnetic effects observed with Zn^{2+} (Table 5). As with the Mn^{2+}-enzyme, the paramagnetic effects of Co^{2+}-enzyme were obtained by subtracting the diamagnetic effects of the Zn^{2+}-enzyme from the total observed effects by means of appropriate control experiments. The large difference between the paramagnetic effects of Mn^{2+}- and Co^{2+}-enzymes arises from differences in the electron spin relaxation times of the two metals. In contrast to the larger paramagnetic effects of the Mn^{2+}-enzyme, the small paramagnetic effects of the Co^{2+}-enzyme on both acetate and p-nitrophenol protons are in the NMR fast exchange region as indicated by their temperature dependencies. As with the Mn^{2+}-enzyme, however, they are removed by the addition of stoichiometric amounts of acetazolamide or by Zn^{2+}. These results one again confirm that the observed paramagnetic effects arise within the enzyme-product complex and are not of outersphere origin.

From the magnitudes of the paramagnetic effects ($1/fT_{1p}$), it is possible to calculate metal-nuclear distances provided the correlation time for the paramagnetic dipolar interaction is known and the relaxation is not exchanged-limited or, if it is exchange-limited, the exchange-time τ_M is known. For the Co^{2+}-enzyme, the τ_S value of 1.2×10^{-11} s (Fabry et al., 1970) was used in distance calculations. The distances obtained from relaxation data for the HCAC-Co^{2+}-p-nitrophenol-acetate complex are included in Table 5.

Table 5. Metal-nuclear distances on Mn (II) HCAC and Co (II) HCAC from paramagnetic effects [a]

Complex [b]	Nucleus	T_{1M}^{-1} s^{-1}	τ_c s	r Å
E·Mn·pnp	^1H (meta)	11,800 ± 400	8×10^{-9}	5.3 ± 0.2
E·Mn·pnp	^1H (ortho)	58,300 ± 1,000	8×10^{-9}	4.1 ± 0.2
E·Mn·pnp·Ac	^1H (meta)	6,200 ± 500	8×10^{-9}	5.9 ± 0.3
E·Mn·pnp·Ac	^1H (ortho)	18,100 ± 1,000	8×10^{-9}	4.9 ± 0.2
E·Mn·pnp·Ac	^1H (methyl)	72,000 ± 18,000	8×10^{-9}	4.0 ± 0.3
E·Mn·pnp·Ac	^{13}C (methyl)	86,000 ± 22,000	8×10^{-9}	3.6 ± 0.3
E·Mn·pnp·Ac	^{13}C (carboxyl)	146,000 ± 37,000	8×10^{-9}	3.3 ± 0.2
E·Co·pnp·Ac	^1H (meta)	190 ± 30	1.2×10^{-11}	5.6 ± 0.3
E·Co·pnp·Ac	^1H (ortho)	230 ± 35	1.2×10^{-11}	5.4 ± 0.3
E·Co·pnp·Ac	^1H (methyl)	80 ± 10	1.2×10^{-11}	6.4 ± 0.4

a T = 15°C (MnHCAC) and 25°C (CoHCAC). NMR frequency = 100 MHz (^1H) and 25.1 MHz (^{13}C)

b pnp, p-nitrophenol; Ac, acetate

For the HCAC-Mn^{2+}-p-nitrophenol and the HCAC-Mn^{2+}-p-nitrophenol-acetate complexes, the T_{1p}/T_{2p} ratio for p-nitrophenol protons was used to calculate a correlation time of $\sim 8 \times 10^{-9}$ s for the metal-nuclear dipolar interaction using Eq. (3) and (4). This dipolar correlation time is close to the rotational correlation time τ_R of the carbonic anhydrase molecule but may also be a combination of τ_R and τ_S. The distance calculations are, of course, independent of the detailed nature of the correlation process(es). Knowing the correlation time and the magnitude of paramagnetic effects, the distances from Mn^{2+} to p-nitrophenol protons shown in Table 5 were obtained.

The calculations of distances to protons and ^{13}C atoms of acetate in the HCAC-Mn^{2+}-acetate-p-nitrophenol complex is more complicated since acetate levels used were nonsaturating and also because the paramagnetic effects are exchange-limited in this case. However, since the inhibition constant for acetate in the p-nitrophenylacetate hydrolysis reaction is known and the correlation time for the dipolar interactions of the enzyme-bound Mn^{2+} with acetate nuclear spins must be the same as for Mn^{2+}-p-nitrophenol interactions in the same enzyme complex (the latter being in the NMR fast exchange region) it is possible to calculate τ_M and T_{1M} values for acetate nuclei according to Eq. (1) and (2) (Table 5). Knowledge of T_{1M} and τ_c values permitted a calculation of the distances from Mn^{2+} to various parts of the acetate molecule shown in Table 5.

Calculated distances indicate direct coordination of the phenolic oxygen of p-nitrophenol to the enzyme-bound Mn^{2+} in the ternary HCAC-Mn^{2+}-p-nitrophenol complex. The Mn^{2+}-p-nitrophenol distances in the quaternary HCAC-Mn^{2+}-acetate-p-nitrophenol and HCAC-Co^{2+}-p-nitrophenol-acetate complexes are also consistent with coordination of p-nitrophenol to the enzyme-bound metal via the phenolic oxygen. However, the orientation of the aromatic ring of p-nitrophenol must be different for the Co^{2+} and Mn^{2+}-enzymes since the relative distances of o- and m-ring protons are different in the two cases. The carboxyl carbon of acetate is in the range of innersphere distances expected for direct coordination of a carboxyl oxygen to the enzyme-bound Mn^{2+}. The distances of 3.3, 3.6, and 4.0 Å from Mn^{2+} to the carboxyl and methyl carbons and protons of acetate, however, indicate that the acetate molecule is folded around the metal. Although direct coordination of acetate to the metal in the HCAC-Co^{2+}-p-nitrophenol-acetate complex cannot be inferred from our data, weak innersphere complexation with a somewhat longer Co^{2+}-carboxyl oxygen bond distance is not excluded. However, there is a significant difference in the metal-H_3C proton distances in the HCAC-Mn^{2+}-acetate-p-nitrophenol and HCAC-Co^{2+}-p-nitrophenol-acetate quaternary complexes. The various distances obtained by us are shown in Fig. 1. The observed small but significant differences in the arrangement of the products at the active site of the Co^{2+}- and Mn^{2+}-enzymes may be responsible for the differences in their reactivities in the hydrolysis reaction. It should be pointed out that although anions displace the metal-bound water in the Co^{2+}-enzyme (Fabry et al., 1970), they do not alter the paramagnetic effects of the Mn^{2+}-enzyme on water relaxation rates (Lanir et al., 1973). Even in the presence of both products of the hydrolysis reaction, p-nitrophenol and acetate, the water relaxation is not affected. Since both products appear to be located in the innersphere of the metal, it would appear that the enzyme-bound Mn^{2+} has a tendency to become octahedral, while the enzyme-bound Co^{2+} may retain a distorted tetrahedral or pentacoordinate structure in the presence of substrates.

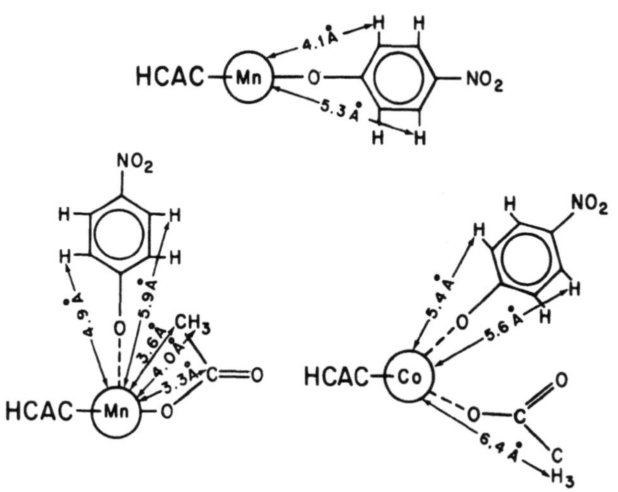

Fig. 1. Metal-nuclear distances for the products of p-nitrophenylacetate hydrolysis on Mn^{2+} and Co^{2+}-substituted human carbonic anhydrase C

Much of our present knowledge about the active site of carbonic anhydrase and other metalloenzymes has been obtained by substituting for the naturally occurring metal ion other transition metal ions whose spectroscopic properties make them more amenable to study. Our present work, showing different active site configurations for the products of p-nitrophenylacetate hydrolysis in two metallo-enzyme derivatives of HCAC, suggests the need for caution in using the results of such experiments to characterize the native enzyme.

2 A Proposal for the Catalytic Mechanism of Carbonic Anhydrase

Although carbonic anhydrase was initially purified almost twenty years ago and its reversible hydration of carbon dioxide involves only six substrate atoms, work by many investogators using often sophisticated methodologies has failed to produce a generally accepted proposal for its mechanism of catalysis (Pocker and Sarkanen, 1978).

Many investigators favor the basic proposal that a zinc-bound water molecule at the active site of the enzyme ionizes around pH 7, producing a metal-bound hydroxide ion that attacks the substrate in the hydration reaction (Khalifah, 1971; Lindskog and Coleman, 1973; Pocker and Sarkanen, 1978). Anion inhibitors are thought to compete with hydroxide for the metal, explaining the apparent increase in the pKa of this group in their presence. Failure of chemical modification studies to identify additional essential groups at the active site provides indirect support for this model (Pocker and Sarkanen, 1978).

While the ionizable zinc-bound water molecule hypothesis is an attractive one, it has several major difficulties: (1) there is no data to indicate that the pKa of water can be reduced from 15.74 in solution to less than 7 when bound to zinc (Prince and Woolley, 1972), (2) there is no explanation for the marked differences in the visible spectra of the high pH and anion-inhibited cobalt-substituted enzymes, both of which would have coordinated anions in this scheme (Dennard and Williams, 1966), (3) the protein itself serves no role that could not be met by a simple metal-chelate complex (Prince and Woolley, 1972; Pesando, 1975b) and (4) magnetic relaxation studies of solvent water protons for both the manganese and cobalt-substituted enzymes suggest that there is no metal-bound hydroxide in the high pH form of the enzyme (Fabry et al., 1970; Lanir et al., 1973).

Several years ago we proposed an alternative model for enzyme-catalysis based on the anomalous behavior of the chemical shifts of a histidine residue at the active site observed by high resolution proton magnetic resonance (PMR) spectroscopy (Gupta and Pesando, 1975; Pesando, 1975b). One of the three histidine ligands to the metal identified by X-ray diffraction studies (Liljas et al., 1972) was thought to simultaneously bind to the metal and ionize with increasing pH, generating an imidazolate anion liganded to the metal by a strained bond. This labile protein form was thought to be the active species in a concerted hydration reaction, serving to generate hydroxide from water while being replaced as a metal-ligand by an oxygen of the substrate. The mechanism appeared to explain the PMR data, and the pKa of metal-bound imidazole based on then available data appeared to be lower than that of metal-bound water (Pesando, 1975b). This model assigns to the protein the essential role of creating strain at the active site. The differences in the visible spectra of the anion-inhibited and high pH forms of the cobalt enzyme occur since the ligand would be constrained by the protein in the latter but not in the former state. This mechanism is also consistent with the PMR relaxivity studies since it does not require the presence of an hydroxide ion on the metal.

Subsequent work has failed to confirm that the pKa of metal-bound imidazole is sufficiently low to identify histidine as the group controlling catalytic activity (Martin, 1974). In addition, the work of Koenig's group (Jacob et al., 1978), if confirmed, indicating that the true pKa of the group controlling enzymatic activity in carbonic anhydrase is closer to 5 than to 7 because of heretofore unappreciated large anion effects, makes both of the two preceding basic models untenable. Neither water nor imidazole when bound to zinc as neutral species would be expected to have a pKa value this low.

Given this new information, we propose the catalytic mechanism shown in Fig. 2. This model incorporates features of both earlier proposals and identifies the group controlling catalytic activity as an active site histidine. The protonated imidazole ring of this residue coordinates to zinc with increasing pH, losing a proton. Geometric constraints on the histidine make this a strained and therefore labile metal-ligand bond. In the hydration reaction this metal-bound neutral histidine serves as a proton acceptor for the *transient* ionization of metal-bound water and as a partial proton donor to carbon dioxide in a concerted mechanism. Zinc-bound hydroxide attacks the carbon of the substrate to generate metal-bound bicarbonate, and the system regenerates itself by losing the elements of carbonic acid.

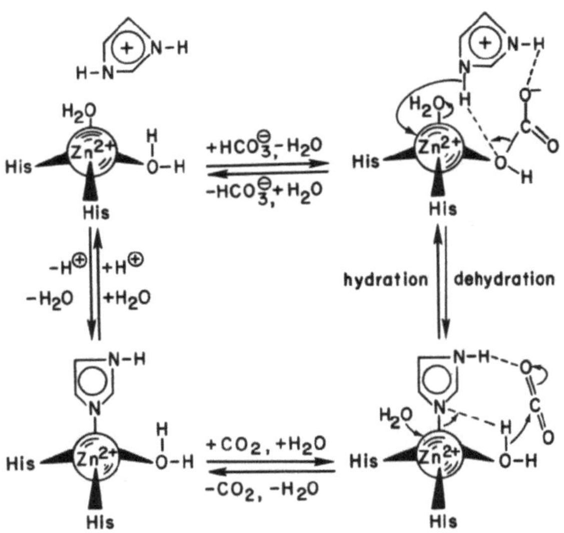

Fig. 2. Proposed catalytic mechanism for carbonic anhydrase

This model appears to be consistent with all available data on the enzyme. Concerted deprotonation and metal-ion coordination of imidazole is chemically reasonable and is consistent with the recent finding of Koenig and co-workers (Jacob et al., 1978) that the pKa of the group controlling catalytic activity is much lower than previously believed. Metal-bound hydroxide exists only as a transient species generated in a concerted mechanism which relieves the strain on the imidazole-metal bond, and its nucleophilicity is enhanced by hydrogen bonding of this imidazole to the substrate. Absence of metal-bound hydroxide as a stable species in the high pH form of the enzyme is consistent with the nuclear magnetic relaxation data (Fabry et al., 1970; Lanir et al., 1973). The protein itself serves to create strained, and therefore labile, species at the active site. The abnormal pH dependence of the imidazole proton resonances observed in PMR studies results from simultaneous ionization and coordination of an imidazole ring to a divalent cation. Differences between the visible spectra of the anion-inhibited and high pH forms of the cobalt-substituted enzyme are explained by protein-induced constraints on the coordination geometry at the metal.

Acknowledgments. This work has been supported by National Institutes of Health Grant AM-19454, by Grants CA-06927 and RR-05539 to The Institute for Cancer Research from the National Institutes of Health and by an appropriation from the Commonwealth of Pennsylvania. J.M.P. is a recipient of a Fellowship from the Cancer Research Institute, New York, New York. R.K.G. is a Research Cancer Development Awardee AM (NIH)-00231 of the United States Public Health Service

References

Bertini I, Luchinat C, Scozzafava A (1977) Carbon-13 longitudinal relaxation times of acetate ion in the presence of metal-substituted bovine carbonic anhydrases. J Chem Soc, Dalton Trans 1962–1965

Dennard AE, Williams RJP (1966) Transition metal ions as reagents in metallo-enzymes. In: Carlin R (ed) Transition metal chemistry, vol II. Marcel Dekker, New York, pp 115–164

Fabry ME, Koenig SH, Schillinger WE (1970) Nuclear magnetic relaxation dispersion in protein solutions. IV. Proton relaxation at the active site of carbonic anhydrase. J Biol Chem 245: 4256–4262

Gupta RK, Pesando JM (1975) Magnetic resonace study of exchangeable protons in human carbonic anhydrases. J Biol Chem 250: 2630–2634

Jacob GS, Brown RD, Koenig SH (1978) Relaxation of solvent protons by cobalt bovine carbonic anhydrase. Biochem Biophys Res Commun 82: 203–209

Khalifah, RG (1971) The carbon dioxide hydration activity of carbonic anhydrases. J Biol Chem 246: 2561–2573

Lanir A, Navon G (1974) NMR studies of the two binding sites of acetate ions to manganese (II) carbonic anhydrase. Biochim Biophys Acta 341: 75–84

Lanir A, Gradsztajn S, Navon G (1973) Proton magnetic relaxation in solutions of manganese carbonic anhydrase. FEBS Lett 30: 351–354

Liljas A, Kannan KK, Bergsten PC, Waara I, Fridborg K, Strandberg B, Carlbom U, Jarup L, Lovgren S, Petef M (1972) Crystal structure of human carbonic anhydrase C. Nature (London) 235: 131–137

Lindskog S, Coleman JE (1973) The catalytic mechanism of carbonic anhydrase. Proc Natl Acad Sci USA 70: 2505–2508

Martin RB (1974) Pyrrole hydrogen ionization of imidazole derivatives in metal ion complexes and carbonic anhydrase. Proc Natl Acad Sci USA 71: 4346–4347

Mildvan AS, Gupta RK (1978) Nuclear relaxation measurements of the geometry of enzyme-bound substrates and analogs. Methods Enzymol 49G: 322–359

Pesando JM (1975a) Proton magnetic resonance studies of carbonic anhydrase. I. Identification of histidine resonances. Biochemistry 14: 675–680

Pesando JM (1975b) Proton magnetic resonance studies of carbonic anhydrase. II. Group controlling catalytic activity. Biochemistry 14: 681–688

Pocker Y, Sarkanen S (1978) Carbonic anhydrase: Structure, catalytic versatility, and inhibition. Adv Enzymol 47: 149–274

Prince RH, Woolley PR (1972) Metal ion function in carbonic anhydrase. Angew Chem Int edn Engl 11: 408–417

CO_2 Exchange and Carbonic Anhydrase in Respiratory Systems

Chairmen: H. H. LOESCHCKE and E. D. CRANDALL

Cl^- and HCO_3^- Movements Across the Red Blood Cell Membrane

R.E. FORSTER, A.L. OBAID, E.D. CRANDALL,[1] and N. ITADA [2]

Although there is a high concentration of carbonic anhydrase inside the mammalian red blood cell, as Booth (1938) has shown the enzyme may be ineffective under circumstances where the HCO_3^- cannot exchange rapidly enough for Cl^- across the cell wall. Under such circumstances the anion exchange becomes rate-limiting. Thus it is that the rate of HCO_3^- flux through the plasma membrane is of importance in blood gas exchange and numerous investigators, some of whom are present, have made measurements of the speed of these processes.

Piiper (1969) at the CO_2 Symposium of 1968 gave an excellent review of the experimental approaches to the measurements of the rate of Cl^--HCO_3^- exchanges between red cells and plasma including the continuous-flow rapid-mixing apparatus with filtration of Dirken and Mook (1931) and Piiper himself (1964) and of the ingenious thin layer technique of Luckner (1939). Overall equlibration including the movements of CO_2 and associated anions had half-times from 0.3 to 0.16 s but the time course was not a simple exponential; the later portions were much slower so that complete equilibration appeared to take more than 1 s. Hemingway et al. (1970) used a continuous-flow filtration apparatus to monitor the extracellular chloride concentration following the rapid mixing of a red cell suspension with Ringer-Locke equilibrated with one atmosphere of CO_2 and found a half time of the order of 0.14 s. However they could not obtain data points earlier than 0.1 s. Klocke (1976) employed continuous-flow filtration to follow Cl^- and HCO_3^- exchanges in human red cells after production of mirror image gradients of the two anions in a rapid mixing chamber at 37°C and physiological pH. He found that HCO_3^- moved out of the red cells with a half-time approximating 0.07 s but moved in with a half-time or approximately 0.13 s.

This paper describes two additional techniques for the measurement of respiratory anion movements, the first of Cl^--HCO_3^- exchange and the second for HCO_3^- self-exchange. Each technique has particular advantages in comparison with earlier methods. The principle of the first technique is outlined in Fig. 1. The hydration-dehydration reactions of CO_2-H_2CO_3 both intra- and extracellular are rapid because of presence of carbonic anhydrase. If H^+ is suddenly added to the extracellular phase, it will convert H_2CO_3 into CO_2 which will then diffuse into the cell where it is rapidly rehydrated to give H^+ and HCO_3^-, a process which will lead to a quasi steady

1 Department of Physiology G4, School of Medicine, University of Pennsylvania, 19104 Philadelphia, Pennsylvania /USA

2 Present address: Kokuritsu Kyoto Hospital, Fushimi-ku, Kyoto, 612, Japan

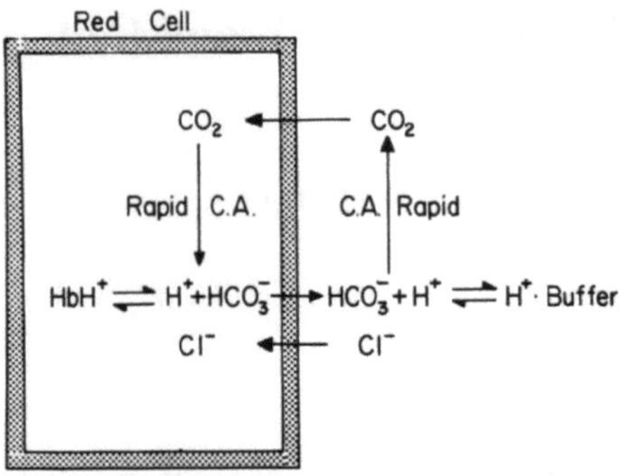

Fig. 1. Schema of the Jacobs-Stewart cycle as modified for the determination of the rate of HCO_3^--Cl^- exchange in red blood cells. *CA* indicates the presence of carbonic anhydrase, that in the extracellular fluid has been added deliberately

state in less 0.05 s with most of the total CO_2 (CO_2 plus HCO_3^-) inside the cell (Chow et al., 1976). Further transfer of H^+ to the interior of the red cell can only take place, the membrane being effectively impermeable to H^+, as fast as HCO_3^- moves into the cell in exchange for Cl^-, thus permitting further formation of CO_2 in the extracellular fluid with an equivalent consumption of H^+. The rate of change of extracellular pH is a measure of the rate of HCO_3^- flux across the cell membrane, quantitatively equaling

$$\frac{dpH}{dt} \text{ x the buffering capacity of the suspending medium.}$$

Experimentally fresh red cells in a concentration to give approximately 16% hematocrit in 146.5 mM NaCl, 3.5 mM mM KCl, 0.025 $NaHCO_3^-$, pH 7.1 to 8.4 and containing 800 units carbonic anhydrase per ml was mixed in a stopped-flow pH apparatus (Crandall et al., 1978) with 30 mM phosphate buffer (temperatures from 5° to 37°C) and the time course of the pH followed with a glass pH electrode. Both solutions were equilibrated with 10% O_2, 0% CO_2 and 90% N_2. pH was adjusted with HCl or NaOH.

Arrhenius plots of effective permeability (k defined as HCO_3^- flux per unit area divided by the HCO_3^- gradient across the membrane, in units of cm/s) at temperatures from 5° to 37°C are shown in Fig. 2. The activation energy between 5° and 17°C is 18.6 kcal per mol but decreases to 11.4 kcal per mol from 17° to 37°C. The temperature at which this change in activation energy appears to take place is considered the "transition temperature". Chow et al. (1976) reported almost the same results except that the variability in their data persuaded them to treat their Arrhenius plot as a continuous curve rather than two straight lines of differing slope. When the cells were incubated with 4-acetamide-4′-isothiocyanostilbene-2-2′-disulfonic acid [SITS], a powerful inhibitor of membrane anion transport, k is decreased approximately one order of magnitude but the shape of the Arrhenius plot is essentially unaltered. That is there is a transition at 17°C from a higher activation energy of around 18 kcal/mol to a lower activation energy of 11 kcal/mol.

Obaid and Crandall (unpublished communication) have gone on to make analogous measurements at pH 5–8 and again found higher activation energies of about

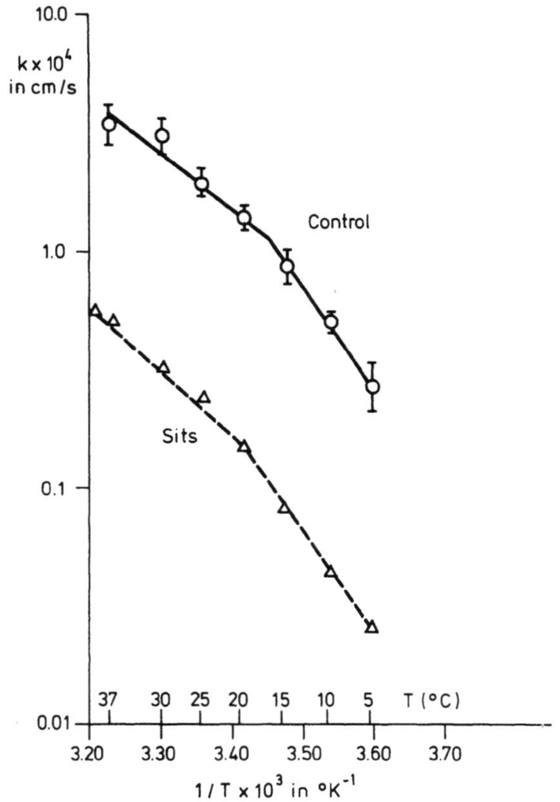

Fig. 2. A graph of the effective permeability, k, of HCO_3^- vs. the reciprocal of the absolute temperature for untreated control cells (*solid line*) and for cells pretreated with SITS (*dashed line*)

19 kcal/mol below a transition temperature of 17°C and a lower activation energy of 12 kcal/mol above this point. Truskey et al. (1979) carried out measurements of acetic acid uptake by red cells using the same apparatus and a similar experimental design. They were able to separate the process into a fast phase (about 1 s), which they considered to represent the movement of unionized acetic acid into the cell and a second slow (700 s or more) phase, which they considered to represent acetate ion exchange. The principle of the separation of these fluxes has been presented by Aubert and Motais (1975). The interesting point is that an Arrhenius plot for the acetate movement showed an activation energy of 25.1 kcal/mol below 19°C and 19.8 kcal/mol above. Surprisingly the unionized acetic acid has a similar transition temperature with an activation energy of 18.8 kcal/mol in the lower range and 11.6 kcal/mol in the upper range. Gunn et al. (1973) have presented convincing evidence that anions, Cl^- at least, move through the red cell membrane via a facilitated transport process, rather than by a simple diffusion mechanism. However, the similarity of the Arrhenius plot for several types of anions and for at least one unionized species suggest that there is a fundamental change in the state of the membrane at the transition temperature and not just an alteration in the anion transport process. It is interesting to note that recently Obaid et al. (1979) determined the effective permeability of dogfish erythrocytes (*Mustelus canis*) as a function of temperature and found a constant activation energy of 13.2 kcal/mol over the entire temperature range from 5°–35°C.

The second technique, actually the only method known to the authors for the measurement of HCO_3^--HCO_3^- exchange, is to follow the disappearance of ^{18}O from CO_2 as described in an earlier paper at this meeting. In a brief review, when red cells are added to a solution of isotopically labeled sodium bicarbonate in which ^{18}O is exchanging with the large pool of unlabeled water via the uncatalyzed chemical reaction of CO_2 and H_2CO_3, the rate of disappearance of labeled CO_2 increases both because labeled CO_2 diffuses into the red cell where it exchanges with water at an accelerated rate and also because labeled bicarbonate diffuses into the red cell. The overall process is thus a function in part of the effective permeability of the red cell to HCO_3^-. On the basis of the theoretical model we discussed earlier, we can calculate an effective permeability, P, for HCO_3^--HCO_3^- exchange and the results are presented in Fig. 3. P at the lowest v is the least reliable and we customarily average the other two results. There is no theoretical basis in our model for a systematic variation in P with v.

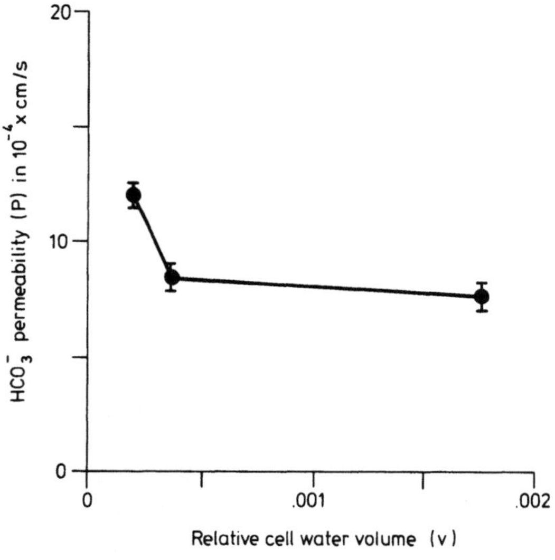

Fig. 3. Graph of the permeability of the red blood cell membrane to HCO_3^- as a function of hematocrit in the reacting mixture, that is cell water volume/extracellular water volume (ν) at 37°C. Blood of HBJ. ***Each point*** is the average of 17 observations; the ***vertical bars*** indicate ± standard error of the mean

Generally the effective permeability of the red cell membrane to anions such as OH^- (Crandall et al., 1971) and Cl^- (Gunn et al., 1973), increases as pH decreases, an experimental finding which has been explained as a result of increasing positive charge on the theoretical hypothetical carrier. HCO_3^--Cl^- exchange is relatively uninfluenced by extracellular pH over the range 6.5–8 (Obaid and Crandall, unpublished results). We also find that HCO_3^--HCO_3^- permeability shows no significant trend as pH increases from 6.5–8.3. Addition of phlorizin, another anion transport inhibitor, reduces HCO_3^--Cl^- exchange within 0.03 s (Chow and Forster, unpublished observations) and also reduces P for HCO_3^--HCO_3^- exchange.

The physiological importance of the absolute value of k and P depend upon the interactions among chemical reactions and diffusion exchanges between cells and plasma within the capillary blood as well as the surrounding tissues, and is not obvious from the magnitude of the value per se. Bidani and Crandall have computed the effects on CO_2 movements out of the blood of decreasing k for HCO_3^--Cl^- exchange to one-tenth of its normal value and found that the exchange is only reduced 25%. Thus one must consider the entire system in attempting to predict the effect of changes in any one parameter.

In conclusion, we find that at pH 7.4 and 37°C, the effective permeability for HCO_3^--Cl^-exchange, k, for normal human red cells is 3.5×10^{-4} cm/s as measured by the stopped-flow rapid-mixing pH apparatus; that the effective permeability of HCO_3^--HCO_3^- exchange, P, under the same conditions is 7.5×10^{-4} cm/s as calculated from the disappearance of ^{18}O from CO_2 in the presence of a very small volume of red cells. The importance of the relative transport of HCO_3^- across the red cell wall upon the speed of blood gas exchanges depends on the dynamic fluxes and reactions of the total cell-plasma tissue system.

Acknowledgments. The authors wish to acknowledge the assistance of Ms. Irina Barinova and Mr. Jack Afford in preparing this manuscript. Supported by Grant HLB-19737 from the National Institutes of Health.

References

Aubert L, Motais R (1975) Molecular features of organic anion permability in ox red blood cell. J Physiol (London) 246: 159–179

Booth VM (1938) Carbonic anhydrase activity inside corpuscles. Enzyme accessibility factors. J Physiol (London) 93: 117–128

Chow EI, Crandall ED, Forster RE (1976) Kinetics of bicarbonate chloride exchange across the human red blood cell membrane. J Gen Physiol 68: 633–652

Constantine HP, Craw MR, Forster RE (1965) Rate of the reaction of carbon dioxide with human red blood cells. Am J Physiol 208: 801–811

Crandall ED, Klocke RA, Forster RE (1971) Hydroxyl ion movements across the human erythrocyte membrane. J Gen Physiol 57: 664–683

Crandall ED, Obaid AL, Forster RE (1978) Bicarbonate-chloride exchange in erythrocyte suspensions. In: Fast chemical reactions in solutions, membranes, and cells. Rockefeller University Press, New York, pp 35–47

Dirken MNJ, Mook HW (1931) The rate of gas exchange between blood cells and serum. J Physiol (London) 73: 349–360

Gunn RB, Dalmark M, Tosteson DC, Wieth JO (1973) Characteristics of chloride transport in human red blood cells. J Gen Physiol 61: 185–206

Hemingway A, Hemingway CJ, Roughton FJW (1970) The rate of the chloride shift of respiration studied with a rapid filtration method. Respir Physiol 10: 1–9

Klocke RA (1976) Rate of bicarbonate-chloride exchange in human red cells at 37°C. J Appl Physiol 40: 707–714

Luckner H (1939) Über die Geschwindigkeit des Austausches der Atemgase im Blut. Pflügers Arch 241: 753–778

Obaid AL, Critz A McE, Crandall ED (1979) Kinetics of bicarbonate exchange in dogfish erythrocytes. Am J Physiol (in press)
Piiper J (1964) Geschwindigkeit des CO_2-Austausches zwischen Erythrocyten und Plasma. Pflügers Arch 278: 500–512
Piiper J (1969) Rates of chloride-bicarbonate exchange between red cells and plasma. In: CO_2: Chemical, biochemical and physiological aspects. Forster RE, Edsall JT, Otis AB, Roughton FJW (eds) NASA, Washington DC, pp 267–273
Truskey G, Deutsch C, Forster RE (1979) Determination of energy of activation for acetic acid and acetate ion transport in the human erythrocyte. Fed Proc 38: 1127

The Rate-Limiting Event in the Depletion of ^{18}O from CO_2 in Red Cell Suspensions

D.N. SILVERMAN and C.K. TU [1]

The measurement of the depletion of ^{18}O from CO_2 was first reported by Mills and Urey (1940) who used this technique to measure the uncatalyzed rate constant for the hydration of CO_2.

$$C^{18}O^{18}O + H_2^{16}O \rightleftharpoons HC^{18}O^{18}O^{16}O^- + H^+ \rightleftharpoons C^{18}O^{16}O + H_2^{18}O \qquad (1)$$

Since the isotope depletion from CO_2 is catalyzed by carbonic anhydrase this method is useful in studying CO_2 kinetics in red cell suspensions. Placing intact red cells into a solution containing ^{18}O-labeled CO_2 in equilibrium with ^{18}O-labeled HCO_3^- will result in rapid depletion of ^{18}O from CO_2 species in the external solution, a result of diffusion into the cell, rapid loss of ^{18}O by Eq. (1) in the cell, and diffusion out of the cell. The purpose of this report is to present three experiments which suggest that the depletion process is limited in rate not by intracellular activity of carbonic anhydrase but by another process, most likely the diffusion of CO_2. The three experiments involve (1) observation of nonrandom distributions of ^{18}O in CO_2, (2) measurement of the rate of oxygen exchange between ^{12}C- and ^{13}C-containing species of CO_2 and (3) measurement of the depletion of ^{18}O from CO_2 when carbonic anhydrase activity is inhibited.

1 Kinetic Methods

The measured variables in this method are the atom fractions of ^{18}O in all CO_2, τ, and in doubly labeled CO_2, c:

$$\tau = \frac{1/2\,(CO^{18}O) + (C^{18}O^{18}O)}{(CO_2) + (CO^{18}O) + (C^{18}O^{18}O)}$$

$$c = \frac{(C^{18}O^{18}O)}{(CO_2) + (CO^{18}O) + (C^{18}O^{18}O)}$$

The random distribution of oxygen atoms is given by the binomial distribution $[(1-\tau)+\tau]^2$, with the atom fraction of ^{18}O in doubly labeled CO_2 given by τ^2.

1 Department of Pharmacology, University of Florida College of Medicine, Gainesville, Florida 32610/USA

Thus a criterion of random distribution is $c/\tau^2 = 1.0$; other values of c/τ^2 indicate nonrandom distributions of oxygens in CO_2.

The atom fractions c and τ were measured in the external solution by mass spectrometry using a CO_2 inlet vessel devised by Hoch and Kok (1963). The vessel consists of a glass cylinder the bottom of which is a membrane permeable to CO_2. The membrane is supported by a porous stainless steel disk. Gas passing across the membrane enters a mass spectrometer (Finnigan 3000) and the system provides a continuous monitoring of isotopic content of CO_2 in solution. The use of this inlet to measure kinetics of isotope depletion in red cell suspensions has been described by Itada and Forster (1977) and Silverman et al. (1976).

The isotope depletion experiments were carried out in solutions containing 15 mM total concentration of all species of CO_2, 4.5 mM glucose, 27 mM glycylglycine, 0.05% albumin and 121 mM NaCl. The pH of this solution was 7.40 ± 0.02 during experiments and temperature was maintained at 25°C. Experimental procedure for performing the depletion experiments has been described (Silverman et al., 1976). Red cells are in the same chemical environment before and after their addition to the inlet vessel, however, the isotopic environment is different. Rat red cells were used to obtain all the data in this report. In experiments using the inhibitors acetazolamide or methazolamide, red cells and inhibitors were incubated for at least one hour prior to measurements.

2 Results

Figure 1 demonstrates the ^{18}O depletion pattern upon addition of lysed rat red cells to a solution containing ^{18}O-labeled, equilibrium mixtures of CO_2 and HCO_3^-. The altered slope represents the enhancement of ^{18}O depletion rate by carbonic anhydrase. The depletion pattern upon addition of intact rat erythrocytes (at the arrow in Fig. 2) is shown in Fig. 2, a pattern described previously by Itada and Forster (1977) and Silverman et al. (1976). The first phase of the biphasic loss of ^{18}O from CO_2 in the external solution is caused by the rapid flux of labeled CO_2 into the red cell. During this phase the ^{18}O content of external CO_2 becomes less than that of external HCO_3^- because of the slower diffusion of HCO_3^- into the cell. In the second phase the rate of formation of labeled CO_2 from HCO_3^- in the external solution is approximately balanced by the loss of ^{18}O from CO_2 caused by diffusion into the red cell. The values for ^{18}O-content of bicarbonate given in Fig. 2 were obtained by rapid acidification of samples from the inlet vessel; the method for calculation of ^{18}O content in bicarbonate in such an experiment is given by Mills and Urey (1940). The values of c/τ^2 given in Fig. 3 and 4 were calculated from the data of Figs. 1 and 2. After the addition of lysed cells there is no variation from the random distribution of ^{18}O in CO_2 which is indicated by $c/\tau^2 = 1.0$. Upon the addition of red cells, under conditions identical to those of Fig. 3 except that cells were intact, it is clear that the distribution of ^{18}O in CO_2 outside the cell becomes nonrandom. The extent to which the distribution is nonrandom is greater for CO_2 than for bicarbonate.

Besides the depletion of ^{18}O from CO_2 there is a second exchange of isotope involving CO_2. This was discovered by Gerster et al. (1973) by mixing ^{18}O-labeled

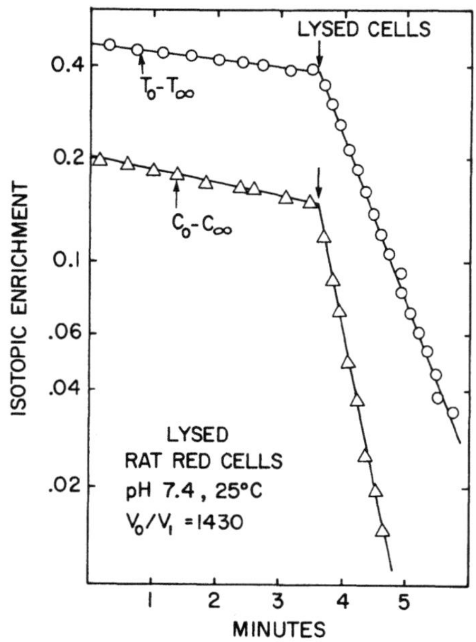

Fig. 1. The atom fractions of ^{18}O in CO_2, $\tau_0-\tau_\infty$ and c_0-c_∞ before and after the addition of lysed rat erythrocytes. The pH was 7.4 at 25°C with a total concentration of all CO_2 species at 15 mM. The number of lysed cells added was equivalent to a ratio of the volume of external solution to the volume of cells V_0/V_i of 1430 had the cells been intact

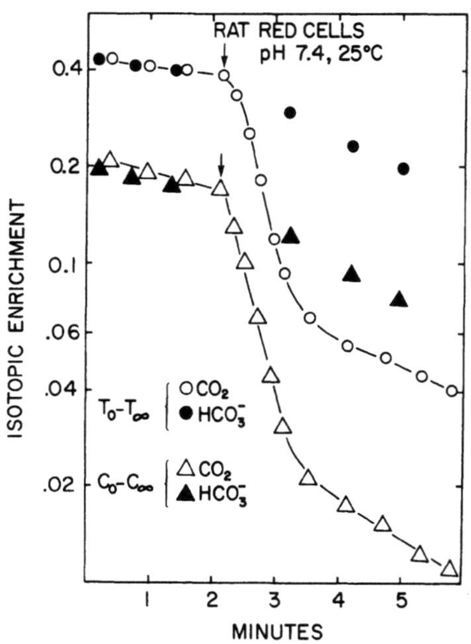

Fig. 2. The atom fractions $\tau_0-\tau_\infty$ for ^{18}O in extracellular CO_2 (○) and in CO_2 derived from extracellular HCO_3^- by rapid acidification (●), and the atom fractions c_0-c_∞ for ^{18}O in CO_2 (△) and in HCO_3^-(▲) before and after the addition of intact rat erythrocytes. V_0/V_i = 1430. The experimental conditions are identical with those of Fig. 1 with the exception that added cells are not lysed

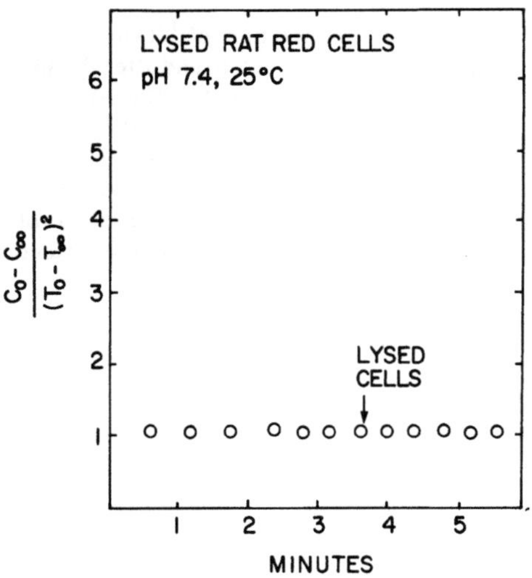

Fig. 3. The data of Fig. 1 showing $(c_0-c_\infty)/(\tau_0-\tau_\infty)^2$ before and after the addition of lysed rat erythrocytes

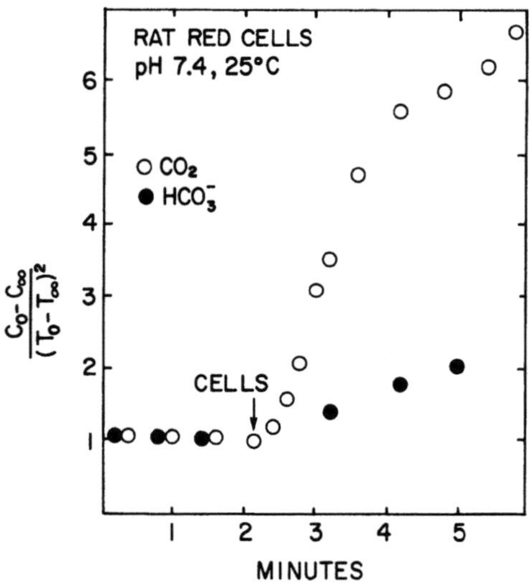

Fig. 4. The data of Fig. 2 showing $(c_0-c_\infty)/(\tau_0-\tau_\infty)^2$ before and after the addition of intact rat erythrocytes

bicarbonate not labeled in ^{13}C with ^{13}C-labeled bicarbonate not labeled in ^{18}O. The appearance of species of CO_2 bearing both ^{13}C and ^{18}O labels is taken as evidence for this second exchange. The uncatalyzed exchange occurs as a result of a chemical reaction between $^{12}CO_2$ and $^{13}CO_3^{2-}$ to exchange oxygens and yield $^{13}CO_2$ and $^{12}CO_3^{2-}$. The rate constant for the exchange of oxygens in this manner in the absence of enzyme is $\phi_{uncat} = 1.5 \times 10^{-4}\ s^{-1}$ under the conditions of these red cell runs. This exchange is catalyzed by carbonic anhydrase by a mechanism in which ^{18}O labels the active site (Silverman and Tu, 1976).

$$H^{12}COO^{18}O^- + E \rightleftharpoons E(^{18}OH) + {}^{12}CO_2$$

$$^{13}CO_2 + E(^{18}OH) \rightleftharpoons E + H^{13}COO^{18}O^-$$

In Table 1 values are given of the rate constant for this exchange measured in the external solution of red cell suspensions and in hemolysate prepared from the same red cell suspension (i.e., containing the same amount of red cells before lysis). Our observation is that ϕ measured in the red cell suspension is nearly identical to ϕ_{uncat}, whereas ϕ measured in the hemolysate is larger by a factor of twenty.

Table 1. Rate constants for the exchange of ^{18}O between ^{12}C- and ^{13}C-containing species of CO_2

	ϕ
Uncatalyzed	$1.5 \times 10^{-4}\ s^{-1}$
Intact red cells [a]	$1.9 \times 10^{-4}\ s^{-1}$
Lysed red cells [b]	$42.0 \times 10^{-4}\ s^{-1}$

a The ratio of external volume to volume of red cells was 1200

b The ratio of external volume to volume of red cells before lysis was 1200. The pH was 7.4 at 25°C

Experiments were carried out in which carbonic anhydrase in the red cell was inhibited with acetazolamide, a potent and specific inhibitor with a K_I of 2×10^{-8} M measured for rat erythrocyte carbonic anhydrase (Maren, 1967). As shown in Fig. 5, the first indication of inhibition of the biphasic depletion pattern is at inhibitor concentrations near 5×10^{-7} M with complete inhibition at 10^{-5} M. At the concentration

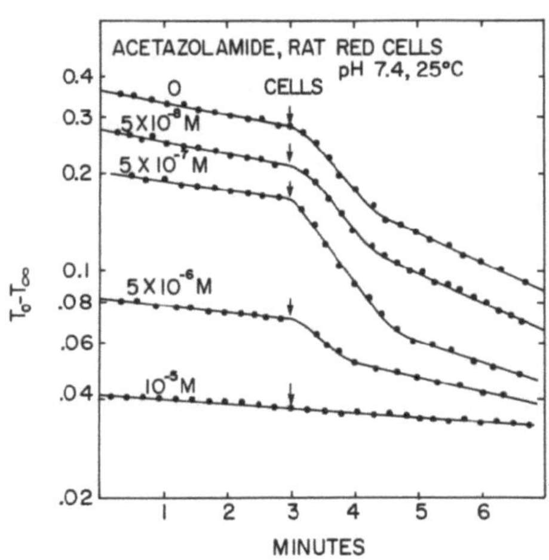

Fig. 5. The effect of inhibition of carbonic anhydrase by acetazolamide on the rate of depletion of ^{18}O from CO_2 in presence of rat erythrocytes. The ratio of volumes of external solution to red cells, V_0/V_i, is 3300 at 25°C and pH 7.4. Total concentration of all CO_2 species is 15 mM

where inhibition is first observed, 5 x 10^{-7} M acetazolamide, we calculate the fraction of enzyme bound by inhibitor to be i = 0.96 using $i = I_{free}/(I_{free} + K_I)$. This fraction has been confirmed by inhibition studies in hemolysate and also by studies in which carbonic anhydrase was added to the external solution of red cell suspensions following which CO_2 hydration activity was measured in the external solution. The data of Fig. 5 indicate that greater than 95% of intracellular carbonic anhydrase must be inhibited before an effect is observed on the rate of depletion of ^{18}O from CO_2 in red cell suspensions. Similar results were obtained with the inhibitor methazolamide.

3 Discussion

It is difficult to obtain a complete kinetic description of the depletion of ^{18}O from CO_2 and HCO_3^- at chemical equilibrium in red cell suspensions, as can be attested by recent publications (Itada and Forster, 1977; Tu et al., 1978). Yet some simple conclusions about the nature of the rate-determining step for ^{18}O depletion can be made. In these experiments there is no net flux of CO_2 or HCO_3^- across the cell membrane, only self-exchange. Most of the loss of ^{18}O from CO_2 in the external solution is caused by labeled CO_2 passing into the red cell and losing ^{18}O in the catalyzed, intracellular hydration-dehydration cycle. The diffusion of HCO_3^- across the red cell membrane is much slower and so is the uncatalyzed rate outside the cell. These comments are verified in ^{18}O depletion studies in which picrate ion, an inhibitor of passive anion diffusion across the red cell, was shown to cause only very small changes in the rate constants for ^{18}O depletion in red cell suspensions (Silverman et al., 1976). Hence, for the purpose of determining the rate-limiting event in the depletion of ^{18}O from CO_2 we consider the following steps where subscript o indicates an ^{18}O content outside the cell:

$$\underset{\tau_o}{(CO_2)_o} \underset{k_i}{\overset{k_e}{\rightleftharpoons}} \underset{\tau_i}{(CO_2)_i} + H_2O \overset{k_c}{\rightleftharpoons} HCO_3^- + H^+ \tag{2}$$

Here k_e and k_i are diffusion constants assuming Fick's law for the diffusion of CO_2 across the cell membrane and k_c is the rate constant for the catalyzed hydration of CO_2 at chemical equilibrium in the cell; $k_c\,(CO_2)$ is the rate of the chemical process of interconversion of CO_2 and HCO_3^- at chemical equilibrium.

It is useful to consider two situations concerning Eq. (2). (1) When $k_i > k_c$ the rate of diffusion of labeled CO_2 out of the cell is faster than the rate of loss of ^{18}O to water. Under this condition $\tau_o \approx \tau_i$ and the rate of depletion of ^{18}O from CO_2 outside the cell is proportional to $k_c \circ \tau_o$, the rate for carbonic anhydrase-catalyzed ^{18}O depletion in the cell; that is, the catalysis is rate limiting. We expect in such a process that the distribution of ^{18}O in CO_2 will remain random because only one oxygen can be lost to water per encounter with enzyme. This results in a cascade in which $C^{18}O_2$ is first reduced to $C^{18}O^{16}O$ then to $C^{16}O_2$, maintaining a random distribution of ^{18}O in CO_2. (2) In the second situation, $k_c > k_i$ and the rate of diffusion of labeled CO_2 out of the cell is much slower than the rate of ^{18}O depletion from CO_2. Under this condition $\tau_o > \tau_i$ and the rate of depletion of ^{18}O from CO_2 outside the cell is approximated by $k_e \cdot \tau_o/(1 + k_i/k_c)$; that is, the rate of access of CO_2 to the interior of the cell, a diffusion step, is rate-limiting. We expect that in such a process the distribution of ^{18}O in the CO_2 outside the cell will be nonrandom. This is because the rate of ^{18}O loss from CO_2 outside the cell is not determined by catalysis but by a diffusion step that does not differentiate between labeled and unlabeled CO_2, neglecting kinetic isotope effects. In this situation, $C^{18}O_2$ enters the cell and then exists as $C^{16}O_2$; this results in $C^{18}O_2$ outside the cell being replaced by $C^{16}O_2$ with no accumulation of intermediate species. In this situation $c_o/\tau_o^2 \neq 1.0$. Further discussion is provided by Gerstet (1971) who developed these techniques to measure CO_2 uptake by leaves.

The experiments in Fig. 3 and 4 demonstrate that hemolysate is unable to alter the random distribution of ^{18}O in CO_2, but intact red cells cause a rapid increase in c_o/τ_o^2, indicative of suppression of singly labeled species of CO_2. Furthermore, the values of c_o/τ_o^2 obtained from CO_2 are greater than c_o/τ_o^2 obtained from bicarbonate. This indicates that the diffusion-limited process is a correct description of ^{18}O depletion from CO_2 since it is not possible for bicarbonate of a certain value of c_o/τ_o^2 to yield by dehydration CO_2 with a greater value of c_o/τ_o^2, which is just to point out that a pool of HCO_3^- bearing a random distribution of ^{18}O cannot yield by dehydration CO_2 with a nonrandom distribution. We conclude from Fig. 3 and 4 that for CO_2 kinetics in the red cell $k_c > k_i$.

Further evidence for this conclusion is given in Table 1. It is clear that in hemolysate the exchange of ^{18}O between ^{12}C- and ^{13}C-containing species of CO_2 is catalyzed by carbonic anhydrase. Yet this catalysis is not seen when the experiment is performed on intact cells (the conditions including total amount of enzyme present were identical in both experiments). The most probable explanation is that CO_2 leaving the red cell is nearly completely depleted of ^{18}O; that is, any $^{13}C^{18}O^{16}O$ formed in the cell is depleted of ^{18}O by other catalytic steps before it returns to the external solution. This is evidence that $\tau_o > \tau_i$.

The inhibition data of Fig. 5 indicate that a decrease is observed in the rate of depletion of ^{18}O from CO_2 outside the cell only after more than 95% of intracellular carbonic anhydrase is inhibited. This evidence also supports the conclusion that $k_c > k_i$. Since the overall depletion of ^{18}O from CO_2 in external solution depends on a diffusion step, inhibition of carbonic anhydrase will not affect depletion until the remaining enzymatic activity is small enough to be comparable in rate to other limiting processes. In this respect the red cell is similar to every other physiological system in which carbonic anhydrase plays a part (Maren, 1967).

These confirmations of the inequality $k_c > k_i$ allow an upper limit to be placed on P, the permeability of the cell to CO_2. From the nature of these experiments we understand that P represents a permeability barrier for CO_2 between external solution and the intracellular sites of carbonic anhydrase. Based on experiments in lysed cells (Tu et al., 1978) and on the fact that rat erythrocytes contain about 25 μM carbonic anhydrase (this is high activity form; a low activity form is present at 10 μM, Wistrand and Wåhlstrand, 1977) we estimate k_c to be 2000 s^{-1}. The diffusion rate constant is $k_i = PA/V_i$ in which $A/V_i \approx 2 \times 10^4$ cm^{-1}, here A is the surface area and V_i the volume of the red cell. These considerations place an upper limit of 0.1 cm s^{-1} on P.

A lower limit for P can be estimated by analyzing the kinetic data assuming that the permeability barrier is so large compared to catalysis in the cell that the ^{18}O enrichments of CO_2 and HCO_3^- inside the cell are equal to natural abundance, which is 0.002, a value negligible compared to enrichments used in this study. This has been the approach of Silverman et al. (1976) who obtained the value $P \cong 0.01$ cm s^{-1} in experiments with rat red cells at 25°C. A complete solution using numerical methods of the kinetic equations describing ^{18}O depletion from CO_2 in suspensions of rat red cells at 25°C has been reported by Tu et al. (1978). They used a Newton-Raphson iterative procedure extended to three variables to find a permeability constant P = 8×10^{-3} cm/s. This value of P represents a permeability barrier for CO_2 to pass from the external solution to the sites of intracellular carbonic anhydrase.

Acknowledgments. We thank Mr. George Wynns for excellent technical assistance. This work was supported by Research Grant HL 18671 from the National Institutes of Health, U.S. Public Health Service.

References

Forster RE (1969) The rate of CO_2 equilibration between red cells and plasma. In: Forster RE, Edsall JT, Otis AB (eds) CO_2: Chemical, biochemical, physiological aspects. NASA SP-188, Washington DC, pp 275–284

Gerster R (1971) Cinétique de l'echange des atomes d'oxygène en phase hétérogene entre $C^{18}O_2$ et H_2O. Int J Appl Radiat Isot 22: 339–348

Gerster R, Maren TH, Silverman DN (1973) A new approach to the study of aqueous solutions of CO_2. Proc 1st Int Conf Stable Isotopes Chem Biol Med, 219–228

Hoch G, Kok B (1963) A mass spectrometer inlet system for sampling gases dissolved in liquid phases. Arch Biochem Biophys 101: 160–170

Itada N, Forster RE (1977) Carbonic anhydrase activity in intact red blood cells measured with ^{18}O exchange. J Biol Chem 252: 3881–3890

Maren TH (1967) Carbonic anhydrase: chemistry, physiology, and inhibition. Physiol Rev 47: 595–781

Mills GA, Urey HC (1940) The kinetics of isotope exchange between carbon dioxide, bicarbonate ion, carbonate ion and water. J Am Chem Soc 62: 1019–1026

Silverman DN, Tu CK (1976) Carbonic anhydrase catalyzed hydration studied by ^{13}C and ^{18}O labeling of CO_2. J Am Chem Soc 98: 978–984

Silverman DN, Tu CK, Wynns GC (1976) Depletion of ^{18}O from $C^{18}O_2$ in erythrocyte suspensions. J Biol Chem 251: 4428–4435

Tu, CK, Wynns GC, McMurray RE, Silverman DN (1978) CO_2 kinetics in red cell suspensions measured by ^{18}O exchange. J Biol Chem 253: 8178–8184

Wistrand PJ, Wåhlstrand T (1977) Rat renal and erythrocyte carbonic anhydrases. Purification and properties. Biochim Biophys Acta 481: 23–31

Kinetics of CO_2 Uptake by Red Cells

R.A.B. HOLLAND [1]

1 Introduction

The uptake and evolution of CO_2 by the blood are complex processes and involve two major phases. These are most easily considered with respect to the uptake of CO_2 and, although one may consider them separately, there is considerable overlap. The first phase is hydration of carbon dioxide to carbonic acid and bicarbonate ion. This takes place almost entirely inside the red cells where it is greatly accelerated by carbonic anhydrase. The second phase is the ionic exchanges which have to take place for the process to be completed, the principal exchange being the outward movement of bicarbonate, which has just been increased by the CO_2 hydration, in exchange for chloride ions.

The first measurement of the rate of CO_2 uptake by red cell suspensions was by Dirken and Mook (1931). Kernohan and Roughton (1968), using a thermal method of measuring the reaction rate, also reported the degree to which the intracellular carbonic anhydrase accelerated the reaction. Both the above groups of workers made their measurements at room temperature and used unphysiological tensions of CO_2 (one atm or more) in the reactant solution.

Constantine et al. (1965) were the first to use a rapidly responding CO_2 electrode to measure the rate of the CO_2 uptake by red cell suspensions and with the introduction of this technique, measurements were made at more physiological temperatures and CO_2 tensions. This enabled the whole of the reaction to be followed, including not only the early rapid phase due to enzymatic hydration of CO_2, but also the later part where the ionic exchange is the most important factor. Holland and Forster (1975) also used the CO_2 electrode to measure the reaction in human adult red cells over a range of temperatures, and Holland (1978) reported measurements in human fetal red cells which are known to have lower concentrations of carbonic anhydrase (Bartels, 1970).

The results of the above work are summarized in this paper and their importance discussed.

1 School of Physiology and Pharmacology, University of New South Wales, Kensington, N.S.W. 2033/Australia

2 Methods

Procedure. In the experiments where the CO_2 electrode was used to measure the progress of the reaction, the apparatus was essentially a modification of the rapid continuous flow apparatus originally introduced by Hartridge and Roughton over 50 years ago. It has been described in detail elsewhere (Constantine et al., 1965; Holland and Forster, 1975). A suspension of red cells at a known P_{CO_2} was held in one syringe and a buffer solution containing a higher P_{CO_2} in another, the syringes and the storage bottles from which they were replenished being kept at a constant temperature. The syringe plungers could be driven by motors; when this occurred their contents entered a mixing chamber and, after mixing, the reacting mixture traveled by turbulent flow down a tube. At the far end of the tube the reaction mixture impinged on a CO_2-sensitive electrode whose output was fed through an electrometer to a pen recorder. Thus the P_{CO_2} of the reacting mixture was found. From a knowledge of the CO_2 tensions in the syringes and their relative rate of delivery, one could calculate what the P_{CO_2} would have been in the absence of any chemical reaction. The difference between this and the measured P_{CO_2} is due to chemical reaction between the mixing point and the observation point. The time taken for the reacting mixture to travel this distance was calculated from the delivery rate of the syringes and the volume of the intervening tubing, so from one delivery from the syringes one could find the change in P_{CO_2} due to chemical reaction in given time. The time was varied by varying the length of tubing between the mixing chamber and the electrode, and in this way a number of points were obtained to give a curve of ΔP_{CO_2} plotted against time. The earliest point obtained was 8 ms after mixing and the latest was 1400 ms after mixing. The end point of the reaction was found by mixing equal volumes of the reactants in a third syringe and, after several minutes, measuring this P_{CO_2} with the same electrode system.

The experiments were done using a 1: 20 suspension of red cells before mixing in the apparatus. In the experiments of Holland and Forster (1975) and Holland (1978) the red cell suspension was normally at a P_{CO_2} of 0.2 torr (in equilibrium with room air) and the buffer in which the cells were suspended was isotonic NaCl-KCl. Phosphate was excluded because it has been reported to catalyze the hydration of CO_2 and bicarbonate was excluded because it would have made the cells too alkaline at this P_{CO_2}. The pH of the red cell suspension was adjusted to a value between 7.4 and 8.0 by slow addition of either dilute hydrochloric acid or sodium hydroxide. The reactant buffer was at about 90 torr P_{CO_2} giving a P_{CO_2} of about 45 torr immediately after mixing in the apparatus; this is close to the physiological tension. Phosphate was also excluded from the reactant buffer, but part of the chloride was replaced by bicarbonate in a concentration of 30 mM. All the measurements made with red cell suspensions and the CO_2 electrode have been with human blood. Adult blood was obtained from normal healthy males aged 20 to 40 years, and fetal blood from the placental end of the cut umbilical cord after delivery. It was normally studied within 8 h of collection; when it had to be kept for 24 h the carbonic anhydrase activity was shown to be the same after this period as it was immediately after the blood was taken. In the fetal blood studied the carbonic anhydrase activity was between one quarter and one eight of the activity in adult blood, the assays being done by the colorimetric method of Philpot and Philpot (1936).

3 Results and Discussion

The Early Phase. In Fig. 1 the esults of Holland and Forster (1975) and Holland (1978) have been summarized. In drawing these curves the end points were taken as 100% and the earlier points expressed as a percentage of the end point. The lines were drawn by eye through the raw data points, all results at the same temperature being pooled. The slowing of the reaction with fall in temperature is apparent, as is the slowness of the reaction in the fetal cells with their low carbonic anhydrase content.

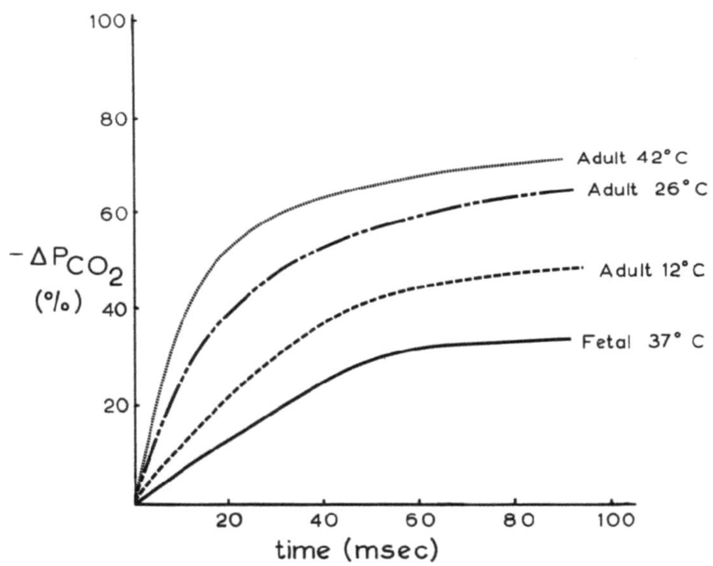

Fig. 1. Time course of early fall of P_{CO_2} after a 1:20 suspension of red cells at a P_{CO_2} of 0.2 torr was mixed with a buffer containing a P_{CO_2} of about 90 torr. All changes are expressed as a percentage of the final change in P_{CO_2}. The original data points are not shown but were in the experiments of Holland and Forster (1975) and Holland (1978)

The rate of the early phase can be expressed in terms of a constant k_{uc} defined by the equation:

$$\frac{d\,P_{CO_2}}{d\,t} = k_{uc}\,(P_{CO_2}\ \text{equil} - P_{CO_2}) \tag{1}$$

where P_{CO_2} equil is the P_{CO_2} with which the red cells were equilibrated and P_{CO_2} is the tension of CO_2 in the mixed reactants. The rate of fall of CO_2 tension was found from the slope of tangents to the curves at zero time and the values for k_{uc} were corrected to the values in cell water, taken to be 70% of the red cell volume. An Arrhenius plot was made of the k_{uc} in adult cells showing an energy of activation (E) for the catalyzed reaction of 7.2 kcal mol^{-1} (30 kJ mol^{-1}), corresponding to a Q_{10} of 1.5 in the physiological range. The regression line gave values for k_{uc} of 915 s^{-1} at 42°C, 761 s^{-1} at 37°C, 476 s^{-1} at 25°C and 274 s^{-1} at 12°C. The value at 37°C is in good agreement with that found by Constantine et al. (1965).

The acceleration factors were calculated for each of the experiments of Holland and Forster (1975) and are shown in Fig. 2. For these calculations the uncatalyzed

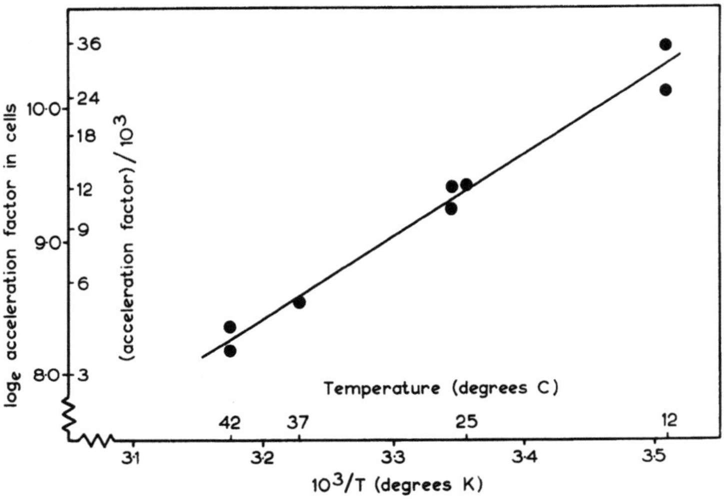

Fig. 2. The acceleration factor by which the red cell carbonic anhydrase multiplied the early stages of the hydration of CO_2 at different temperatures. All experiments were on adult human cells and were reported by Holland and Forster (1975). The equation for the regression line was: ln (accel. factor) = 6.13 (10^3/T) – 11.21

rates were calculated at each temperature as the sum of k_u and k'' $[OH^-]$ where the velocity constants are defined in the following equation (back reaction omitted)

$$\frac{d\,[CO_2]}{d\,t} = -k_u\,[CO_2] - k''\,[OH^-]\,[CO_2]$$

and the values of k_u and k'' were taken from Roughton's review (1964). An Arrhenius plot of the acceleration factor is linear but because temperature has more effect on the uncatalyzed rate than on the catalyzed rate, the acceleration is less at higher temperatures. The value of 11,600 at 25°C given by the regression line of Fig. 2 is in good agreement with the acceleration found by Kernohan and Roughton in bovine red cells at this temperature. They used a thermal method of measuring the extent of the reaction and found acceleration factors varying from 4,900 to 14,000 with a mean of 11,300. In contrast to the acceleration factor of 5,200 given by Fig. 2 for adult cells at 37°C, Holland (1978) found a factor of only 860 for fetal cells at this temperature.

It must be remembered that the acceleration factors given above and in Fig. 2 were based on a value of 0.122 s^{-1} for k_u at 37°C as given by Roughton (1964) and recently Swenson and Maren (1978) have reported this value to be 0.079 s^{-1}. If the latter is the true value, these acceleration factors should be increased appropriately.

The Later Stages of the Reaction. In Fig. 3 are the mean curves for human adult cell suspensions at 12° and 42°C and for fetal cells at 37°C. There were insufficient points at 25°C to show the results for this temperature at the later stages. The adult cells at both high and low temperatures showed the early rapid phase described above followed by a much slower phase. The slowing down was caused by a build up of bicarbonate in the cells which had to exchange across the cell membrane with chloride. This was shown to be so by measuring the rate of P_{CO_2} fall when hemolysates were used instead of cell suspensions: this sped up the first phase a little and later stages of the reaction were much more rapid with no sharp bend or elbow in the curves. This elbow can be calculated to occur when the bicarbonate inside the cells has increased to or above the level outside the cells. At 42°C the change in slope occurred after a drop in P_{CO_2} of 12.7 torr corresponding to a rise in intracellular bicarbonate of 15.3 mM. At 12°C the elbow occurred after an increase in intracellular bicarbonate of 16.6–23 mM. Thus it would appear that up to the elbow there is a nett entry of bicarbonate and after the elbow there is a nett efflux in exchange for chloride.

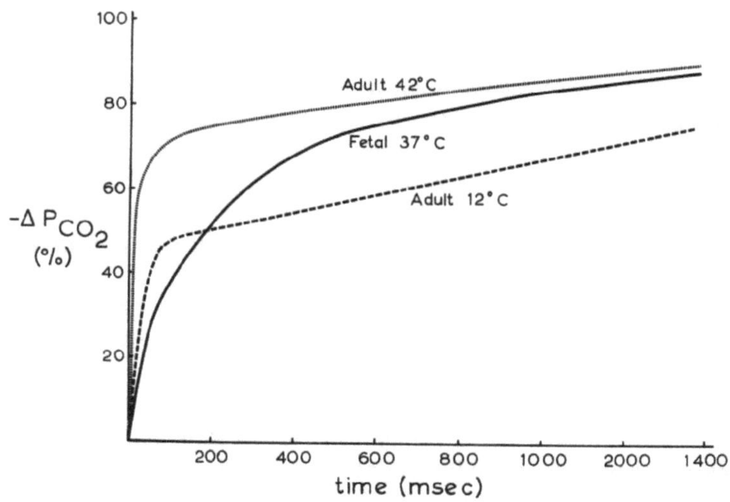

Fig. 3. Early and late course of fall in P_{CO_2} after a 1:20 suspension of red cells at a P_{CO_2} of 0.2 torr mixed with a buffer containing a P_{CO_2} of about 90 torr. The curves were drawn as for Fig. 1, where their first 100 ms has been shown with the X-axis expanded

It had been hoped that the temperature effect on the later stages of the reaction could be treated quantitatively but the scatter of the original points was too great to permit this. More recent work by Chow et al. (1976) has shown that the bicarbonate-chloride exchange is temperature-dependent with the Q_{10} increasing at low temperatures. The results summarized in Fig. 3 show a slowing of the later stages at low temperatures, and a failure of the process to reach completion in the time that blood takes to pass through a capillary at rest (see below).

The curve of uptake in fetal cells at 37°C started even more slowly than in adult cells at 12°C but at 200 ms it had overtaken the adult 12°C curve, and from 800 ms it was nearly as advanced as in adult cells at 42°C. This reflects the initial slowness due to relative lack of carbonic anhydrase while the later stages were presumably able to catch up because ion exchange across the fetal red cell membrane was similar to that in the adult cells.

Physiological Implications. In adult cells the hydration rate of CO_2 appears fast enough at rest and on exercise. The curves of Figs. 1 and 3 are for a 1:40 suspension of red cells and the increased concentration of red cells, and hence of carbonic anhydrase, in whole blood would make the first phase go faster than shown in Fig. 1. At a hematocrit of 0.43, cell water is 30% of the total volume of blood so the values of k_{uc} at 42°C and 37°C would be 275 and 228 s^{-1} respectively, to give half times for the initial stages of CO_2 uptake of less than 5 ms. However it does not appear that the later and slower phase would be sped up by an increase in hematocrit and so this may not be complete within the capillary transit time of blood. The transit time in the lung at rest appears to be 0.75 s (Johnson et al., 1960; Roughton and Forster, 1957) and even at 42°C the reactions were not complete in this time. In severe exercise, where the muscle temperature may rise to 42°C, the likely capillary transit time in muscles is 0.3 s (Folkow and Neil, 1971) and in the lung the transit time is about 0.5 s when the cardiac output is tripled (Johnson et al., 1960). In the experiments of Fig. 3 where the equilibrated P_{CO_2} was zero, the process was only about 80% completed after 0.3 s and in other experiments of Holland and Forster (1975) where the equilibrated red cell P_{CO_2} and bicarbonate were physiological, the reaction had only gone to 70% completion in this time. Although experiments have not been reported on the rate of release of CO_2 from red cell suspensions, one may reasonably apply the uptake data to release of CO_2 in the lungs as the evolution of CO_2 should be symmetrical to the loading process. Thus it appears that in the adult at rest and on exercise the uptake and evolution of CO_2 may be incomplete in tissue and pulmonary capillaries.

The significance of the fetal results is harder to assess as the transit times are less well known. However when Hill et al. (1973) computed the course the course of placental CO_2 exchange they used a value for the initial rate of CO_2 hydration in the mother over twice as great as that reported by Holland and Forster and by Constantine et al., and a value for the initial rate in the fetus about eight times as great as that reported by Holland (1978). As their model gave a small gradient of CO_2 tension at the end of the capillaries, it is possible that there would be a more significant failure of equilibration if the slower values reported here were used.

References

Bartels H (1970) Prenatal respiration. North-Holland Publishing Company, Amsterdam London. Wiley Interscience Division, John Wiley & Sons Inc, New York, p 57

Chow E I-H, Crandall ED, Forster RE (1976) Kinetics of bicarbonate-chloride exchange across the human red blood cell membrane. J Gen Physiol 68: 633–652

Constantine HP, Craw MR, Forster RE (1965) Rate of the reaction of carbon dioxide with human red blood cells. Am J Physiol 208: 801–811

Dirken MJN, Mook NW (1931) The rate of gas exchange between cells and serum. J Physiol (London) 73: 349–360

Folkow B, Neil E (1971) Circulation. Oxford Univ Press, New York, p 41

Hill EP, Power GG, Longo LD (1973) A mathematical model of carbon dioxide transfer across the placenta and its interaction with oxygen. Am J Physiol 224: 283–299

Holland RAB (1978) Rate of carbon dioxide uptake by suspensions of human fetal red cells at 37°C. In: Longo LD, Reneau DD (eds) Fetal and newborn cardiovascular physiology, vol II. Fetal and newborn circulation. Garland STPM Press, New York London, pp 475–482

Holland RAB, Forster RE (1975) Effect of temperature on rate of CO_2 uptake by human red cell suspensions. Am J Physiol 228: 1589–1596

Johnson RL, Spicer WS, Bishop JM, Forster RE (1960) Pulmonary capillary blood volume, flow, and diffusing capacity during exercise. J Appl Physiol 15: 893–902

Kernohan JC, Roughton FJW (1968) Thermal studies of the rates of reaction of carbon dioxide in concentrated haemoglobin solutions and in red blood cells. A. The reactions catalyzed by carbonic anhydrase. B. The carbamino reactions of oxygenated haemoglobin. J Physiol (London) 197: 345–361

Philpot FJ, Philpot JStL (1936) A modificated colorimetric estimation of carbonic anhydrase. Biochem J 30: 2191–2193

Roughton FJW (1964) Transport of oxygen and carbon dioxide. In: Rahn H, Fenn WO (eds) Handbook of Physiology, Respiration Sect. 3, vol I. The American Physiological Society, pp 767–825

Roughton FJW, Forster RE (1957) Relative importance of diffusion and chemical reaction rates in determining rate of exchange of gas in the human lung, with special reference to true diffusing capacity of pulmonary membrane and volume of blood in the lung capillaries. J Appl Physiol 11: 290–302

Swenson ER, Maren T (1978) A quantitative analysis of CO_2 transport at rest and during maximal exercise. Respir Physiol 35: 129–159

Intravascular End-Capillary CO_2-H^+ Disequilibrium

E.D. CRANDALL, A. BIDANI, and J.E. O'BRASKY [1]

1 Introduction

In recent years, our understanding of the anion exchange properties of the red cell membrane have led to greater insight into the kinetics of CO_2 elimination from blood in the lung and CO_2 uptake by blood in body tissues. When metabolically produced carbon dioxide enters the blood in tissue capillaries, most of the gas diffuses rapidly into the red blood cell. It is then quickly hydrated to carbonic acid and bicarbonate ions under the influence of the carbonic anhydrase present inside the erythrocytes. The HCO_3^- thus formed leaves the red cells, flowing down its electrochemical potential gradient into plasma and thereby promoting the entry of more carbon dioxide into the red blood cells. It is through this mechanism that the plasma bicarbonate ion pool has access to intra-erythrocyte carbonic anhydrase. These processes occur in the reverse direction when CO_2 is eliminated in the lungs.

In the sequence of events described above, the slowest step is the exchange of bicarbonate for chloride ions across the red cell membrane. It is possible that CO_2 uptake and elimination in capillary beds is limited by the speed of the bicarbonate-chloride exchange, despite the fact that its half-time at physiological temperature is about 100 ms. In preliminary calculations, it has been determined that if the bicabonate-chloride exchange were infinitely fast, CO_2 elimination in the lungs might be increased by as much as 10% (Bidani and Crandall, unpublished observations). One result of these considerations is that the entire sequence of events might not be in electrochemical equilibrium by the time the blood leaves the capillary bed after gas exchange.

Roughton [16] noted as early as 1935 that a disequilibrium of hydrogen ions across the red cell membrane would be expected in blood after CO_2 exchange. This consideration, largely ignored in most modern textbooks of physiology, has recently been rediscovered. Carbon dioxide uptake by red cell suspensions was shown to have a half-time of about 45 ms [4]. When identical experiments were performed while observing pH instead of P_{CO_2} in the cell suspension, it was found that equilibration of hydrogen ions took about 50 times as long as CO_2 uptake [5, 10]. After carbonic anhydrase was added to the extracellular fluid of the cell suspension, pH equilibration took place almost instantaneously. These data showed that the rate-limiting step in H^+ equilibration was probably the extracellular hydration-dehydration reactions. Theoretical models which included reaction rates and transport across the red cell

1 University of California, Los Angeles, California 90024 /USA and University of Pennsylvania, Philadelphia, Pennsylvania 19104/USA

membrane within blood after gas exchange agreed well with these experimental results [10]. Extension of the calculation to in vivo simulations suggested that red cells and plasma are not in equilibrium by the time blood leaves a capillary bed in which gas exchange takes place, resulting in the prediction that blood pH and P_{CO_2} continue to change after the blood has left the gas exchanging site [10, 11].

The reason that post-capillary changes in plasma pH are predicted to take place after gas exchange in a capillary bed can be appreciated by considering the events occurring within blood. When carbon dioxide enters blood in the tissue capillaries, it is rapidly converted to hydrogen and bicarbonate ions inside the red cell. Bicarbonate ions leave the red cells in exchange for chloride ions and enter the plasma. However, the hydrogen ions produced inside the red cells, though largely buffered, remain intracellular, and the hydrogen ion concentration there increases. On the other hand, the plasma hydrogen ion concentration changes little while blood is in the pulmonary capillary since no process occurs extracellularly that would lead to a significant change within the 0.75 s transit time. The result is that hydrogen ions must leave the red cell and enter the plasma (causing a fall in pH) in order for electrochemical equilibrium within blood to be reestablished. This equilibration process takes place largely after the blood has left the capillary bed. Although O_2 exchange complicates these events somewhat [10, 11], the sequence is essentially as described.

The process by which hydrogen ions equilibrate across the red cell membrane after the blood has left the tissue capillary is thought to be the Jacobs-Stewart cycle [13]. In this cycle, the excess hydrogen ions inside the red blood cell combine with bicarbonate ions there to form CO_2. CO_2 then diffuses into the plasma where it slowly (uncatalyzed) rehydrates into carbonic acid and hydrogen and bicarbonate ions. The bicarbonate ions then diffuse back into the red cell in exchange for chloride ions and the cycle is complete. A net result of one cycle is the transfer of one hydrogen ion and one chloride ion from red cells to plasma with water following. The Jacobs-Stewart cycle represents a reversal of the usual chloride shift that takes place while the blood is within the capillary bed. It is obvious that if carbonic anhydrase is present in the plasma, the equilibration process via the Jacobs-Stewart cycle will be accelerated, since the rate-limiting step in the Jacobs-Stewart cycle is the extracellular hydration-dehydration reactions. Because these reactions have a half-time under physiological conditions that is measured in seconds, it is possible that the pH and P_{CO_2} changes that take place in blood after it has left the capillary bed may take enough time to be measurable. Note that the above processes are reversed after CO_2 elimination in the lung capillaries.

The purpose of the present paper is to describe recent experimental and theoretical insights into H^+ disequilibrium. Attempts to measure slow pH changes in arterial blood in vivo met with limited success at first, presumably because of the excessive hemolysis that was present in the blood of the animals tested [7]. However, the use of variable doses of carbonic anhydrase inhibitors overcame some of these problems and allowed for considerable insight into the mechanisms by which H^+ equilibration was taking place [1]. After a good deal of effort, two laboratories have been able to demonstrate (smaller than expected) slow pH changes in arterial blood after the exchange of CO_2 and O_2 in the pulmonary capillaries [2, 12]. Because of the difficulty in demonstrating slow pH changes in the early animal studies, however, several laboratories began to use bloodless, isolated perfused lungs to look for the slow post-capillary pH changes. The results from three such studies have suggested that carbonic anhydrase known to be present in lung tissue is capable of eliminating at least part, if not all, of the disequilibrium during passage through the pulmonary capillaries [6, 9, 14]. The discrepancies between the in vivo and the in vitro studies have yet to be resolved. The possibility that species and organ capillary bed differences may be involved is supported by two recent studies on guinea pig liver and cat hindlimb.

2 Methods

2.1 Apparatus

Fluid to be tested was withdrawn into a temperature-controlled, rapidly responding pH electrode assembly, incorporating a pressure- and flow-insensitive reference electrode liquid junction. The output of the electrode was amplified (Transidyne General MPA6) and recorded on a storage oscilloscope (Tektronix 5103N). Simultaneously recorded on the oscilloscope screen was the temperature of the blood in the measuring chamber (Yellow Springs Instruments 954-C42). The response time of the electrode was less than 10 ms. Calibration of the electrode assembly was performed using standard pH buffers. A detailed description of the experimental apparatus has been given previously [7].

2.2 Procedure for Animal Experiments

The animal (dog or cat) was anesthetized with 25 mg/kg of freshly prepared sodium pentobarbitol solution given intravenously. Cannulas were placed in the right common carotid artery, and in the central veins via the femoral vein. The animals were allowed to breathe spontaneously through a tracheostomy tube. 750 units/kg of heparin was administered intravenously after surgery was complete, and an additional dose of 200 units/kg was administered hourly thereafter. Blood from the carotid arterial cannula was drawn through the electrode apparatus into a syringe at a constant rate, using a withdrawal pump. Blood flow was then suddenly stopped and the time course of plasma pH and temperature of the blood in the electrode chamber was observed thereafter. Any experiment in which the temperature of the blood in the chamber changed more than 0.25°C after stopping flow was rejected. Transit time from lung capillary to electrode chamber was estimated to be about 4–6 s.

Some animals had enough carbonic anhydrase in their plasma due to hemolysis to prevent the observation of any slow pH changes that might exist. Carbonic anhydrase inhibitors were administered to these animals intravenously and the above experiments repeated between 1 and 60 min thereafter. Benzolamide was given at 2 mg/kg, while acetazolamide was used at dosages varying from 0.1 to 100 mg/kg.

2.3 Procedure for Isolated Organ Experiments

Rats or guinea pigs were anesthetized with 50 mg/kg of sodium pentobarbitol intraperitoneally. In the rat lung experiments, a tracheostomy was performed and the lungs were mechanically ventilated (Harvard 680 Respirator) with room air. The abdomen was opened and blood cleared from the pulmonary circulation using Krebs-Ringer-Bicarbonate (KRB) perfusate at 37°C infused into the portal vein. After opening the chest and cannulating the pulmonary artery, perfusate was infused via that vessel at about 20 ml/min (Cole Palmer 7104 Roller Pump). The lungs were removed from the chest and mounted in a chamber maintained at 37°C. Fluid from a left atrial cannula was drawn as described above through the electrode chamber at a constant rate. Flow was then suddenly stopped and perfusate pH and temperature were monitored thereafter. The perfusing fluid was then switched to a KRB + acetazolamide (0.25 mg/ml) perfusate and the experimental procedure repeated. Perfusates were preequilibrated with gas mixtures to a pH of about 7.4, P_{CO_2} of about 39 mm Hg and P_{O_2} of about 48 mm Hg. Hemoglobin concentration and carbonic anhydrase activity in the effluent perfusate were both measured by techniques described previously [6].

Perfusions of guinea pig livers were performed after opening the abdomen, cannulating the portal vein, and perfusing with KRB solution. Perfusate from the inferior vena cava above the diaphragm (cannulated after opening the thorax) was withdrawn into the pH elelctrode apparatus. The livers were maintained at 37°C by performing the experiments in situ and by using a heat lamp. Experimental runs were performed using KRB, KRB + carbonic anhydrase (860 U/ml), or KRB + acetazolamide (0.25 mg/ml) as perfusates. In all isolated organ studies, inflowing and effluent perfusates were sampled for equilibrated pH, P_{CO_2} and P_{O_2} in order to assess the amount of carbon dioxide transferred during a single pass through the capillary bed.

A computational model described previously [3] was used to simulate the experimental animal results under the conditions studied.

3 Results

Eight animals (5 dogs, 3 cats) were studied under conditions when no carbonic anhydrase inhibitor had been administered. In these animals, a pH rise of about 0.01 unit was observed with a half-time of 4.4 s after arterial blood entered the electrode chamber [2]. The average hemoglobin concentration in these animals was 5.6 mg%. The magnitude of this pH change was about half that predicted by the mathematical model for the situation when no carbonic anhydrase is present in plasma. When carbonic anhydrase was purposely added to the plasma, by returning hemolysed blood to the animal, no pH change was seen after the blood entered the pH elelctrode. These data suggest that slow pH changes are present after blood leaves the pulmonary capillary in vivo.

Figure 1 shows the results from an animal given variable doses of acetazolamide after showing that sufficient carbonic anhydrase was present in plasma to prevent pH changes from being seen (0 mg/kg). With a small dose of acetazolamide (1.2 mg/kg),

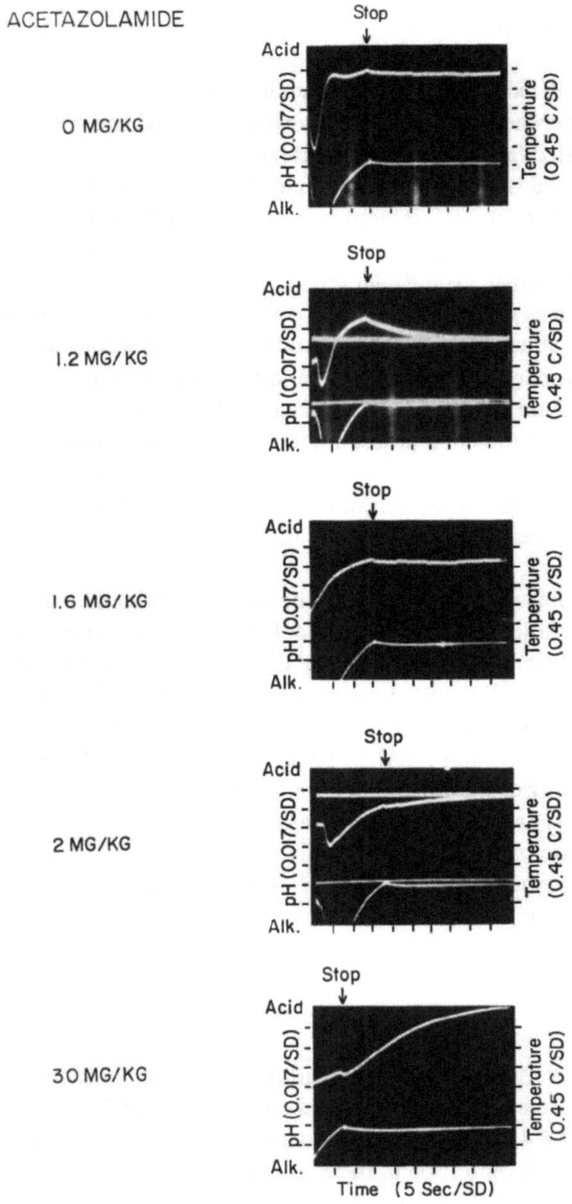

Fig. 1. Oscilloscope tracings of plasma pH (*upper traces*) and temperature (*lower traces*) vs. time. Corresponding total dosage of acetazolamide (0 to 30 mg/kg) is shown next to each tracing. The time at which the withdrawl of blood through the electrode apparatus was suddenly stopped is indicated. Experimental animal: dog

an alkaline pH change is noted. Identical results were obtained with 2 mg/kg benzolamide. At a slightly higher dose of acetazolamide (1.6 mg/kg), once again no pH change is seen. As the dose is increased further to 2–30 mg/kg, it is seen that the pH change observed in the blood in the electrode chamber was in an acidward direction. These data show that differing degrees of inhibition of plasma and red cell carbonic anhydrase lead to qualitatively different slow post-capillary pH changes [1].

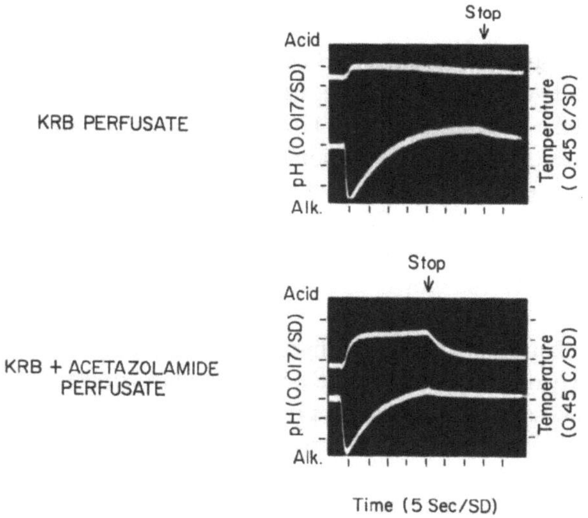

Fig. 2. Oscilloscope records of effluent perfusate pH (*upper traces*) and temperature (*lower traces*) vs. time for an isolated perfused rat lung. The time at which the flow of perfusate through the pH electrode chamber was suddenly stopped is indicated. The corresponding perfusate for each record is shown next to each tracing

Figure 2 shows a sequential record obtained using KRB perfusate alone and KRB plus acetazolamide perfusate in an isolated rat lung [6]. In eleven lungs the mean rise in perfusate pH after flow was stopped in the electrode chamber, when the lungs were perfused with KRB alone, was not significantly different from zero. When acetazolamide was in the perfusate, the pH rise was about 0.05 unit. These data suggest that the carbonic anhydrase in the isolated rat lung is accessible to the perfusate, and is capable of accelerating the CO_2 reactions almost to completion before the perfusate reaches the pH electrode (~ 4 s). These results differ quantitatively from those in the intact animals.

The results of similar experiments on isolated perfused guinea pig livers showed a pH fall of about 0.004 unit in livers perfused with KRB alone [15]. When they were perfused with KRB plus carbonic anhydrase, no pH change was observed after the fluid entered the pH electrode. When acetazolamide was added to the KRB perfusate, the pH fall was about 0.005 unit. Since metabolically produces CO_2 is entering the perfusate in the liver, these data suggest that carbonic anhydrase known to be present in liver tissue is not able to accelerate CO_2 reactions in the perfusate. The results from the liver and lung perfusion studies, therefore, lead to opposite conclusions. In both the liver and lung perfusion studies, no carbonic anhydrase activity was found in the effluent perfusates.

4 Discussion

The major mechanism underlying slow post-capillary pH changes in blood is the disequilibrium for hydrogen ions between red cells and plasma at the end of the pulmonary capillary. Three reactions take place inside the red blood cell that contribute to changes in hydrogen ion concentration there: dehydration of carbonic acid, carbamate dissociation, and hemoglobin oxygenation. The first two reactions lead to a decrease in intracellular hydrogen ion concentration after CO_2 elimination, while the third causes an increase due to the release of Bohr protons during O_2 uptake. Outside the cells, hydrogen ion concentration falls due only to dehydration of carbonic acid there.

The consequences of these changes in intracellular hydrogen ion concentration are best appreciated from the experiments using variable doses of acetazolamide (Fig. 1). When significant carbonic anhydrase activity is present both inside and outside the red cells, any disequilibrium of hydrogen ions across the cell membrane is rapidly dissipated through the Jacobs-Stewart cycle. This occurs because the Jacobs-Stewart cycle is rate-limited by the extracellular reactions; when they are speeded up (by the activity of carbonic anhydrase due to hemolysis or other source), no pH change is observed in our pH electrode apparatus (top panel in Fig. 1). When 1.2 mg/kg of acetazolamide is present, however, extracellular carbonic anhydrase is essentially totally inhibited, while intracellular carbonic anhydrase activity remains high. In this case, the three reactions occurring inside the red blood cells lead to a decrease in hydrogen ion concentration there, while in the plasma, very little change in hydrogen ion concentration takes place since the dehydration reaction there is uncatalyzed. As a result, some of the plasma hydrogen ions must enter the red blood cells via the Jacobs-Stewart cycle in order to attain electrochemical equilibrium, and a rise in pH of the blood within the electrode chamber was observed. This situation is closest to the case of the normal animal not given any carbonic anhydrase inhibitors but in whom little hemolysis has taken place. When the animal is given 1.6 mg/kg acetazolamide, intracellular dehydration of carbonic anhydrase is somewhat slowed. As a result, the three reactions inside the red blood cell lead to almost no change in hydrogen ion concentration there. Since no change in plasma hydrogen ion concentration takes place either, no H^+ flux is required and no change in plasma pH was observed after blood entered the electrode chambers. At higher doses of acetazolamide, the intracellular dehydration of carbonic acid is slowed even further. This results in an increase in intracellular hydrogen ion concentration during transit through the pulmonary capillary, due primarily to the release of Bohr protons. Consequently, hydrogen ions must now leave the red cells in order to reestablish electrochemical equilibrium within blood. As a result, a fall in the pH of plasma was observed for the blood in the electrode chamber at concentrations of acetazolamide of 2 mg/kg and greater. Computational results agreed very well with these experimental findings.

In those animals in which hemolysis had been reduced to a minimum, a small plasma pH rise was observed after blood entered the electrode chamber. This rise occurs for the same reasons that the rise in pH takes place in the animals given 1.2 mg/kg of acetazolamide, discussed above. However, computations suggested that these pH changes should have been about twice as large as actually found. Possible explanations for this observation are that the hemolysis that was present caused the release of enough carbonic anhydrase to account for the smaller pH change, or that carbonic anhydrase available to plasma within the lung could have partially catalyzed the Jacobs-Stewart cycle toward equilibrium.

The isolated rat lung experiments were undertaken in order to investigate the possibility that the lung tissue carbonic anhydrase is available to a blood-free perfusate. The results (Fig. 2) indicate that in our preparation the dehydration of carbonic acid in the perfusate was catalyzed almost to completion by the lung tissue carbonic anhydrase. These results are similar to those obtained by Effros et al. [9]. Klocke [14], however, suggested that the reactions are accelerated only about 20–30% toward equilibrium. Depending upon the extent of catalysis, these results together suggest that the in vivo normal animal studies should show an even smaller post-capillary pH change than was observed. The quantitative differences between these in vitro perfused lung studies and the in vivo animal studies have yet to be resolved.

One possibility that might help expalin the discrepancy is that different capillary beds may have different carbonic anhydrase activities available to their perfusates. The guinea pig liver perfusions, undertaken in an attempt to investigate this question, indicate that liver carbonic anhydrase is not available to perfusate. These data suggest that different capillary beds in different species have variable levels of carbonic anhydrase activity available to their perfusates. Effros and Weissman [8], in a preliminary study, suggested that the carbonic anhydrase activity of cat hindlimb is accessible to perfusate. It has been suggested [6, 9, 14] that the carbonic anhydrase in capillary beds may be present on the endothelial cell surface in the lumen.

At the present time, a large number of questions remain to be answered for further understanding of the existence, mechanism, and importance of slow intravascular pH changes. The presence of these changes in humans, in whom hemolyis can be kept at very low levels, should be verified. The importance of these changes to peripheral chemoreceptor control of respiration, and to total CO_2 exchange in lung and tissue capillary beds, should be explored. The reasons for the apparent quantitative discrepancies between the perfused lung preparations and the in vivo animal studies requires explanation. Specific localization of carbonic anhydrase in lung tissue and in systemic organs is an important new consideration that has grown out of this work. The fate and activity of carbonic anhydrase in plasma in vivo has itself never been studied and must be ascertained. The effects of exercise on slow pH changes in vivo may have important consequences. These, and a number of other problems related to CO_2 exchange and kinetics both in vivo and in vitro, are likely to be the subject of intense investigation over the next several years.

Acknowledgments. This work was supported in part by research grants HL 19737 and AHA 75-992. Edward D. Crandall is the recipient of RCDA HL 00668. James E. O'Brasky was supported in part by an ATS Fellowship. Akhil Bidani was supported in part by NRSA 7027. We thank, in particular, Ms. Theodora Mauro for expert technical assistance and enthusiasm.

References

1. Bidani A, Crandall ED (1978) Slow post-capillary pH changes in blood pH in vivo: titration with acetazolamide. J Appl Physiol 45: 565–573
2. Bidani A, Crandall ED (1978) Slow post-capillary pH changes in the blood in anesthetized animals. J Appl Physiol 45: 674–680
3. Bidani, A, Crandall ED (1978) Analysis of post-capillary pH changes in blood in vivo after gas exchange. J Appl Physiol 44: 770–781
4. Constantine HP, Craw MR, Forster RE (1965) Rate of the reaction of carbon dioxide with human red blood cells. Am J Physiol 208: 801–811
5. Crandall ED, Forster RE (1973) On some rapid ion exchanges across the red cell membrane. Am Chem Soc Adv Chem Ser 118: 65–87
6. Crandall ED, O'Brasky JE (1978) Direct evidence for participation of rat lung carbonic anhydrase in CO_2 reactions. J Clin Invest 62: 618–622
7. Crandall ED, Bidani A, Forster RE (1977) Post-capillary changes in blood pH in vivo during carbonic anhydrase inhibition. J Appl Physiol 43: 582–590
8. Effros RM, Weissman ML (1978) Carbonic anhydrase activity of cat hindleg. Physiologist 21: 33
9. Effros RM, Chang RSY, Silverman P (1978) Acceleration of plasma bicarbonate conversion to carbon dioxide by pulmonary carbonic anhydrase. Science 199: 427–429
10. Forster RE, Crandall ED (1975) Time course of exchanges between red cells and extracellular fluid during CO_2 uptake. J Appl Physiol 38: 710–718
11. Hill EP, Power GG, Longo LD (1973) Mathematical simulation of pulmonary O_2 and CO_2 exchange. Am J Physiol 224: 904–917
12. Hill EP, Power GG, Gilbert RD (1977) Rate of pH changes in blood plasma in vitro and in vivo. J Appl Physiol 42: 928–934
13. Jacobs MH, Stewart DR (1942) The rôle of carbonic anhydrase in certain ionic exchanges involving the erythrocyte. J Gen Physiol 25: 539–552
14. Klocke RA (1978) Catalysis of CO_2 reactions by lung carbonic anhydrase. J Appl Physiol 44: 882–888
15. O'Brasky JE, Mauro T, Crandall ED (1979) Slow pH equilibration in perfusate after gas exchange in the isolated liver. Am Rev Respir Dis 119: 341
16. Roughton FJW (1935) Recent work on carbon dioxide transport by the blood. Physiol Rev 15: 241–296

Changes in pH and P_{CO_2} with Time in Pulmonary Post-Capillary Blood in Cats

J. PONTE and M.J. PURVES [1]

Crandall's difficulty (this vol) is this. The hypothesis proposed by Hill et al (1973) and Forster and Crandall (1975) predicts that as blood flows through the pulmonary capillaries and CO_2 is lost, a disequilibrium between $[H^+]$ in plasma and $[H^+]$ in the red cell will be set up because the dehydration reaction will be catalyzed in the red cell but not in plasma. They further predict that the resolution of this disequilibrium will be reflected by a slow rise of pH and P_{CO_2} in post pulmonary capillary blood. In favour of this hypothesis is the observation that pH does, in fact, rise in pulmonary post-capillary blood (Hill et al., 1977; Bidani and Crandall, 1978) but against the hypothesis is the compelling evidence that carbonic anhydrase is available to plasma as blood passes through the pulmonary capillaries (Effros et al., 1978; Crandall and O'Brasky, 1978; Hanson et al., 1979). Thus, no H^+ disequilibrium should occur. Nor should pH in post-capillary blood change. Yet, it patently does. It is therefore likely that there must be an alternative explanation for the changes in pH_a.

We have studied this question in cats, using a technique similar to that described by Hill et al. (1977) and Bidani and Crandall (1978). In addition to a small pH electrode with through-flow cuvette, we have devised a CO_2 sensor using mass spectrometry which responds linearly to changes in P_{CO_2} in plasma with a time constant of 0.5 s or less and which has been shown not to be sensitive to changes in flow or pressure. Therefore if the CO_2 sensor and pH electrode are placed in series in a carotid artery-jugular venous loop and blood flow is suddenly stopped, any changes in CO_2 and pH are likely to be those which would occur in naturally flowing blood as it passes to the periphery. Both pH electrode and CO_2 sensor are sensitive to changes in temperature and they have therefore been enclosed in a water jacket which is kept at 37°C.

In Fig. 1 we show the changes which occur when blood flow in such an arterio-venous loop was suddenly stopped in an anaesthetized cat being artificially ventilated with air at a frequency of 15 min^{-1}. The traces indicate changes in pH_a, $PaCO_2$, together with end-tidal oxygen and CO_2 measured with a second mass spectrometer. When blood flow is stopped, pH rises but $PaCO_2$ decays and it is of additonal interest that the amount and rate of these changes depends upon whether blood flow was stopped at the peak or the trough of the pH and CO_2 fluctuation which has the same period as respiration. If blood flow was stopped at the peak of the CO_2 fluctuation the decay is slow and slight: if at the trough the decay is more rapid and more obvious.

1 Department of Physiology, School of Veterinary Sciences, Park Row, Bristol BS1 5LS Avon/GB

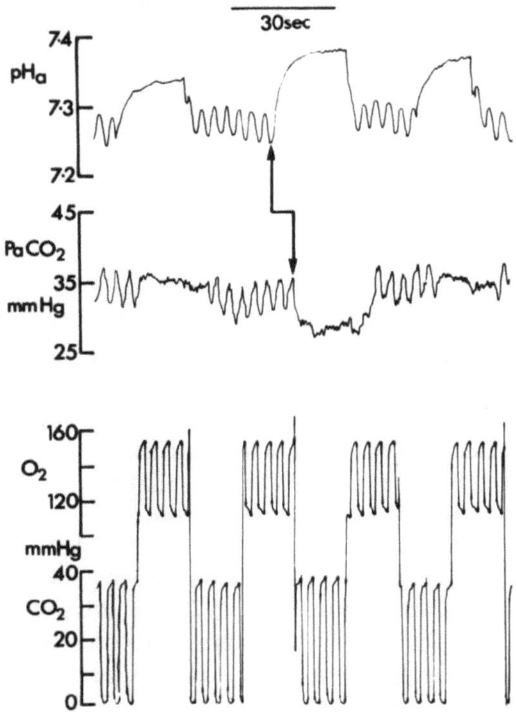

Fig. 1. Traces showing *from above down,* the output of the pH electrode, the output of the CO_2 sensor, tidal O_2 and CO_2 in an anaesthetized cat artificially ventilated with air at a frequency of 15 min $^{-1}$. The *arrows* show the pen displacement of the recorder. Blood flow in the cuvette was stopped, in sequence, at the peak, the trough and towards the peak of CO_2 respiratory fluctuations

The magnitude of the changes in $PaCO_2$ and pH_a in pulmonary post-capillary blood with stop-flow could be affected by various factors as is shown in Fig. 2, where changes in $PaCO_2$ alone are shown. When the cat was given 4–5% CO_2 in air to inhale, as end-tidal and mean $PaCO_2$ rose, the magnitude of CO_2 decay with stop flow also increased. Similarly when the volume of CO_2 to be eliminated with time was increased by infusing equal volumes of 0.1 M HCl and 0.1 M $NaHCO_3$ intravenously, the size of the CO_2 decay also increased. When the cats were made hypoxic (PaO_2 45–55 mm Hg), the CO_2 decay was increased and it was reduced or abolished when the cats were given high O_2 ($PaO_2 > 400$ mm Hg) to inhale. When the animals were given acetazolamide 25–30 mg/kg^{-1} intravenously, end-tidal P_{CO_2} fell, and when arterial blood flow was stopped, $PaCO_2$ now rose by 10 mm Hg or more as is shown in Fig. 3. This also shows that pH_a also rose initially but then fell to a steady, intermediate level. The changes in pH_a or $PaCO_2$ with stop-flow were not affected by the infusion of carbonic anhydrase of up to 20,000 Roughton-Booth units.

These results indicate that under these experimental conditions in the cat, $PaCO_2$ decayed variably in pulmonary post-capillary blood to reach equilibrium which would be a few mm Hg lower than $PaCO_2$ in the first seconds after leaving the lung and also lower than alveolar P_{CO_2}. It would be this value at equilibrium which would be measured on the bench using a standard CO_2 electrode. In parallel experiments, we have been able to show that under these experimental conditions a small significant decay occurs in P_{CO_2} with time and that this decay can be varied by various factors.

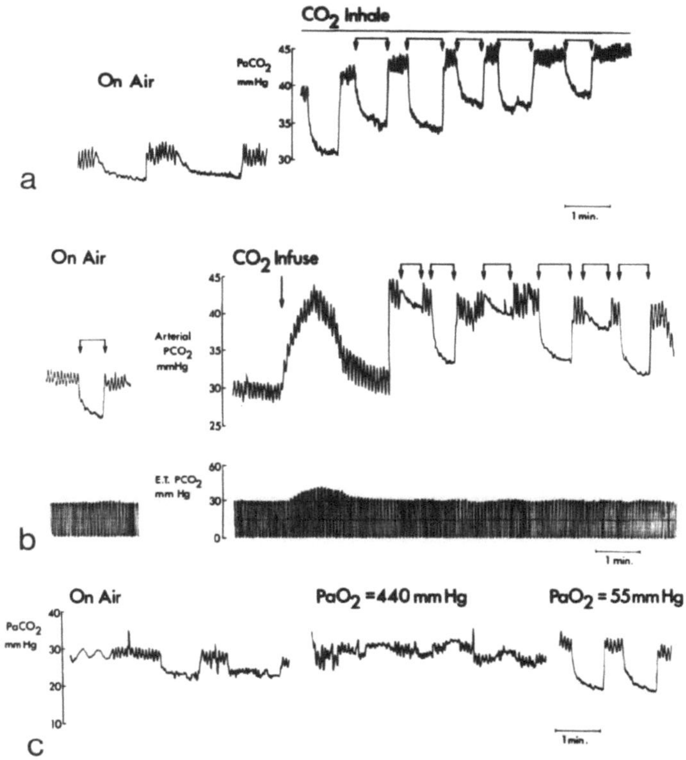

Fig. 2. **a** The effect of stopping blood flow through the arterial cuvette in a cat breathing air, on the *left*, and on the *right*, 3 min after it had been given 5% CO_2 in air to breathe. **b** Traces of $PaCO_2$ above and end-tidal P_{CO_2} below when, on the *left*, the cat was artificially ventilated with air and, on the *right*, during the infusion of CO_2 intravenously. As CO_2 was infused and $PaCO_2$ and end-tidal P_{CO_2} rose, the stroke volume of the pump was increased to bring these variables back towards control levels. The trace of $PaCO_2$ was then shifted by the equivalent of 12 mm Hg. **c** Traces of $PaCO_2$ in a spontaneously breathing cat on the *left* when it breathed air: in the *middle* when it inhaled a high O_2/N_2 mixture and on the *right*, when PaO_2 was reduced to 55 mm Hg. The periods of stop-flow are indicated by *arrows*

In the first group of experiments a delay coil which was enclosed in a water bath at 37°C was interposed into the carotid artery circuit in cats anaesthetized and artificially ventilated. The effect of this coil was to impose a delay of approximately 10 s, the system being arranged such that blood from the heart could flow directly or via the delay coil and P_{CO_2} measured at steady state under these two conditions with a standard Radiometer electrode. At the same time, end-tidal CO_2 was measured continuously and samples of blood were taken from the carotid artery for measurement of $PaCO_2$ with a standard CO_2 electrode which was identically calibrated to that electrode used in vivo. When the cat was ventilated with air, end-tidal P_{CO_2} being approximately 30 mm Hg, $PaCO_2$ measured by the direct circuit in vivo was approximately 2 mm lower than end-tidal and after the delay of 10 s had been imposed, $PaCO_2$ was approximately 3 mm Hg lower and was virtually identical to that $PaCO_2$

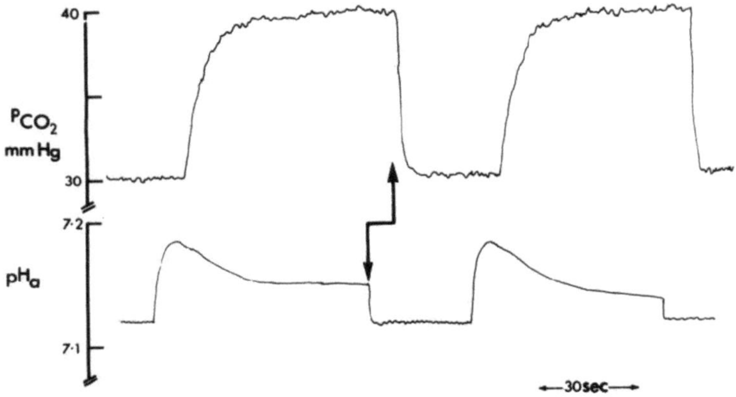

Fig. 3. Traces to show the effect of stopping flow in the cuvette upon *above*, $PaCO_2$ and *below*, pH_a after the administration of acetazolamide, 30 mg/kg^{-1}. The *arrows* indicate corresponding times on the traces. Now, $PaCO_2$ rises by 10 mm Hg or more while pH_a rises initially and then adapts to an intermediate level

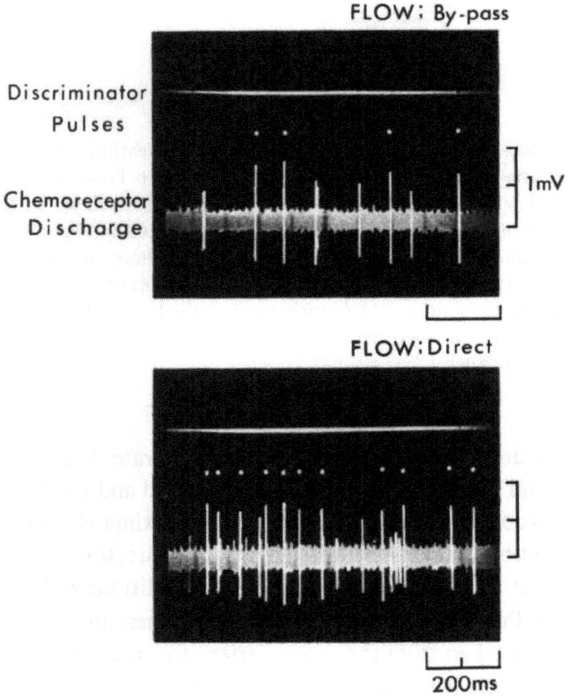

Fig. 4. Neurograms showing afferent carotid body chemoreceptor discharge *above*, when flow bypassed the carotid body and *below*, when arterial blood flowed directly to the carotid body. In each trace *below*, action potentials from the sinus nerve; *above*, standard pulses from the discriminator

measured at equilibrium on the bench. When the cat was given high CO_2 to inhale and end-tidal P_{CO_2} was raised to approximately 45 mm Hg, the gradient between end-tidal and arterial P_{CO_2} now increased to between 5 and 7 mm Hg and again, P_{CO_2} measured after the imposition of the 10 s delay was 2 or 3 mm Hg lower. An even more striking gradient was observed when CO_2 was infused intravenously when the difference between end-tidal and direct P_{CO_2} now increased to 10 mm Hg or more, P_{CO_2} after the 10 s delay or measured at equilibrium being a further 3 or 4 mm lower.

In a second group of experiments, a similar delay was imposed upon blood flowing in the carotid artery, this time the carotid body chemoreceptors were used effectively as a CO_2 electrode and in six cats studied in this way in which flow was alternated between a direct route and a delay or bypass route, the chemoreceptor discharge, as is shown in Fig. 4, was invariably lower when blood flow was delayed, compared to the direct route.

Discussion

We are tolerably certain that the observed changes in P_{CO_2} and pH are not artefactual but their cause remains uncertain. Since their magnitude could be affected by O_2, we propose that they follow oxygenation of Hb, the release of the Bohr proton and carbamino CO_2 and the subsequent binding of H^+ to the sites so vacated. The fall in $[H^+]_i$ could be substantial, since the ratio H^+ binding/CO_2 released is 1.5 or greater (Rossi-Bernardi and Roughton, 1967; Bauer and Schröder, 1972; Klocke, 1973). Further, part of the H^+ binding could well take place in post-capillary blood since Klocke (1973) has shown that the evolution of carbamino CO_2 is not complete in under 480 ms and we have shown that the pulmonary capillary transit time is 0.5 s or as little as 0.1 s (Vogel, 1947). We further propose that the fall in $[H^+]_i$ can only be rectified by an inward flux of CO_2 from plasma which will (a) lead to a slow uncatalyzed dehydration reaction in plasma (reflected by the rise in pH) and (b) a catalyzed hydration reaction in the red cell, HCO_3^- diffusing into plasma in exchange for Cl^-. This explanation is consistent with the observation that the decay of CO_2 is enhanced by hypercapnia (when loss of carbamino CO_2 becomes more important) or in hypoxia (when the sequence of reactions might be expected to be slowed) and is reversed with acetazolamide when plasma bicarbonate is virtually unchanged in the pulmonary capillaries.

We therefore suggest that the changes in pH_a observed by Hill et al. (1977), Bidani and Crandall (1978) and in this series are genuine but do not arise for the reasons proposed by Hill et al. (1973) and Forster and Crandall (1975). The physiological significance of this decay of CO_2 and rise of pH can at present only be guessed at. Clearly, if as shown in these experiments, the decay can be sensed by the peripheral arterial chemoreceptors, the CO_2 signal in arterial blood becomes, among other things, a function of cardiac output and further, changes in the signal may not be detected by measurements on the bench at equilibrium. This could have relevance to the long-debated question of how ventilation rises in voluntary exercise and is maintained at high levels without obvious change in the arterial chemical signal.

Acknowledgments. These studies were supported by N.I.H. Grant No. RO1 HD 09457.

References

Bauer C, Schröder E (1972) Carbamino compounds of haemoglobin in human adult and foetal blood. J Physiol (London) 227: 457–471

Bidani A, Crandall ED (1978) Slow post-capillary changes in blood in vivo: titration with acetazolamide. J Appl Physiol 45: 565–573

Crandall ED, O'Brasky JE (1978) Direct evidence for participation of rat lung carbonic anhydrase in CO_2 reactions. J Clin Invest 62: 618–622

Effros RM, Silverman P, Chang RSY (1978) The role of pulmonary carbonic anhydrase in CO_2 transport. Fed Proc 37: 868, 3432

Forster RE, Crandall ED (1975) Time course of exchanges between red cells and extracellular fluid during CO_2 uptake. J Appl Physiol 38: 710–718

Hanson MA, Nye PCG, Torrance RW (1979) Pulmonary capillary luminal surface carbonic anhydrease. J Physiol (London) 293: 25 P

Hill EP, Power GG, Longo LD (1973) Mathematical simulation of pulmonary O_2 and CO_2 exchange. Am J Physiol 224: 904–917

Hill EP, Power GG, Gilbert RG (1977) Rate of pH changes in blood plasma in vitro and in vivo. J Appl Physiol 42: 928–934

Klocke RA (1973) Mechanism and kinetics of the Haldane effect in human erythrocytes. J Appl Physiol 35: 673–681

Rossi-Bernardi L, Roughton FJW (1967) The specific effect of CO_2 and carbamate compounds on the buffer power and Bohr effects in human haemoglobin solutions. J Physiol (London) 189: 1–29

Vogel H (1947) Die Geschwindigkeit des Blutes in den Lungenkapillaren. Helv Physiol Pharmacol Acta 5: 105–121

Kinetics of Acid-Base Parameters of Blood After Respiratory and Metabolic Disturbances

A. LUTTMANN, K. MÜCKENHOFF, and H.H. LOESCHCKE [1]

1 Introduction

Theoretical considerations lead to the result that the time course of the equilibration process after a disturbance of the acid-base balance of blood is mainly determined by the uncatalyzed CO_2 hydration in the plasma, if no carbonic anhydrase is available to the plasma (Roughton, 1935; Jacobs and Stewart, 1942; Forster, 1972; Hill et al., 1973; Forster and Crandall, 1975; Bidani et al., 1978). The time course of the pCO_2 was measured in red cell suspensions first by Constantine et al. (1965) using a continuous flow method. Measurements of the pH time course with a stopped flow method were performed first by Crandall et al. (1971). Our own determinations of the time courses of the acid-base parameters in untreated bovine blood with a continuous flow method (Luttmann, 1977) show a faster decrease of the plasma pH than is predicted by the theoretical considerations. Therefore, H^+-ion transfer between the plasma and the cells is proposed, which could be mediated by a transfer of weak acids or bases. Some experiments to test the influence of the diffusion of H_2CO_3, lactic acid, and ammonia are described in this paper. The concentrations of all these substances are small at normal pH due to the great differences between their pK and the blood pH. Beside the small actual concentration it has been shown by Jacobs and Stewart (1942) and Loeschcke and Sugioka (1963) for ammonia, that the diffusion of NH_3 plays an important role in the equilibration process between the intra- and extracellular phase of blood and tissue. For lactate Moll et al. (this vol.) have demonstrated that the transfer process across the placenta is partly mediated by an exchange of lactic acid.

2 Material and Methods

The pH was measured by an H^+-ion sensitive glass electrode. The reference electrode was connected with a liquid KCl-bridge. The flow- and pressure sensitivity was found to be very small.

The CO_2-partial pressure was measured with a membrane covered pH-electrode. The calibration curve was linearized by using an antilogarithmic amplifier. The response time (T_{95}) of the electrode was smaller than 1 s due to adding carbonic anhydrase (10^5 Wilbur-Anderson units 1^{-1}) to the aqueous layer in front of the

1 Institut für Physiologie, Ruhr-Universität Bochum, Postfach 10 21 48, 4630 Bochum/FRG

H^+-ion sensitive area and due to linearizing the frequency response of the electrode electronically as described by Luttmann et al. (1978).

The electrical resistance of blood depends upon the cell content since the erythrocyte membrane has isolating properties. The resistance was measured by applying an alternating current (10 kHz, 45μA) on a sample of blood using two platinum electrodes and determining the voltage drop in the liquid with two other electrodes. In previous studies, by comparison of the resistance of blood and true plasma, it was observed that during respiratory disturbances the change of the electrical resistance in blood is mainly caused by a change of the cell volume (Luttmann, 1977).

The chloride concentration was measured with a chlorinated silver electrode. Since the stability of the electrode was not good enough to measure the absolute values of the chloride concentration due to the high protein of the blood, only its relative changes were obtained.

All experiments were carried out with bovine blood within the first 6 h after taking the blood (mean value of hematocrit HT = 40.4 ± 0.66%, n = 57; hemoglobin concentration [Hb] = 140 ± 4.3 g · 1^{-1}, n = 21; lactate concentration [Lac] = 4.4 ± 0.3 mM · 1^{-1}, n = 17). The blood was warmed to 310 K and equilibrated with an O_2-CO_2-mixture (F_{CO_2} = 0.03). Thus the hemoglobin was saturated and further results could not be influenced by the Bohr effect. To produce a respiratory disturbance CO_2 gas was added using a modified Hartridge-Roughton continuous flow method: The blood was pressed into a tube by a pump (Fig. 1). At point (a) a CO_2-O_2-mixture (F_{CO_2} = 0.7) was added. After about 200 ms the blood-gas-mixture leaves the tube at point (b). During passage of the tube CO_2 diffuses into the liquid and the CO_2-content increases. At point (b) a sample of the liquid is drawn through the electrodes with a second pump. In some experiments, instead of the CO_2-electrode a chloride electrode was installed. The time courses of the diverse signals were determined by varying the velocity of the suction pump and by varying the length of the tube between point (b) and the electrodes.

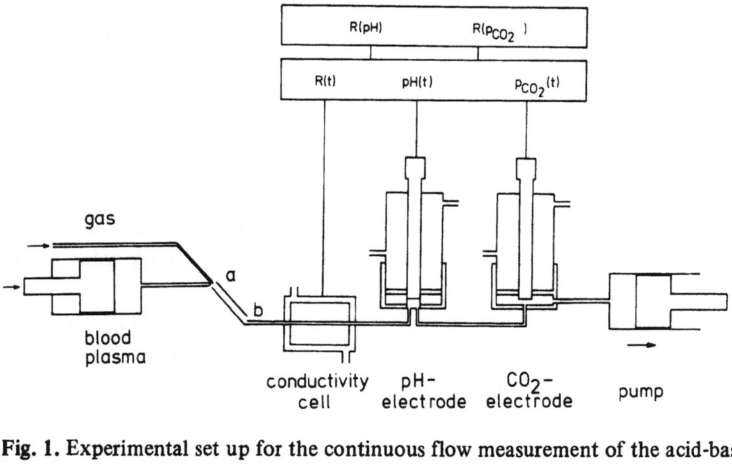

Fig. 1. Experimental set up for the continuous flow measurement of the acid-base parameters after respiratory disturbance (see text)

To measure the time courses of pH and pCO_2 after metabolic disturbances the blood was filled into a thermostated container and covered with paraffin. It was drawn through the electrodes with constant velocity. Isoosmotic solutions of acids or bases were added swiftly to the blood container and well mixed with the blood by a magnetic stirrer.

3 Results

3.1 Time Course of the Acid Base Parameters After Respiratory Disturbance

Figure 2 demonstrates the time courses of pCO_2, pH and the changes of the electrical resistance and the chloride concentration in whole blood (mean values ± SEM). Due to the addition of CO_2 the pCO_2 increases abruptly from the starting value (3.3 kPa). It then decreases because the physically dissolved CO_2 is transformed into bound CO_2 (HCO_3^-, H_2CO_3, carbamate). During this process the H^+-ion concentration in the plasma increases rather fast to its final value. The chloride concentration decreases because of the Hamburger Shift. The increase of the electrical resistance due to the water shift into the cells is found to be the slowest process. In control experiments the influence of a possible hemolysis was tested. Therefore we intended to inhibit the carbonic anhydrase in the plasma only. This was done with benzolamine (10 mg $\cdot\ 1^{-1}$) and with acetazolamine (200 mg $\cdot\ 1^{-1}$), both added to the blood just before doing the experiment. The time courses of pH and pCO_2 are not markedly changed in these experiments, indicating that during the experiments there is no marked influence of any extracellular carbonic anhydrase activity.

The time courses may be characterized by the time needed to reach 90% respectively 110% (overshooting response) of the final change (T_{90} for pH, R and $\Delta\ [Cl^-]$ and T_{110} for pCO_2). The mean values of several series of experiments did not exceed the following values: T_{110} = 10.5 s for pCO_2; T_{90} = 5 s for pH; T_{90} = 2.5 s for $\Delta\ [Cl^-]$; T_{90} = 12 s for R. In another series of experiments the influence of the hematocrit upon the time courses was determined. Therefore blood was diluted in several steps with its own plasma. Figure 3 gives the results. The dilution factor c_B (ratio between the actual Hb and the Hb of whole blood) was varied from 0 (isolated plasma) to 1 (whole blood). It turns out that the speed of the equilibration process is strongly dependent on the cell content. The time course of pCO_2 for isolated plasma can be described by a monoexponential function. Increasing the cell content yields a more and more biexponential time course. The time course of pH for whole blood shows a bend which disappears when the cell content is decreased.

Comparison of the experimental data with model calculations (Bidani et al., 1978; Hill et al., 1973, and own calculations) assuming an uncatalyzed CO_2-hydration in the plasma shows that the formation of H^+-ions by CO_2-hydration is not fast enough to explain the measured time course of the plasma pH. The discrepancy cannot be explained by a catalytic effect in the plasma, since the experiments with isolated plasma (c_B = 0) are in good agreement with the theoretical calculations and because in the experiments with whole blood the inhibition of the plasma carbonic anhydrase had no marked effect on the time courses. To explain the discrepancy

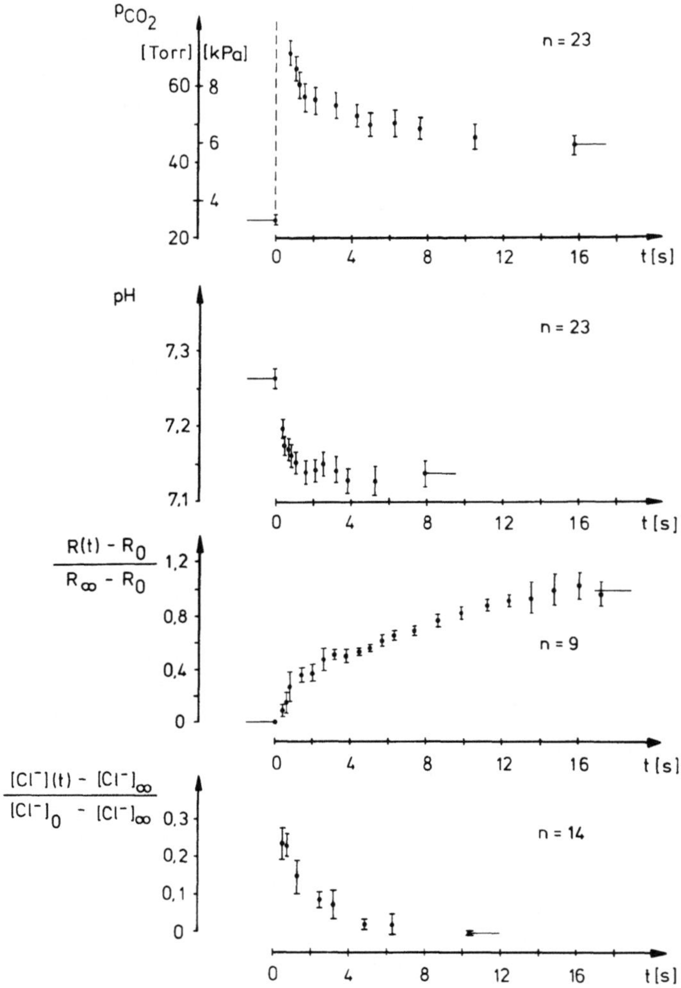

Fig. 2. Time courses of pCO_2, pH and the relative changes of the resistance and the chloride concentration in whole blood after respiratory disturbace (mean values ± SEM)

between the experimental result and the theoretical considerations an H^+-ion transfer mechanism from inside the cells is hypothesized. This H^+-ion transfer could be mediated by a transfer of undissociated weak acids or bases. To test this hypothesis experiments were performed; their results are described later.

3.2 The Time Courses of pH and pCO_2 After Metabolic Disturbances

Figure 4 shows the original recording of the pH- and the pCO_2-time course measured in whole blood after addition of 1.5 mM · 1^{-1} NaOH. Addition of the base causes a fast alteration of the plasma pH. After this steep displacement the pH slowly

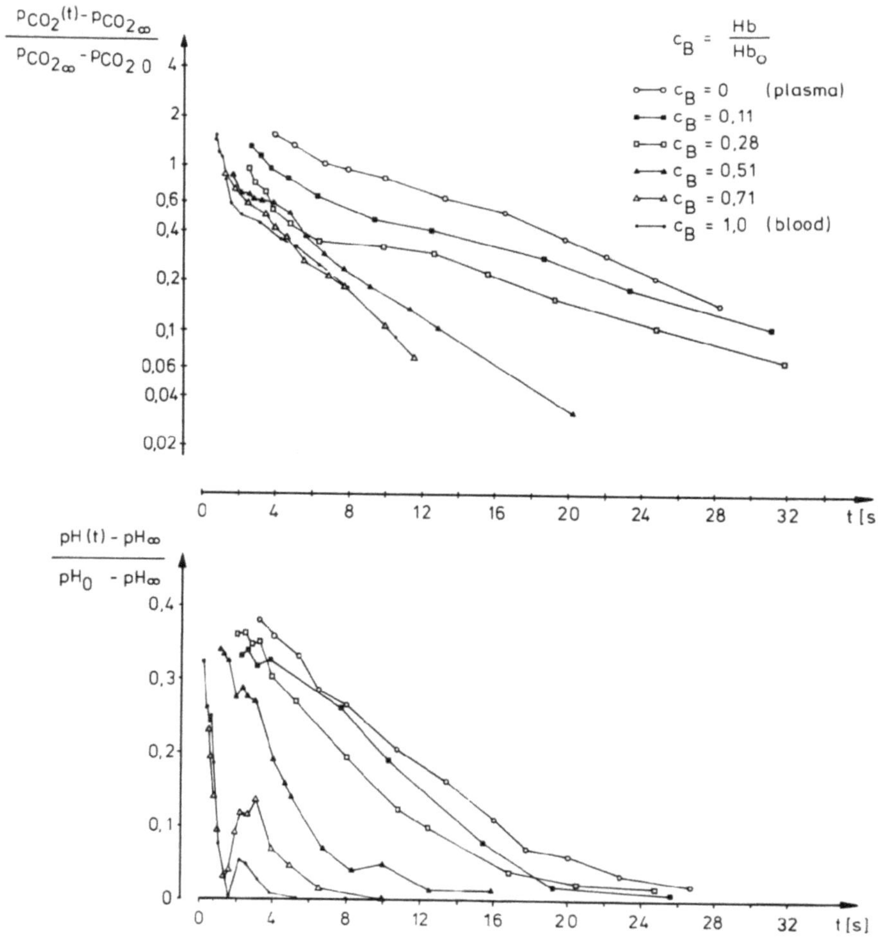

Fig. 3. Influence of the cell content on the pCO_2- and pH-time courses after adding CO_2

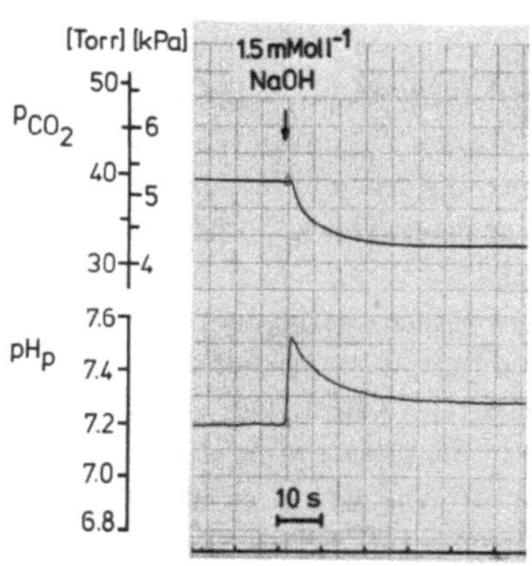

Fig. 4. Original recording of the pH- and pCO_2-time course in whole blood after additon of 1.5 mM • 1^{-1} NaOH

approaches its final value. Simultaneously a slow alteration of the CO_2 partial pressure is observed. (The pCO_2-recording is delayed in comparison to the pH-recording due to serial connection of the electrodes.) The pH- and pCO_2-time courses can be described by monoexponential time functions; the time constants are summarized in Table 1. In untreated blood the time constants are smaller than in the experiments with benzolamide or acetazolamide due to the presence of carbonic anhydrase in the plasma from hemolysis. There are small differences between the time constants after adding lactic acid, HCL or H_2SO_4. This may be explained by noting that the $SO_4^{=}$-permeability is smaller than the Cl^--permeability and that lactic acid may permeate the membrane in an undissociated form too (Deuticke, 1973; Motais, 1977) or may be transported by a specific H^+-lactate symport mechanism as recently demonstrated by Dubinsky and Racker (1978). The time constants are found to be smaller after addition of acids than after addition of NaOH. This agrees with the findings of Jacobs and Stewart (1942) who assessed the equilibration process after addition of NaOH and HCl by monitoring the concomitant volume change of the red cells. An explanation for the different time courses is given by Hladky and Rink (1977).

Table 1. Time constants of the pH- and the pCO_2-time course after addition of acid or base

	Time constants [s]					
	Blood without carbonic anhydrase inhibitor		Blood with acetazolamide [a]		Blood with benzolamine [b]	
	pH	pCO_2	pH	pCO_2	pH	pCO_2
Acid						
HCl	7.0	7.7	12.9	13.8	12.8	13.0
lactic acid	6.1		11.5			
H_2SO_4	13.5		14.1			
Base						
NaOH	9.3	8.1	18.6	14.2	18.7	13.8

a 200 mg · 1^{-1}
b 10 mg · 1^{-1}, both added just before experiment

3.3 Influence of Weak Acids or Bases on the Time Courses of pH and pCO_2

As pointed out above, a fast H^+-ion transfer between the cell- and plasma-phase could be mediated by a transport of undissociated acids or bases. This hypothesis is experimentally tested for H_2CO_3, lactic acid and ammonia.

After a respiratory disturbance the pH- and pCO_2-time courses are measured in untreated blood ([Lac] = 4 mM · 1^{-1}) and in blood from the same animal, to which isotonic solutions of Na-lactate or HN_4Cl (+ 6.6 mM · 1^{-1}) are added. The result is shown in Fig. 5. In both cases a fast initial change of the plasma pH is obtained. The results are not markedly influenced by any plasma carbonic anhydrase activity as could be demonstrated in experiments with benzolamide (10 mg · 1^{-1}).

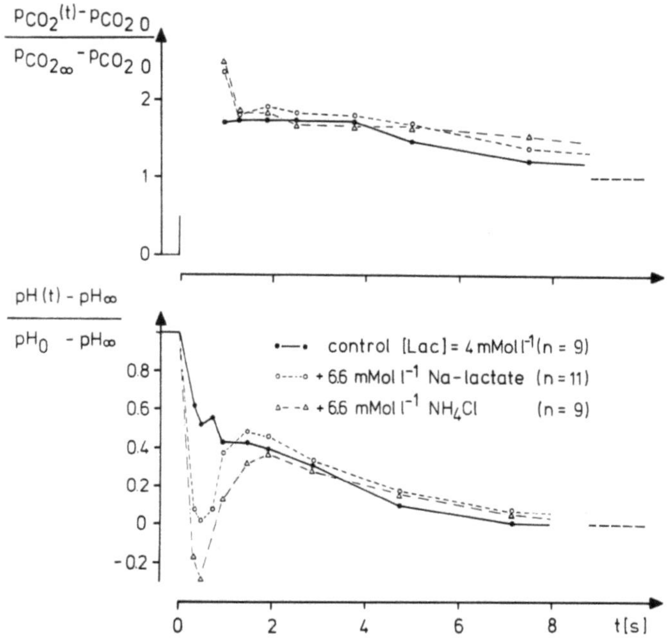

Fig. 5. Influence of the addition of Na-lactate or NH_4Cl on the pH- and pCO_2 time courses after addition of CO_2 (mean values). The final values are indicated by *dotted lines*

The influence of an increased lactate concentration was also tested after metabolic disturbances: Therefore we measured the pH- and pCO_2-time courses after addition of HCl or NaOH in untreated blood ([Lac] = 4.4 mM · 1^{-1}) and in blood to which Na-lactate was added (+ 6 mM · 1^{-1} and + 10 mM · 1^{-1}). The influence of plasma carbonic anhydrase activity was prevented by benzolamide (10 mg · 1^{-1}). The results are shown in Fig. 6. It demonstrates that the equilibration process is accelerated by lactate.

Investigations on the influence of nonionic diffusion of H_2CO_3 are more complicated because of the interconversion of H_2CO_3 to HCO_3^- and to CO_2. A test was carried out in blood with diminished anion permeability [blood with 10^{-4} M · 1^{-1} 4-acetamido-4′-isothiocyanatostilben-2,2-disulfonic acid (SITS)]. The idea of this experiment was that if there is any H_2CO_3-flux across the membrane, this flux should go on even in the case of diminished anion permeability. After addition of CO_2 this H_2CO_3-flux should produce a fast rise of the H^+-ion concentration in the plasma, since HCO_3^- cannot get out of the cell and cannot act as a buffer. This result could not be obtained in SITS-treated blood. On the contrary, the equilibration process was markedly delayed (time constants τ = 8.7 s for pH, τ = 11.8 s for pCO_2), because the HCO_3^--Cl-shift was retarded. So the result of this experiment cannot be regarded as a proof of a nonionic H_2CO_3 diffusion.

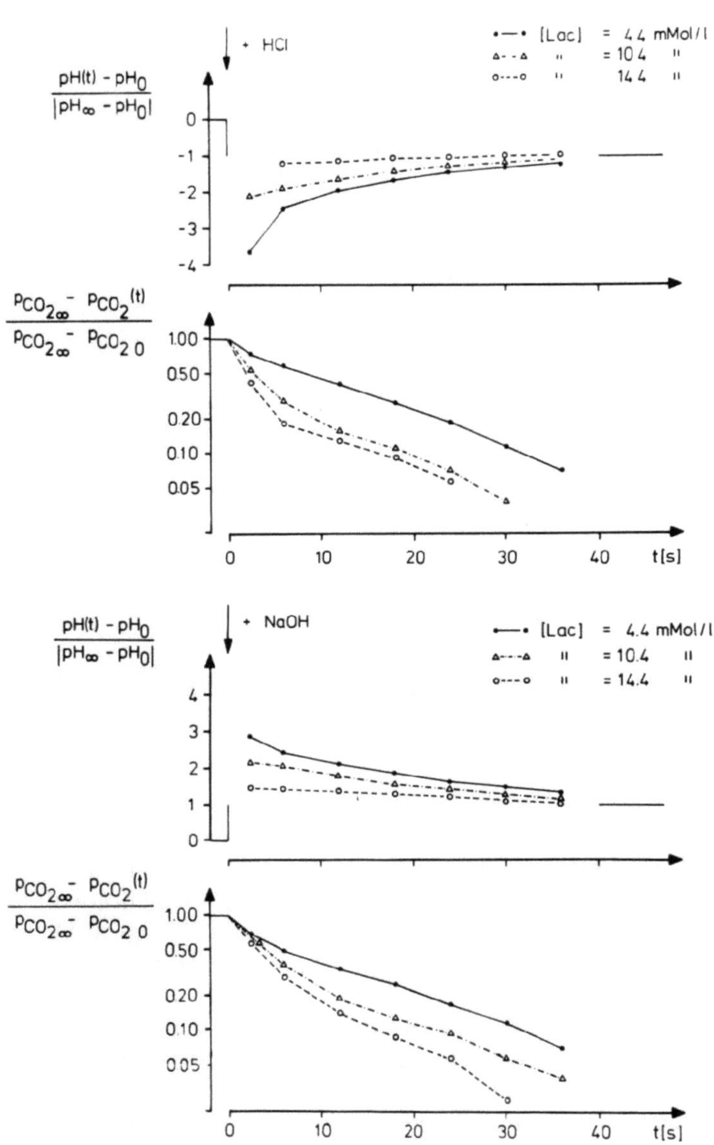

Fig. 6. Influence of increased lactate concentration on the pH- and pCO_2-time courses after addition of HCl (*upper diagram*) or NaOH (*lower diagram*) (mean values of 4–8 experiments)

4 Discussion

The respiratory disturbance of the blood was produced by adding CO_2 in a gaseous form in order to prevent any other initial disequilibria. Therefore the blood was in contact with the gas during 200 ms, and so the begin of the reaction and all measured time courses are poorly defined within this time interval.

In the continuous flow method the response times of the electrode systems do not play a role, in the measurements after metabolic disturbance however the response times of the electrodes directly affect the measured time courses of pH and pCO_2. The response time (T_{95}) of the CO_2-electrode was < 1 s, that of the pH-electrode was much smaller. After addition of acid or base the pH changes abruptly. This initial displacement will be damped by the wash-in process into the measuring chambers. This will affect only the time course within the first second as it was demonstrated in control experiments in which sudden changes of the composition of a test liquid were produced by mixing phosphate buffers.

The described pH time courses after respiratory disturbances as measured with the continuous flow method are faster than those measured with the stopped flow technique (Crandall et al., 1971, 1977; Hill et al., 1977). Two facts may explain this difference:

a) The lactate concentration of the untreated bovine blood used here is relatively high (mean value: 4.4 mM · l^{-1}). Therefore, the H^+-ion transfer by transport of lactic acid may play an important role.
b) After stopping the flow diffusion gradients for all substances, which are exchanged between the cells and the plasma, may be formed due to the diminished convection between the cells and the surrounding plasma. In some experiments we stopped the flow too with the result that the equilibration process was delayed, T_{90} for pH rose from 5 s to 15 s, T_{110} for pCO_2 from 10.5 s to 18.5 s (mean values from 10 measurements).

After adding CO_2 the pCO_2 shows a fast initial drop followed by a slower decrease. This biphasic time course was also found by Constantine et al. (1965) and Holland and Forster (1975) using a rapid mixing continuous flow technique.

Measurements of the Cl^--time course indicating the Hamburger shift have been obtained by several authors (for ref. see Piiper, 1964, and Chow et al., 1976). The half times scatter widely between 0.04 and 0.2 s. In our experiments a T_{90} of 2.5 s is obtained.

The measurements of the electrical resistance may indicate that the time course is mainly determined by the water permeability of the cell membrane. This is confirmed by the finding that after inhibition of the carbonic anhydrase with 250 mg · l^{-1} acetazolamide added 3 h before the experiment, the time for reaching 75% of the total change is the same as in untreated blood in spite of the different time course of pH and pCO_2 (Luttmann, 1977).

The mechanism of lactic acid exchange across the red cell membrane is not completely understood. The findings of Deuticke (1973) suggest that nonionic diffusion of lactic acid is small. Dubinsky and Racker (1978) demonstrated recently a specific H^+-lactate symport mechanism. From our experiments we cannot distinguish between these two mechanisms. The dependency of the equilibration process upon the lactate concentration may play a role during severe muscular exercise, in which the lactate concentration in the blood can increase up to 18 mM · l^{-1} (Koepchen, 1972). This will accelerate the equilibration processes after uptake of acids or CO_2. So exchange processes in the tissue and the lung may be facilitated at the same time as the passage time is shortened.

Acknowledgments. Supported by the Deutsche Forschungsgemeinschaft (Sonderforschungsbereich 114 Bionach).

References

Bidani A, Crandall ED, Forster RE (1978) Analysis of postcapillary pH changes in blood in vivo after gas exchange. J Appl Physiol 44: 770–781

Chow EI, Crandall ED, Forster RE (1976) Kinetics of bicarbonate-chloride exchange across the human red blood cell membrane. J Gen Physiol 68: 633–652

Constantine HP, Craw MR, Forster RE (1965) Rate of reaction of carbon dioxide with human red blood cells. Am J Physiol 208: 801–811

Crandall ED, Klocke RA, Forster RE (1971) Hydroxyl ion movement across the human erythrocyte membrane. J Gen Physiol 57: 664–683

Deuticke B (1973) Transfer of monovalent organic anions across the red cell membrane: Mechanisms and experimental alterations. In: Gerlach E, Moser K, Deutsch E, Willmanns W (eds) Erythrocytes, thrombocytes, leukocytes. Thieme, Stuttgart, pp 81–87

Dubinsky WP, Racker E (1978) The mechanism of lactate transport in human erythrocytes. J Membr Biol 44: 25–36

Forster RE (1972) Interaction of CO_2 and O_2 exchange in red blood cells. In: Astrup P, Rørth M (eds) Oxygen affinity of hemoglobin and red cell acid-base status. Munksgard, Copenhagen, pp 494–506

Forster RE, Crandall ED (1975) Time course of exchanges between red cells and extracellular fluid during CO_2 uptake. J Appl Physiol 38: 710–718

Hill EP, Power GG, Longo LD (1973) Mathematical simulation of pulmonary O_2 and CO_2 exchange. Am J Physiol 224: 904–917

Hill EP, Power GG, Gilbert RD (1977) Rate of pH changes in blood plasma in vitro and in vivo. J Appl Physiol 42: 928–934

Hladky SB, Rink TJ (1977) pH equilibrium across the red cell membrane. In: Ellory JC, Lew VL (eds) Membrane transport in red cells. Academic Press, London New York, pp 115–135

Holland RAB, Forster RE (1975) Effect of temperature on rate of CO_2 uptake by human red cell suspensions. Am J Physiol 228: 1589–1596

Jacobs MH, Stewart DR (1942) The role of carbonic anhydrase in certain ionic exchanges involving the erythrocyte. J Gen Physiol 25: 539–552

Koepchen HP (1972) Atmungsregulation. In: Gauer OH, Kramer K, Jung R (eds) Physiologie des Menschen, vol VI, Atmung. Urban und Schwarzenberg, München

Loeschcke HH, Sugioka K (1963) Über die Natur der Ammoniumchlorid-Acidose. Schweiz Med Wochenschr 33: 1030–1048

Luttmann A (1977) Steady-state- und transientes Verhalten von Säuren-Basen-Parameters und Erythrozytenvolumen bei der Bindung von CO_2 in Blut (Entwicklung eines schnell ansprechenden CO_2-Meßsystems). Diss Abt Biol Ruhr-Univ Bochum

Luttmann A, Mückenhoff K, Loeschcke HH (1978) Fast measurement of the CO_2 partial pressure in gases and fluids. Pflügers Arch 375: 279–288

Motais R (1977) Organic anion transport in red blood cells. In: Ellory JC, Lew VL (eds) Membrane transport in red cells. Academic Press, London New York, pp 197–220

Piiper J (1964) Geschwindigkeit des CO_2-Austausches zwischen Erythrocyten und Plasma. Pflügers Arch 278: 500–512

Roughton FJW (1935) Recent work on carbon dioxide transport by the blood. Physiol Rev 15: 241–296

Carbonic Anhydrase in Lung Tissue

R.A. KLOCKE [1]

Carbonic anhydrase (EC 4.2.1.1) activity has been demonstrated in lung tissue (Berfenstam, 1952), but the enzyme has not been associated with any major function in the adult. In the fetus the enzyme may play a role in the secretion of lung fluid (Adamson and Waxman, 1976), possibly remaining in the mature lung as a vestige of intrauterine events.

Du Bois (1969) has concluded that storage of carbon dioxide in lung tissue dampens variations in alveolar and arterial P_{CO_2} resulting from cyclic ventilation. Lung carbonic anhydrase enhances this capacative effect by facilitating interconversion of CO_2 and bicarbonate ion in pulmonary tissue. In the presence of active enzyme, uptake of CO_2 by lung tissue following a step change in P_{CO_2} is 90% complete in 10 s (Plewes et al., 1976). Without enzymatic activity this time increases to over 25 s. However, the overall effect of lung tissue stores on P_{CO_2} oscillations is small and lung carbonic anhydrase enhances dampening of CO_2 transients by only 10%–15% (Farhi et al., 1976).

Recent evidence (Effros et al., 1978; Klocke, 1978; Crandall and O'Brasky, 1978) has suggested another possible physiologic effect of pulmonary carbonic anhydrase. Enzyme located in the lung may be important in attainment of chemical equilibrium in blood during pulmonary capillary transit.

1 Slow Postcapillary pH Changes

Carbonic anhydrase activity is absent from plasma (Maren, 1967). This led Roughton (1935) to postulate that confinement of carbonic anhydrase to the red cell interior would lead to disequilibrium between end-capillary plasma and red cell pH. Dissolved CO_2 rapidly leaves both plasma and erythrocyte as blood enters the pulmonary capillary. This upsets the equilibrium between bicarbonate and carbon dioxide, leading to conversion of bicarbonate to CO_2 in both the plasma and red cell. The latter reaction occurs rapidly under the influence of carbonic anhydrase, but in the absence of enzyme the plasma reaction proceeds very slowly at the uncatalyzed natural rate. At cellular bicarbonate is depleted, plasma bicarbonate enters the cell in exchange for chloride ion and combines with hydrogen ions to form carbonic acid. The acid is then dehydrated to carbon dioxide and excreted. End-capillary plasma is depleted of both bicarbonate and CO_2, but plasma pH is unchanged because appreciable trans-

1 Department of Medicine, State University of New York at Buffalo, Buffalo, New York/USA

membrane hydrogen ion movement does not occur during the short capillary transit time (Crandall et al., 1971). Roughton (1935) recognized that failure to catalyze plasma CO_2 reactions would result in an excess of hydrogen ions in plasma relative to the erythrocyte. Equilibrium can be reached only by transferring large quantities of hydrogen ion into the erythrocyte with CO_2 acting as a proton carrier (Jacobs and Stewart, 1942). Since the rate of transfer is dependent upon uncatalyzed CO_2 reactions in plasma, equilibrium is reached long after blood has left the capillary. Indeed, without catalysis of plasma CO_2 reactions, blood will never reach chemical equilibrium as it circulates between the tissues which produce CO_2 and the lung which excretes the gas (Bidani et al., 1978).

Three different groups of investigators (Sirs, 1970; Hill et al., 1973; Forster and Crandall, 1975) independently investigated the theoretical aspects of this disequilibrium and concluded that plasma pH will continue to change for 20–30 s after blood leaves the lung if plasma CO_2 reactions are not catalyzed. Slow plasma-red cell pH adjustments following CO_2 exchange have been observed in vitro in red cell suspensions (Forster and Crandall, 1975). In vivo confirmation of the disequilibrium theory is less convincing. Initial efforts were frustrated due to hemolysis with spillage of red cell enzyme into the plasma of the experimental animal. Slow pH changes could only be demonstrated in the presence of carbonic anhydrase inhibitors (Crandall et al., 1977). Later efforts were successful in detecting slow pH changes in blood leaving the lung in the intact animal, but the observed changes were much smaller than predictions from theoretical models (Hill et al., 1977; Bidani and Crandall, 1978). Observed changes were 0.01 pH units or less with expected changes of 0.03–0.04 pH units.

2 Lung Carbonic Anhydrase Activity

Simultaneous experiments from three different laboratories using isolated, perfused lung preparations have provided evidence that plasma may have access to lung carbonic anhydrase in the pulmonary capillary. Removal of erythrocytes and perfusion of the lung with bicarbonate solutions allows study of CO_2 reactions which otherwise might be masked by the overwhelming effect of red cell carbonic anhydrase.

Effros et al. (1978) injected a bolus of radioactive CO_2 or bicarbonate and measured the volume of tracer distribution. Under control circumstances the volume of bicarbonate distribution was almost as large as CO_2, and was much greater than the vascular volume. In the presence of acetazolamide, a potent carbonic anhydrase inhibitor, bicarbonate distribution was limited essentially to the vascular space. CO_2 distribution was affected but to a much lesser degree. These authors concluded that enzyme localized to the vascular endothelium catalyzed conversion of bicarbonate to CO_2 under control conditions, but was inhibited when acetazolamide was added to the perfusate.

Crandall and O'Brasky (1978) measured the rate of pH change in perfusate after it left the lung. Slow pH changes were absent unless carbonic anhydrase inhibitors were added to the perfusate. In these experiments perfusate could not be sampled during the first 4–6 s after leaving the pulmonary capillary, but this lag should not have completely obscured the expected pH changes.

In our laboratory steady-state CO_2 excretion was quantitated in isolated rabbit lungs which were ventilated with air and perfused with bicarbonate buffers equilibrated with CO_2 (Klocke, 1978). Dissolved carbon dioxide entered the alveolar space as perfusate reached the pulmonary capillaries. The decrease in perfusate P_{CO_2} disrupted chemical equilibrium, leading to production of additional CO_2. The quantity of CO_2 produced from bicarbonate was proportional to the rate of CO_2 reactions during capillary transit. Addition of acetazolamide to the perfusate caused a pronounced decrease in CO_2 excretion, indicating that the reactions had been catalyzed by carbonic anhydrase. As in the experiments of Effros et al. (1978) and Crandall and O'Brasky (1978), analysis of perfusate leaving the lungs revealed no significant carbonic anhydrase activity, ruling out the possibility of artifact secondary to contamination with red cells.

These three series of experiments demonstrate that bicarbonate ion in perfusate has access to lung carbonic anhydrase. Two possible mechanisms could be responsible for the observations. Bicarbonate could diffuse across the capillary endothelium and be converted to CO_2 in the presence of carbonic anhydrase located in the pulmonary interstitium. More likely, the enzyme is localized to the vascular endothelium and catalyzes CO_2 reactions inside the exchange vessels. The bolus experiments performed by Effros et al. (1978) are particularly supportive of this possibility. A precedent for endothelial localization of an enzyme exists in the demonstrated association of angiotension-converting enzyme with the pulmonary endothelium (Ryan et al., 1972). Carbonic anhydrase has been localized to reptilian vascular endothelium (Fain and Rosen, 1973), but the histochemical technique utulized has been criticized (Muther, 1972).

3 End-Capillary pH, P_{CO_2}

The degree of catalysis of CO_2 reactions in the capillary will determine the extent, if any, that slow pH changes occur in vivo. This problem is currently being investigated in our laboratory by studying pulmonary CO_2 excretion in the presence of an excess of carbonic anhydrase (Klocke, 1980).

3.1 Methods

Lungs from New Zealand white rabbits (mean weight 1.0 kg) were washed free of blood and perfused with buffer (102.5 mM NaCl, 23.5 mM $NaHCO_3$, 15.5 mM Na_2HPO_4, 4.5 mM NaH_2PO_4, 4.0 mM KCl) using previously reported methods (Klocke, 1978). Buffer was equilibrated with 5.4% carbon dioxide, 6.9% acetylene, 40% oxygen, balance nitrogen. Buffer and lungs were maintained at 37°C. Lungs were perfused serially with three buffers identical in pH (7.407) and P_{CO_2} (36.8 torr) (Fig. 1). After control measurements, the lungs were perfused with buffer containing 20 mg/l bovine carbonic anhydrase (Sigma Chemical Co.), followed by perfusion with buffer containing 250 mg/l acetazolamide. There was no recirculation of buffer at any time. Lungs were ventilated with a rodent respirator (Harvard Apparatus Co.).

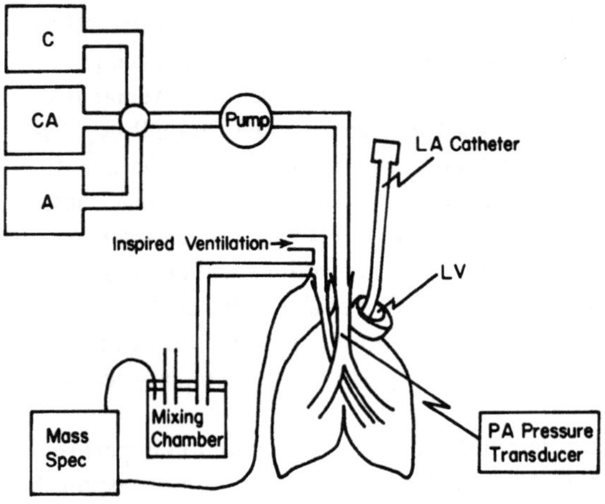

Fig. 1. Experimental apparatus. Buffer reservoirs are labeled *C* for control, *CA* for carbonic anhydrase and *A* for acetazolamide. Venous effluent drained from the transected left ventricle (*LV*) and was sampled by the left atrial (*LA*) catheter. Pulmonary artery (*PA*) pressure was monitored by pressure transducer and mixed expiratory and end-tidal CO_2 concentrations by mass spectrometry (*Mass Spec*)

Mean expired and end-tidal CO_2 concentration was monitored by mass spectrometry. Acetylene concentration in mixed expired gas was quantitated by gas chromatography. The quantity of dissolved CO_2 excreted in the lungs was predicted from simultaneous measurements of acetylene excretion (Klocke, 1978). The remainder of the observed CO_2 excretion was attributed to conversion of bicarbonate to carbon dioxide. Gas excretion was corrected for loss by diffusion through the pleural surfaces (Klocke, 1978). In six experiments a left atrial catheter was placed for anaerobic sampling of drainage from the pulmonary veins. Hemoglobin concentration in the effluent was calculated from optical density readings at 412 nm.

3.2 Results

Steady-state CO_2 excretion in ten isolated lungs is illustrated in Fig. 2. In the upper panel, observed and predicted CO_2 excretions during control and carbonic anhydrase experiments are plotted as a function of perfusion rate. The difference between the paired predicted and observed excretions represents the quantity of CO_2 formed from bicarbonate ion. In controls the volume of carbon dioxide produced from bicarbonate in the pulmonary capillaries averaged 97.2% of the carbonic anhydrase values. This difference is statistically significant ($p < 0.02$, paired t test). CO_2 production in both control and carbonic anhydrase experiments was significantly greater ($p < 0.001$) than the paired acetazolamide values (Fig. 2, lower panel). Hemoglobin concentration in the venous effluent averaged 9.8 (± 5.8 SD) x 10^{-5} mM (tetramer).

The end-capillary P_{CO_2} values expected if equilibrium were reached were determined from in vitro tonometry. Known quantities of equilibrated buffer and air were mixed in 30 ml syringes and agitated at 37°C for 90 min. The ratios of gas to liquid were chosen to encompass the span of $\dot{V}_A/\dot{Q}$ ratios present during the experiments. Alveolar ventilation in the experimental lungs was calculated from minute ventilation, and mixed expired and end-tidal CO_2 tensions. The latter was used as an approximation of alveolar P_{CO_2}. The solid line in Fig. 3 is fitted to the P_{CO_2} values

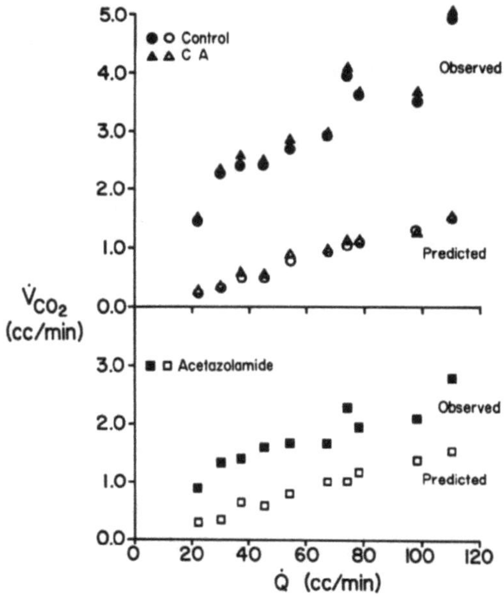

Fig. 2. Carbon dioxide excretion, $\dot{V}_{CO_2}$, as a function of perfusion, $\dot{Q}$. *Upper panel,* control and carbonic anhydrase (*CA*) experiments are indicated by *circles* and *triangles* respectively. *Lower panel,* acetazolamide experiments are represented by *squares. Open symbols* indicate predicted CO_2 excretion; *closed symbols* represent observed CO_2 excretion (Data in figure from Klocke, 1980)

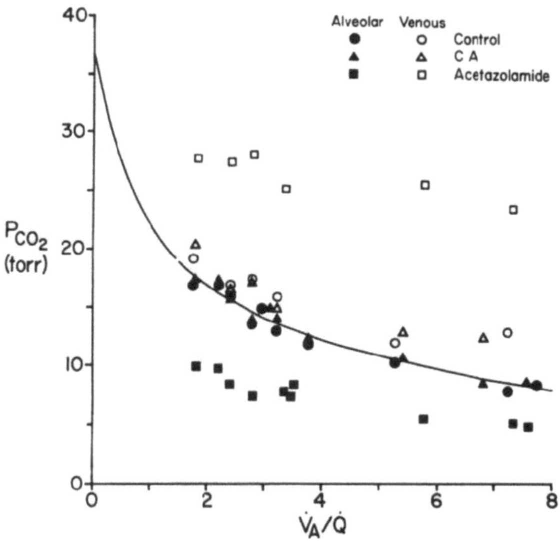

Fig. 3. Alveolar and pulmonary venous P_{CO_2} values as a function of ventilation-perfusion ratio, $\dot{V}_A/\dot{Q}$. The *solid line* is fitted to nine in vitro determinations (not shown) of the equilibrium relationship of buffer P_{CO_2} to $\dot{V}_A/\dot{Q}$. Symbols for experimental data are given in the legend in the figure (Data in figure from Klocke, 1980)

from the tonometered samples; none of the nine measurements (not shown) deviated more than 0.5 torr from this line. Alveolar (end-tidal) samples are shown by the filled symbols; pulmonary venous values are indicated by the open symbols. Alveolar samples for both the control (circles) and carbonic anhydrase (triangles) experiments lie along the line of equilibrium, indicating that CO_2 exchange was essentially complete as perfusate left the capillary bed. Points from the acetazolamide experiments (squares) are well below the line of equilibrium. The pulmonary venous points lie quite close, but slightly above, the equilibrium line in all but the acetazolamide experiments. The alveolar points were obtained during exchange and are representative of actual tensions present in perfusate at the end of the capillary. The pulmonary venous points were determined by electrode measurement after completion of the experiments, and are the values present in perfusate which has reached equilibrium after leaving the capillary. Perfect end-capillary equilibrium would result in both alveolar and venous points lying on the line of equilibrium. As the degree of equilibrium attained in the capillary decreases, the alveolar and venous points move away from each other and the line of equilibrium.

3.3 Discussion

The decrement in steady-state CO_2 excretion produced by the addition of acetazolamide to the perfusate is similar to previous results (Klocke, 1978). This indicates that the perfusate has access to lung carbonic anhydrase during CO_2 exchange. Since the quantity of CO_2 excreted in the expired ventilation decreases with enzyme inhibition, the enzyme must be present in the capillary bed. The possibility that enzyme activity is derived from trapped red cells is unlikely in view of previous measurements of enzyme activity and the low concentration of hemoglobin in the venous effluent. The latter measurements indicate that the maximum possible contamination of perfusate with blood would only increase CO_2 reaction rates by 10%. This would account for only 1/15 of the CO_2 production observed in the control experiments.

Addition of carbonic anhydrase to the perfusate increased CO_2 excretion significantly from a statistical viewpoint. However, the increase over control values was less than 3% and is of no physiologic importance. Under the conditions of these experiments, for practical purposes CO_2 reactions can be considered to have reached equilibrium without addition of carbonic anhydrase. This conclusion is strengthened by comparison of CO_2 tensions in the isolated lung experiments to those obtained in the in vitro equilibrium work. The alveolar CO_2 tensions in the control and carbonic anhydrase experiments are essentially the same as those obtained under equilibrium conditions (Fig. 3). The comparable pulmonary venous values lie slightly above the equilibrium line, but these values are subject to the effect of intrapulmonary shunting or nonuniform matching of ventilation and perfusion. A shunt of a few percent of the total perfusion is all that is required to produce such minor discrepancies since the initial P_{CO_2} of the perfusate is much higher than the alveolar value.

The tensions obtained with acetazolamide in the perfusate are quite distant from the equilibrium line, indicating that carbonic anhydrase activity is necessary to attain chemical equilibrium during capillary transit. The alveolar points are well below the line of equilibrium because the small quantity of CO_2 dissolved in the perfusate was diluted in a comparatively large volume of alveolar ventilation. The slow natural rate of bicarbonate dehydration did not permit enough generation of CO_2 to raise the alveolar tension to the equilibrium value. As a result, relatively little CO_2 was excreted, and the P_{CO_2} gradually rose after the perfusate left the lung until it reached a value relatively close to the initial P_{CO_2} in the perfusate.

Other estimates of the degree of equilibrium reached in the pulmonary capillary have varied widely. Calculations from our previous work indicated that only 20%–59% of total exchange was attained by the end of the capillary at normal pH and P_{CO_2} (Klocke, 1978). Effros et al. (1978) used a simple model and concluded that CO_2 exchange was 86% complete in the lung. Both approaches required critical assumptions in the calculations. The data of Crandall and O'Brasky (1978) support the present findings. They observed no slow pH changes in the absence of carbonic anhydrase inhibition, but the lag in transit time between the exchange bed and the measuring pH electrode did not permit a definition of the end-capillary situation. In contrast, both Hill et al. (1977) and Bidani and Crandall (1978) observed small slow pH changes in intact dogs and cats. However, the changes observed averaged only 0.01 or less, well below the expected alterations. It is not clear whether this represents a species difference, the result of different experimental conditions, or possible artifact despite the precautions taken. Certainly under the conditions of our experiments, it appears that slow pH changes do not occur in the pulmonary vasculature. However, this does not preclude slow pH alterations in the venous drainage of other organs.

Acknowledgments. This work was supported by Grant HL-15194 from the National Heart, Lung and Blood Institute. The author appreciates the technical assistance of Mrs. Anne Coe, Miss Maureen Cannon and Mrs. Kate Fields. Mrs. Patricia Baier and Mrs. Marsha Barber aided in the preparation of the manuscript. Acetazolamide was supplied as Diamox by Lederle Laboratories.

References

Adamson TM, Waxman BP (1976) Carbonate dehydratase (carbonic anhydrase) and the fetal lung. In: Lung liquids. American Elsevier, New York, pp 221–234

Berfenstam R (1952) Carbonic anhydrase activity in fetal organs. Acta Paediatr 41: 310–315

Bidani A, Crandall ED (1978) Slow postcapillary pH changes in blood in anesthetized animals. J Appl Physiol 45: 674–680

Bidani A, Crandall ED, Forster RE (1978) Analysis of postcapillary pH changes in blood in vivo after gas exchange. J Appl Physiol 44: 770–781

Crandall ED, O'Brasky JE (1978) Direct evidence for participation of rat lung carbonic anhydrase in CO_2 reactions. J Clin Invest 62: 618–622

Crandall ED, Klocke RA, Forster RE (1971) Hydroxyl ion movements across the human erythrocyte membrane. Measurement of rapid pH changes in red cell suspensions. J Gen Physiol 57: 664–683

Crandall ED, Bidani A, Forster RE (1977) Postcapillary changes in blood pH in vivo during carbonic anhydrase inhibition. J Appl Physiol 43: 582–590

DuBois AB (1969) Significance of carbonic anhydrase in lung tissue. In: Forster RE, Edsall JT, Otis AB, Roughton FJW (eds) CO_2: Chemical, biochemical and physiological aspects. NASA, Washington DC, pp 257–259

Effros RM, Chang RSY, Silverman P (1978) Acceleration of plasma bicarbonate conversion to carbon dioxide by pulmonary carbonic anhydrase. Science 199: 427–429

Fain W, Rosen S (1973) Carbonic anhydrase activity in amphibian and reptilian lung: a histochemical and biochemical analysis. Histochem J 5: 519–528

Farhi LE, Plewes JL, Olszowka AJ (1976) Lung carbonate dehydratase (carbonic anhydrase). CO_2 stores and CO_2 transport. In: Lung liquids. Elsevier, Amsterdam, pp 235–249

Forster RE, Crandall ED (1975) Time course of exchanges between red cells and extracellular fluid during CO_2 uptake. J Appl Physiol 38: 710–718

Hill EP, Power GG, Longo LD (1973) Mathematical simulation of pulmonary O_2 and CO_2 exchange. Am J Physiol 224: 904–917

Hill EP, Power GG, Gilbert RD (1977) Rate of pH changes in blood plasma in vitro and in vivo. J Appl Physiol 42: 928–934

Jacobs MH, Stewart DR (1942) The role of carbonic anhydrase in certain ionic exchanges involving the erythrocyte. J Gen Physiol 25: 539–552

Klocke RA (1978) Catalysis of CO_2 reactions by lung carbonic anhydrase. J Appl Physiol. 44: 882–888

Klocke RA (1980) Equilibrium of CO_2 reactions in the pulmonary capillary. J Appl Physiol (in press)

Maren TH (1967) Carbonic anhydrase: chemistry, physiology, and inhibition. Physiol Rev 47: 597–781

Muther T (1972) A critical evaluation of the histochemical methods for carbonic anhydrase. J Histochem Cytochem 20: 319–330

Plewes JL, Olszowka AJ, Farhi LE (1976) Amount and rates of CO_2 storage in lung tissue. Respir Physiol 28: 359–370

Roughton FJW (1935) Recent work on carbon dioxide transport by the blood. Physiol Rev 15: 241–296

Ryan JW, Smith U, Niemeyer RS (1972) Angiotensin I: metabolism by plasma membrane of lung. Science 176: 64–66

Sirs JA (1970) The interaction of carbon dioxide with the rate of exchange of oxygen by red blood cells. In: Hershey D (ed) Blood oxygenation. Plenum Press, New York, pp 116–136

Pulmonary Capillary Carbonic Anhydrase

R.M. EFFROS, L. SHAPIRO, P. SILVERMAN, and M. LIEBER [1]

During the past two years, we have reported a series of studies which indicate that the conversion of plasma HCO_3^- to CO_2 in the pulmonary vasculature is accelerated by carbonic anhydrase (c.a.) in the lung tissues [5, 6, 8]. In these experiments, injections of $H^{14}CO_3^-$ or $^{14}CO_2$ were made into the pulmonary artery of isolated or in situ rabbit lungs which were perfused with solutions that did not contain carbonic anhydrase. The outflow patterns of the ^{14}C labels in fluid collected from the left atrium were quite similar under control circumstances. However, when the carbonic anhydrase inhibitor, acetazolamide, was added to the perfusion solution, transit times of $H^{14}CO_3^-$ through the lungs were decreased and approached those of ^{22}Na and ^{125}-I-albumin, whereas the transit times of $^{14}CO_2$ through the lungs were considerably delayed and greater than those of tritiated water.

We calculated that equilibration between HCO_3^- and CO_2 is 90% complete during a single circulation. These observations led us to suggest that c.a. is associated with the pulmonary endothelium in a manner similar to that of angiotensin converting enzyme [12].

Klocke [10] found that acetazolamide diminished the excretion of CO_2 relative to acetylene in a similar preparation but concluded that equilibration between HCO_3^- and CO_2 was only 14%–26% complete at normal flows. Subsequently Crandall and O'Brasky, using a pH electrode, confirmed our observation that equilibration in the plasma is nearly complete in the isolated lung [3]. However, both Crandall et al. [9] and Hill et al. [4] have continued to report that a pH disequilibrium persists in the arterial blood of anesthetized dogs, which would suggest that conversion of HCO_3^- to CO_2 is by no means complete in the intact animal.

We have recently completed studies which indicate that (1) the excretion of CO_2 from the lungs is not normally limited by the amount of c.a. in the lung tissues and (2) c.a. is present in the lungs of intact rabbits. In the first set of experiments, injections of $H^{14}CO_3^-$ or $^{14}CO_2$, or an equilibrated mixture at a pH of 7.4 were made into the pulmonary artery line of an in situ lung in a manner similar to that described previously [5, 6, 8]. However, on this occasion we collected expired gas and left atrial outflow in rubber bags and calculated the excretion fraction by dividing the amount of $^{14}CO_2$ in the gas bag by the sum of that in the gas and left atrial bags. The excretion fractions of $H^{14}CO_3^-$ and $^{14}CO_2$ were then divided by the excretion fraction of the equilibrated mixture obtained under control conditions in each experiment to obtain the "relative excretion fraction" of each. Under control circumstances, the

1 Harbor-U.C.L.A. Medical Center, 1000 West Carson, Torrance, California 90509/USA

relative excretion fractions (R.E.F.) of $H^{14}CO_3^-$ and $^{14}CO_2$ were very similar (R.E.F.$_{\cdot HCO_3^-}$ = 1.12 ± 0.07, R.E.F.$_{\cdot CO_2}$ = 1.17 ± 0.05, r = 9, S.E.M.). After acetazolamide (20 mg/dl), R.E.F.$_{\cdot HCO_3^-}$ = 0.60 ± 0.26 and R.E.F.$_{\cdot ^{14}CO_2}$ = 1.96 ± 0.22. Furthermore, addition of bovine c.a. (10 mg/dl) had no effect upon the excretion fraction of the equilibrated mixture (R.E.F.$_{\cdot e}$ = 1.06 ± 0.07, n = 4). These data indicate that there is enough accessible c.a. in the lungs to ensure that equilibration between HCO_3^- and CO_2 is essentially complete in a single passage. The apparent deficiency in the amount of CO_2 excreted and that expected from the excretion of acetylene in the experiments of Klocke may therefore be due to some factor other than the amount of c.a. in the lungs.

In a separate group of studies, we replaced the $H^{14}CO_3^-$ in the injection solution with a cocktail containing ^{125}I-albumin, $^{22}Na^+$, ^{3}H-acetazolamide and ^{14}C-antipyrine (a "water" indicator). The activity of ^{3}H-acetazolamide was sufficiently high to permit us to use a dose of only 70 picomoles in each injection. In addition, 5 g/dl albumin in the perfusion fluid was replaced with 5 g/dl dextran (M.W. 70,000) to avoid plasma protein binding of ^{3}H-acetazolamide. As indicated in Fig. 1, the mean transit time of ^{3}H-acetazolamide was considerably greater than that of the water indicator (^{14}C-antipyrine) under control conditions. This delay is consistent with binding to some site within the tissue and since tissue binding to c.a. is considered quite specific, this provides additional evidence that c.a. is accessible to plasma in the isolated lung. After addition of only 0.1 mg/dl of unlabeled acetazolamide to the perfusion solution, the outflow pattern of the labeled inhibitor became quite similar to that of $^{22}Na^+$ as well as ^{125}I-albumin, indicating that when the tissue binding sites are saturated, the volume of distribution of acetazolamide is only slightly greater than vascular volume (Fig. 2).

Of particular interest was the observation that ^{3}H-acetazolamide is also bound to pulmonary tissue sites in the intact, anesthetized rabbit (Fig. 3), despite the presence

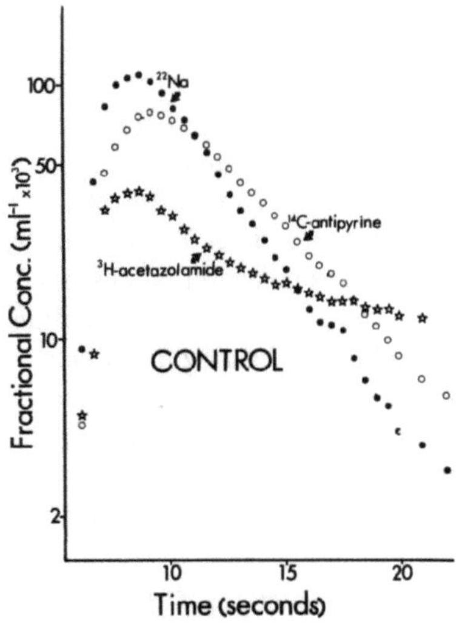

Fig. 1. Pulmonary outflow patterns of ^{22}Na, ^{14}C-antipyrine (a "water" indicator) and ^{3}H-acetazolamide from perfused, in situ rabbit lung. The outflow of ^{3}H-acetazolamide is considerably delayed in comparison to the other indicators

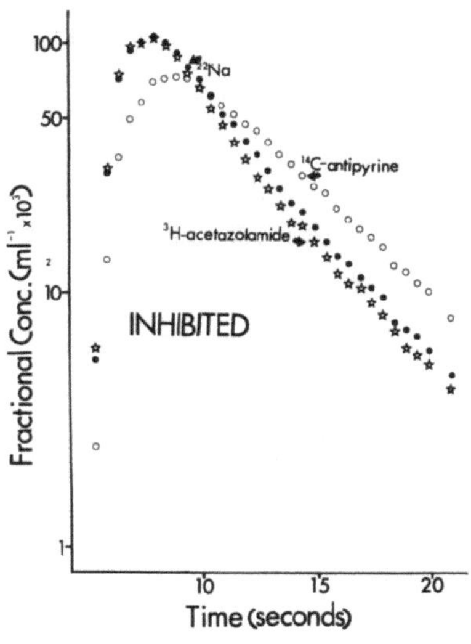

Fig. 2. Pulmonary outflow patterns of the same preparation after 7 mg/l unlabeld acetazolamide added to the perfusion solution. Note that the ^{3}H-acetazolamide now resembles that of ^{22}Na

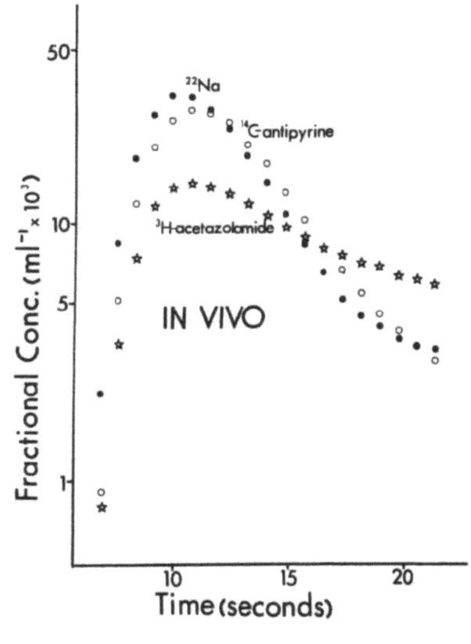

Fig. 3. Arterial outflow patterns of the same indicators in an anesthetized rabbit. Despite the presence of red cells in the rabbit vasculature, binding of ^{3}H-acetazolamide to the pulmonary tissue, presumably endothelium, is evident

of red cells. The entry of acetazolamide into red cells is known to be relatively slow [11], requiring more than 5 min to be 50% complete in studies we have performed with the labeled compound. Access to the tissue c.a. appears to be significantly faster than access to c.a. in erythrocytes, and it is likely that the tissue enzyme is bound to the outer surface of cells, whereas red cell enzyme is strictly intracellular. When hemolyzed rabbit red cells or bovine c.a. was added to the injection, binding of ^{3}H-acetazolamide to the tissue site could no longer be measured, presumably because most had bound to c.a. in the perfusate. Additional support for the hypothesis that c.a. is bound to the external surface of endothelial cells has recently been provided by U. Ryan and W. Ryan (personal communication) who have found c.a. associated with the membrane fractions of pure bovine pulmonary artery endothelial cell cultures.

It is possible that the disequilibrium described by Hill et al. [9] and Crandall et al. [4] is related to c.a. inhibitors which have been described in sheep and dogs but are absent in rabbits and man [11]. Another factor which could contribute to the slow changes in arterial pH observed by these investigators would be mixing of blood from various regions of the lung with different CO_2 tensions in large vessels where access to endothelial enzyme is limited.

Since c.a. appears to be present in the vasculature of the leg as well as the lung [7], some consideration must be given to the role of the enormous quantities of c.a. within red cells. By catalyzing equilibration between HCO_3^- and CO_2, erythrocyte c.a. should accelerate movement of HCO_3^- across the erythrocyte membrane in exchange for Cl^-. In this way, the considerable buffering capacity of hemoglobin can be effectively used to aid in the transport of CO_2 between peripheral tissues and the lung. However, rapid stabilization of pH, P_{CO_2}, and HCO_3^- in the plasma requires the presence of the endothelial enzyme.

References

1. Bidani A, Crandall ED (1978) Analysis of P_{CO_2} differences during rebreathing due to slow pH equilibration in blood. J Appl Physiol Resp Environ Exercise Physiol 45: 666–673
2. Bidani A, Crandall ED, Forster RE (1978) Analysis of postcapillary pH changes in blood in vivo after gas exchange. J Appl Physiol Respir Environ Exercise Physiol 44: 770–781
3. Crandall ED, O'Brasky JE (1978) Direct evidence for participation of rat lung carbonic anhydrase in CO_2 reactions. J Clin Invest 70: 618–622
4. Crandall ED, Bidani A, Forster RE (1977) Postcapillary changes in blood pH in vivo during carbonic anhydrase inhibition. J Appl Physiol Respir Environ Exercise Physiol 43: 582–590
5. Effros RM (1978a) Pulmonary carbonic anhydrase and the conversion of plasma bicarbonate to carbon dioxide. In: Fitzgerald RS, Gautier H, Lahiri S (eds) The regulation of respiration during sleep and anesthesia. Plenum Press, New York, pp 255–268
6. Effros RM (1978b) Pulmonary carbonic anhydrase and the release of carbon dioxide from the blood. Trans Assoc Am Phys Phil 91: 186–196
7. Effros RM, Weissman ML (1978) Carbonic anhydrase activity of cat hindleg. Physiologist 21: 33
8. Effros RM, Chang RSY, Silverman P (1978) Acceleration of plasma bicarbonate to carbon dioxide by pulmonary carbonic anhydrase. Science 199: 427–429
9. Hill EP, Power GG, Gilbert RD (1977) Rate of pH changes in blood plasma in vitro and in vivo. J Appl Physiol Respir Environ Exercise Physiol 42: 928–934
10. Klocke RA (1978) Catalysis of CO_2 reactions by lung carbonic anhydrase. J Appl Physiol Respir Environ Exercise Physiol 44: 882–888
11. Maren TH (1967) Carbonic anhydrase: chemistry, physiology, and inhibition. Physiol Rev 47: 595–781
12. Ryan JW, Smith U, Niemeyer RS (1972) Angiotensin I: Metabolism by plasma membrane of lung. Science 176: 64–66

Can CO_2 Be Actively Eliminated by the Lungs?

G.H. GURTNER and R.J. TRAYSTMAN [1]

A few years ago we presented our observation that in dogs, during rebreathing under conditions of no gas exchange, the alveolar P_{CO_2} was higher than the P_{CO_2} in the mixed venous blood (Gurtner et al., 1969). The phenomenon had been originally described by Bohr in 1891 who found that alveolar P_{CO_2} was higher than arterial P_{CO_2} when 5% to 10% CO_2 was present in the inspired air. A similar observation had also been made by Jones et al. (1967) during rebreathing experiments in exercising man a few years prior to our observations. Subsequently the pehnomenon has been reported by several groups of investigators including Guyatt et al. (1973) who did experiments similar to our own, and by Jennings and Chen (1975) who found results similar to those of Bohr. Recently, the phenomenon has been demonstrated to exist during normal gas exchange by Robertson and Hlastala (1977).

Two hypotheses have been proposed to explain the P_{CO_2} differences. These are, our model, the charged membranes hypothesis (Gurtner et al., 1969; Gurtner, 1972, 1977) and the delayed equilibration hypothesis (Forster and Crandall, 1975).

We proposed the charged membrane hypothesis to explain the P_{CO_2} differences between alveolus and mixed venous blood during rebreathing. In brief, the mechanism of the step-up in P_{CO_2} involves a transient chemical disequilibrium between H^+ and HCO_3^- and resultant production of CO_2 in the vicinity of a negatively charged capillary wall. The disequilibrium is caused by the H^+ dissociation of polyelectrolytes, such as proteins, when exposed to an electrical field. This phenomenon, called the Wien effect, is due to a rapid movement of counterions such as H^+ away from the protein molecule due to the electrostatic attraction of the negatively charged capillary wall. Since proteins are complex weak acids, the loss of H^+ in the microenvironment of the protein molecule causes the protein, acting as a buffer, to dissociate more H^+ ions. The dissociated H^+ combines with HCO_3^-, producing CO_2 in the vicinity of the wall. This higher P_{CO_2}, at the capillary wall, is the CO_2 tension which is important in gas exchange across the alveolar capillary membrane. The theory predicts that the P_{CO_2} differences should be directly related to the H^+ activity and HCO_3^- concentration of mixed venous blood. These predictions agree well with the experimental results (Gurtner et al., 1969). The proposed mechanism is not specific for CO_2 but should cause differences in the uncharged form of other weak acids and bases as well; such differences did occur for several weak acids and bases (Gurtner, 1972).

---

1 Depts. of Medicine and Environmental Health Sciences, The Johns Hopkins Medical Institutions, Baltimore, Maryland 21205/USA

The delayed equilibration hypothesis describes the relatively slow changes in P_{CO_2}, H^+, and HCO_3^- which occur after blood leaves the tissue or pulmonary capillary. These changes result from the absence of the enzyme carbonic anhydrase in the plasma causing the rate-limiting process in equilibration of H^+, HCO_3^-, and CO_2 across the red blood cell membrane to be the uncatalyzed reaction between CO_2 and water outside the red blood cell. No attempt will be made here to describe the sequence of events; however, these reactions and ionic exchanges result in a small decrease in blood P_{CO_2} after the blood leaves the pulmonary capillary. The time required to reach equilibrium is on the order of seconds and it has been argued that CO_2 equilibrium is not achieved in the time taken for blood to travel from tissue to the lung (Forster and Crandall, 1975).

The charged membrane hypothesis can explain all of the experimental results which were described above, however the delayed equilibration hypothesis can only explain the results of the rebreathing experiments. The latter hypothesis predicts that during gas exchange, in the presence or absence of CO_2 in the inspired air, the P_{CO_2} of blood increases after leaving the pulmonary capillary. Therefore, since the hypothesis predicts that there will always be a negative A-a gradient, it cannot explain the positive A-a gradient observed by Bohr (1891) and by Jennings and Chen (1975) during CO_2 breathing or the results of Robertson and Hlastala (1977). These results would appear to falsify the delayed equilibration hypothesis as being the sole cause of the phenomenon; however even though the authors went to some length to verify the accuracy of their measurements, the magnitude of the CO_2 gradients was small. Because of this we felt it worthwhile to carry out an experiment similar to that of Jennings and Chen, but with much higher inspired CO_2 tensions. In these experiments we found very large gas to blood P_{CO_2} differences, the magnitude of which were similar to the predictions of the charged membrane hypothesis.

In the course of these experiments we found that the HCO_3^- in arterial blood did not increase during exposure to CO_2 as much as predicted from the results for blood exposed to CO_2 in vitro. This is further evidence that in vivo CO_2 buffering is different than in vitro buffering, and may indicate movement of HCO_3^- from extracellular to intracellular fluid during hypercapnia.

1 Methods

The experiments were carried out on mechanically ventilated dogs (tidal volume 300–400 ml, frequency 15–20 breaths per min.) which had been anesthetized with pentobarbital (30 mg/kg initially, with 2 mg/kg/h subsequently) and paralyzed with succinyl choline (2 mg/kg). The animals were made hypercapnic by stepwise addition of CO_2 in O_2 to the inspired air.

Catheters had been placed in the femoral artery and pulmonary artery for blood sampling and pressure measurement. An electromagnetic flow probe was placed around the thoracic aorta for measurement of aortic blood flow. It was possible to increase the inspired CO_2 fraction to about 40% without adverse effects on the heart. At higher levels cardiac failure occurred manifested by decreasing cardiac output, decreased arterial pressure and increased pulmonary artery pressure. The cardiac failure was reversed by administration of epinephrine. The cardiovascular response to severe hypercapnia has been reported elsewhere (Traystman et al., 1976).

Temperature was measured in the pleural space and in the rectum using approximately calibrated mercury thermometers, no temperature differences were observed between the two sites in any of the experiments.

Gas and blood P_{CO_2} and pH measurements were made on an Instrumentation Laboratories electrode (model 113). The P_{CO_2} electrode was calibrated in the range of 100–700 torr. The P_{CO_2} electrode was calibrated with 100% CO_2 and known gas mixtures containing approximately 80%, 60%, 40%, and 20% CO_2. Over the entire range the electrode output was linearly related to the log P_{CO_2} as would be expected. Samples of dog blood were also tonometered using the calibration mixtures at the temperature of the electrode; there was no measured differences in P_{CO_2} between blood and gas at each level. The pH electrode was calibrated in the range 6.5 to 7.0 pH units. HCO_3^- was calculated from the measured P_{CO_2} and pH using a Severinghaus slide rule (1963). This is discussed in detail below. Blood gas measurements were made at 37°C; the dog's body temperature was maintained within 1/2 a degree of the measurement temperature by using a heat lamp and an electric heating pad under the animal. Each P_{CO_2} level was maintained for at least 30 min; the blood and gas measurements were made at the end of the interval. Three ml blood samples were drawn in 5 ml glass syringes, the dead space of which had been filled with a heparin solution (Upjohn 1000 IU/ml). Previous experiments had shown that dilution of the blood with the heparin contained in the dead space did not measurably affect the P_{CO_2} of the sample. The gas samples were collected in 50 ml syringes from the expired line of the ventilator at end expiration, the syringe was flushed two to three times with end tidal gas. The final sample was collected over several breaths.

2 Results

Figure 1 shows the steady-state P_{CO_2} differences divided by the arterial HCO_3^- plotted against the arterial pH. The solid lines indicate the magnitude of the normalized P_{CO_2} differences, predicted by the charged membrane hypothesis, for two values of electrical potential difference between the wall phase and bulk phase of capillary blood. As discussed in an earlier paper, these values represent the lower limit of possible potentials which could give the P_{CO_2} differences at any certain pH (Gurtner, 1972). The data from our rebreathing experiments are plotted as triangles; the present data is plotted as closed circles. Note that the magnitudes of the P_{CO_2} differences are similar to the predictions of the charged membrane hypothesis and appear to agree with our earlier published results using rebreathing techniques.

The blood gas data in Table 1 shows an interesting relationship between the arterial (HCO_3^-) concentration and P_{CO_2}. This is plotted in Fig. 2. The solid lines indicate different values of base excess and were taken from the nomogram of Siggard-Andersen (1963) for blood exposure to CO_2 in vitro. In our experiments the blood HCO_3^- did not increase as much as it would if the CO_2 equilibration had taken place in vitro. In the control situation the blood base excess was near zero. As the P_{CO_2} is increased base excess becomes more and more negative, reaching a value of – 30 mEq at the highest P_{CO_2} levels.

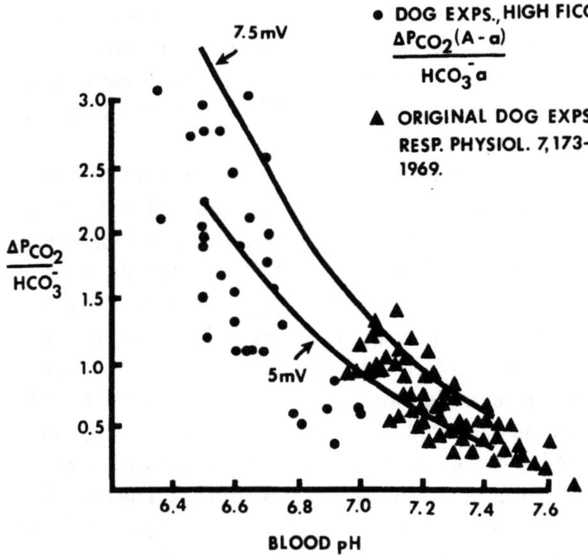

Fig. 1. the relationship between P_{CO_2} normalized by dividing by the blood HCO_3^- and the blood pH. In the present study (*closed circles*) ΔP_{CO_2} represents the A-a P_{CO_2} difference and the HCO_3^- and pH used are values for arterial blood. In our previous work (Gurtner et al., 1969) (*solid triangles*) ΔP_{CO_2} represents alveolar-mixed venous differences and HCO_3^- and pH values are those of mixed venous blood. The *solid lines* represent the relationship predicted by the charged membrane hypothesis for two different values for potential difference between wall and bulk phase of capillary blood. The experimentally determined relationship is similar to the predictions of the theory

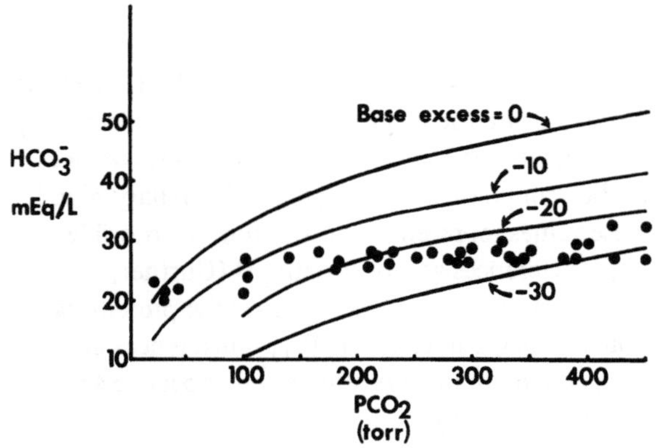

Fig. 2. Arterial HCO_3^- concentration is plotted aginst arterial P_{CO_2}. The *circles* represent our measurements. The *solid lines* give the relationship between for blood esposed to CO_2 in vitro. The observed increase in blood HCO_3^- was less than the in vitro increase in HCO_3^-. This data supports the hypothesis that in vivo and in vitro buffering of CO_2 by blood is markedly different (see text)

Table 1. Steady-state results during hypercapnia

	P_ACO_2	$P_{\bar{V}}CO_2$	$PaCO_2$	$\frac{PaCO_2}{PACO_2}$	pHa	HCO_{3a}	$\frac{\Delta P_{CO_2}\ (A\text{-}a)}{HCO_{3a}}$
			32.5		7.385	20	
(1)	117	117	103	0.8803	6.994	21	0.67
	195	180	180	0.923	6.789	25.3	0.59
	255	236	227	0.890	6.689	25.5	1.1
	315	274	286	0.908	6.608	26.4	1.1
	325	311	295	0.908	6.579	26.5	1.13
	380	350	342	0.9	6.530	26.5	1.43
	455	415	391	0.859	6.478	27	2.0
	465	425	425	0.913	6.471	27	1.5
(2)	35.8	51.6	43.3		7.302	21	
	122	116	107	0.877	6.981	24.0	0.625
	195	179	180	0.923	6.801	26.2	0.534
	278	250	230	0.827	6.706	27	2.7
	335	305	298	0.889	6.625	28.8	1.52
	364	324	320	0.88	6.591	28	1.57
	385	380	350	0.909	6.578	28.6	1.22
	520	480	457	0.878	6.498	32.5	1.91
(3)	16	39	31.8		7.444	21.8	
	210	157	165	0.78	6.870	28.2	0.627
	400	250	245	0.6125	6.688	27.4	2.55
	350	275	263	0.751	6.699	28	3.1
	330	245	315	0.955	6.590		0.058
	350	293	290	0.829	6.650	28.0	2.11
	440	322	330	0.750	6.592	30	2.47
	485	395	399	0.823	6.500	28.8	2.9
	475	392	395	0.832	6.499	28.8	2.75
(4)		27.5	21.5		7.640	23.3	
	130	129	121	0.930	6.979	27	0.333
	247	214	207	0.838	6.725	25.4	1.575
	332	286	291	0.876	6.619	22.8	1.79
	390	360	345	0.885	6.546	27.2	1.65
	456	380	405	0.888	6.470	27.5	1.85
	500	495	415	0.83	6.351	27	3.14
(5)	159		138	0.866	6.920	27	0.82
	249		214	0.859	6.751	27.4	1.296
	315		284	0.901	6.620	27	1.13
	407		333	0.818	6.555	27	2.74
	556		482	0.867	6.438	27	2.74
	560		498	0.889	6.345	29	214

$\bar{x} = 0.861 \pm 0.011$

3 Discussion

The factors which can cause spurious P_{CO_2} differences are well known and include differences in calibration between instruments and differences in calibration characteristics between blood and gas on the same instrument. We believe that we have experimentally eliminated these possibilities as an explanation for the observed P_{CO_2} gradient.

Temperature differences between the animal and the measuring electrode can lead to large apparent gas to blood P_{CO_2} differences if the electrode temperature is lower than the body temperature. We were careful to control the animals' body temperature within 0.5°C of the electrode temperature. It is interesting to calculate the temperature difference required to produce the sort of gradient which we observed. The mean ratio of arterial P_{CO_2} to alveolar P_{CO_2} in Table 1 was 0.861 ± 0.011. In order to produce an apparent P_{CO_2} difference of this magnitude a temperature difference of 3.4°C would be required (Severinghaus, 1966). This sort of temperature difference did not occur in our experiments.

The present experiments confirm the findings of Jennings and Chen (1975) and the results shown in Fig. 1 indicate that the magnitude of the normalized P_{CO_2} differences during severe hypercapnia are consistent with our earlier observations and can be explained by the charged membrane hypothesis.

During the control measurements arterial and venous P_{CO_2} was higher than mixed expired P_{CO_2} as would be expected; however even under these circumstances the P_{CO_2} is higher than would be predicted by the excretion of inert gases. The experimental results of Robertson and Hlastala (1977), those of Jennings and Chen and the present experiments indicate that the mechanism causing the P_{CO_2} differences is operating to move CO_2 from blood to gas during gas exchange as well as under circumstances, such as rebreathing, in which there is no net CO_2 flux. As mentioned above, all of these observations are consistent with the predictions of the charged membrane hypothesis but the results of the measurements made during gas exchange are exactly opposite to the prediction of the delayed equilibration hypothesis. The present experimental results add further evidence that delayed equilibration cannot solely explain the blood to gas P_{CO_2} differences observed by several different groups to investigators.

Recently the experiments of Jennings and Chen (1975) have been repeated in anesthetized dogs with different results (Scheid et al., 1978). The authors failed to demonstrate a negative a-A gradient and concluded that the results of Jenning and Chen (1975) might be in error, even though no specific errors could be identified in the methodology of Jennings and Chen (1975) which appeared to be of similar accuracy to that of Scheid et al. (1978). An alternative explanation seems more likely. It is well known that anesthetized animals and humans are proven to develop atelectasis which is reversible with periodic hyperinflation of the lung. It is also known that controlled respiration in an anesthetized animal or human is associated with increased dispersion in the distribution of pulmonary $\dot{V}/\dot{Q}$ ratios. Since it seems likely that there would be an increased physiological dead space in Scheid's anesthetized dogs, it follows that the failure of these authors to observe a negative a-A gradient could be explained on this basis. Because Scheid et al. had no independent method to measure the effect of $\dot{V}/\dot{Q}$ distribution on expired P_{CO_2}, their results do not represent convincing evidence that such P_{CO_2} differences do not occur in unanesthetized animals (such as those used by Jennings and Chen) in which the dispersion of $\dot{V}/\dot{Q}$ should be minimal.

Robertson and Hlastala (1977) divised their ingenious inert gas method to measure independent the effects of $\dot{V}/\dot{Q}$ distribution on the expired and arterial P_{CO_2}. In anesthetized dogs they consistently demonstrated negative a-A P_{CO_2} gradients when the effects of $\dot{V}/\dot{Q}$ distribution were accounted for.

The results of the present experiments also demonstrate large negative a-A P_{CO_2} differences during severe hypercapnia which are too large to explain by experimental error. Therefore, even though we have no doubt that the measurements of Scheid et al. were accurate, we conclude that their experimental design was not appropriate for the demonstration of the gas to blood P_{CO_2} differences observed by the other groups of investigators.

The failure of the arterial HCO_3^- concentration to increase as much as expected is another demonstration that acid base relationships of blood are markedly different when CO_2 exposure is done in vitro or in vivo. This phenomenon has been previously observed by several investigators and discussed in detail (Brown and Michel, 1968; Cohen et al., 1964). To our knowledge no other measurements have been made at the very high CO_2 levels reported here. The relationship between base excess and arterial P_{CO_2} in our experiments is similar to previously reported results, in that base excess becomes more negative as arterial P_{CO_2} is increased. The explanation may involve movement of HCO_3^- from extracellular to intracellular space. Since the buffering capacity of blood is higher than that of tissue more HCO_3^- is generated with exposure to hypercapnia. The HCO_3^- so formed could cause disequilibrium between HCO_3^- in the extra- and intracellular space and movement of HCO_3^- down to electrochemical gradient into the tissues could cause the changes in blood HCO_3^-. The present experiments do not demonstrate movement of HCO_3^-, into tissue, only HCO_3^- disappearances from blood. Because of the experiment's design, the animal was killed after the highest CO_2 levels was reached, therefore the present data is only suggestive of such HCO_3^- movement but not conclusive.

We believe that the HCO_3^- concentration calculated at the high P_{CO_2} levels are accurate. In some experiments we tonometered dog blood, withdrawn during normocapnia with gas containing large fractional concentrations of CO_2. We found that the calculated HCO_3^- agreed well with the Siggard-Andersen nomogram.

References

Bohr C (1891) Über die Lungenathmung. Scand Arch Physiol 2: 236–268

Brown EB, Michel CC (1968) Whole body CO_2 exchange. Proc Int Union Physiol Sci 6: 185–186

Cohen JJ, Brackett Jr NC, Schwartz WB (1964) The nature of the carbon dioxide titration curve in the normal dog. J Clin Invest 43: 777–786

Forster RE, Crandall ED (1975) Time course of exchanges between red cells and extracellular fluid during CO_2 uptake. J Appl Physiol 38: 710–718

Gurtner GH (1972) Nonequilibrium steady-state differences in partial pressure of CO_2 and in concentration of weak acids and bases between blood and tissue. Biophys J 12: 597–608

Gurtner GH (1977) Controversy: Can alveolar PCO_2 exceed pulmonary end capillary CO_2? J Appl Physiol 42: 323–328
Gurtner GH, Song SH, Farhi LE (1969) Alveolar to mixed-venous PCO_2 difference under conditions of no gas exchange. Respir Physiol 7: 173–187
Guyatt AR, Yu CJ, Lutherer B, Otis AB (1973) Studies of alveolar-mixed venous CO_2 and O_2 gradients in the rebreathing dog lung. Respir Physiol 17: 178–194
Hill EP, Power GG, Longo LD (1973) Mathematical simulation of pulmonary O_2 and CO_2 exchange. Am J Physiol 224: 904–947
Jennings DB, Chen CC (1975) Negative arterial mixed expired PCO_2 gradient during acute and chronic hypercapnia. J Appl Physiol 38: 382–388
Jones NL, Campbell EJM, McHardy GJR, Higgs BE, Clode M (1967) The estimation of carbon dioxide pressure of mixed venous blood during exercise. Clin Sci 32: 311–327
Robertson HT, Hlastala MP (1977) Elevated alveolar PCO_2 relative to predicted values during normal gas exchange. J Appl Physiol 43: 357–364
Scheid P, Meyer M, Piiper J (1978) Gas/blood CO_2 equilibration in dog lungs in hypercapnia. Pflügers Arch Suppl 37: R 21
Severinghaus JW (1966) Blood gas calculator. J Appl Physiol 21: 1108–1116
Siggard-Andersen O (1963) Blood acid base alignment nomogram. Scand J Clin Lab Invest 15: 211–217
Traystman RJ, Gurtner GH, Permutt S (1976) Effect of severe hypercapnia on the circulatory system. Circulation 54: 11–13

Experimental Results on Gas-Blood CO_2 Equilibrium in Lungs

J. PIIPER, M. MEYER, and P. SCHEID [1]

Three experimental studies were undertaken in our laboratory with the aim of reproducing and checking previous experimental findings which had indicated that blood-gas CO_2 equilibration in lungs led to a condition in which P_{CO_2} in alveolar gas was higher than in blood.

1. Since 1968, several groups have reported that in rebreathing equilibrium of CO_2 (and of O_2) P_{CO_2} in rebreathing gas is higher than in mixed venous or arterial blood (reviewed by Piiper, 1979). Scheid et al. (1972) attempted to reproduce these findings in anesthetized dogs and in isolated lung lobes of dogs, using the same methods which had been applied in previous investigations. The unexpected result was that with all methods, and in all experimental conditions, blood P_{CO_2} was found to be identical with gas P_{CO_2} in rebreathing equilibrium.

2. Jennings and Chen (1975) reported negative P_{CO_2} differences between arterial blood and mixed expired gas in dogs subjected to acute or chronic hypercapnia. We reproduced identical hypercapnia levels in anesthetized dogs in our laboratory, but found that P_{CO_2} in arterial blood was never lower than P_{CO_2} in mixed expired or end-expired (alveolar) gas (Scheid et al., 1978, 1979).

Since a very high accuracy is required in such measurements and a number of directional errors tending to produce negative blood-gas P_{CO_2} differences, both in rebreathing and steady state conditions, can be listed, we are inclined to believe that the findings of "paradoxical" P_{CO_2} differences are artifacts.

3. Davies and Dutton (1975) reported data showing that in chickens breathing hypercapnic mixtures in steady state both arterial and mixed venous P_{CO_2} were lower than end-tidal P_{CO_2}. The authors believed that this finding could only be explained by assuming that blood-gas equilibration in lungs led to an equilibrium in which P_{CO_2} in blood was lower than in gas. This result was reproduced in our laboratory by Meyer et al. (1976). Further experimentation and model calculations revealed, however, that these findings were explainable on the basis of a particular role of the Haldane effect in the cross-current-type parabronchial lungs of birds. Assumptions of a negative P_{CO_2} difference between end-capillary blood and gas of single elements of the system was unnecessary and contrary to findings in rebreathing equilibrium.

1 Abteilung Physiologie, Max-Planck-Institut für experimentelle Medizin, 34 Göttingen/FRG

References

Davies DG, Dutton RE (1975) Gas-blood P_{CO_2} gradients during avian gas exchange. J Appl Physiol 39: 405–410

Jennings DB, Chen CC (1975) Negative arterial-mixed expired P_{CO_2} gradient during acute and chronic hypercapnia. J Appl Physiol 38: 382–388

Meyer M, Worth H, Scheid P (1976) Gas-blood CO_2 equilibration in parabronchial lungs in birds. J Appl Physiol 41: 302–309

Piiper J (1979) Blood-gas equilibration of CO_2 in pulmonary gas exchange of mammals and birds. Physiologist, in press

Scheid P, Teichmann J, Adaro F, Piiper J (1972) Gas-blood CO_2 equilibrium in dog lungs during rebreathing. J Appl Physiol 33: 582–588

Scheid P, Meyer M, Piiper J (1978) Gas/blood CO_2 equilibration in dog lungs in hypercapnia. Pflügers Arch Suppl 377: R21

Scheid P, Meyer M, Piiper J (1979) Arterial-expired P_{CO_2} differences in the dog during acute hypercapnia. J Appl Physiol, submitted

Intracapillary CO_2 Gradients

R.M. EFFROS [1]

The appearance of a number of reports that CO_2 tensions in the alveoli may exceed those in the pulmonary artery and vein have led to several hypotheses which are intended to explain such a gradient [17, 11, 18, 8]. Much attention has been given to the proposal of Gurtner et al. that this phenomenon could be explained on the basis of a charged membrane hypothesis [11, 13]: Hydrogen ions are dissociated from serum protein molecules under the influence of the negative charge on the endothelial cell surfaces and they are attracted to the capillary wall. Bicarbonate ions are repelled from the endothelium but at a slower rate. Consequently carbonic acid concentrations are elevated and CO_2 is formed in the region near the capillary wall and diffuses into the alveoli, raising alveolar P_{CO_2}.

Our original objection to this hypothesis was based on the amount of energy which would be required to maintain a gradient of CO_2 in the pulmonary capillaries [1]. Any increase in CO_2 tensions near the capillary wall should be rapidly dissipated in the plasma phase. Gurtner proposed that diffusion of CO_2 away from the endothelium could be slowed by red cells passing through the vessels [12]. Work by Roughton, Forster, and Silverman seems to show that the red cell membranes are less permeable to CO_2 than might be expected in aqueous or lipid solutions [20, 7, 23]. However, even if the red cells provide a barrier to diffusion, numerous problems remain:

1. Red cells flowing through the capillaries appear to remain about one μ away from the capillary wall and could not maintain a P_{CO_2} gradient in the intervening plasma [10, 19]. CO_2 diffusion across this layer of plasma would quickly eliminate the proposed gradient [2].

2. If, on the contrary, it were assumed that the red cell membrane very closely approached the endothelial cell, the amount of HCO_3^- in the volume between these surfaces could become a limiting factor [2]. Since the hypothesis assumes that CO_2 does not readily escape from this region, it would not be appropriate to assume that additional bicarbonate would be transferred from between red cells to this same volume.

3. Convection near the capillary wall and between and around red cells may be sufficient to eliminate local H^+, HCO_3^-, or CO_2 gradients.

4. It is by no means clear that the charge on the endothelium is any greater than that on red cells [21, 22]. Since convection near the red cell surface may be as great as that near the endothelial surface, CO_2 generation by the proposed mechanism might be equal at both surfaces.

1 Harbor-U.C.L.A. Medical Center, 1000 West Carson, Torrance, California, 90509/USA

5. If red cell membranes were sufficiently impermeable to impede CO_2 movement, then CO_2 generated within red cells should be released to the plasma slowly and a rise in P_{CO_2} might therefore occur in the pulmonary veins.

6. There is no evidence that CO_2 diffusion across the alveolar capillary wall is more rapid than diffusion back into the center of the vessel. Consequently, even if P_{CO_2} is greater near the endothelial surface, this would not necessarily elevate alveolar P_{CO_2} above capillary P_{CO_2}.

7. Calculations by Forster suggest that although the movement of HCO_3^- away from the wall is only one-eighth as rapid as the movement of H^+ toward the wall, an equilibrium distribution of H^+ and HCO_3^- near the wall would be obtained within 10^{-9} s and excess CO_2 generation at the periphery of the capillary would not be significant [8].

Forster also argued that there was no evidence for carbonic anhydrase near the endothelial surface to accelerate the production of CO_2 and the uncatalyzed rate of conversion would be too slow to produce an elevated P_{CO_2} in this region [8]. In addition, Forster and Crandall and several other investigators have suggested that the absence of carbonic anhydrase on red cell surfaces, plasma and endothelium should result in a slow equilibration of hydrogen ion in concentrations between red cells and plasma following gas exchange [6, 9, 14, 15, 24]. One consequence of this disequilibrium would be a slow rise in the P_{CO_2} of venous blood draining the peripheral tissues. In those experimental preparations in which gas exchange was avoided, this gradual increase in P_{CO_2} could continue in the blood traversing the pulmonary vasculature and the arterial P_{CO_2} could, therefore, exceed alveolar P_{CO_2}.

This mechanism would not explain those experiments in which gas exchange was permitted and alveolar P_{CO_2} appeared to exceed pulmonary arterial or venous P_{CO_2} since a decline in P_{CO_2} would be predicted in blood leaving the lungs after CO_2 exchange had occurred. Furthermore, recent studies suggest that carbonic anhydrase is readily accessible to plasma in both the lungs and legs and may be associated with the external surface of the endothelial cells [3a, 3b, 4, 5]. However, the location of this enzyme suggests an alternative manner in which CO_2 tensions might be elevated at the endothelial wall. If the conversion of plasma HCO_3^- arriving in the pulmonary capillaries to CO_2 is specifically catalyzed by the endothelial enzyme, P_{CO_2} would increase near the wall. This could only occur if gas exchange were permitted and would not explain the gradients reported in which exchange was avoided. Furthermore, the considerations outlined above make it unlikely that a significant difference between alveolar and capillary P_{CO_2} would be created in this fashion. Further speculation in this regard would only be justified if a consensus could be reached that such gradients do exist in some experimental situation.

References

1. Effros RM (1972) Pulmonary capillary carbon dioxide gradients and the Wien effect. J Appl Physiol 32: 221–222
2. Effros RM (1973) Red cells and CO_2 gradients within pulmonary capillaries. J Appl Physiol 34: 740

3a. Effros RM (1978) Pulmonary carbonic anhydrase and the conversion of plasma bicarbonate to carbon dioxide. In: Fitzgerald RS, Gautier H, Lahiri S (eds) The regulation of respiration during sleep and anesthesia. Plenum Press, New York, pp 255–268
3b. Effros RM (1978) Pulmonary carbonic anhydrase and the release of carbon dioxide from the blood. Trans Assoc Am Phys Phil 91: 186–196
4. Effros RM, Weissman ML (1978) Carbonic anhydrase activity of cat hindleg. Physiologist 21: 33
5. Effros RM, Chang RSY, Sliverman P (1978) Acceleration of plasma bicarbonate to carbon dioxide by pulmonary carbonic anhydrase. Science 199: 427–429
6. Filley GF, Heineken FG (1976) A blood gas disequilibrium theory. Br J Dis Chest 70: 223–245
7. Forster RE (1969) The rate of CO_2 equilibration between red cells and plasma in CO_2. In: Forster RE, Edsall RT, Otis AB, Roughton FJW (eds) Chemical, biochemical and physiological aspects. NASA SP-188, Washington DC, pp 275–284
8. Forster RE (1977) Can alveolar P_{CO_2} exceed pulmonary end-capillary P_{CO_2}? J Appl Physiol 42: 326–328
9. Forster RE, Crandall ED (1975) Time course of exchanges between red cells and extracellular fluid during CO_2 uptake. J Appl Physiol 38: 710–718
10. Goldsmith HL (1968) Microrheology of blood cell suspension. J Gen Physiol 52: 5s–28s
11. Gurtner GH (1972) Nonequilibrium steady-state differences in partial présence of CO_2 and in concentration of weak acids and bases between blood and tissue. Biophys J 12: 597–608
12. Gurtner GH (1973) Energy requirements for intracapillary P_{CO_2} gradients. J Appl Physiol 34: 740
13. Gurtner GH, Song SH, Farhi LE (1969) Alveolar to mixed venous P_{CO_2} difference under conditions of no gas exchange. Respir Physiol 7: 173–287
14. Hill EP, Power GG, Lonzo LD (1973) Mathematical simulation of pulmonary O_2 and CO_2 exchange. Am J Physiol 224: 904–917
15. Hill EP, Power GG, Gilbert RD (1977) Rate of pH changes in blood plasma in vitro and in vivo. J Appl Physiol 42: 928–934
16. Hochmuth RM, Marple RN, Sutera SP (1970) Capillary bllod flow. I. Erythrocyte deformation in glass capillaries. Neurovasc Res 2: 409–419
17. Jones NL, Campbell EJM, Edwards RHT, Wilkoff WG (1969) Alveolar-to-blood P_{CO_2} difference during rebreathing in exercise. J Appl Physiol 27: 356–360
18. Lazlo G, Clark TJH, Pope H, Campbell EJM (1971) Differences between alveolar and arterial P_{CO_2} during rebreathing experiments in resting human subjects. Respir Physiol 12: 36–52
19. Miyamoto T, Moll W (1971) Measurements of dimensions and pathway of red cells in rapidly frozen lungs in situ. Respir Physiol 12: 141–156
20. Roughton FJW (1959) Simultaneous chemical reaction velocity in hemoglobin solutions and red cell suspensions. Prog Biophys Biophys Chem 9: 55–104
21. Sawyer PN, Himmelfarb EH (1965) Studies of streaming potentials in large mammalian blood vessels in vivo. In: Sawyer PN (ed) Biophysical mechanisms in vascular haemeostasis and intravascular clotting. Appleton-Century Crafts, New York, pp 69–79
22. Seaman GVP (1975) Electrokinetic behavior of red cells. In: MacN Surgenor F (ed) The red blood cell, 2nd ed. Academic Press, London New York, pp 1135–1229
23. Silverman DN, Tu C, Wynns GC (1976) Depletion of ^{18}O from $C^{18}O_2$ in erythrocyte suspensions. The permeability of erythrocyte membrane to CO_2. J Biol Chem 251: 4428–4436
24. Sirs JA (1970) The interaction of carbon dioxide with the rate of exchange of oxygen by red cells. In: Hershey D (ed) Blood oxygenation. Plenum Press, New York, pp 116–136

Contribution of Red Cell-Plasma H^+ Disequilibrium to Alveolar-Blood P_{CO_2} Differences During Rebreathing

E.D. CRANDALL and A. BIDANI [1]

In recent years, a number of investigators have observed that arterial and/or mixed venous P_{CO_2} can be lower than alveolar P_{CO_2} during conditions when no CO_2 is exchanged in the lung capillaries. The range of values reported for these differences has been exceptionally wide. For example, in one laboratory, careful studies with several different preparations failed to reveal any P_{CO_2} differences between alveolar gas and blood [7]. Other laboratories have reported that the P_{CO_2} differences can become quite large during exercise [6] or acidosis [5]. The wide range of reported values probably reflects the many technical difficulties involved in obtaining such data. Several explanations have been proposed for these differences, but none has been entirely accepted to date.

In vitro and in vivo experiments [3, 4] have suggested the possibility that electrochemical equilibration of hydrogen ions within blood might not be complete during pulmonary and systemic capillary transit. As a result, slow changes in blood P_{CO_2} would continue to occur after capillary exit. A mathematical model describing reaction and transport events in blood after CO_2 exchange was developed (assuming no carbonic anhydrase activity in plasma) and yielded results that agreed well with the experimental observations on the slow pH changes [2]. Computations predicted that P_{CO_2} in arterial blood falls if no CO_2 exchange takes place within the lung, but oxygen uptake continues. This occurs because the Bohr protons released upon oxygenation of hemoglobin cause a disequilibrium for hydrogen ions across the red cell membrane within the pulmonary capillaries, and the equilibration process leads to a fall in P_{CO_2} [1]. These computational results are illustrated in Fig. 1 for normal man at rest. The rise in P_{CO_2} between mixed venous blood and alveolar gas is due predominantly to the production of the Bohr protons, leading to some dehydration of carbonic acid intracellularly. The difference between alveolar P_{CO_2} and arterial P_{CO_2} is due to the subsequent equilibration process of hydrogen ions across the red cell membrane [1]. For the case where no CO_2 or O_2 is exchanged across the lung capillaries, the results of the computations are shown in Fig. 2. Note that once again alveolar gas P_{CO_2} is predicted to be greater than either mixed venous or arterial blood P_{CO_2}. This observation is due to the fact that P_{CO_2} of venous blood falls after oxygen leaves the blood and carbon dioxide enters it in the systemic capillaries [1]. When no gas is exchanged in the lung, the fall in P_{CO_2} continues to completion (equilibration) after the blood has passed through the lung capillaries. As a result, alveolar P_{CO_2} ends up higher than both arterial and mixed venous blood P_{CO_2}, which are equal to

1 University of Pennsylvania, Philadelphia, PA. 19104/USA

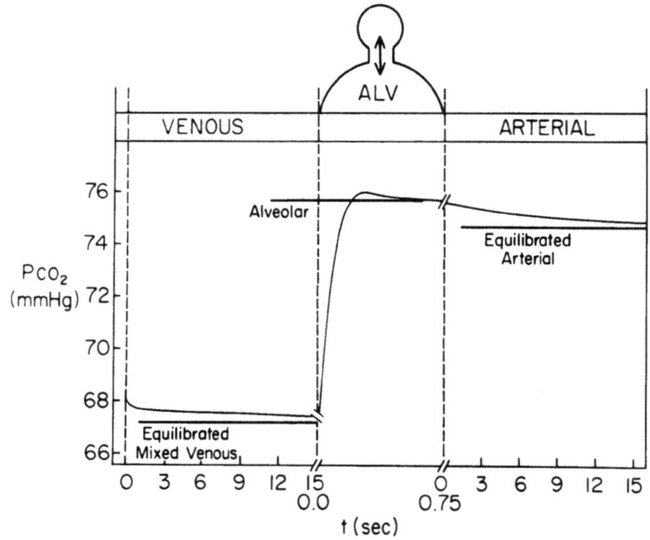

Fig. 1. Time course of P_{CO_2} in a unit volume of blood during one passage through veins, lung capillaries, and arteries in a resting animal. Blood does not exchange CO_2 in the lungs, although oxygen uptake takes place. Equilibrated mixed venous P_{CO_2}, equilibrated arterial P_{CO_2}, and alveolar P_{CO_2} are all indicated. Alveolar P_{O_2} equals 100 torr

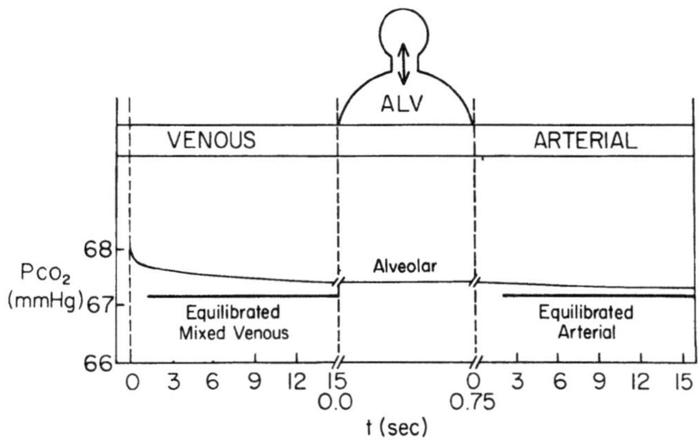

Fig. 2. Time course of P_{CO_2} in a unit volume of blood during one passage through veins, lung capillaries, and arteries in a resting animal. Blood does not exchange CO_2 or O_2 in the lungs. Equilibrated mixed venous P_{CO_2}, equilibrated arterial P_{CO_2}, and alveolar P_{CO_2} are all indicated. Alveolar P_{O_2} equals 100 torr

each other. Computations were also performed for the case of exercise rebreathing with and without oxygen uptake, resulting in larger predicted P_{CO_2} differences. The results of the computations and the reported values for P_{CO_2} differences summarized from the literature are shown in Table 1.

Table 1. Predicted alveolar gas-blood P_{CO_2} differences during rebreathing

	Rest		Exercise	
	O_2 uptake	No O_2 uptake	O_2 uptake	No O_2 uptake
AV P_{CO_2} difference, torr	8.5	0.3	25.7	1.3
Range of reported measured mean values		0–7		0–6
Aa P_{CO_2} difference, torr	1	0.3	3.2	1.3
Range of reported measured mean values	0–3	0–1	3–10	4–6

From Table 1, it can be seen that for almost every case studied, the predicted P_{CO_2} differences lie within the range of experimentally measured values for corresponding conditions. However, the magnitudes of predicted differences are always at the low end of the range. If all the experimentally measured values are valid, then the disequilibrium phenomenon can only account for a fraction of the reported P_{CO_2} differences. Since the range of reported values is quite wide and some controversy still exists about their meaning, the contribution of other mechanisms to the creation of the P_{CO_2} differences remains to be determined. In any case, it is important to recognize that the disequilibrium phenomenon leads to P_{CO_2} differences that are always positive between alveolar gas P_{CO_2} and either mixed venous P_{CO_2} or arterial P_{CO_2} measured at equilibrium in a blood-gas machine. The differences are predicted to be greater with exercise, and are considerably greater when oxygen uptake takes place. However, the disequilibrium phenomenon, at least as currently understood, cannot explain positive P_{CO_2} differences under most conditions of net carbon dioxide elimination in the lung capillaries.

References

1. Bidani A, Crandall ED (1978) Analysis of pCO_2 differences during rebreathing due to slow pH equilibration in blood. J Appl Physiol 45: 666–673
2. Bidani A, Crandall ED, Forster RE (1978) Analysis of post-capillary pH changes in blood in vivo after gas exchange. J Appl Physiol 44: 770–781
3. Crandall ED, Bidani A, Forster RE (1977) Post-capillary changes in blood pH in vivo during carbonic anhydrase inhibition. J Appl Physiol 43: 582–590
4. Forster RE, Crandall ED (1975) Time course of exchange between red cells and extracellular fluid during CO_2 uptake. J Appl Physiol 38: 710–718
5. Gurtner GH, Song SH, Farhi LE (1969) Alveolar to mixed venous pCO_2 difference under conditions of no gas exchange. Respir Physiol 7: 173–187
6. Jones NL, Campbell EJM, Edwards RHP, Wilkoff WG (1969) Alveolar-to-blood pCO_2 difference during rebreathing in exercise. J Appl Physiol 27: 356–360
7. Scheid P, Teichmann J, Adaro F, Piiper J (1972) Gas-blood CO_2 equilibration in dog lungs during rebreathing. J Appl Physiol 33: 582–588

Diffusion Limitation in Alveolar-Capillary CO_2 Transfer in Human Lungs: Experimental Evidence from Rebreathing Equilibration

J. PIIPER, M. MEYER, and P. SCHEID [1]

1 Introduction

It is generally held that diffusional equilibration of CO_2 is much faster than that of O_2 as Krogh's diffusion constant is about 20 times higher for CO_2 than for O_2. Hence, since in normoxia pulmonary O_2 exchange is generally considered to be only slightly limited by diffusion, there should be no significant diffusion limitation in alveolar-capillary CO_2 transfer.

However, gas equilibration in a model capillary is not determined by the diffusive conductance (or diffusing capacity), D, alone, but by the ratio of D to convective conductance $\dot{Q}\beta$ ($\dot{Q}$ = flow rate, β = effective solubility or slope of blood dissociation curve), $D/\dot{Q}\beta$ (Piiper et al., 1971; Piiper and Scheid, 1975). For CO_2, the β value is much higher than for O_2, due to the much steeper dissociation curve of CO_2. Therefore, the $D/\dot{Q}\beta$ ratio must be less than 20 times higher for CO_2 than for O_2. Moreover, the release of CO_2 from blood is a complex, composite process involving at least two elemental processes known to be inherently slow: (a) the bicarbonate/chloride exchange between red cells and plasma, and (b) the very slow dehydration of carbonic acid in plasma due to absence of carbonic anhydrase (Hill et al., 1973; Forster and Crandall, 1976; Bidani and Crandall, 1978). The limitations would further reduce the effective $D/\dot{Q}\beta$ ratio.

Therefore, it is not unreasonable to look for diffusion (+ reaction) limitation in capillary-alveolar CO_2 transfer. In steady-state gas exchange, however, it is not reasonable to try to identify diffusion limitation in face of potential alveolar dead space ventilation giving rise to sizable arterial-to-alveolar P_{CO_2} differences. A more promising approach is provided by the analysis of the equilibration kinetics of alveolar gas with mixed venous blood on the basis of a model comprising both diffusion and perfusion limitation.

Hyde et al. (1968) applied a breath-holding method in which the kinetics of $^{13}CO_2$ equilibration was compared with that of acetylene. Both at rest and during exercise the diffusion-reaction limitation effect was either very small or not discernible, the diffusing capacity for CO_2 being estimated to exceed 200 ml $\cdot$ min^{-1} $\cdot$ torr^{-1}. A clear diffusion-reaction limitation for $^{13}CO_2$ equilibration could only be established after inhibition of carbonic anhydrase by Diamox, when effective D_{CO_2} fell to 42 ml $\cdot$ min^{-1} $\cdot$ torr^{-1}.

The functional inhomogeneity of lungs, due to unequal distribution of ventilation, perfusion, and diffusing capacity to lung volume leading to local diversity of

1 Abteilung Physiologie, Max-Planck-Institut für experimentelle Medizin, D-3400 Göttingen/FRG

alveolar P_{CO_2} and P_{O_2}, is generally recognized to be one of the major obstacles to an accurate assessment of alveolar-capillary diffusion. Since rebreathing is expected to reduce local differences in alveolar gas composition, we adapted a rebreathing technique, previously elaborated for measurement of D_{O_2} (Adaro et al., 1973; Cerretelli et al., 1974; Veicsteinas et al., 1976), for simultaneous measurement of the disappearance of acetylene and of $^{13}CO_2$ from lungs, allowing calculation of the diffusing capacity for CO_2, D_{CO_2}.

2 Methods

2.1 Principle

The two-compartment model comprising an alveolar compartment (volume V_A, gas partial pressure P_A) and a rebreathing bag (volume V_R, gas partial pressure P_R) is ventilated in closed system (effective ventilation $\dot{V}_{eff}$). The alveolar compartment has gas transfer contact across a diffusion barrier (diffusive conductance or diffusing capacity D) with blood (flow rate $\dot{Q}$). Further important parameters are the "capacitance coefficients" (β_g for gas phase; β for blood, denoting solubility for inert gases, and slope of the dissociation curve for O_2 or CO_2; see Piiper et al., 1971).

For the alveolar-mixed venous blood transfer conductance (transfer rate/alveolar-to-mixed venous partial pressure difference), $G = \dot{M}/(P_A - P_{\bar{v}})$, the following equation is obtained:

$$G = V_A\beta_g \cdot k \cdot \left(1 + \frac{V_R/V_A}{1 - k \cdot V_R/\dot{V}_{eff}}\right) \tag{1}$$

in which k is the rate constant of exponential approach of P_A to $P_{\bar{v}}$:

$$k = -\frac{d\,[\ln (P_A - P_{\bar{v}})]}{dt} \tag{2}$$

For gases whose transfer is exclusively limited by perfusion (assumedly all inert gases), the conductance G is

$$G_{(perf)} = \dot{Q}\beta \tag{3}$$

For gases with transfer limited both by perfusion and diffusion (assumedly CO_2), the conductance G is

$$G_{(perf + diff)} = \dot{Q}\beta \cdot (1 - e^{-\frac{D}{\dot{Q}\beta}}) \tag{4}$$

The procedure consists in determining $\dot{Q}$ from the rate constant for acetylene, $k_{C_2H_2}$, using Eqs (1) and (3), and in solving Eqs. (1) and (4) for D_{CO_2}, using k_{CO_2} and $\dot{Q}$.

For soluble gases, the effective alveolar dilution volume, V_A, is considerably higher than the alveolar gas volume (determinable e.g., from insoluble gas dilution) due to solution of gas in lung tissue and in pulmonary capillary blood. Furthermore, for CO_2 the tissue solubility is much higher than the physical solubility of CO_2 as buffering leads to formation or decomposition of bicarbonate. The effective dilution volume for each gas can be determined by the extrapolation procedure of Cander and Forster (1959).

2.2 Advantages of Isotopic Gases

In principle, D can be determined from the equilibration kinetics of the naturally abundant isotopes of CO_2 (and O_2). However, use of rare (stable) isotopes offers important technical advantages (cf. Piiper et al., 1979) which are particularly important, and practically indispensable, in the case of CO_2 mainly for the following reasons.

1. Partial pressure in mixed venous blood, $P_{\bar{v}}$, which is the asymptote for equilibration of P_A, can be made practically zero.

2. The slope of the blood dissociation curve, β, is generally variable. But the effective β for transfer of a rare isotope (e.g., $^{13}CO_2$) is constant and equal to the ratio $C_{\bar{v}}/P_{\bar{v}}$ of the abundant isotope, provided this major abundant isotope is in rebreathing equilibrium and P_A for the rare isotope is small compared with P_A ($= P_{\bar{v}}$) of the abundant isotope.

3. Determination of the effective dilution volume for CO_2 would be much too inaccurate when based on changes of the normal, abundant CO_2 isotope.

2.3 Experimental Procedure

The subjects were sitting on a bicycle ergometer, resting or pedalling against a work load of 75 W. They were allowed a pre-period (4–5 min) of open circuit breathing prior to the rebreathing maneuver. After two rapid priming breaths, taken within less than 2 s from a spirometer containing an oxygen-free hypercapnic mixture (about 8% CO_2 in N_2), the subject rebreathed for 15–18 s a mixture of 0.07% $^{13}CO_2$, 1% He, 1% C_2H_2, 3% O_2, and about 7.5% $^{12}CO_2$ in N_2, starting at or below FRC. The concentrations of O_2 and $^{12}CO_2$ in the rebreathing mixture were adjusted to rapidly bring alveolar P_{O_2} and P_{CO_2} close to their respective values in mixed venous blood ($P_{\bar{v}}$) and to provide equilibrium between lung gas and inflowing blood for both gases over the rebreathing period before onset of recirculation. Rebreathing was performed at a frequency of 50/min aided by a metronome. On exercise the pedalling rate imposed was equal to the rebreathing rate. The mean tidal volume was kept at 2 l.

2.4 Technical Design

The stable isotope $^{13}CO_2$ (^{13}C enrichment 94.5 atom %) was obtained in a pressurized cylinder (Prochem B.O.C. Ltd., London, England). A septum-sealed injector was fitted to the cylinder allowing uncontaminated withdrawal of small amounts of gas with a gas-tight Hamilton syringe. About 1.8 ml of isotopic gas was added to the rebreathing gas mixture supplied by precision gas mixing pumps from ordinary sources to yield the appropriate concentration when the bag was filled up to 2.5 l.

Gas partial pressures were continuously monitored by a 5-channel variable magnetic sector respiratory mass spectrometer (a redesigned model of Varian M3) sampling at a rate of 6 ml/min next to the mouth of the subject. The collector plates were set to read the following masses: 45 ($^{13}CO_2$), 32 (O_2), 26 (C_2H_2), 22 ($^{12}CO_2$), 4 (He).

The relative natural abundance of ^{13}C, 1.11 atom % of total carbon isotopes, produced a signal at mass 45 that could not be neglected. It was taken into account by recording $^{12}CO_2$ at mass 22 to derive an equivalent electrical signal to compensate for its contribution at mass 45. The mass spectrometer signal from $^{13}CO_2$ thus corrected was displayed on a separate recorder channel.

The mass spectrometer output was stored on magnetic tape and subsequently recorded, using an X-Y recorder, on graph paper in a format allowing convenient reading.

3 Results

Measurements were performed on three healthy male subjects (28 to 33 of age, body weight 72 to 83 kg). The following mean D_{CO_2} values were found: at rest, 180 ml · min^{-1} · torr^{-1}; during exercise (75 W), 300 ml · min^{-1} · torr^{-1}.

4 Discussion

4.1 Comparison of D_{CO_2} with D_{O_2}

In a comparable (in part identical) group of subjects the pulmonary diffusing capacity for O_2, D_{O_2}, was measured by a similar rebreathing technique, both at rest and during exercise, using the oxygen isotope $^{18}O_2$. The following average values for D_{O_2} were found (in ml · min^{-1} · torr^{-1}): at rest, 55; during exercise, 70. Thus one obtains for the ratio,

$$D_{CO_2}/D_{O_2} \begin{cases} = 3.3 \text{ at rest,} \\ = 4.3 \text{ during exercise.} \end{cases}$$

The values of this conductance ratio may be compared to values expected on theoretical grounds considering as mechanisms underlying the conductances: (1) diffusion through tissue barriers, (2) diffusion in alveolar gas, and (3) chemical reactions and ionic exchange between red cells and plasma.

1. If the resistance corresponding to the diffusing capacity represented mainly a resistance to diffusion of CO_2 and O_2 molecules across a tissue barrier between alveolar gas and pulmonary capillary blood, the ratio D_{CO_2}/D_{O_2} should be close to the ratio of Krogh's diffusion constants, i.e., about 20. Since the actual D ratio is much smaller, some other processes are expected to be involved in limiting CO_2 transfer.

2. The role of diffusion limitation in lung gas has recently received much attention in investigation into alveolar gas exchange. There is experimental evidence for incomplete mixing of inspired gas with lung gas (= stratification) limiting the efficiency of alveolar gas exchange (cf. Piiper, 1979). It is difficult to predict the effects of stratification in rebreathing experiments. However, if there is stratification, all gases should be involved. In particular, removal of inert gases (like acetylene) from lungs would be a mixed diffusion and flow-limited process, and the $\dot{Q}$ calculated therefrom by neglecting diffusion limitation would be an "effective" $\dot{Q}$ (smaller than the true $\dot{Q}$). The diffusion coefficients of $^{13}CO_2$, $^{18}O_2$ and C_2H_2 should not differ by more than 25% as the molecular weights for these gases are similar. Therefore, one may conclude that at least a large part of stratification effects is expected to be compensated by the procedure of comparing the simultaneously measured rate constants of CO_2, O_2, and C_2H_2.

3. The next limiting factors to be considered are chemical reactions and exchange processes between red cells and plasma, initiated by uptake of O_2 and release of CO_2.

The kinetics of oxygenation of hemoglobin in red cells has been characterized by a specific conductance, Θ_{O_2}, defined as O_2 uptake rate of red cells from well-stirred plasma per unit P_{O_2} difference per unit blood volume. There is no agreement as to the value of Θ_{O_2} which has been reported to equal (in ml $\cdot$ min^{-1} $\cdot$ torr^{-1} $\cdot$ ml^{-1}) 2.7 (Staub et al., 1962), 1.5 (Holland et al., 1977) and 0.9 (Mochizuki, 1966). Multiplication with the pulmonary capillary volume, Q_c, yields the O_2 uptake conductance of pulmonary capillary blood (= $\Theta_{O_2} \cdot Q_c$). Pulmonary capillary blood volume in man has been estimated at 100–150 ml. Thus $\Theta_{O_2} \cdot Q_c$ is expected to be in the range of 90–400 ml $\cdot$ min^{-1} $\cdot$ torr^{-1}.

Using these values one can calculate according to the well-known formula of Roughton and Forster (1957)

$$\frac{1}{D} = \frac{1}{D_M} + \frac{1}{\Theta \cdot Q_c} \tag{5}$$

the "membrane" diffusing capacity for O_2 (D_{MO_2}). Furthermore, assuming that $D_{MCO_2}/D_{MO_2} \approx 20$, one can apply the equation to equilibration of CO_2 and, with the measured D_{CO_2} and assumed Q_c, solve the equation for Θ_{CO_2}. Values ranging from 1.6 to 2.2 ml $\cdot$ min^{-1} $\cdot$ torr^{-1} $\cdot$ ml^{-1} are obtained.

These values are much lower than the value 5.5 (same units) determined in experiments on red cells with the continuous-flow-rapid-reaction technique by Constantine et al. (1965). These authors point out, however, that their Θ_{CO_2} value is based on the initial rate and that later the kinetics slows down. Since in lungs the equilibration may be estimated as 98% to 85% complete (see below), a lower "effective" Θ_{CO_2} is expected.

Furthermore, partitioning of the resistance to CO_2 elimination into a membrane and a blood component shows that the blood component clearly prevails, the membrane component accounting for only 8% to 17% of the total resistance, whereas for O_2 uptake, the membrane resistance amounts to 45%–87% of the total uptake resistance (according to the Θ_{O_2} and Q_c values assumed).

4.2 Overall Diffusion Limitation

By "overall diffusion limitation" we designate the effect of all diffusion, reaction, and ion exchange processes in limiting the attainment of equilibrium between alveolar gas and end-capillary blood.

The degree of diffusion limitation (L_{diff}) may be defined as the relative difference between the transfer rate without diffusion limitation, $\dot{M}$ (D = ∞), and the actual transfer rate, $\dot{M}$:

$$L_{diff} = \frac{\dot{M}_{(D=\infty)} - \dot{M}}{\dot{M}_{(D=\infty)}} \tag{6}$$

For the simplest pulmonary capillary model (perfused homogeneous tube surrounded by alveolar gas) the following relationship holds:

$$L_{diff} = e^{-\frac{D}{\dot{Q}\beta}} \tag{7}$$

Using standard data on cardiac output ($\dot{Q}$), β_{CO_2} [$= (C_{\bar{v}} - C_a)_{CO_2}/P_{\bar{v}} - P_a)_{CO_2}$] and the measured D_{CO_2} the following values are calculated:

1. $D/\dot{Q}\beta$: 4.0 at rest; 1.7 at very heavy exercise ($\dot{M}_{O_2}$ = 3 l/min);
2. the corresponding L_{diff} values are: 0.02 and 0.18;
3. the end-capillary-to-alveolar P_{CO_2} difference, $(P_{c'} - P_A)_{CO_2}$, equals: 0.1–0.2 torr at rest, about 4 torr at very heavy exercise.

These calculations show that pulmonary CO_2 transfer is very slightly diffusion-limited in resting conditions, but may be appreciable diffusion-limited in heavy exercise.

References

Adaro F, Scheid P, Teichmann J, Piiper J (1973) A rebreathing method for estimating pulmonary D_{O_2}: theory and measurements in dog lungs. Respir Physiol 18: 43–63

Bidani A, Crandall ED (1978) Analysis of P_{CO_2} differences during rebreathing due to slow pH equilibration in blood. J Appl Physiol 45: 666–673

Cander L, Forster RE (1959) Determination of pulmonary parenchymal tissue volume and pulmonary capillary flow in man. J Appl Physiol 14: 541–551

Cerretelli P, Veicsteinas A, Teichmann J, Magnussen H, Piiper J (1974) Estimation by a rebreathing method of pulmonary O_2 diffusing capacity in man. J Appl Physiol 37: 526–532

Constantine HP, Craw MR, Forster RE (1965) Rate of the reaction of carbon dioxide with human red blood cells. Am J Physiol 208: 801–811

Forster RE, Crandall ED (1976) Pulmonary gas exchange. Annu Rev Physiol 38: 69–93

Hill EP, Power GG, Longo LD (1973) Mathematical simulation of pulmonary O_2 and CO_2 exchange. Am J Physiol 224: 904–917

Holland RAB, Van Hezewijk W, Zubzanda J (1977) Velocity of oxygen uptake by partly saturated adult and fetal human red cells. Respir Physiol 29: 303–314

Hyde RW, Puy RJM, Raub WF, Forster RE (1968) Rate of disappearance of labeled carbon dioxide from the lungs of humans during breath holding: a method for studying the dynamics of pulmonary CO_2 exchange. J Clin Invest 47: 1535–1552

Mochizuki M (1966) Study on the oxygenation velocity of the human red cell. Jp J Physiol 16: 635–648

Piiper J (1979) Series ventilation, diffusion in airways and stratified inhomogeneity. Fed Proc 38: 17–21

Piiper J, Scheid P (1975) Gas transport efficacy of gills, lungs and skin: theory and experimental data. Respir Physiol 23: 209–221

Piiper J, Dejours P, Haab P, Rahn H (1971) Concepts and basic quantities in gas exchange physiology. Respir Physiol 13: 212–304

Piiper J, Meyer M, Scheid P (1979) Pulmonary diffusing capacity for O_2 and CO at rest and during exercise. Advantages of rebreathing techniques using stable isotopes. Bull Eur Physiopathol Respir 15: 145–150

Roughton FJW, Forster RE (1957) Relative importance of diffusion and chemical reaction rates in determining rate of exchange of gases in the human lung, with special reference to true diffusing capacity of pulmonary membrane and volume of blood in the lung capillaries. J Appl Physiol 11: 291–302

Staub NC, Bishop JM, Forster RE (1962) Importance of diffusion and chemical reaction rates in O_2 uptake in the lung. J Appl Physiol 17: 21–27

Veicsteinas A, Magnussen H, Meyer M, Cerretelli P (1976) Pulmonary O_2 diffusing capacity at exercise by a modified rebreathing method. Eur J Appl Physiol 35: 79–88

Physiological Consequences of Carbonic Anhydrase Inhibition During Rest and Exercise, and on the Bohr Effect

T.H. MAREN and E.R. SWENSON [1]

The questions which led to the present work are also those that led to the discovery of carbonic anhydrase, and thus set in motion the events which led to this conference. It was 50 years ago that the Danish physiologist and physician, Henriques, showed that the uncatalyzed conversion of HCO_3^- to CO_2 could not occur rapidly enough to account for the elimination of CO_2 in the one second that blood traversed the lungs. He showed that in the presence of blood, however, this reaction was extremely rapid; he speculated that this was due either to a catalytic process or a direct combination of CO_2 with hemoglobin. Because it seemed to him that the reaction did not conform to the pattern of an enzymic rate (it was so rapid that he missed readings in the first 5 s) he thought that the carbamino-hemoglobin process was the more likely [4, 5].

The repetition of these experiments in 1932 by Meldrum and Roughton showed their enzymic nature, and so carbonic anhydrase was discovered [10]. Soon after, Roughton returned to the problem set by Henriques, and calculated that for the dehydration of blood HCO_3^- to reach 90% of equilibrium with alveolar CO_2 without a catalyst, some 200 s would be required [12] [2]. It was therefore thought that inhibition of red cell carbonic anhydrase would be lethal.

However, quite the contrary was found during the development of acetazolamide. Animals lived for years with very large daily doses of the drug [9], which (it is now clear, see below) totally inhibit the red cell enzyme, as well as carbonic anhydrase elsewhere in the body.

The paradox was resolved by the work of Cain and Otis in anesthetized dogs [2], who showed that there were profound changes in CO_2 gradients across the lungs following carbonic anhydrase inhibition, and that near-equilibrium of the species of CO_2 was not necessary. The equations of Henriques and of Roughton did not take any physiological adjustments into account, and their conclusions were based on considerations of an equilibrium state.

In returning to the problem, with particular attention to man at rest and during exercise, we no longer consider it primarily in terms of equilibria, but of quanta of CO_2 evolved per unit time. In the presentation to follow, we summarize our published data on this question [13], and also give new data on the related question of the rates of the Bohr off-shift during carbonic anhydrase inhibition [8], a subject introduced by Forster and Steen [3].

---

1 Department of Pharmacology and Therapeutics, University of Florida, College of Medicine, Gainesville, Florida/USA

2 This calculation was based on the equation for a first-order unopposed reaction, using the rate constant for dehydration of HCO_3^-. If we use the expression for opposing first-order reaction, which incorporates the hydration rate constant, the time for 90% equilibrium would be 10–20 s

1 An Analysis of CO_2 Transport in Man During Rest and Exercise

We have analyzed six situations, three for rest and three for exercise. The three units in each case are: normal, following clinical doses of acetazolamide, and complete absence of the enzyme. A very large number of factors must be taken into account, and these underlying data are given in the original publication [13]. They include kinetics of the catalyzed and uncatalyzed hydration and dehydration of CO_2 at 37°C; blood levels and inhibition kinetics of acetazolamide; CO_2, HCO_3^- and H^+ concentrations throughout the vascular system; physiological and chemical measurement of men during exhausting exercise; computer-generated plots of HCO_3^- loss at ms intervals during passage through the lungs, and catalytic CO_2 output at varying levels of carbonic anhydrase activity.

In assembling these data, we also used relevant material form certain animal experiments, and earlier work in resting man. To some degree, therefore, each of the six situations may be regarded as a model. In each case we give CO_2 outputs due to (a) the uncatalyzed reaction, $\dot{V}_{uncat}$, (b) the calazyed reaction, $\dot{V}_{cat}$, (c) carbaminohemoglobin, and (d) CO_2 diffusion. In the first of the situations analyzed, the bases for the calculations will be given briefly. These apply, with some variation, to the other five.

1.1 Resting Man

Table 1, left frame, shows how the 12 mmol/min of normal CO_2 output is partitioned among the four modes of excretion. (a) The uncatalyzed rate is calculated from the net dehydration of HCO_3^-, using the known rate constants and the concentrations of HCO_3^- in the pre-pulmonary capillaries and CO_2 in the post-pulmonary capillaries. It is evident that this is a negligible quantity. This, of course, agrees with the Henriques-Roughton idea that the uncatalyzed rate is inadequate (by a large margin) to deal with the situation. (b) The catalyzed rates entered under "Chem" are calculated from

Table 1. Evolution of CO_2 in resting man and in response to carbonic anhydrase inhibition. Output 12 mmol/min

		Normal		Clinical inhibiton [a] mmol/min		Total inhibition (model)
$\dot{V}_{uncat}$		0.07		0.3		0.6
		Chem	*Obs*	*Chem*	*Obs*	
$\dot{V}_{cat}$	Enzyme C	915	8	1	1	0
	Enzyme B	216	2	9	7	0
$\dot{V}_{carbamino\ Hb}$		1.3		1.3		1.3
$\dot{V}_{diffusion}$		1.2		3		10

a 12 mg/kg acetazolamide orally. Free plasma conc = 33 μM, whence i = 0.9994 for enzyme C and 0.983 for B

the concentrations and activities of the two isozymes (B and C, or I and II) in human red cells, and the HCO_3^- and CO_2 concentrations in the capillaries as described above. These show, as in other physiological systems, a great excess over observed rates. Note also that the ratio of the calculated catalyzed/uncatalyzed rates is about 16,000, in agreement with other work by Roughton [12] and recently by Itada and Forster [7]. The catalyzed rates entered under "Observed", totaling 10 mmol/min, are obtained by difference between the total rate and the noncatalytic components. The partition between isozymes B and C is based on their relative rates, integrating their concentrations and activities in the red cell. (c) $\dot{V}_{carbamino}$ is calculated from the known A-V carbamino hemoglobin difference x cardiac output. (d) CO_2 diffusion is calculated from the A-V pCO_2 difference x cardiac output.

1.2 Resting Man, with High Clinical Doses of Acetazolamide

These data are taken from the work of Pocidalo et al., in which 12.5 mg/kg of acetazolamide was given by mouth, and measurements were made 1–2 h later [11]. From our own studies on the pharmacology of this drug, we estimate that the unbound concentration in plasma was 33 μM, from which we calculate that the fractional inhibition (i) was 0.9994 for enzyme C and 0.983 for enzyme B. In these experiments, enzyme inhibition led to a large increase of the pCO_2 difference between pre- and post-pulmonary capillary blood, from 6 mm Hg (normal) to 23 mm Hg. Integrating all of these facts leads to the middle frame of Table 1. The uncatalyzed and carbamino contributions are still small. The chemical basis for the enzymic rate has been greatly lowered, but the residual rate is only slightly less than the usual physiological contribution of the enzyme, 10 mmol/min. This modest deficit, of about 2 mmol/min, is now contributed by molecular diffusion of CO_2, a function of the new raised gradient of the gas.

This analysis gives the basis for the finding that patients treated with carbonic anhydrase inhbitors do not ordinarily show any deficit in pulmonary function. The elimination of CO_2 is subserved by residual enzyme activity and the raised gradient for diffusion.

1.3 Rest. No Carbonic Anhydrase

We now ask if adequate CO_2 elimination occurs in the complete absence of carbonic anhydrase. The data are extrapolated from animal experiments [2] in which 50–100 mg/kg acetazolamide were given, and a few human data following 25–50 mg/kg intravenously [1]. At these doses there is less than 1 part in 10^4 remaining of enzyme C, and about 1 part in 10^3 of enzyme B. It may be shown that these residues yield rates scarcely different from that of the uncatalyzed reaction.

The right frame of Table 1 shows that diffusion of CO_2 alone is able to substitute for the absent enzyme. This is because the CO_2 gradient from pre- to post-pulmonary capillary is now very large – 51 mm Hg [2]. It is, then, quite clear why our animals [9] survived very high doses of acetazolamide at rest. The questions posed by Henriques and Roughton are answered: while the uncatalyzed rate cannot nearly approach physiological needs, a second means of CO_2 elimination is available when new gradients appear. Carbonic anhydrase inhibition induces the new situation that makes function possible in its absence.

1.4 Normal Exercise

Three men trained in good physical condition exercised on the Balke treadmill to exhaustion. Their heart rate reached 190 beats/min and $\dot{V}_E$ reached 115 l/min, 20 min into the exercise. Chemical findings were as expected, and included a sharp drop in plasma HCO_3 and widening of the A-V pCO_2 gradient to 13 mm Hg.

Figure 1 shows CO_2 and O_2 exchange during the exercise. Table 2 (left frame) shows the partition of CO_2 elimination, calculated in the same way as described in Sect. 1.1 above. The 15-fold increase in CO_2 output during exercise in easily accomplished by drawing on the catalytic reserve. Other elements are increased also due to increased A-V CO_2 gradients and cardiac output.

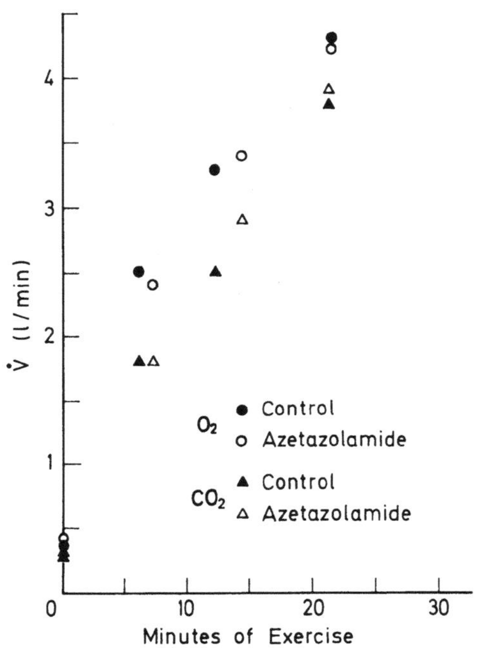

Fig. 1. Mean gas exchange in three trained men exercising to exhaustion. Acetazolamide in plasma (unbound) = 8 μM

1.5 Exercise During Clinical Inhibition of Carbonic Anhydrase

The exercise routine described in Sect. 1.4 above was repeated exactly, except that the men had received 10 mg/kg of acetazolamide 10 h before and 2.0–2.5 mg/kg of the drug 2 h before, all by mouth. The resulting mean plasma concentration at the time of the test was 105 μM, of which 7.5 μM was not bound to plasma protein.

Table 2. Evolution of CO_2 in maximally exercising man. Effect of carbonic anhydrase inhibition. Output = 172 mmol/min

		Normal		Clinical inhibition [a] mmol/min		Total inhibition
$\dot{V}_{uncat}$		0.25		1.5		1.7
		Chem	*Obs*	*Chem*	*Obs*	
$\dot{V}_{cat}$	Enzyme C	3050	124	24	14	0
	Enzyme B	713	30	137	105	0
$\dot{V}_{carbamino\ Hb}$		6		6		6
$\dot{V}_{diffusion}$		12		45		75

a Plasma conc. = 8 μM whence i = 0.997 for enzyme C and 0.927 for B

The concentration of drug in red cells was 150 μM, which shows near stoichiometric binding to the carbonic anhydrase present.

Exercise tolerance was not impaired by the drug. The chief finding was that total ventilation, $\dot{V}_E$, was increased 27% over that for the drug-free experiment, to 146 l/min. From this and the fact that total CO_2 output was unchanged (Fig. 1) the lowering of alveolar pCO_2 could be computed. The result was that the pCO_2 gradient from pre- to postpulmonary capillary was 65 mm Hg; in exercise without drug this figure was 13 mm Hg.

From these data the numbers in the middle frame of Table 2 were generated. It is seen (as found also for rest, Table 1) that clinical (and not quite complete) inhibition of carbonic anhydrase is compatible with severe exercise because of two factors: some residual enzyme activity, and elevated gradients for the diffusion of molecular CO_2. Thus even the 15-fold increase of CO_2 output is accomplished, with no untoward chemical or physiological signs.

1.6 Exercise Without Carbonic Anhydrase

The right frame of Table 2 shows the (hypothetical) situation in which exercise as described here would be done in the absence of carbonic anhydrase. This would be the case if man were to take some 50 mg/kg of acetazolamide (see Table 3 of [13]). The data are assembled from experiments in rats which yielded pCO_2 concentrations during swimming [13], and also extrapolation from experiments in resting man in which very large doses of drug were given (Table 3 and discussion in [13]). In this model the pCO_2 of the tissues is taken to be 120 mm Hg and the gradient between pre- and post-pulmonary capillary pCO_2 is 89 mm Hg (Table 7 of [13]). Despite these maximal values, the resulting contribution of CO_2 diffusion is seen to be 75 mmol/min, and the sum of all the elements of CO_2 excretion only 83 mmol/min, far short of the 172 that is necessary. It is therefore concluded that severe exercise in man is impossible without carbonic anhydrase.

Table 3. Summary of relations between chemical catalysis of HCO_3^- dehydration and physiological CO_2 output

	Rest	Maximal exercise
$\dot{V}$ observed, total (mmol CO_2/min)	12	172
$\dot{V}$ observed catalyzed [a] HCO_3^- dehydration (mmol CO_2/min)	10	154
$\dot{V}$ maximal catalyzed [b] HCO_3^- dehydration (mmol CO_2/min)	1131	3763
Enzyme "excess", i.e., can be drawn on before any perturbations occur [c]	17.4	5.8

a $\dot{V}$ observed minus noncatalytic components, see Tables 1 and 2
b From Tables 1 and 2
c From data of Dr. Esther Hill, Fig. 2

Table 3 summarizes these relations. The first row gives the total CO_2 output, and the second row shows that normally most of this output is due to the catalytic dehydration of HCO_3^-. The third row gives the chemically calculated rates of dehydration, using the known kinetic constants, the blood volume of the lung, and the total gradients of the various species of CO_2 across the lung. This large number is misleading, however, because the full gradients are never realized; rather the enzyme works over small gradually changing gradients within the blood. The true relation between CO_2 outputs using 1 ms intervals, and catalysis, is obtained from Fig. 2. This depends

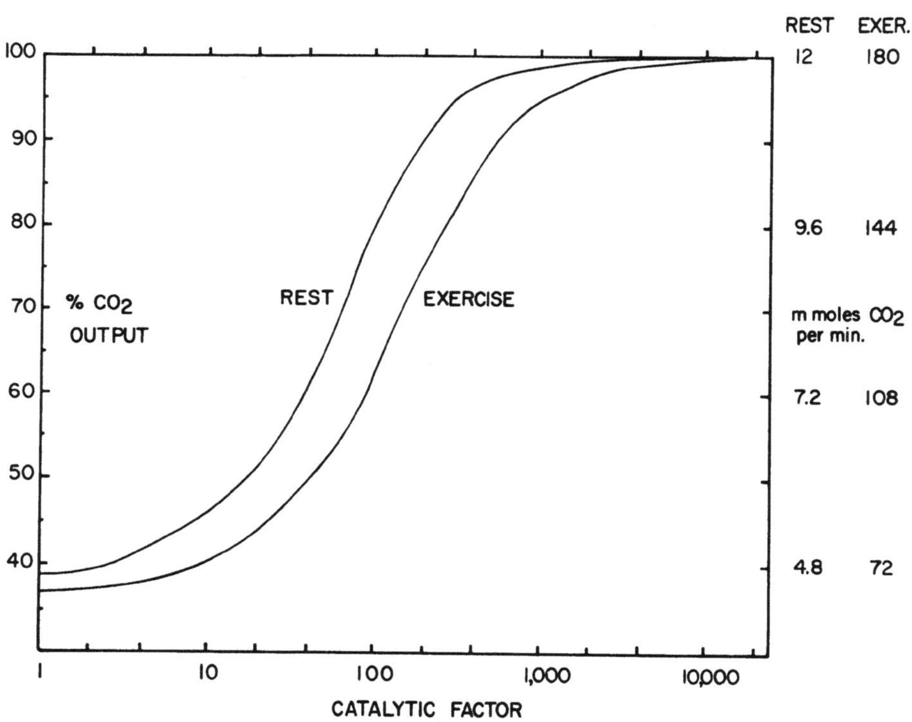

Fig. 2. The relation between carbonic anhydrase activity and CO_2 output at rest, and in present exercise experiments. *Abscissa* is the factor by which carbonic anhydrase increases the dehydration of HCO_3^-. Curves kindly supplied by Dr. Esther Hill according to [6]

upon computer-generated rates using the known rate and concentration factors for the various species involved, according to the model of Hill et al. [6]. It is seen that at rest, CO_2 output would decline when the catalytic factor is reduced from the normal of 17,400 to about 1000. At exercise, this decline sets in at a catalytic factor of 3000. Thus the "excess" enzyme – if no perturbations are allowed – is 17.4 and 5.8 for rest and exercise respectively; these are entered in the fourth row of Table 3.

Thus it is seen that the question of the relation of carbonic anhydrase activity or inhibition to function has several different sorts of answer. If we ask how much enzyme is needed to ensure normal CO_2 output at rest, with no perturbations, the answer is 1/17 of the normal amount. But if we ask how much enzyme is needed to ensure normal CO_2, allowing the organism to adjust and use other means of CO_2 excretion, the anwer is that no enzyme is needed at rest. For exercise, the margin is quite narrow if no perturbations are made, 1/6 of the normal amount is needed. Finally and importantly, severe exercise would be impossible in the absence of carbonic anhydrase, since the chief compensatory mechanism, CO_2 diffusion, would not be adequate for the excretion of CO_2 at some 15 times resting rates.

2 Relation of the Kinetic Bohr Effect to Carbonic Anhydrase

The rate of oxygen release following the acidification of hemoglobin (Bohr off-shift) was found to depend in part on the carbonic anhydrase in red cells [3]. In the present study we quantify these relations and compare the rates in the red cells of fish, amphibia, bird, and man. The experiments were carried out with a modified Hartridge-Roughton reaction apparatus [3, 8]. A 1% suspension of red cells in appropriate Ringer's solution is mixed "instantly" with the same Ringer's solution of different pH or pCO_2. The mixture travels at known speed down a thin observation tube to a modified Clark electrode; the change in pO_2 is displayed on a Microsensor (Model 1201, Transidyne General Co.). Details may be found in our full paper [8].

Figure 3 (left) shows the rate of the Bohr "off-shift" in five representative vertebrates. The differences in rate are probably a function of the temperatures of the tests, which were carried out at the temperature of the animals. Note that in man the effect is complete well within the time of blood flow through the lungs (1 s). The same is true of the dogfish, in which blood transverses the gill in ~ 2 s. The right frame of Fig. 3 shows the very large reduction in rate (some 90%) when carbonic anhydrase is completely inhibited by 10^{-3} M methazolamide. It is clear that the Bohr effect would be reduced to an insignificant contribution in the absence of carbonic anhydrase.

Table 4 shows the relations in human red cells among chemical rates of proton generation, and measured oxygen release, and the effects of graded doses of methazolamide upon these rates. Column 2 gives the fractional carbonic anhydrase activity, for the different concentrations of drug, based on K_I for the drug enzyme complex of 10^{-8} M and the simplifying assumption of noncompetitive inhibition. Column 3 gives the calculated rates of proton generation, from the known kinetic parameters for the enzyme, the concentrations of CO_2 and HCO_3 in the system, and the equation for the net velocity of opposing reactions. As with all other carbonic anhydrase

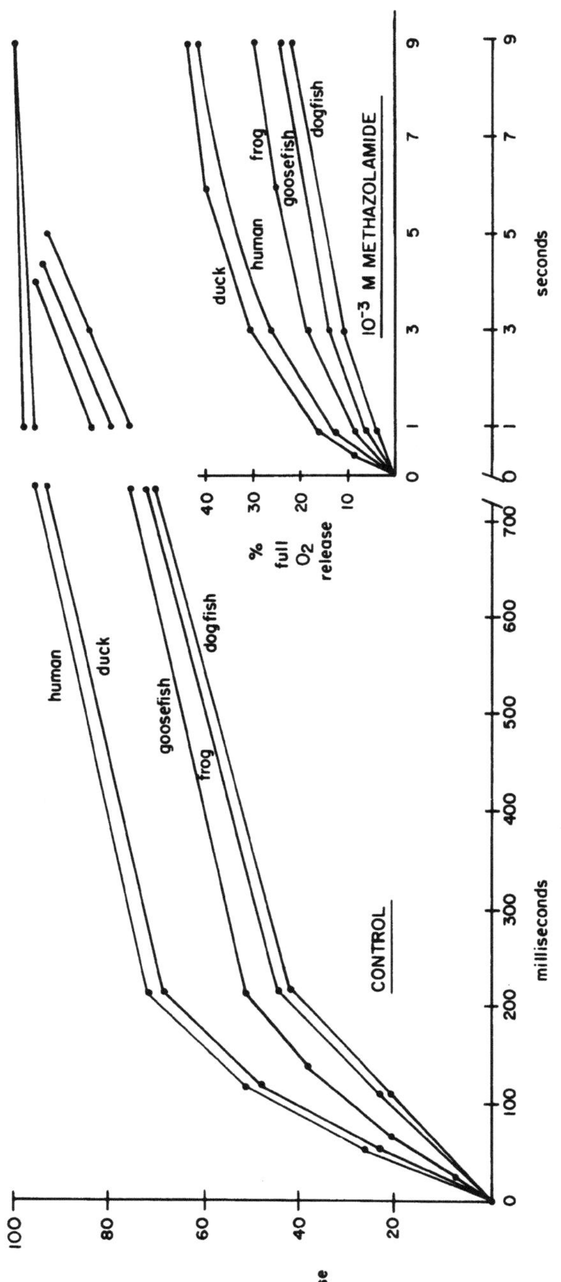

Fig. 3. Rate of the Bohr "off-shift" in suspensions of red cells from representative vertebrates. At 0 time red cell suspensions are mixed with Ringer's solution containing 2.2 mM (man, bird) or 4.4 mM (fish, frog) dissolved CO_2

Table 4. Carbonic anhydrase inhibition on proton formation and oxygen release following CO_2 (Bohr effect) in human red cells

	Fractional carbonic anhydrase activity $a = 1 - i$ [b]	Initial H^+ generation (calc.) [a] mM/s	Initial O_2 release (observed) mM/s	t 1/2 O_2 release s	% completion of Bohr effect in 1 s
Control	1	2850	16.5	0.12	99.7
Methazolamide μM					
10	1.6×10^{-3}	4.7	7.6	0.22	96
30	5.5×10^{-4}	1.7	3.5	0.54	74
100	1.6×10^{-4}	0.6	1.3	1.3	42
300	5.5×10^{-5}	0.3	0.6	2.9	21
1000	1.6×10^{-5}	0.2	0.4	4.2	15

a From catalytic constants and concentrations [8], and uncatalyzed rate of 0.16 mM/s
b i = fractional inhibition, calculated in [8, 13]

mediated reactions this calculated rate, 2850 mM/s, is very much greater than the observed rate, which in this case (Col. 4) is oxygen release, 16.5 mM/s. When enzyme is inhibited, the calculated rate drops greatly, but at 10^{-6} M methazolamide it is still enough to sustain normal oxygen release. Not until enzyme activity drops to about 0.1% of normal (at 10^{-5} M methazolamide) is the release affected. When enzyme is completely inhibited by 10^{-3} M methazolamide and proton generation falls to the level of the uncatalyzed rate, we observe the true stoichiometry between H production and O_2 release – that binding of one proton induces the release of 2 O_2 – the Bohr effect factor of 0.5.

The concentration range of 10^{-4}–10^{-5} M for methazolamide is equivalent to plasma levels of patients receiving this and other carbonic anhydrase inhibitors for various diseases, notably glaucoma. Clearly the Bohr effect must be at least somewhat impaired, as shown by columns 5–6. The physiological implications of this surely deserves study.

Acknowledgment. This work was supported by National Institutes of Health grants GM 16934 and HL 22258-02.

References

1. Burton GC, Collier CR, Hackney JD, Bauer C (1960) In vivo inhibition of erythrocytic carbonic anhydrase in man. Fed Proc 19: 379
2. Cain SM, Otis AB (1961) Carbon dioxide transport in anesthetized dogs during inhibition of carbonic anhydrase. J Appl Physiol 16: 1023–1028
3. Forster RE, Steen JB (1968) Rate limiting processes in the Bohr shift in human red cells. J Physiol (London) 196: 541–562
4. Henriques OM (1928a) Über die Geschwindigkeiten der Anhydrierung bzw. der Hydratisierung der Kohlensäurekomponenten im Blute. Biochem Z 200: 1–4

5. Henriques OM (1928b) Der experimentelle Nachweis schnell reagierenden, gebundenen CO_2 im Hämoglobin. Biochem Z 200: 5–9
6. Hill EP, Power GG, Longo LD (1973) Mathematical simulation of pulmonary O_2 and CO_2 exchange. Am J Physiol 224: 904–917
7. Itada N, Forster RE (1977) Carbonic anhydrase activity in intact red blood cells measured with ^{18}O exchange. J Biol Chem 252: 3881–3890
8. Maren TH, Swenson ER (1980) A comparative study of the kinetics of the Bohr effect in vertebrates. J Physiol (London) 302
9. Maren TH, Mayer E, Wadsworth BC (1954) The pharmacology of Diamox ® 2-acetylamino-1,3,4-thiadiazole-5-sulfonamide. Bull Johns Hopkins Hosp 95: 199–243
10. Meldrum NU, Roughton FJW (1933) Carbonic anhydrase: Its preparation and properties. J Physiol (London) 80: 113–142
11. Pocidalo JJ, Corcket F, Amiel JL, Lissac J, Finetti P, Blayo MC (1960) Action respiratoire de l'acétazolamide. Etude chez l'homme à poumons sains ou emphysématieux maintenu sous ventilation constante. Rev Fr Etud Clin Biol 6: 582–588
12. Roughton FJW (1935) Recent work on carbon dioxide transport by the blood. Physiol Rev 15: 241–296
13. Swenson ER, Maren TH (1978) A quantitative analysis of CO_2 transport at rest and during maximal exercise. Respir Physiol 35: 129–159

CO_2 Transport and Carbonic Anhydrase in Placenta, Muscle and Calcium-Mobilizing Tissues. Localization of Carbonic Anhydrase

Chairman: C. ALBERS

Placental CO_2 Exchange

E.P. HILL [1] and L.D. LONGO [2]

1 Placental CO_2 Exchange

In the fetus in utero, carbon dioxide must be eliminated by diffusing across the placenta into maternal blood. Despite a number of studies on placental CO_2 exchange, several vital problems relating to this process remain unsolved.

1.1 Species Differences

Both oxygen and carbon dioxide transport have been studied extensively in the placentas of numerous species, but generalizing the results is difficult because of large species differences. Figure 1 summarizes some of these differences in terms of functional efficiency for gas exchange and the morphology of the placental barrier. Rabbits and guinea pigs have efficient placentas, with a relatively thin "hemochorial" barrier consisting of a fetal endothelial layer, mesenchymal tissue, and chorionic membranes arranged in a labyrinthic pattern. Fetal umbilical vein O_2 tension usually exceeds uterine vein O_2 tension (Faber and Hart, 1966), suggesting a countercurrent type exchange pattern for oxygen, although this does not seem to explain the CO_2 measurements.

Sheep and other ruminants have much less efficient placentas, with a thicker "epitheliochorial" barrier. Although the vascular anatomy suggests a countercurrent (Barcroft and Barron, 1946) or mixed flow pattern (Makowski, 1968), the efficiency of transport suggests a much less efficient system. On that basis, it is often more useful to treat these placentas as having concurrent patterns functionally. Even assuming the rather inefficient concurrent vascular arrangement, the large O_2 tension difference between exiting maternal and fetal blood suggests that there must be a large diffusion barrier or the presence of sizable shunting of blood past the exchange region in both maternal and fetal placental circulations.

Human and other primate placentas appear to be intermediate in efficiency. The microvascular structure is hemochorial, similar to that found in guinea pigs, but the vascular structure is often described in terms of a maternal pool with fetal vessels into it, forming a "multivillous" flow arrangement.

---

1 Department of Medicine, University of California, San Diego, La Jolla, California 92093/USA
2 Division of Perinatal Biology, Loma Linda University, Loma Linda, California 92354/USA

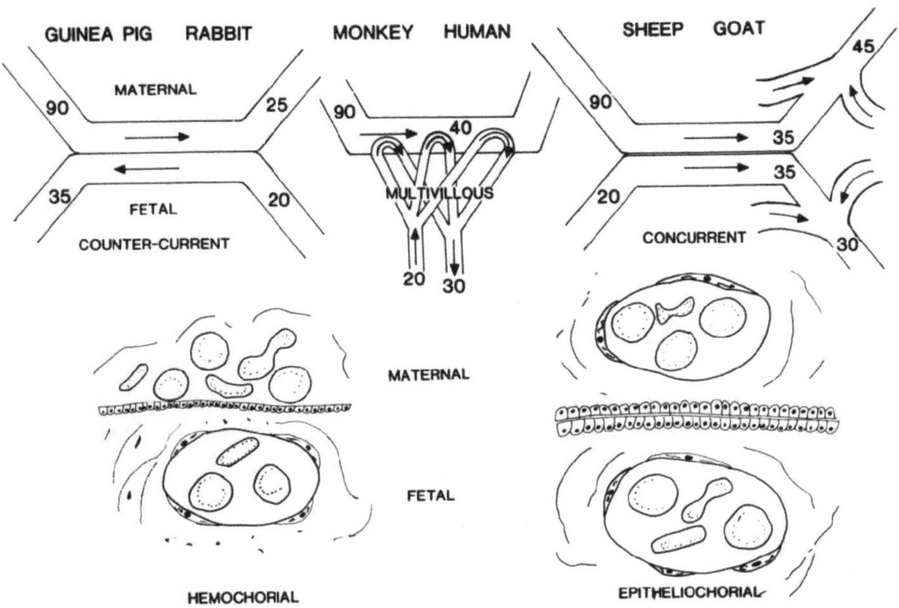

Fig. 1. Highly schematic illustrations of functional anatomy and morphology of the placental barrier of various species. Numerical values indicate typical O_2 values (P_{O_2} in torr) for inflowing and outflowing maternal and fetal blood

1.2 P_{O_2} and P_{CO_2} Differences Across the Placenta

Figure 1 includes representative P_{O_2} values in inflowing and outflowing blood for the different types of placenta. Note that the maternal-to-fetal vein difference ranges from − 10 torr in the guinea pig to about + 15 torr in sheep (Fig. 1). Typical P_{CO_2} values in the various species are listed in Table 1. The surprising results indicate that although the uterine-to-umbilical vein P_{O_2} difference varies greatly with species, the umbilical-to-uterine vein (fv-mv) P_{CO_2} difference equals 4 to 5 torr in all species listed, with the exception of the mare, in which the P_{CO_2} difference is only 1 torr. (The mare also has a very small fv-mv P_{O_2} difference, suggesting an especially efficient exchange, despite a thick, epitheliochorial placental barrier.) Why placentas which show very efficient exchange for oxygen (guinea pigs and rabbits) do not show a similar efficiency for carbon dioxide exchange, is puzzling. Bartels et al. (1967) suggested that it might be related to the low carbonic anhydrase activity in fetal blood. However the reduced carbonic anhydrase activity appears to explain only a small part of the observed (fv-mv) P_{CO_2} difference (see below).

Table 1. Summary of P_{CO_2} values in uterine artery (ma), uterine vein (mv), umbilical vein (fv), umbilical artery (fa), and fetal-to-maternal vein P_{CO_2} difference (fv-mv) in various species

Species	ma	mv	fv	fa	fv-mv	Authors
Sheep	34	38	42	46	4	Comline and Silver, 1970a
	36	40	43	48	3 [a]	Comline and Silver, 1970a
		38	43		5 [a]	Meschia et al., 1970
	27	36	40	46	4 [a]	Barron and Meschia, 1954
	34	40	45	55	5 [a]	Kaiser and Cummings, 1957
	34	40	46	55	6 [a]	Battaglia et al., 1965
	34	38	40	47	2 [a]	Baillie et al., 1971
	35	41	48	53	7 [a]	Longo et al., 1974
	35	40	44	52	4	Longo et al., current studies
Goat	36	39	46	49	7 [a]	Kaiser and Cummings, 1958
		43	50		7 [a]	Bartels et al., 1967
	29	33	40	42	7	Blechner et al., 1968
Cow	33	37	41	48	4	Gahlenbeck et al., 1968
	36	40	44	49	4	Comline and Silver, 1974
Mare	35	40	41	50	1 [a]	Comline and Silver, 1970b
	35	38	39	46	1	Comline and Silver, 1970b
	38	42	43	50	1	Comline and Silver, 1974
Pig	36	41	45	52	4 [a]	Comline and Silver, 1974
Rabbit		43	48		5 [a]	Bartels et al., 1967
Guinea pig		57	60		3 [a]	Bartels et al., 1967
Human	26	29 [i]	33	39	4 [i] N	Bruns et al., 1961
		37 [i]	40	49	3 [i] N	Rooth and Sjöstedt, 1962
		37 [i]	42	46	5 [i a]	Prystowsky et al., 1961
	30	33	42		9	Stenger et al., 1964
	36	46	49	60	3 [a]	Wulf, 1962

a denotes anestisized subjects, i denotes sample from intervillous space in humans rather than uterine vein, and N denotes samples from newborn rather than in utero

1.3 Possible Explanations of Umbilical-to-Uterine Vein (fv-mv) P_{CO_2} Differences

Many investigators have interpreted the P_{O_2} and P_{CO_2} differences across the placenta as the result of diffusion limitation. Lamport (1954) calculated the placental diffusing capacity for sheep, using the Bohr integration procedure (Bohr, 1909) and assuming that the uterine-to-umbilical P_{O_2} difference is due to diffusion limitation rather than to shunts or other factors. Bissonnette and Gurtner (1972) suggest that some gases may be diffusion-limited, but Longo et al. (1970) found evidence that diffusion does not limit gas exchange in sheep since H_2 and SF_6 equilibrate to the same extent. On the other hand, equilibration depends on the ratio of diffusing capacity to effective solubility of the gas in the blood (Wagner, 1977, and Piiper, this symposium). Because of the large capacity of blood for O_2 and CO_2, it is possible that O_2 or CO_2 may be diffusion-limited, even though SF_6 is not. We made calculations of O_2 and CO_2 equilibration rates, based on measured diffusing capacity of carbon monoxide (Hill et al., 1973a). We used primarily human values, but assumed a concurrent flow pattern for simplicity. Figure 2 shows the calculated P_{CO_2} changes during the placental

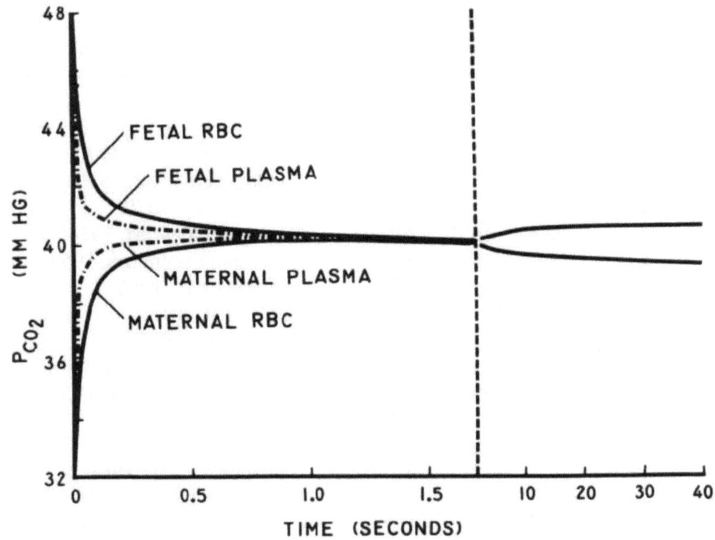

Fig. 2. Calculated P_{CO_2} changes during a placental capillary transit. *Solid lines* represent P_{CO_2} changes in fetal and maternal erythrocytes; *broken lines* represent plasma P_{CO_2} changes. Changes occurring after blood leaves exchange area due to slow continuing reactions are shown on *right* with the scale compressed 40 times. (From Hill et al., 1973a)

capillary transit in maternal and fetal blood. These calculations indicate that P_{CO_2}. equilibrates between maternal and fetal capillary blood, but that slow chemical reactions continue in the blood after it leaves the exchange area because the plasma does not contain carbonic anhydrase. Reactions continue in the veins and produce the shifts in P_{CO_2} illustrated on the right side of Fig. 2, and are similar to the slow chemical reactions postulated in the pulmonary system (Hill et al., 1973b; Forster and Crandall, 1975). Evidence presented at this symposium by Effros, Klocke, Crandall and Purves indicates that the importance of these slow chemical reactions is still in dispute. Nevertheless, according to these calculations, this effect can produce at most 1 torr fv-mv P_{CO_2} difference. These calculations were initially made assuming that fetal red cells had carbonic anhydrase activity which catalyzes the CO_2 hydration rate by a factor of 6500 (compared to a factor of 13 000 in adult red cells). Even if the fetal catalysis factor is as low as 800 (Berfenstam, 1952; Holland, 1978), our calculations indicate that the post-capillary P_{CO_2} change is only about 1 torr. This agrees with studies by Swenson and Maren (1978) that indicate the catalysis can be reduced substantially without interfering with gas exchange in the lung at rest.

Shunts have been estimated to range from nearly zero in rabbits and guinea pigs (Moll and Kastendieck, 1977) to about 20% on both sides of the placenta in sheep and goats (Metcalfe et al., 1967; Rankin and Peterson, 1969). If we assume a 20% shunt on both sides of the sheep placenta, this can explain a P_{O_2} difference of 15 torr as shown in Fig. 1. This degree of shunting would also produce a fv-mv P_{CO_2} difference of 3 torr, and thus also explain most of the fv-mv P_{CO_2} difference observed.

Power and co-workers measured distributions of maternal and fetal blood flows in sheep placentas, using radioactive-labeled macroaggregates of albumin (Power et al., 1967) and microspheres (Power et al, 1978). Their calculations (Longo and Power, 1969) indicate that the degree of mismatching of blood flows can explain a vein-to-vein P_{O_2} difference of 12 torr and a vein-to-vein P_{CO_2} differece of 5 torr, accounting for most of the differences observed experimentally. It is likely that measurements previously interpreted as indicating large shunts were actually the result of nonuniform distribution of maternal and fetal blood flows.

Another possible explanation of the vein-to-vein P_{O_2} and P_{CO_2} differences is placental tissue metabolism. Our calculations (Hill et al., 1973a) considered this possibility. Figure 3 indicates that regardless of whether we assume the CO_2 is added to fetal capillary blood, maternal capillary blood or some to each, P_{CO_2} equilibration occurs at a higher P_{CO_2} value, but equilibration is still complete by the end of the capillary transit, and post-capillary chemical reactions can account for at most a 1 torr P_{CO_2} difference. An additional fv-mv P_{CO_2} difference could occur if CO_2 from placental metabolism were added to the venous blood after it left the exchange area.

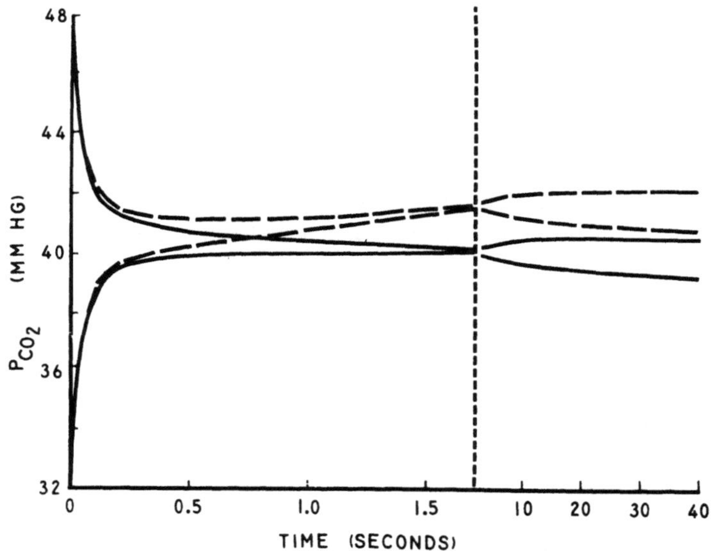

Fig. 3. Calculated effect of placental tissue CO_2 production (4 ml/min) on P_{CO_2} changes during capillary transit. (From Hill et al., 1973a)

1.4 CO_2 Dissociation Curves

Figure 4 shows CO_2 dissociation curves for human adult deoxygenated blood (curve a) and oxygenated blood (curve b), with the heavier lines indicating the positions one would expect for maternal and fetal blood on the basis of the degree of oxygenation. On just that basis, the total fetal CO_2 at the end-capillary position would be less than that of maternal end-capillary blood because the left shifted O_2 dissociation curve of

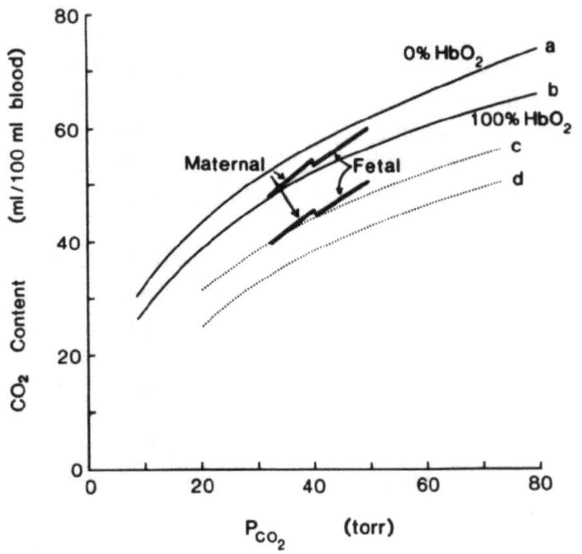

Fig. 4. CO_2 dissociation curves for human maternal and fetal blood, showing effects of oxygen saturation and base deficit. *a* adult deoxygenated blood; *b* adult oxygenated blood; *c* curve for pregnant women and fetuses; *d* curve for fetal blood according to Eastman et al., 1933

the fetus produces a higher fetal than maternal O_2 saturation at the same P_{O_2}. However, this effect is probably minor compared to other factors shifting the CO_2 dissociation curves. First, pregnant women hyperventilate. This produces a respiratory alkalosis which, when compensated, leads to a base deficit and reduces the total CO_2 (curve c). One may also consider the effect of a lowered hematocrit in pregnant women relative to the normal adult. The lower hemoglobin concentration should produce less carbamino, which would decrease the CO_2 content slightly. On the other hand, total bicarbonate is inversely related to hamatocrit, so total bicarbonate should be elevated by the hematocrit effect. Of all these effects, the most important is almost certainly that due to the base deficit produced by hyperventilation.

The fetal curve is also shifted relative to the adult curve. Fetal blood is acidotic, having a base deficit of 1 to 6 meq/l (Bruns et al., 1962; Stenger et al., 1964; and Rooth and Sjöstedt, 1962). Eastman et al. (1933) show a fetal CO_2 dissociation curve considerably below that of pregnant women (curve d), but that blood (having a base deficit of 6 meq/l) was obtained from the umbilical cord after delivery when there may have been considerable hypoxia. More recent data indicate an average fetal base deficit of about 4 meq/l, which places the fetal dissociation curve close to the maternal curve (curve c). Again, one might consider the effects of the high hematocrit of fetal blood, producing more carbamino, but less bicarbonate than blood with normal hematocrit. The net result of all these considerations is that the maternal and fetal curves are nearly superimposed (curve c). Total CO_2 and bicarbonate concentrations are higher due to the higher average P_{CO_2} values in the fetus. but not significantly different on the basis of acid-base status.

1.5 Placental CO_2 Diffusion

Longo et al. (1974) studied the relationships between CO_2 and bicarbonate ion in the placenta in more detail. They added THAM (tris hydroxymethyl amino methane) to the fetal blood to purposely increase pH. Fetal H^+ concentration decreased, bicarbonate concentrations increased, and P_{CO_2} decreased. In some of the experiments, fetal P_{CO_2} dropped below maternal P_{CO_2} and CO_2 moved from the mother to the fetus, despite a higher bicarbonate concentration in fetal blood. This provided evidence that CO_2 moves across the placenta rather than bicarbonate, and that CO_2 can move against a bicarbonate gradient.

1.6 CO_2 and Placental Water Movement

Wilbur et al. (1978) investigated the possibility that CO_2 and bicarbonate differences between fetal and maternal blood might influence maternal-fetal water balance. They developed a mathematical model (Fig. 5) which calculated changes in O_2 and CO_2 along the capillary transit, similar to results illustrated in Fig. 2. In addition, they allowed for transport of glucose across the placenta by facilitated diffusion, active

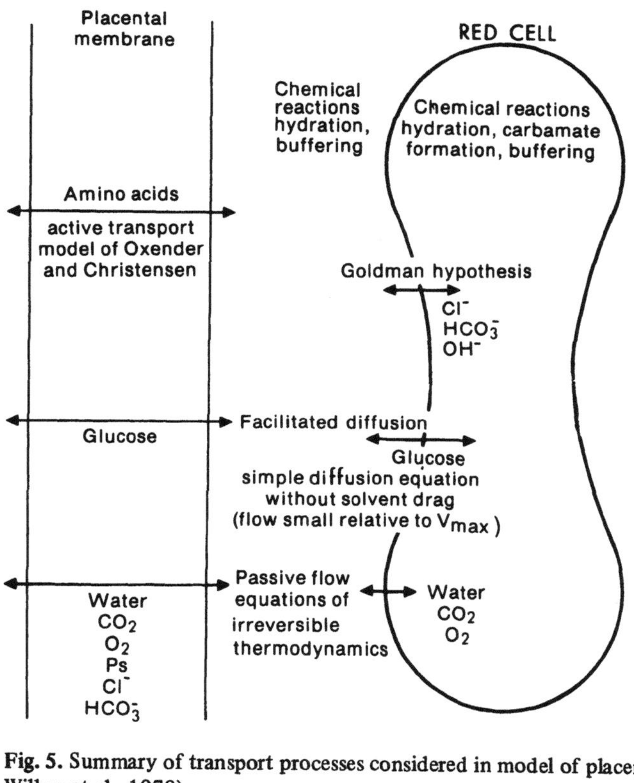

Fig. 5. Summary of transport processes considered in model of placental water exchange. (From Wilbur et al., 1978)

transport of amino acids and passive movement of other ions and water. Finally, they calculated the movement of water in response to osmotic pressure differences produced by the concentrations of the other ions and metabolites. Their calculations included red cell exchanges with plasma, because they found that the degree of water movement in and out of red cells during CO_2 exchange produces a significant change in osmotic pressure of the plasma which, in turn, can affect placental water transfer.

Figure 6 illustrates the osmotic gradients calculated at the beginning, middle, and end of the placental capillaries, using this model. They assumed the bicarbonate ion concentration was initially higher in the fetal blood because of the higher CO_2. This concentration difference tends to move water across the placenta from maternal blood.

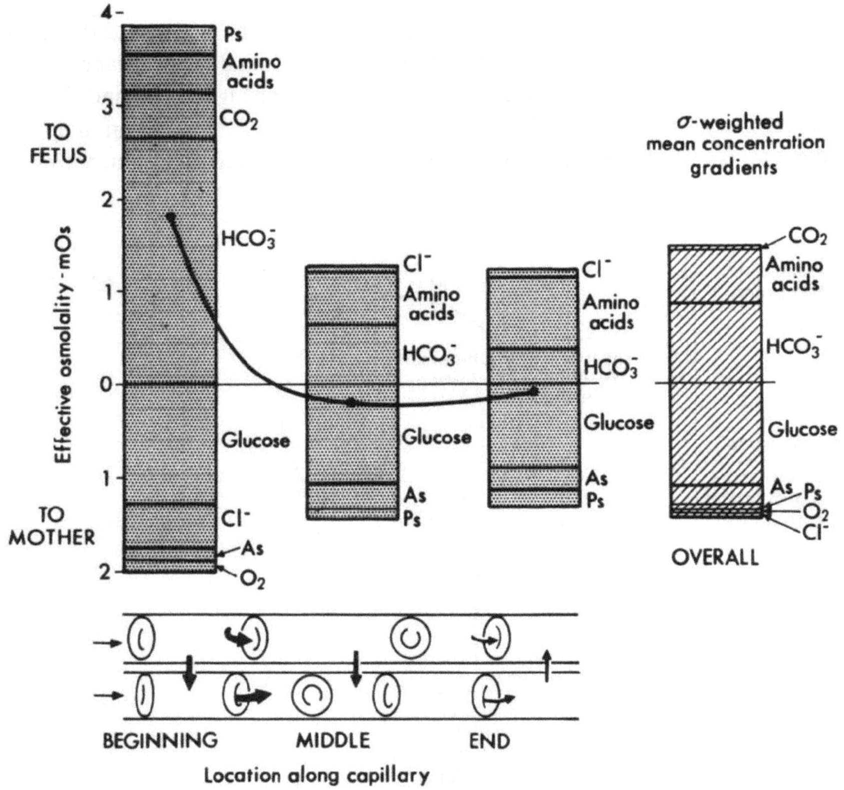

Fig. 6. Calculated transplacental osmotic forces at the beginning, middle, and end of the placental capillaries and the overall effective gradients (shown at *right*). (From Wilbur et al., 1978)

Toward the end of the capillary, this effect becomes much less important. Amino acids, which are actively transported from mother to fetus, increase their concentration in fetal blood later in the capillary, and produce a larger drive to move water from the mother to the fetus. Glucose has a higher concentration in maternal blood and tends to move water toward the mother, but that tendency decreases as the glucose gradient decreases along the capillary. Other ions have less effect. The net effect is that shown by the solid curved line. Initially, the osmotic gradient tends to move water into the fetus, but later during the capillary transit the osmotic gradient reverses, and water returns to the mother. These calculations predict that a large amount of water flows from mother to fetus, but that most of it moves back to the mother again so that the net water flux amounts to 0.014 ml/min or the 20 ml/day, which accounts for the fetal weight gain. The results suggest that the large water flux can produce substantial movement of ions and solutes by solvent drag.

It should be noted that certain unknown constants needed for these calculations were adjusted in order to obtain the net water flux which agrees with known fetal weight gain. In addition, some dispute exists as to the relative importance of osmotic pressure vs hydrostatic pressure differences in determining placental water balance (Armentrout et al., 1977; Conrad and Faber, 1977). Nevertheless, this analysis of the forces determining water balance provides some interesting insight into the complex interactions between the various substances exchanging in the placenta. It also suggests a possible mechanism for control of fetal weight gain and oxygen delivery to tissues. Fetal metabolism alters fetal bicarbonate ion concentration, hence osmotic balance and transplacental water movement. The fetal water gain affects fetal blood volume, venous return, cardiac output, and oxygen delivery (Longo and Power, 1973).

1.7 Conclusions

Despite major differences in placental architecture, morphology, and patterns of maternal and fetal blood flows among different species, the fetal to maternal P_{CO_2} tension difference of about 4 torr is remarkably constant. Post-exchange-capillary changes in P_{CO_2} tensions, a matter of theoretical interest, probably account for no more than 1 torr difference between umbilical and uterine venous blood; while placental CO_2 production probably accounts for none of this difference. Most of the umbilical to uterine vein difference can be explained on the basis of uneven distributions of maternal and fetal blood flow, at least in sheep. Although maternal blood has lower hemoglobin and higher bicarbonate ion concentration than fetal blood, the carbon dioxide dissociation curves of these bloods appear to be superimposed. Carbon dioxide apparently crosses the placenta as molecular CO_2 rather than as bicarbonate ion. Carbon dioxide helps establish fetal homeostasis, as the bicarbonate ion seems to play an important role in transplacental movements, fetal blood volume, cardiac output, and the ultimate oxygen delivery to fetal tissues.

References

Armentrout T, Katz S, Thornburg KL, Faber JJ (1977) Osmotic flow through the placental barrier of chronically prepared sheep. Am J Physiol 233: H466–H474

Baillie P, Dawes GS, Merlet CL, Richards R (1971) Maternal hyperventilation and foetal hypocapnia in sheep. J Physiol (London) 218: 635–650

Barcroft J, Barron DH (1946) Observations upon the form and relations of the maternal and fetal vessels in the placenta of the sheep. Anat Rec 94: 569–595

Barron DH, Meschia G (1954) A comparative study of the exchange of the respiratory gases across the placenta. Cold Spring Harbor Symp Quant Biol 19: 93–101

Bartels H, El Yassin D, Reinhardt W (1967) Comparative studies of placental gas exchange in guinea pigs, rabbits and goats. Respir Physiol 2: 149–162

Battaglia FC, Behrman RE, Bruns PB, Hellegers AE, Meschia G, Schruefer JJ, Seeds AE, Barron DH (1965) Gestational changes in arteriovenous differences of oxygen across the uterus of the rhesus monkey. Am J Obstet Gynecol 93: 32–36

Berfenstam R (1952) Studies in carbonic anhydrase activity in children. I. Enzyme activity in the blood of infants and children of different ages, particularly in premature infants. Acta Paediatr Scand 41: 32–52

Bissonnette JM, Gurtner GH (1972) Transfer of inert gases and tritiated water across the sheep placenta. J Appl Physiol 32: 64–69

Blechner JN, Cotter JR, Hinkley CM, Prystowsky H (1968) Observations on pregnancy at altitude. Am J Obstet Gynecol 102: 794–804

Bohr C (1909) Über die spezifische Tätigkeit der Lungen bei der respiratorischen Gasaufnahme und ihr Verhalten zu der durch die Alveolarwand stattfindenden Gasdiffusion. Scand Arch Physiol 22: 221–280

Bruns PD, Cooper WE, Drose VE (1962) Maternal-fetal oxygen and acid-base studies and their relationships to hyaline membrane disease in the newborn infant. Am J Obstet Gynecol 82: 1079–1088

Comline RS, Silver M (1970a) Daily changes in foetal and maternal blood of conscious pregnant ewes, with catheters in umbilical and uterine vessels. J Physiol (London) 209: 567–586

Comline RS, Silver M (1970b) P_{O_2}, P_{CO_2} and pH levels in the umbilical and uterine blood of the mare and ewe. J Physiol (London) 209: 587–608

Comlime RS, Silver M (1974) A comparative study of blood gas tensions, oxygen affinity and red cell 2, 3 DPG concentrations in foetal and maternal blood in the mare, cow and sow. J Physiol (London) 242: 805–826

Conrad EE, Faber JJ (1977) Water and electrolyte acquisition across the placenta of the sheep. Am J Physiol 233: H475–H487

Eastman NJ, Geiling EMK, DeLawder AM (1933) Foetal blood studies. IV. The oxygen and carbon dioxide dissociation curves of foetal blood. John Hopkins Hosp Bull 53: 246–254

Faber JJ, Hart FM (1966) The rabbit placenta as an organ of diffusional exchange. Circ Res 19: 816–833

Forster RE, Crandall ED (1975) Time course of exchanges between red cells and extracellular fluid during CO_2 uptake. J Appl Physiol 38: 710–718

Gahlenbeck H, Frerking H, Rathschlag-Schaefer AM, Bartels H (1968) Oxygen and carbon dioxide exchange across the cow placenta during the second part of pregnancy. Respir Physiol 4: 119–131

Hill EP, Power GG, Longo LD (1973a) A mathematical model of carbon dioxide transfer in the placenta and its interaction with oxygen. Am J Physiol 224: 283–299

Hill EP, Power GG, Longo LD (1973b) Mathematical simulation of pulmonary O_2 and CO_2 exchange. Am J Physiol 224: 904–917

Holland RAB (1978) Rate of carbon dioxide uptake by suspensions of human fetal red cells. In: Longo LD, Reneau DD (eds) Fetal and newborn cardiovascular physiology, vol II. Fetal and newborn circulation. Garland STPM Press, New York

Kaiser IH, Cummings JN (1957) Hydrogen ion and hemoglobin concentration, carbon dioxide and oxygen content of blood of the pregnant ewe and fetal lamb. J Appl Physiol 10: 484–492

Kaiser IH, Cummings JN (1958) pH, carbon dioxide, oxygen, hemoglobin and plasma electrolytes in blood of pregnant goats and their fetuses. Am J Physiol 195: 481–486

Lamport H (1954) The transport of oxygen in the sheep's placenta: the diffusion constant of the placenta. Yale J Biol Med 27: 26–34

Longo LD, Power GG (1969) Analysis of P_{O_2} and P_{CO_2} differences between maternal and fetal blood in the placenta. J Appl Physiol 26: 48–55

Longo LD, Power GG (1973) Long-term regulation of fetal cardiac output. Gynecol Invest 4: 277–287

Longo LD, Delivoria-Papadopoulos M, Power GG, Hill EP, Forster II RE (1970) Diffusion equilibration of inert gases between maternal and fetal placental capillaries. Am J Physiol 219: 561–569
Longo LD, Delivoria-Papadopoulos M, Forster II RE (1974) Placental CO_2 transfer after fetal carbonic anhydrase inhibition. Am J Physiol 226: 703–710
Makowski EL (1968) Maternal and fetal vascular nets in placentas of sheep and goats. Am J Obstet Gynecol 100: 283–288
Meschia G, Makowski EJ, Battaglia FC (1970) The use of indwelling catheters in the uterine and umbilical veins of sheep for a description of fetal acid-base balance and oxygenation. Yale J Biol Med 42: 154–165
Metcalfe J, Bartels H, Moll W (1967) Gas exchange in the pregnant uterus. Physiol Rev 47: 782–838
Moll W, Kastendieck E (1977) Transfer of N_2O, CO and HTO in the artificially perfused guinea-pig placenta. Respir Physiol 29: 283–302
Power GG, Longo LD, Wagner Jr HN, Kuhl DE, Forster II RE (1967) Uneven distribution of maternal and fetal placental blood flow, as demonstrated using macroaggregates, and its response to hypoxia. J Clin Invest 46: 2053–2063
Power GG, Nelson PS, Dale PS (1978) Microdistribution of maternal and fetal blood flows in the sheep placenta. Fed Proc 37: 615
Prystowsky H, Hellegers AE, Bruns P (1961) Fetal blood studies XV. The carbon dioxide concentration gradient between the fetal and maternal blood of humans. Am J Obstet Gynec 81: 372–376
Rankin JHG, Peterson EN (1969) Application of the theory of heat exchangers to a physiological study of the goat placenta. Circ Res 24: 235–250
Rooth G, Sjöstedt S (1962) The placental transfer of gases and fixed acids. Arch Dis Child 37: 366–370
Stenger V, Eitzman D, Andersen T, De Padua C, Gessner I, Prystowsky H (1964) Observations on the placental exchange of the respiratory gases in pregnant women at cesarean section. Am J Obstet Gynecol 88: 45–47
Swenson ER, Maren TH (1978) A quantitative analysis of CO_2 transport at rest and during maximal exercise. Respir Physiol 35: 129–159
Wagner PD (1977) Diffusion and chemical reaction in pulmonary gas exchange. Physiol Rev 57: 257–312
Wilbur WJ, Power GG, Longo LD (1978) Water exchange in the placenta: a mathematical model. Am J Physiol 235: R181–R199
Wulf H (1962) Der Gasaustausch in der reifen Placenta des Menschen. Z Geburtshilfe Gynäkol 158: 117–134

Transfer of Bicarbonate in a Hemochorial Placenta

W. MOLL[1], E. KASTENDIECK[2], and H. GIRARD[3]

1 Introduction

1.1 The Placental Barrier in the Pregnant Uterus

The pregnant organism may be considered – as first approximation – as a two-compartment system, consisting of the maternal space and the fetal space (Fig. 1). The two spaces are separated by the placental barrier. This placental barrier is a multi-layered membrane mediating countless transport processes between the two spaces. This membrane may contain different numbers of tissue layers in the various species. In this paper we would like to focus on the bicarbonate transport in a hemochorial placenta which also occurs in man. Bicarbonate transport is likely to occur frequently in the hemochorial placenta in response to transplacental concentration gradients arising during birth. These gradients will be considered first.

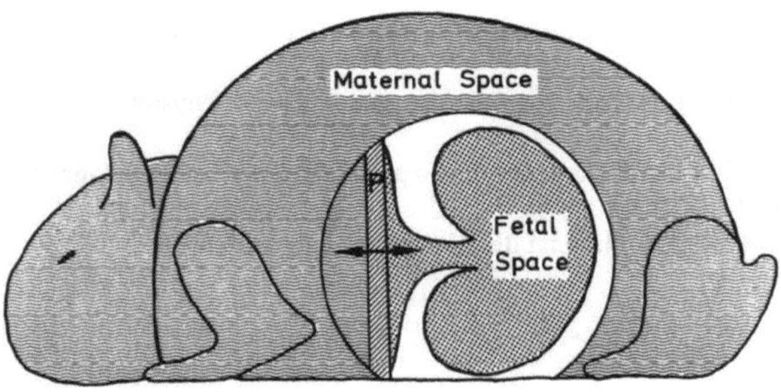

Fig. 1. The pregnant organism as a two-compartment system consisting of a maternal space and a fetal space separated by the placental membrane (*P*). The placental membrane mediates the maternal-fetal exchange. It shows large species difference in histological structure and permeability

1.2 Transplacental Gradients of Bicarbonate and Lactate Due to Production and Buffering of Lactic Acid in the Pregnant Organism

Labor, i.e., uterine contractions and eventually contractions of skeletal muscles lead to lactic acid accumulation as do most forms of muscular activity. Clinical reports show a steady rise of lactate in maternal blood during labor (Schmid, 1973).

1 Universität Regensburg, Institut für Physiologie, 8400 Regensburg/FRG
2 Universität Würzburg, Frauenklinik, 8700 Würzburg/FRG
3 CNRS, Laboratoire de Physiologie Respiratoire, 67087 Strasbourg/F

Simultaneously protons are produced which are predominantly buffered by bicarbonate, so that bicarbonate falls and lactate rises in maternal blood. In this way a maternofetal gradient of lactate and an opposite gradient of bicarbonate may develop across the placental barrier in the course of delivery. During fetal anoxia gradients in the opposite direction may occur: When fetal arterial O_2 saturations falls, lactate accumulates and bicarbonate falls at an equivalent speed of nearly 1 mmol/l/min when fetal anoxia is complete (Moll and Kastendieck, 1978). In these conditions, a maternofetal gradient for bicarbonate and a fetomaternal gradient for lactate occur in the placental barrier.

1.3 Present Evidence of Bicarbonate Transfer in the Hemochorial Placenta

What is the rate of placental bicarbonate transfer which occurs in response to the transplacental bicarbonate concentration gradients? Bartels and his coworkers found that during delivery standard bicarbonate changes in fetal blood during passage through the placenta (Beer et al., 1955). Rooth (1964) and Newman et al. (1967) found that bicarbonate infusion into the maternal blood results in a rise of fetal bicarbonate so that it appears firmly established that the human placenta transfers bicarbonate. However, there has been still a lack of quantitative data on the rate of bicarbonate transfer in the placenta. We therefore tried to measure the placental bicarbonate transfer for various transplacental concentration gradients in order to determine the placental bicarbonate permeability (Kastendieck and Moll, 1977). We used the guinea-pig for these studies since the histological structure of the placental membrane in this animal is very similar to that in humans (Enders, 1965). In the present paper the experimental approaches will be described and the placental bicarbonate transfer will be presented and characterized in quantitative terms so that some conclusion on the transport mechanism may be drawn.

2 Experimental Approaches for Studying Placental Transfer in the Guinea-Pig: "Open Circuit" and "Closed Circuit" Perfusion Techniques

For determining the relationship between transfer and transplacental concentration gradients in the placenta it is advantageous to perfuse the placenta artificially on one or both sides. In our studies we used a preparation where the blood flow is maintained on the maternal side and the fetal side only is artificially perfused. This technique yields stable permeability values which are similar to those found in the intact animal. The technique of artificial perfusion of the umbilical circulation has originally been described by Dancis and Money (1960) and by Metcalfe et al. (1966). It was improved by us by developing a method for controlling maternal placental flow. In Fig. 2 two alternative ways of placental perfusion are shown. In Fig. 2A, the placenta is perfused applying an "open circuit" perfusion system. In this preparation, the fetal arterial concentrations may be adjusted to a given value. The release or uptake of substances by the perfusate may be measured and yield the rate of placental transfer since placental metabolism seems not to interfere with the transfer in these conditions.

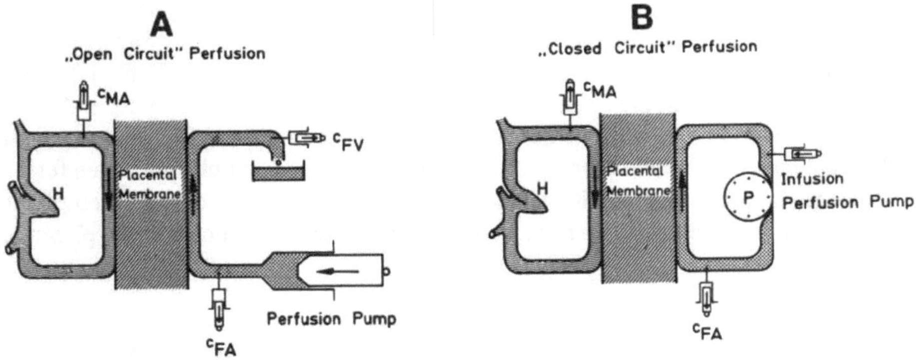

Fig. 2 A, B. Experimental approaches for studying the relation between transplacental gradients and transfer. The umbilical circulation is perfused applying (**A**) an "open circuit" perfusion technique, or (**B**) a "closed circuit" perfusion technique. Maternal blood flow is maintained and controlled. Samples are drawn from maternal arteries (*MA*), fetal arteries (*FA*) and fetal vein (*FV*)

In the "closed circuit" perfusion technique (Fig. 2B) the perfusion fluid is recirculating. Bicarbonate or any other substance may be continuously infused on the fetal side. In the steady state the rate of transfer becomes equal to the rate of infusion. In this way, the transfer may be adjusted to a given value in this preparation and the concentration difference may be measured for these conditions. The "closed circuit" technique has a number of additional advantages when compared with the "open circuit" technique: The perfusate equilibrates automatically with the maternal partial pressures and the permeability values are not affected by short transient changes of blood flow. The data show a smaller scatter than those obtained with the "open circuit" system. With both techniques the data yield the transfer per concentration difference between the maternal and fetal arteries, i.e., the placental clearance as defined by Meschia et al. (1967). Since the guinea-pig placenta is known to have the functional behavior of a quasi-ideal countercurrent system (Moll and Kastendieck, 1977) the transfer per mean concentration gradient, i.e., the placental permeability can be calculated from the clearance value. Finally, since the exchange surface area is known for the guinea-pig placenta near term (0.4 m^2 according to Müller and Fischer, 1968; Kaufmann and Davidoff, 1977) even the transfer per mean concentration gradient and surface area, i.e., the true permeability of the placental barrier can be estimated.

3 Characteristics of Placental Bicarbonate Transfer in the Guinea-Pig Placenta

3.1 The Rate of Bicarbonate Transfer Across the Placental Barrier

Applying an "open circuit" perfusion technique we perfused the fetal side of the placenta with a lactate-free solution containing 20 mM bicarbonate. Maternal bicarbonate concentration was 16 mM and maternal lactate concentration was 5 mM, i.e., there was a fetomaternal concentration gradient of bicarbonate and a lactate concentration gradient in the opposite direction (Fig. 3). In these conditions, bicarbonate was transferred at a rate of about 60 μmol/h per g of placental weight. Placental bicarbonate transfer was proportional to the transplacental concentration difference over a wide range of concentrations. The value of placental permeability for bicarbonate may be illustrated best by the rate with which fetal and maternal bicarbonate spaces equilibrate: The half time of this process is in the range of around 20 min, i.e., there is apparently a fairly rapid bicarbonate transfer in the guinea-pig placenta.

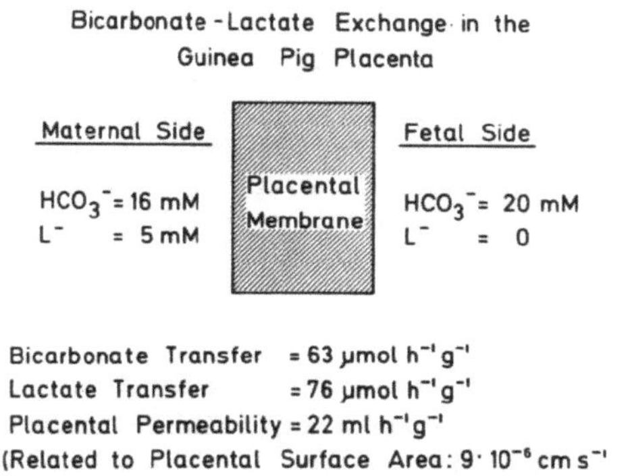

Fig. 3. Bicarbonate transfer and lactate transfer across the guinea-pig placenta (the values are taken from Kastendieck and Moll, 1977)

The high bicarbonate permeability seems to be only true one for the hemochorial placenta, i.e., the placenta of man, most primates, rodents and rabbits, where maternal blood immediately contacts fetal tissue. The bicarbonate permeability in the hemochorial placenta is much higher than in sheep where the placental barrier includes also maternal epithelium and endothelium. As a result, the placental clearance, i.e., transfer per arterial concentration difference, is 20 times higher in guinea-pigs than in sheep (Kastendieck et al., 1979). The placental bicarbonate permeability appears to be extraordinarily high in the hemochorial placenta.

3.2 Free Diffusion of Bicarbonate Ions Across the Placental Barrier

We were interested in the characteristics of placental bicarbonate transfer in the hemochorial placenta. In the above studies, bicarbonate transfer in fetomaternal direction occurred along with a lactate transfer in the opposite direction, i.e., actually a bicarbonate-lactate exchange took place. We wondered whether the bicarbonate transfer is strongly coupled with lactate transfer in the other direction or whether bicarbonate transfer can also proceed in exchange for other ions such as chloride in the form of free ionic diffusion across hydrophilic pores as indicated in Fig. 4A. If such a free ionic diffusion is true, placental bicarbonate transfer should also occur when only chloride is available for exchange against bicarbonate. In order to test for bicarbonate transfer when no lactate is available for exchange, we perfused the fetal

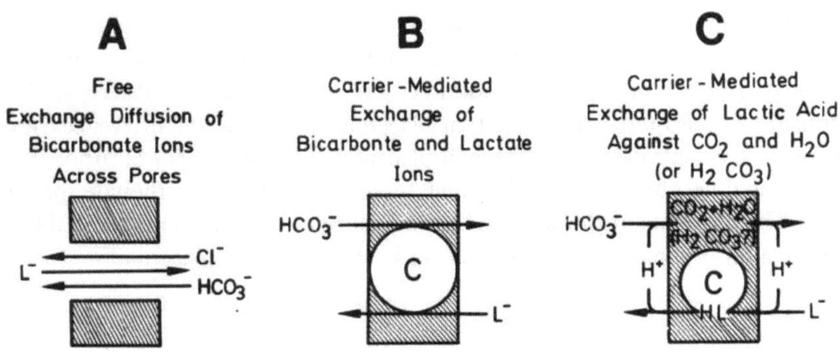

Fig. 4 A–C. Schematic diagram illustrating different ways of placental bicarbonate transfer. **A** Free diffusion of bicarbonate in exchange against another diffusable anion via hydrophilic pores. **B** Carrier-mediated direct exchange of lactate ions and bicarbonate ions (*C* indicates the carrier). **C** Carrier-mediated exchange of CO_2 and water or possibly carbonic acid, against lactic acid. *C* indicates the carrier which transfers lactic acid. The lactic acid transfer may occur in any form of proton-lactate symport

side of the placenta with a lactate-free, low-bicarbonate solution. As indicated in Fig. 5, there was a maternofetal concentration gradient and a parallel lactate gradient. In these conditions, we found that a parallel bicarbonate transfer and lactate transfer occurred across the placenta. This observation indicates that bicarbonate may exchange even against chloride as it is expected on the basis of a free ionic diffusion of bicarbonate across hydrophilic pores.

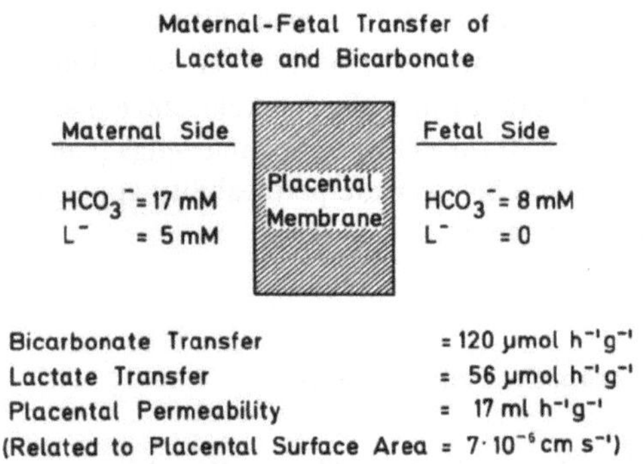

Fig. 5. Simulatenous transfer of lactate and bicarbonate from the maternal to the fetal side demonstrating that bicarbonate may also be transferred when no lactate is available for exchange. The data are taken from Kastendieck and Moll (1977)

3.3 Lactate-Coupled Bicarbonate Transfer

Whereas the above observations clearly show that bicarbonate may penetrate the placental barrier even independently of lactate transfer, they do not exclude that part of placental bicarbonate transfer is coupled with lactate transfer under normal conditions. Actually, in the present experiments bicarbonate transfer per concentration difference was less than found when lactate and bicarbonate gradients were in opposite directions. This indicates some extent of coupling between placental lactate transfer and placental bicarbonate transfer. The coupling became finally evident by an observation we made in experiments where the fetal perfusate was recirculating as shown in Fig. 2B: At the beginning of such a perfusion study maternal and fetal bicarbonate concentrations were fairly close as indicated in Fig. 6. After 30 min sodium lactate was infused

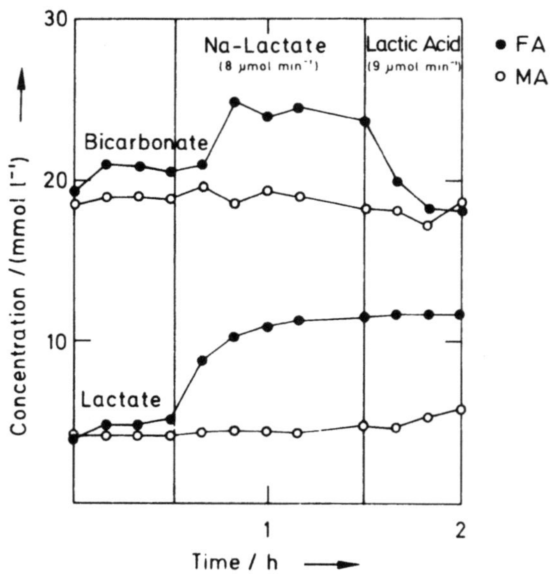

Fig. 6. Bicarbonate and lactate concentrations in recirculating fetal perfusate and maternal plasma before and during infusion of sodium lactate and lactic acid, respectively. During infusion of sodium lactate a concentration gradient of bicarbonate develops, which disappears when lactic acid is infused. The *points* indicate means of data obtained in three experiments (unpublished results). P_{CO_2} was around 40 mm Hg

into the fetal circulation at a rate of 8 μmol/min. As expected, the lactate concentration on the fetal side rose and reached a plateau, indicating that the amount of lactate that was infused was also transferred across the placenta. In this condition, bicarbonate concentration in the umbilical circulation showed an unexpected, remarkable increase. Apparently, bicarbonate was transferred across the placental barrier from the maternal to the fetal side against a concentration difference. This means counter-transport which strongly indicates some sort of coupling between bicarbonate transfer and lactate transfer. We finally could show that just as a lactate gradient may drive bicarbonate ions, bicarbonate gradients are able to drive lactate ions: When a bicarbonate gradient was established across the placental barrier by infusing NaOH or HCl into the umbilical circulation a gradient of lactate developed across the placental barrier (see Fig. 7A and B), indicating that lactate was transferred from the maternal blood to the perfusate against a concentration gradient. This kind of interaction between lactate transfer and bicarbonate transfer could not be explained by linkage due to diffusion potentials as can be shown theoretically by applying the Goldman equations.

What a kind of transport system is mediating the uphill transfer? First, we may think of a transport mechanism which directly exchanges bicarbonate and lactate ions as such allowing for mediating bicarbonate flux in addition to ionic flux (see Fig. 4B).

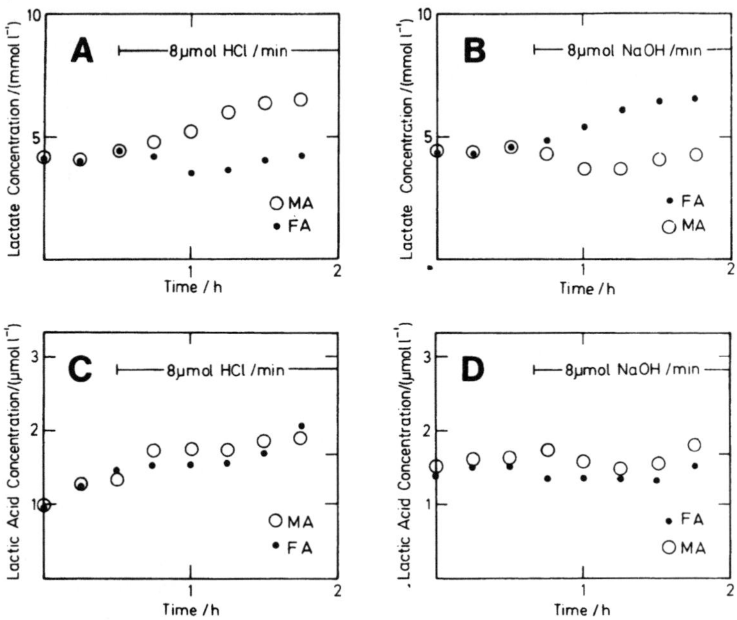

Fig. 7 A–D. Lactate concentrations in recirculating fetal perfusate before and during infusion of HCl (**A**) and NaOH (**B**) into the fetal circulation and calculated concentrations of lactic acid during infusion of HCl (**C**) and NaOH (**D**). The *points* are based on data obtained in three experiments (according to Moll et al., 1978). P_{CO_2} was around 40 mm Hg

Such a system would be capable of maintaining a counterflow of bicarbonate, when a lactate gradient is present. However, other models are also possible: There may be an exchange of lactic acid against CO_2 and water. CO_2-H_2O transfer means bicarbonate transfer when protons are fetched back by lactic acid transfer, i.e., by proton lactate symport (see Fig. 4C). Both models (Fig. 4B and C) postulate a specific transport system for lactate.

3.4 Mediated Transfer of Lactate in the Placenta

The placental transport system which produces the coupling of bicarbonate transfer and lactate transfer could be characterized by observing the lactate transfer across the placenta under various conditions. It turned out that the transport of lactate exhibits some of the well-known identifying characteristics of mediated transfer (facilitated diffusion):

Stereospecifity. Placental transfer per concentration difference shows stereospecifity as evidenced by the fact that the placental permeability for the physiological L-lactate is more than twice that for D-lactate. Stereospecifity is generally considered as an identifying characteristic for facilitated (mediated) transport (see Lehninger, 1975).

Saturation Kinetics. When L-lactate concentration is increased in the fetal perfusate, L-lactate transfer increases but not proportionally. The rate of transfer of L-lactate approaches a maximum value as indicated in Fig. 8. There is evidence that placental L-lactate transfer exhibits saturation kinetics.

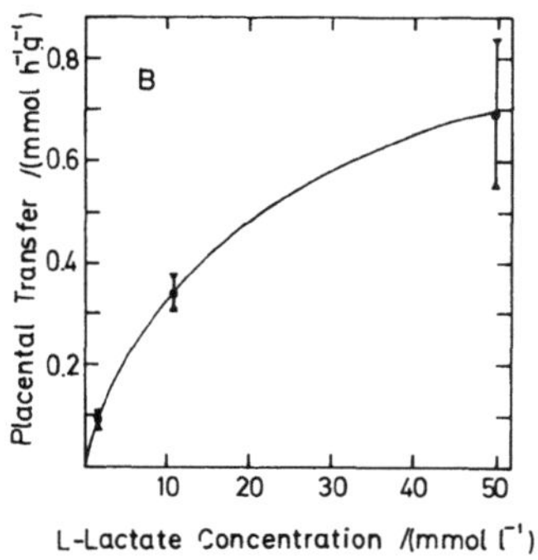

Fig. 8. Maternofetal lactate transfer related to L-lactate concentration in the perfusate. The transfer exhibits saturation kinetics (according to Moll et al., 1978)

Competitive Inhibition. Placental transfer of L-lactate is inhibited by pyruvate. In the presence of pyruvate in the perfusate at a concentration of 50 mM placental transfer of L-lactate was reduced to about 13% (SE = 3%).

Thus, placental lactate transfer shows three identifying characteristics for mediated transport and we believe therefore that indeed facilitated diffusion of lactate occurs in the placenta. A similar facilitated diffusion has been described for ascites tumor cells by Spencer and Lehninger (1976).

3.5 Carbon Dioxide – Lactic Acid Exchange in the Placenta?

The final discussion will focus on the question whether direct mediated bicarbonate-lactate exchange or CO_2-lactic acid exchange occurs in the placenta. Both models explain the bicarbonate-lactate interactions. How can these two processes be distinguished? The bicarbonate-lactate exchange system as described in Fig. 2B should depend on bicarbonate concentration. When the bicarbonate concentration is decreased on the fetal side, feto-maternal transfer of lactate should be accelerated. Actually, we found no change of feto-maternal lactate transfer when the umbilical circulation was perfused with a phosphate-buffered solution of low bicarbonate concentration. On the other hand, a characteristic phenomenon which is common to both systems (Fig. 4B and C) can be demonstrated: Both systems promote the equilibration of undissociated lactic acid on the maternal and fetal side. Such an equilibration of lactic acid could be observed: In the steady state there is an equilibrium of lactic acid concentrations on the fetal and maternal side even in conditions of large lactate concentrations differences as they occur during infusion of HCl and NaOH (see Fig. 7C and D). Thus, on the basis of present evidence, we may explain the characteristics of placental bicarbonate transfer by assuming that CO_2 and water (possibly H_2CO_3) penetrate the placental barrier and protons are transported back by lactic acid transfer so that bicarbonate is formed on the other side of the membrane.

4 Conclusions

The present evidence shows that the hemochorial placenta of the guinea-pig is highly permeable for bicarbonate. The observations indicate that bicarbonate has two mechanisms of penetrating the placental barrier. One mechanism seems to be based on an apparent free ionic diffusion, possibly across hydrophilic pores. The other mechanism is related to a specific transport system of lactate. The results may be explained by assuming an exchange of CO_2 and water against lactic acid.

References

Beer R, Bartels H, Raczkowski H (1955) Die Sauerstoffdissoziationskurve des fetalen Blutes und der Gasaustausch in der menschlichen Placenta. Pflügers Arch 260: 306–319

Dancis J, Money WL (1960) Transfer of sodium and iodoantipyrine across guinea-oig placenta with an in situ perfusion technique. Am J Obstet Gynecol 80: 215–220

Enders AC (1965) A comparative study of the fine structure of the trophoblast in several hemochorial placentas. Am J Anat 116: 29–68

Kastendieck E, Moll W (1976) The transfer of lactic acid across the guinea-pig placenta. Abstr 5th Eur Congr Perinatal Med, Uppsala, Sweden, June 9–12

Kastendieck E, Moll W (1977) The placental transfer of lactate and bicarbonate in the guinea-pig. Pflügers Arch 370: 165–171

Kastendieck E, Künzel W, Kurz CS (1979) Placental clearance of lactate and bicarbonate in sheep. Gynec Obstet Invest 10: 9–22

Kaufmann P, Davidoff M (1977) Adavances in anatomy, embryology and cell biology, vol 53. Springer, Berlin Heidelberg New York

Lehninger AL (1975) Biochemistry. Worth, New York

Meschia G, Battaglia FC, Bruns PD (1967) Theoretical and experimental study of transplacental diffusion. J Appl Physiol 22: 1171–1178

Metcalfe J, Moll W, Bartels H, Hilpert P, Parer JT (1966) Transfer of carbon monoxide and nitrous oxide in the artificially perfused sheep placenta. Circ Res 16: 95–101

Moll W, Kastendieck E (1977) Transfer of CO, N_2O and HTO in the artificially perfused guinea-pig placenta. Respir Physiol 29: 283–302

Moll W, Kastendieck E (1978) Accumulation and disappearance of lactate in a fetus with a hemochorial placenta. The role of placental transfer and fetal metabolism. J Perinat Med 6: 246–254

Moll W, Girard H, Gros G (1978) Facilitated diffusion of lactic acid across the guinea-pig placenta. Pflügers Arch 373: R50

Müller G, Fischer WM (1968) Über den fetalen und maternalen Blutkreislauf in der Meerschweinchenplacenta. Verh Anat Ges 62: 231–239

Newman W, Mitchell P, Wood C (1967) Fetal acid-base status. II. Relationship between maternal and fetal blood bicarbonate concentrations. Am J Obstet Gynecol 97: 52–57

Rooth G (1964) Early detection and prevention of foetal acidosis. Lancet Febr 1964: 290–293

Schmid J (1973) Glucose, Laktat und Pyruvat in der Schwangerschaft und unter der Geburt. Fortschr Geburtshilfe Gynaekol 50: 1–99

Spencer TL, Lehninger AL (1976) L-lactate transport in Ehrlich Ascites-Tumour cells. Biochem J 154: 405–414

Recent Developments in Carbonic Anhydrase Localization: Occurrence of Carbonic Anhydrase in Calcium-Mobilizing Tissues

C.V. GAY [1]

1 Introduction

There are currently three approaches to carbonic anhydrase localization: the cobalt precipitation method of Häusler [15] as modified by Hansson [14], autoradiography of the specific carbonic anhydrase inhibitor, ^{3}H-acetazolamide [7, 8], and immunocytochemistry [9, 13, 17, 27, 32].

A number of investigators consider that Häusler's method is unreliable [6, 18, 23–25] primarily because of the long incubation times required (20–120 min) and the high levels (millimolar) of acetazolamide required for inhibition of the enzyme. Hansson's method has stirred considerable controversy [18, 24, 25, 29] although it continues to be used. The major reasons for believing the validity of Hansson's method are that localizations are frequently found in tissues known to contain carbonic anhydrase and that the staining reaction is prevented by low concentrations of acetazolamide. However, there are several questionable aspects of Hansson's method. One bothersome finding is that when carbonic anhydrase catalyzes the dehydration of bicarbonate, enzyme activity is rapidly lost as the pH rises to 8 [26]; Hansson's unbuffered medium rises to pH 8 within about 3 min after mixing [19], yet 10–15 min are recommended for adequate staining [14, 18]. Anions such as bisulfate and phosphate, both constituents of Hansson's medium, alter the pKa of the activity-linked group [16], and it is possible that enzyme activity is reduced in the presence of these ions. Nuclei stain with Hansson's method [14, 18, 25] while biochemical assay of nuclear fractions of tissue homogenates are devoid of carbonic anhydrase activity. It is also curious that carbonic anhydrase in muscle, recently observed by biochemical techniques [12, 21, 28], has not been detected by Hansson's method [18]. Loveridge [19] reports staining in muscle, but believes it is due to nonspecific adsorption of cobalt.

Other than inhibition by acetazolamide, controls for Hansson's method appear to be inadequate. When staining is carried out in a CO_2 atmosphere, no reaction product develops. Presumably, the catalytic reaction has been driven in reverse. However, high CO_2 would cause the pH of the medium to remain low and cobalt salts would not precipitate. Heating of sections also prevents staining, presumably by loss of enzyme activity. However, the boiling may alter or wash away molecules so that binding of cobalt salts no longer occurs. When the substrate, bicarbonate, is omitted staining is reduced, but not abolished [19].

1 Department of Biochemistry and Biophysics, The Pennsylvania State University, University Park, PA 16802/USA

Since there is reason to question Hansson's method, we have applied several purified substances to filter paper and stained them by Hansson's procedure.

2 Filter Paper Experiment

As shown in Table 1, glycogen, hyaluronic acid, bovine hemoglobin, chicken erythrocyte carbonic anhydrase purified in this laboratory [2], pepsin and cytochrome c were spotted and dried on glass fiber filter paper (Whatman GF/C). Each spot contained 20 μg of material. In addition, carbonic anhydrase had 20 units of activity using Maren's micro-method [20]. The filter paper was approximately 200 μm thick, but was very porous. The filter papers were stained for 10 min by floating them on freshly mixed Hansson's medium which contained 1.75 mM $CoSO_4$, 53 mM H_2SO_4, 11.7 mM KH_2PO_4, and 157 mM $NaHCO_3$. The filters were rinsed three times in 1% NaCl for a total of 10 min, exposed to 0.5% $(NH_4)_2$ S, rinsed in 1% NaCl and mounted in glycerol. Initially all substances were stained, but after 24 h the glycogen, hyaluronic acid and pepsin spots faded. Acetazolamide (10 μM) added to the reaction medium diminished staining only slightly, indicating that enzyme activity was not the basis for staining of any spots.

Table 1. Staining intensities of several substances spotted on filter paper, dried and stained by Hansson's method (see text)

Substance	Immediate staining intensity	Intensity after 24 h
Carbonic anhydrase	+ + + +	+ + +
Glycogen	+ + +	–
Hyaluronic acid	+ + +	–
Hemoglobin	+ + +	+ +
Pepsin	+ +	–
Cytochrome c	+ + +	+ + +

While this preliminary experiment requires further work, it suggests that several substances, including carbonic anhydrase, can bind reaction product. The fading observed may be due to solvation of glycogen, hyaluronic acid, and pepsin since no fixation was used. Using a similar approach, others have found that carboxypeptidase A [25] and cytochrome c [6] stain.

Because of the confusion associated with the cobalt precipitation method, we have explored, over the past several years, alternate procedures for localizing carbonic anhydrase.

3 Autoradiography of ^{3}H-Acetazolamide

For localization of carbonic anhydrase by labeled inhibitor autoradiography, 25 mCi/100 mg body weight ^{3}H-acetazolamide (sp. act. = 4.1 Ci/mM, ICN, Irvine, CA) was added to unlabeled acetazolamide to provide a total concentration of 5–10 mg/kg body weight. The drug was injected intravenously and the dosage level was sufficient to saturate the active sites throughout the body. After allowing time for renal clearance of loosely bound and free inhibitor (3 h for quail, 24 h for chickens), tissues were excised and processed for microscopy. Originally, we embedded freeze-dried tissues directly into epoxy resin to minimize dissolution and relocation of the radioactive label. Subsequently we have found that with glutaraldehyde fixation followed by dehydration in ethylene glycol and embeddment in epoxy resin, 80% of the label was retained (unpublished observation). Localization was the same as when freeze-drying was used.

The requirements for successful labeled inhibitor autoradiography have been described in general by E.A. Barnard [1] and for carbonic anhydrase by us [7, 8].

1. The inhibitor must enter the cell and bind to enzyme. There is ample evidence that acetazolamide enters cells in a matter of minutes.
2. The inhibitor must be highly specific. Carbonic anhydrase is the only known high affinity binding site for acetazolamide. Acetyl CoA carboxylase [4] and pyruvate carboxylase [5] are inhibited by acetazolamide, but only when levels of drug 100- to 1000-fold greater than those needed to inhibit carbonic anhydrase are used.
3. The binding of the inhibitor to the enzyme must have a small dissociation constant ($\leqslant 10^{-8}$ M). This is true for both avian [2] and mammalian carbonic anhydrase [3].
4. The inhibitor must remain bound to enzyme throughout tissue preparation. Evidence for this is the observation of discrete localizations in tissues known to contain carbonic anhydrase. Also, there was no detectable diffusion into adjacent embedding medium.
5. The inhibitor must not be metabolized.

Further validation was provided by finding that skeletal muscle, which contains an acetazolamide-resistant isoenzyme of carbonic anhydrase [21, 28], had no more grains than background.

Table 2 gives a comparison of enzyme activities and autoradiographic grain counts for several tissues. Pancreas had relatively low activity, while shell gland and

Table 2. Comparison of enzyme activity with autoradiographic grain density in tissues from the hen

Tissue	Carbonic anhydrase (μg/g wet weight tissue) [a]	Grain counts per 1000 μm^2
Pancreas	155 ± 7.8 [b]	21.5 ± 1.8
Shell gland mucosa	225 ± 14.4	39.4 ± 2.5
Kidney	270 ± 21.4	40.1 ± 2.1
Gastric mucosa	2820 ± 152.2	89.5 ± 5.1
Red blood cells	3580 ± 205.3 [c]	40.3 ± 4.4

a Measured by Maren's micromethod [20] using purified chicken erythrocyte carbonic anhydrase as a basis

b Means ± S.E

c Based on a hematocrit of 30%

kidney had moderate amounts. Gastric mucosa and red blood cells had high levels of activity. The grain counts reflect enzyme concentration, not activity. If we make the reasonable assumption that all tissues were maximally labeled, pancreas had about half the enzyme concentration of shell gland and kidney, while stomach had twice as much. It is particularly interesting that grain counts of red blood cells and gastric mucosa do not reflect the high levels of activity in these tissues. In red blood cells, the catalytic activity of carbonic anhydrase may be enhanced by closely apposed hemoglobin serving as a proton donor to the active site, as Silverman has suggested [31]. Hence, enzyme activity could be very high while enzyme concentration is low. Possibly gastric mucosa also contains a hydrogen donor.

4 Immunocytochemistry

The immunocytochemical method has been described previously by us [9]. In brief, frozen sections are cut in a cryostat, thawed on a microscope slide, air-dried and stained with the indirect antibody method. Antibodies specific for chicken erythrocyte carbonic anhydrase, which we produced in rabbits, were applied to the sections; the excess was washed away with phosphate buffered saline. A secondary antibody, specific for rabbit immunogammaglobulin and labeled with fluorescein was applied and the excess washed away.

For controls, we replaced primary antibody with saline or nonspecific rabbit immunogammaglobulin, applied nonfluorescent gamma globulin prior to fluorescein-labeled IgG (the blocking test), or replaced the secondary antibody with saline. All controls had little or no fluorescence when appropriate antibody dilutions were used. This will vary with each antibody preparation and proper dilutions must be determined with each new preparation. It was also important to wash rigorously between antibody applications and to avoid fixation. Considerable nonspecific staining occurred when tissues were fixed with glutaraldehyde.

The lack of staining in muscle is presumably due to the antibodies employed being specific for the high-activity enzyme.

More recently, antibodies to human erythrocyte carbonic anhydrase B have been used to localize carbonic anhydrase in human red blood cells and brain where specific fluorescence was closely related to blood vessels [17]. The human antibodies were reported to cross-react with carbonic anhydrase from rat, mouse, and guinea pig. In our studies, we found no cross-reactivity between antibodies to chicken enzyme and rat erythrocyte carbonic anhydrase. The studies of Kumpulainen and Korhonen [17] and of Spicer et al. [32] are important since they allow localization of mammalian carbonic anhydrase.

5 Comparison of Methods

Both the autoradiographic and antibody methods reflect enzyme concentration, not activity. There is no amplification factor built into the autoradiographic method since acetazolamide and carbonic anhydrase bind mole for mole. In the antibody method, amplification is achieved, for example, when several secondary antibodies attach to the primary specific antibody. Regarding sensitivity, approximately 1000 molecules/site must be present for detection by autoradiography [1], whereas in the antibody method, under the right conditions, single molecules may be detected [33]. The resolution of autoradiography is around 1 μm for light and 100 nm for electron microscopy. Resolution for immunocytochemistry appears to be somewhat better than autoradiography. The major disadvantage of both methods is long preparation time. Weeks to months are required for autoradiographs to be exposed or for antibodies to be produced. Once one has the antibodies, however, staining time is short. The major attraction of both methods is high specificity.

In tissues examined by both autoradiography and immunocytochemistry, identical localizations have been obtained (Table 3). The enzyme was distributed throughout the cytoplasm of most cells, as might be expected since carbonic anhydrase is largely found in the soluble fraction of cell homogenates. One exception to this is gastric mucosa where it is found at the luminal surface of the cells.

Table 3. Comparison of carbonic anhydrase localization in avian tissues by autoradiography [7, 8] and antibody [9] methods

	Autoradiography	Antibody
Red blood cells	+ [a]	+
Erythroblasts	+	+
Gastric surface epithelium	+	+
Pancreatic acinar cells [b]	+	
Shell gland columnar cells	+	+
Shell gland tubular gland cells	+	+
Kidney tubules and collecting ducts [b]	+	
Osteoclasts	+	+
Muscle, connective tissue, nuclei	–	–

a Carbonic anhydrase present (+) or absent (–)
b Not analyzed by the antibody method

6 Carbonic Anhydrase in Bone and Chorioallantoic Membrane

Our interst in mineral mobilization has led us to examine bone and chorioallantoic membrane for sites of carbonic anhydrase. In bone, the enzyme occurs in osteoclasts [8]. By biochemical assay we have determined that osteoclast-rich bone, namely, metatarsi of rapidly growing chicks, contains 7.7 units carbonic anhydrase/g wet weight bone [10]. When 50 units parathyoid extract (Lilly) were given to the birds 2 h before killing, carbonic anhydrase activity increased by 32% [11]. While the mechanism of hormonally induced enzyme activation is not known, it seems possible that the role of carbonic anhydrase in bone is secretion of H^+ at sites of bone resorption. Blood exiting from bone predominantly in a state of resorption has been found to be enriched in H^+ and diminished in HCO_3^-, while lactate and citrate remained unchanged [22].

In the chorioallantoic membrane of the developing chick embryo, carbonic anhydrase is barely detectable at 8 days of development [30]. Enzyme activity rises rapidly and continuously from 0.3 units/mg protein at 12 days to 1.13 units at 20 days. Similar findings have been reported by others [34] but our values were approximately ninefold higher since we included a sulfhydryl protecting agent (10 mM dithiothreitol) in the assay system [2]. Localization by autoradiography is in progress. The villous-cavity cell has a particularly high affinity for ^{3}H-acetazolamide (unpublished observation). This cell bears striking resemblance to osteoclasts in that its apical side contains numerous microvilli beneath which are numerous vesicles. There are also numerous mitochondria present.

A likely role for carbonic anhydrase in both osteoclasts and villous cavity cells is elaboration of a H^+-enriched fluid for dissolution of bone mineral or the calcareous egg shell.

We are in the process of applying the autoradiographic and immunocytochemical methods at the ultrastructural level.

Acknowledgment. This work was supported by grant DE 04345 from the National Institutes of Health.

References

1. Barnard EA (1970) Location and measurement of enzymes in single cells by isotopic methods, Part I. Principles and light microscope applications. Int Rev Cytol 29: 213–243
2. Bernstein RS, Schraer R (1972) Purification and properties of an avian carbonic anhydrase from the erythrocytes of *Gallus domesticus*. J Biol Chem 247: 1306–1322
3. Bundy HF (1977) Carbonic anhydrase. Comp Biochem Physiol 57B: 1–7
4. Cao TP, Rous S (1978) Inhibitory effect of acetazolamide on activity of acetyl CoA carboxylase of mouse liver. Life Sci 22: 2067–2072
5. Cao TP, Rous S (1978) Action of acetazolamide on liver pyruvate carboxylase activity, glycogenolysis and gluconeogenesis of mice. Int J Biochem 9: 603–605
6. Churg A (1973) Carbonic anhydrase histochemistry: Evidence for nonenzymatic reaction and artifact production. Histochemie 36: 293–302
7. Gay CV, Mueller WJ (1973) Cellular localization of carbonic anhydrase in avian tissues by labeled inhibitor autoradiography. J Histochem Cytochem 21: 693–792
8. Gay CV, Mueller WJ (1974) Carbonic anhydrase and osteoclasts: Localization by labeled inhibition autoradiography. Science 183: 432–434
9. Gay CV, Faleski EJ, Schraer H, Schraer R (1974) Localization of carbonic anhydrase in avian gastric mucosa, shell gland and bone by immunohistochemistry. J Histochem Cytochem 22: 819–825
10. Gay CV, Sharkey DJ, Schraer H (1978) Carbonic anhydrase levels in osteoclast-rich bone. Fed Proc 37: 272A
11. Gay CV, Sharkey DJ, Shanabrook VM, Schraer H (1979) Parathyroid hormone induced increase in skeletal carbonic anhydrase. Fed Proc 38: 187A
12. Gros G, Siffert W, Schmid A (1979) Activity and properties of carbonic anhydrase in skeletal muscle. In: Symp Biophys Physiol Carbon Dioxide (this volume)
13. Hansson HPJ (1965) Demonstration of carbonic anhydrase by means of fluorescent antibodies in human erythrocytes. Life Sci 4: 965–968
14. Hansson HPJ (1967) Histochemical demonstration of carbonic anhydrase activity. Histochemie 11: 112–128
15. Häusler G (1958) Zur Technik und Spezifität des histochemischen Carboanhydrasennachweises im Modellversuch und in Gewebsschnitten von Rattennieren. Histochemie 1: 29–47

16. Koenig SH, Brown RD, Jacob GS (1979) The pH independence of carbonic anhydrase activity: Apparent pKa due to inhibition of HSO_4^-. Symp Biophys Physiol Carbon Dioxide (this volume)
17. Kumpulainen T, Korhonen LK (1978) Immunohistochemical demonstration of carbonic anhydrase. Histochemistry 58: 183–192
18. Lönnerholm G (1976) Carbonic anhydrase histochemistry, a critidal study of Hansson's cobalt-phosphate method. Acta Physiol Scand Suppl 418: 1–43
19. Loveridge N (1978) Quantitative cytochemical method for measuring carbonic anhydrase activity. Histochem J 10: 361–372
20. Maren TH (1960) A simplified micromethod for the determination of carbonic anhydrase and its inhibitors. J Pharmacol Exp Ther 130: 26–29
21. Moynihan JB (1977) Carbonic anhydrase activity in mammlian skeletal and cardiac muscle. Biochem J 168: 567–569
22. Mueller WJ, Brubaker RL, Gay CV, Boelkins JN (1973) Mechanisms of bone resorption in laying hens. Fed Proc 32: 1951–1954
23. Mustakallio KK, Raekallio J, Raekallio E (1960) The histochemical demonstration of carbonic anhydrase. Ann Med Exp Biol Fenn 38: 247–251
24. Muther TF (1972) A critical evaluation of the histochemical methods for carbonic anhyddrase. J Histochem Cytochem 20: 319–330
25. Muther TF (1977) On the lack of specificity of the cobalt-bicarbonate method for carbonic anhydrase. J Histochem Cytochem 25: 1043–1050
26. Pocker Y, Sarkanen S (1978) Carbonic anhydrase structure, catalytic versatility and inhibition. Adv Enzymol 47: 149–274
27. Poortmans JR (1969) Immunological identification of human plasma $3Sg_1$-globulin as carbonic anhydrase. Experientia 25: 133–134
28. Register AM, Koester MK, Noltmann EA (1978) Discovery of carbonic anhydrase in rabbit skeletal muscle and evidence for its indentity in "Basic Muscle Protein". J Biol Chem 253: 4143–4152
29. Rosen S, Musser GL (1972) Observations on the specificity of newer histchemical methods for the demonstration of carbonic anhydrase activity. J Histochem Cytochem 20: 951–954
30. Sharkey DJ, Schraer H, Gay CV (1979) Carbonic anhydrase in the chorioallantoic membrane. Biochem J (submitted)
31. Silverman DN (1979) Role of buffer in protein transfer to the active site of carbonic anhydrase. Symp Biophys Physiol Carbon Dioxide (this volume)
32. Spicer SS, Stoward PJ, Tashian RE (1979) The immunohistolocalization of carbonic anhydrase in rodent tissies. J Histochem Cytochem 27: 820–831
33. Sternberger LA (1974) Immunocytochemistry. Englewood Cliffs, Prentice-Hall, p 169
34. Tuan RS, Zrike J (1978) Functional involvement of carbonic anhydrase in calcium transport of the chick chorioallantoic membrane. Biochem J 176: 67–74

Arguments for and Against the Histochemical Carbonic Anhydrase Test

O. LEDER [1]

Localization studies of enzymes, carried out by means of enzyme-catalyzed reactions, are difficult both in theory and practice. For the evaluation of such methods it is not only indispensable (1) to have a consistent idea of the mechanism, but also (2) to start control experiments (e.g., with inhibitors) in order to rule out interference and (3) to compare the results with those of other methods. These requirements are of special importance for the histochemical carbonic anhydrase test, which has been regarded as questionable since its introduction by Kurata (1953). The following account begins with the mechanism of Häusler's (1958) and Hansson's (1967, 1968) tests. It should, however, not be forgotten that Kurata's method can also produce correct results (Braun-Falco and Rathjens, 1955, 1955/56; Kurata, 1953).

1. All histochemical carbonic anhydrase methods use the dehydration reaction. In the method of Häusler (cf. Table 1) a basic cobaltous carbonate is precipitated on structures containing carbonic anhydrase. If the specimen (pure enzyme on filter paper or tissue section) is floating on the surface of the incubation solution, CO_2 can escape and the solution alkalinizes locally on the surface layer of the structures containing carbonic anhydrase. The precipitate is made visible by $(NH_4)_2S$ as a brownish or black deposit. The incubation solution has a pH of about 7.3 at the beginning and this rises to approximately 8.4 (the equilibrium pH of a $NaHCO_3$ solution without effervescence) within 90 min. The alkalinization is not followed by a general precipitation of cobalt and does not interfere with the enzyme-specific reaction, the pH of which is higher. Clearly, proper calculation of this equlibrium pH must take into account the balance between dissolved and gaseous CO_2, and we know that the maximum

Table 1. Incubation medium (Häusler, 1958)

$CoSO_4$	$1.75 \cdot 10^{-3}$ M
H_2SO_4	$5.26 \cdot 10^{-3}$ M
Na_2SO_4	$8.77 \cdot 10^{-2}$ M
$NaHCO_3$	$2.09 \cdot 10^{-1}$ M [a]

a Without correction for expulsion of CO_2. For a slight modification see Waldeyer and Häusler (1959)

1 Anatomisches Institut II der Universität Freiburg, Albertstr. 17, 7800 Freiburg/FRG

possible value of the enzyme-catalyzed alkalinization corresponds to pH 9.1 (Leder, 1971) The high rise of the pH may explain the long incubation time of Häusler's test, since at the pH > 8 enzyme dehydration catalysis is very slow, as is the uncatalyzed dehydration of bicarbonate. Hansson (1967) overcame this difficulty. He added KH_2PO_4 ($1.17–11.7 \cdot 10^{-3}$ M) for the precipitaiton of Co^{2+} at a much lower pH, raised the concentration of H_2SO_4 ($5.3 \cdot 10^{-2}$ M) and omitted the Na_2SO_4 which serves to stabilize Häusler's medium.

2. It is well known that low concentrations of acetazolamide do not completely inhibit Häuser's test if a long incubation time is used. A complex of cobalt and inhibitor has been suggested as the cause of this phenomenon (Muther, 1972). In Hansson's method there is a similar correspondence between the dose of acetazolamide required to produce inhibition and the duration of incubation. However, a cobalt inhibitor complex can be ruled out, since only acetazolamide (10^{-6} M) is used. Moreover, the dependence of the necessary inhibitor concentration on the concentration of KH_2PO_4 suggests that the difficulty of bringing about inhibition is caused by the extreme sensitivity of the method. A 10^{-6} M concentration of acetazolamide is also sufficient in Häusler's test for incubation of 15 to 60 min (the time is inversely related to the enzyme activity) (Leder, 1971). However, Hansson's method should have an advantage over Häusler's, in view of the pH dependence of acetazolamide binding on the enzyme, which is highest at pH 7.2 (Coleman, 1967). Although Hansson's method seems to be more suitable for carbonic anhydrase studies than Häusler's test, our own experience with the latter before 1967 (Leder, 1967a, b, 1971; Leder and Tritschler, 1966), and later comparisons between the two, have revealed essentially the same results as Hansson's (1968). Histochemistry first demonstrated the enrichment of carbonic anhydrase within cell borders (infoldings of basal cell membranes and brushborders (cf. Fig. 1; Leder, 1967; Hansson, 1968). Rosen's findings (1972a)

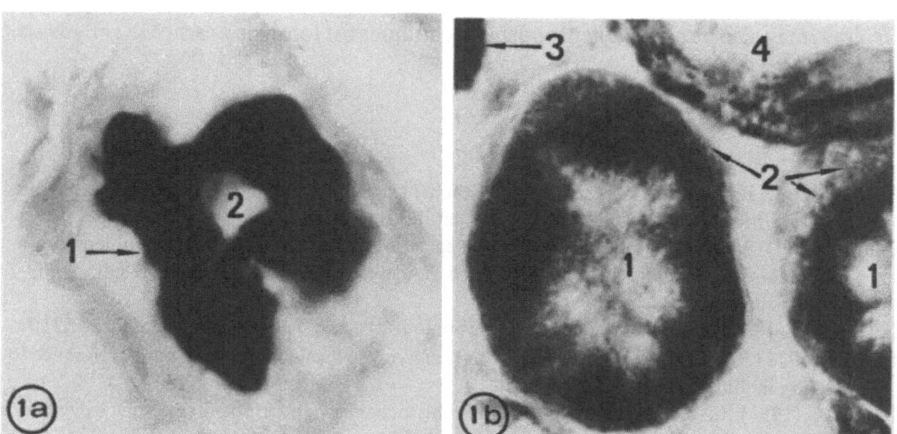

Fig. 1 a, b. Kidney sections 6 μm thick. Incubation 45 min. Häusler's test. Scale 1600:1 (after Leder, 1967a). **a** Straight portion of proximal tubule with strongly stained brushborder (*1*). *2* Collapsed lumen. **b** Strong carbonic anhydrase activity within the infoldings of the basal cell membrane (*2*) of thick ascending parts of limbs (*1*) of Henle's loop. *3* Brushborder of the straight part of a proximal tubule. *4* Collecting duct

are at variance with these results. He interprets the pictures of rat kidney published by Hansson as a diffuse cell staining (which is the case in his own experiments) and neglects all earlier work. Between Hansson's original experiments on kidneys and Rosen's method there is a significant difference. Hansson (1968) exposed the surface of the mixed solution supporting the floating tissue sections to a stream of CO_2. Hence the histochemical reaction could begin at the lowest possible pH after removal of this stream. Rosen (1970, 1972b) left the incubation medium to stand for 20 min before use. In these circumstances a lot of CO_2 is lost and the reaction proceeds at a higher pH.

3. Although biochemists pay scant attention to histochemistry, the biochemical findings corrobate the histochemical work on carbonic anhydrase (Leder, 1971). Of particular interest (in view of CO_2 transport) is the activity of carbonic anhydrase in the placental labyrinth of the rat (Leder and Groh, 1965) and in lung tissue (Leder, 1967a, 1971). For the low carbonic anhydrase activity of muscle the test of Häusler, on the sensitivity of which too much reliance should not be placed, is not well suited. Nevertheless, incubation with Hansson's medium shows some precipitation after between 20 and 40 min of incubation even with a low KH_2PO_4 concentration ($1.17 \cdot 10^{-3}$ M; frog rectus). Recent biochemical work, which is at variance with the older findings, has shown that brushborders and basal cell infoldings are indeed carbonic anhydrase-rich structures (Wistrand and Kinne, 1977). Because this is in accordance with our histochemical results of 1967 and 1968, it seems correct to say that the histochemical carbonic anhydrase test (Häusler's method with low sensitivity, Hansson's procedure without this limitation) works quite well. Moreover, recent histochemical investigations with Hansson's method on the electron microscopic level also confirm carbonic anhydrase activity close to cell membranes (Lütjen-Drecoll and Mohr, 1979) and produce such an attractive result that further work should be encouraged. On the other hand there is no doubt that immunohistochemistry and autoradiography of ^{3}H-labeled inhibitors, as is demonstrated in this volume by Dr. Gay, may also be especially promising techniques for localizing carbonic anhydrase.

References

Braun-Falco O, Rathjens B (1955) Über die histochemische Darstellung der Kohlensäureanhydratase in normaler Haut. Arch Klin Exp Dermatol 201: 73–82

Braun-Falco O, Bathjens B (1955/56) Zur Frage spezifischer Hemmung der Nierenkohlensäureanhydratase im histologischen Schnitt durch 2-Acetylamino-1,3,4-thiodiazol-5-sulfonamid. Acta Histochem 2: 39–46

Coleman JE (1967) Mechanism of action of carbonic anhydrase. J Biol Chem 242: 5212–5219

Gay CV (1979) Recent developments in carbonic anhydrase localization with special reference to calcium mobilizing tissues. This volume

Häusler G (1958) Zur Technik und Spezifität des histochemischen Carboanhydrasenachweises im Modellversuch und in Gewebsschnitten von Rattennieren. Histochemie 1: 29–47

Hansson HPJ (1967) Histochemical demonstration of carbonic anhydrase activity. Histochemie 11: 112–128

Hansson HPJ (1968) Histochemical demonstration of carbonic anhydrase activity in some epithelia noted for active transport. Acta Physiol Scand 73: 427–434

Kurata Y (1953) Histochemical demonstration of carbonic anhydrase activity. Stain Technol 28: 231–233

Leder O (1967a) Die Organ-Verteilung der Carboanhydrase, ein Beitrag zur Physiologie des CO_2-und Ionen-Transports. Habilitationsschr Med Fak Freiburg
Leder O (1967b) Die intrazelluläre Verteilung der Carboanhydrase in der Nieren von Ratten und Mäusen. Pflügers Arch 297: R54
Leder O (1971) Untersuchungen über die Spezifität des histochemischen Karboanhydrasetest. Acta Histochem Suppl 10: 369–377
Leder O, Groh R (1965) Histochemische Untersuchungen über die Carboanhydrase-Aktivität der Rattenplacenta. Pflügers Arch 285: 19–25
Leder O, Tritschler P (1966) Nachweis höchster Carboanhydrase-Aktivitäten im Epithel der Speicheldrüsengänge. Pflügers Arch 292: 229–238
Lütjen-Drecoll E, Mohr M (1979) Funktionelle Unterschiede an Epithelien mit aktivem Flüssigkeitstransport. Verh Anat Ges 74: in press
Muther TF (1972) A critical evaluation of the histochemical methods for carbonic anhydrase. J Histochem Cytochem 20: 319–330
Rosen S (1970) Localization of carbonic anhydrase activity in transporting urinary epithelia. J Histochem Cytochem 18: 668–670
Rosen S (1972a) Localization of carbonic anhydrase activity in the vertebrate nephron. Histochem J 4: 35–48
Rosen S (1972b) Localization of carbonic anhydrase activity in turtle and toad urinary bladder mucosa. J Histochem Cytochem 20: 696–702
Waldeyer A, Häusler G (1959) Histochemische Studien über Carboanhydraseaktivit der Samenwege und ihrer Anhangdrüsen von Mus rattus. Acta Biol Med Germ 2: 568–589
Wistrand PJ, Kinne R (1977) Carbonic anhydrase activity of isolated brush border and basal-lateral membranes of renal tubular cells. Pflügers Arch 370: 121–126

Activity and Properties of Carbonic Anhydrase in Striated Muscle

G. GROS, W. SIFFERT, and A. SCHMID [1]

Skeletal muscle has long been considered to be one of the few tissues that contain no carbonic anhydrase. Roughton (1935) held the view that in a tissue such as skeletal muscle the presence of carbonic anhydrase should even be disadvantageous for the elimination of the CO_2 produced inside the cells.

More recently, however, evidence has been accumulated that carbonic anhydrase does occur in skeletal muscle. Indications to its presence can be derived from the results of Zborowska-Sluis et al. (1974), and from the demonstration of a facilitation of the diffusion of CO_2 across intact muscle layers by Scheid and his coworkers (Kawashiro and Scheid, 1976; Scheid et al., this volume). It can be shown that for diffusion paths as short as the diameter of a muscle fiber, facilitated diffusion of CO_2 is only possible when the hydration of CO_2 is speeded up by carbonic anhydrase. Direct evidence that carbonic anhydrase occurs in the cytoplasm of rat, rabbit, sheep, chicken, human, and ox skeletal muscle has been obtained by Tashian et al. (this volume), by Moynihan (1977), Koester et al. (1977), Holmes (1977), and by Carter et al. (1978). These authors isolated a carbonic anhydrase from homogenates of unperfused muscles which was different from the red cell carbonic anhydrases with respect to its amino acid composition, immunochemical properties, specific CO_2 hydrase activity, and its affinity for acetazolamide.

Our aim was mainly to quantitate the catalytic activity of muscle carbonic anhydrase under physiological conditions, and furthermore, to give a description of its kinetic properties under these conditions.

1 Methods

The hind limbs of rabbits were artificially perfused with a dextrane solution via a catheter inserted into the abdominal aorta until they were practically free of red cells. Cardiac muscles were perfused through a catheter inserted into the Aorta ascendens, the opening of the catheter pointing toward the aortic valves. The perfused muscles were frozen in liquid nitrogen, homogenized in a "Mikro-Dismembrator" (Braun Comp., Melsungen), thawed and centrifuged for 1/2 h at 20 000 *g*. The velocity of CO_2 hydration was then determined, after suitable dilution, in the supernatants of the homogenates using a rapid reaction stopped flow apparatus.

1 Institut für Physiologie, Universität Regensburg, 8400 Regensburg/FRG, and the Institut für Physiologie, Universitätsklinikum Essen, Hufelandstr. 55, 4300 Essen/FRG

To be able to estimate how many erythrocytes had remained in the muscles, the red cells of the rabbits' circulating blood were labeled with ^{51}Cr about 1 h before the perfusion was started. By comparing the radioactivity of the perfused muscles with that of the blood, the contamination of the tissues with red cells could be evaluated. Examples from two preparations are given in Table 1. It may be seen that the radioactivity of the homogenates was four to five orders of magnitude lower than that of the circulating red cells. The carbonic anhydrase activity (i.e., the acceleration factor of CO_2 hydration minus 1) due to residual red cells is estimated from these figures to be about 1 or less. In view of the high activities found in these homogenates we may conclude that red cell carbonic anhydrase is practically absent in the perfused muscles.

Table 1. Radioactivities of circulating red cells and of the homogenates of artificially perfused skeletal muscles. The data were obtained from two different rabbits (1, 2). The radioactivities of the red cells were measured on samples taken from the A. carotis and the V femoralis 1 h after the labeled erythrocytes had been injected into the rabbits' circulation. Carbonic anhydrase activity of rabbit red cells was determined to be 20,000; from this figure, and the radioactivities of red cells and muscle, the carbonic anhydrase activity due to red cells remaining in the muscle after perfusion was calculated

	Preparation 1	Preparation 2
Radioactivity of the circulating red cells (CpM/ml)	$1.5 \cdot 10^7$	$1.6 \cdot 10^7$
Radioactivity of homogenates of perfused muscles (CpM/ml)	$9 \cdot 10^2$	$1.1 \cdot 10^2$
Estimated carbonic anhydrase activity in the homogenates due to residual red cells	1.2	0.14

Another argument against the presence of significant amounts of red cell carbonic anhydrase in our preparations derives from the low concentration of heme in the white muscle homogenate. From a small absorbance peak found at 410 nm the heme concentration in the supernatant from white muscle was estimated to be $3 \cdot 10^{-6}$ M. Part of this heme may represent myoglobin, but if all of it is assumed to be hemoglobin, an enzyme activity of ~ 3 due to red cell carbonic anhydrase is predicted. This compares with an enzyme activity of ~ 300 that was actually measured in this supernatant (Table 2).

To determine CO_2 hydration velocities the supernatants of the homogenates were usually diluted 1:5 in a solution containing 0.06 M imidazole, 0.14 M NaCl, 0.4 mM EDTA, and 0.1% bovine serum albumin. After the diluted supernatants had been freed from CO_2 by equilibration with nitrogen, they were rapidly mixed with a CO_2 solution (0.15 M NaCl equilibrated with CO_2-N_2-mixtures) in a pH stopped-flow apparatus such as has been described by Crandall et al. (1971). From the initial slope of the recorded pH trace and the buffer capacity of the solutions the initial rate of CO_2 hydration was calculated. Subtracting the independently determined uncatalyzed rate, the CO_2 hydration rate due to enzyme catalysis was obtained.

2 Carbonic Anhydrase Activity in Muscle Tissue

Determining catalyzed CO_2 hydration rates at various CO_2 concentrations allowed us to produce Lineweaver-Burk diagrams and calculate Michaelis constants, K_M, and maximal catalytic rates, V_{max}. An example obtained with a supernatant from white muscle homogenate (final dilution after mixing was 1:10) at pH 6.8 and 37°C is shown in Fig. 1. K_M is calculated from the slope of the regression line to be 3.8 mM,

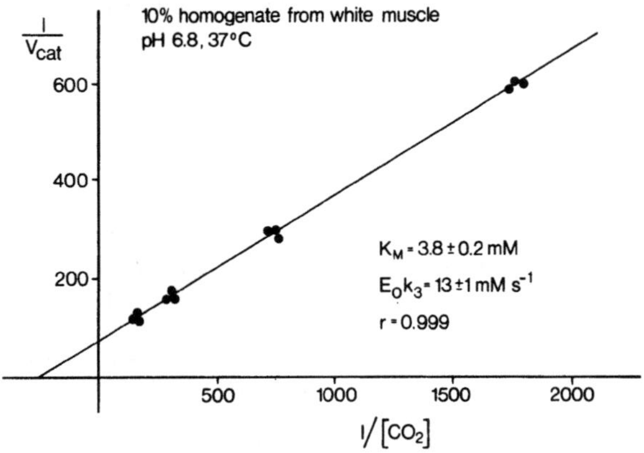

Fig. 1. Lineweaver-Burk plot of catalyzed CO_2 hydration rates, v_{cat}, obtained at various CO_2 concentrations, $[CO_2]$, with supernatant of white muscle homogenate (dilution after mixing 1:10) at pH 6.8 and 37°C. K_M and V_{max} (= $E_0 \cdot k_3$) were obtained from a regression analysis. Correlation coefficient, r, was 0.999

V_{max}, as estimated from the intercept, is 13 mM s^{-1}. Similar data were collected for red skeletal muscle and for cardiac muscle, and K_M and V_{max} were obtained in the same fashion. The results are compiled in Table 2. Significant carbonic anhydrase activity was present in white as well as in red skeletal muscle, but was not detectable in cardiac muscle. This is in qualitative agreement with Zborowska-Sluis (1974, 1975) and Moynihan (1977), who found evidence for carbonic anhydrase in skeletal but not in cardiac muscle. K_M and V_{max} for red and white skeletal muscle were identical within experimental error. After correcting for the dilution of the supernatants and

Table 2. Michaelis constant, K_M, maximal catalytic rate, V_{max}, and carbonic anhydrase activity, a, in white and red skeletal muscle of rabbit. No carbonic anhydrase activity significantly different from that expected from red cell contamination was detectable in rabbit cardiac muscle. Carbonic anhydrase activities are calculated from K_M and V_{max} for pCO_2 = 40 mm Hg. pH was 6.8, temperature 37°C. V_{max} and a refer to the fluid space of whole muscle tissue (cytoplasm + extracellular space)

	K_M (mM)	V_{max} (mM s^{-1})	a (pCO_2 = 40 mm Hg)
White muscle	3.8 ± 0.2	160 ± 10	280
Red muscle	3.5 ± 0.4	160 ± 10	320
Cardiac muscle		No activity detectable	

for a perfusion edema of ~ 20%, V_{max} is found to be 160 mM s^{-1} for both types of skeletal muscle. Using the values of K_M and V_{max} given in Table 2 we can estimate the activity of carbonic anhydrase in muscle tissue at roughly physiological conditions of pCO_2 = 40 mm Hg, 37°C, and pH = 6.8: the hydration of CO_2 accordingly is speeded up by a factor of ~ 300. It should be noted that this figure represents the carbonic anhydrase activity related to the fluid space of whole muscle tissue. To obtain the true intracellular activity one has to correct for the volume fraction of the extracellular space (which has not been determined here), and would then arrive at an activity of about 400 to 500. It can be shown that this roughly corresponds to the enzyme activity that has to be postulated from the facilitation of CO_2 diffusion across muscle layers described by Kawashiro and Scheid (1976).

3 Sensitivity to Inhibitors and Turnover Number

Figure 2 shows that the catalysis of CO_2 hydration in muscle homogenate can be almost completely inhibited by acetazolamide. At an acetazolamide concentration of $1 \cdot 10^{-4}$ M there is a residual activity of 4.5 that could not be further reduced by raising the acetazolamide concentration to $5 \cdot 10^{-3}$ M. It appears likely that this residual activity represents unspecific catalysis, for example, by proteins and phosphate present in the homogenate. From a least squares fit to the data points of Fig. 2 the inhibition constant K_I for acetazolamide was estimated to be $1.8 \cdot 10^{-8}$ M (the calculation is based on the assumption of noncompetitive inhibition).

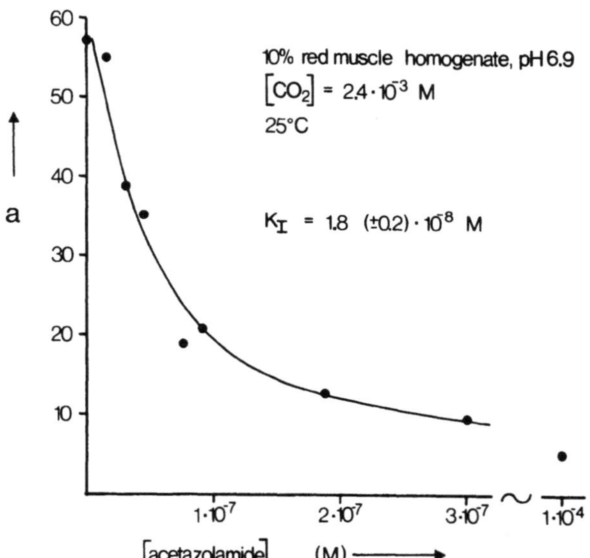

Fig. 2. Inhibition of red muscle carbonic anhydrase by acetazolamide. Carbonic anhydrase activity, a, is plotted versus acetazolamide concentration. The inhibition constant, K_I, for acetazolamide was obtained from a second-order regression analysis (catalytic rate, v, and the total inhibitor concentration, $[I_{tot}]$, were fitted to the following polynomial: $v\,[I_{tot}] = a + b \cdot v + c \cdot v^2$; K_I was obtained from the intercept, a)

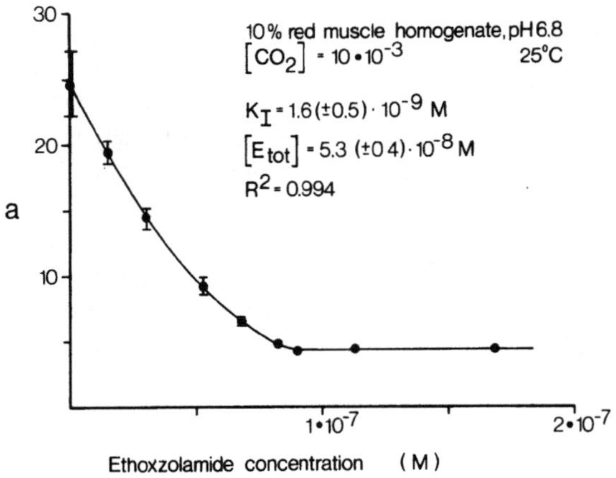

Fig. 3. Inhibition of red muscle carbonic anhydrase by ethoxzolamide. Carbonic anhydrase activity, a, is plotted versus ethoxzolamide concentration. The absolute values of a are lower than in Fig. 2 because an about four times greater substrate concentration was used for these experiments ($[CO_2] = 10 \cdot 10^{-3}$ M). Inhibition constant, K_I, and total enzyme concentration, $[E_0]$, were obtained from the coefficient a and b of a second-order regression analysis (see legend to Fig. 2). The *bars* indicate the standard deviations of the activity determinations

Figure 3 shows that ethoxzolamide is an even more potent inhibitor of muscle carbonic anhydrase. From a least squares fit to these data an inhibition constant of $1.6 \cdot 10^{-9}$ M was calculated. In addition, an estimate of the total enzyme concentration, E_0, was obtained. In a 10% supernatant from red muscle (without correction for the perfusion edema) E_0 was found to be $5.3 \cdot 10^{-8}$ M. Using this latter figure for E_0, and the V_{max} value ($16 \cdot 10^{-3}$ M s^{-1}) obtained from a Lineweaver-Burk plot for the same preparation at identical conditions of pH and temperature, we calculate a turnover number, k_3, of muscle carbonic anhydrase of 302,000 s^{-1} (temperature 25°C).

When compared to N-methyl-imidazole and phosphate in the presence of ~ 0.15 M NaCl, imidazole did not appear to have an inhibitory effect on the muscle enzyme. Replacing imidazole by identical concentrations of N-methyl-imidazole or phosphate, values of K_M and V_{max} were obtained that were not significantly different from those found with imidazole.

4 Conclusion

Table 3 gives a compilation of some of the properties of muscle carbonic anhydrase and a comparison with the properties of human carbonic anhydrases B and C. It may be seen that the properties of muscle carbonic anhydrase resemble rather closely those of the high activity human enzyme C. This conclusion is in sharp contrast to the results of Koester et al. (1977), Holmes (1977), and Carter et al. (1978), who found the muscle enzyme after purification by ion exchange chromatography or affinity chromatography to have a low specific activity (k_3 about 50 times lower than for the C enzyme) and a very low affinity for acetazolamide (K_I ~ 1000 times higher than for the C enzyme). The reason for this discrepancy is not clear. One possibility may be an alteration of the properties of the enzyme due to the purification procedures used by these authors. Another, more likely, explanation may be the presence of a high *and* a low activity enzyme in skeletal muscle. The low activity form may then be undetectable in our preparation due to the dominance of the high

Table 3. Comparison of some properties of muscle carbonic anhydrase (from this study) and of the low activity human carbonic anhydrase B and the high activity human carbonic anhydrase C. Temperature 25°C

	Human carbonic anhydrase B	Human carbonic anhydrase C	Rabbit muscle carbonic anhydrase	Source (human data)
K_M (M)	$4 \cdot 10^{-3}$	$9 \cdot 10^{-3}$	$4 \cdot 10^{-3}$	Khalifah (1971)
k_3 (s^{-1}) pH = 7	30,000	500,000	300,000	Khalifah (1971)
K_I (M) for acetazolamide	$10 - 30 \cdot 10^{-8}$	$0.4 - 2 \cdot 10^{-8}$	$2 \cdot 10^{-8}$	Maren (1967), Maren et al. (1976), Coleman (1967)
K_I (M) for ethoxzolamide	$2 \cdot 10^{-9}$	$2 \cdot 10^{-9}$	$1.6 \cdot 10^{-9}$	Maren (1967)
pH dependence of K_M	–	–	–	Khalifah (1971)
pH dependence of V_{max}	+	+	+	Khalifah (1971)

activity form, and the latter may have been removed, together with red cell carbonic anhydrase, from the unperfused muscle homogenates in the course of the purification used by the above authors. Work to resolve this question is currently in progress.

Note Added in Proof: After submitting this manuscript we have been able to show that in the homogenates of perfused muscles there are indeed two carbonic anhydrase isoenzymes: one with a high affinity for ethoxzolamide (CA type C), and one with a very low affinity for this inhibitor (probably CA type A). The latter isoenzyme is responsible for only about 5–10% of the total carbonic anhydrase activity in red skeletal muscle.

References

Carter N, Schiels A, Tashian R (1978) Carbonic anhydrase III isoenzyme from human and bovine muscle. Biochem Soc Trans 6: 552–553

Coleman JE (1967) Mechanism of action of carbonic anhydrase. J Biol Chem 242: 5212–5219

Crandall ED, Klocke RA, Forster RE (1971) Hydroxyl ion movements across the human erythrocyte membrane. J Gen Physiol 57: 664–684

Holmes RS (1977) Purification, molecular properties and ontogeny of carbonic anhydrase isozymes. Eur J Biochem 78: 511–520

Kawashiro T, Scheid P (1976) Measurements of Krogh's diffusion constant of CO_2 in respiring muscle at various CO_2 levels: evidence for facilitated diffusion. Pflügers Arch Eur J Physiol 362: 127–133

Khalifah RG (1971) The carbon dioxide activity of carbonic anhydrase. I. Stop-flow kinetic studies on the native human isozymes B and C. J Biol Chem 246: 2561–2573

Koester MK, Register AM, Noltman EA (1977) Basic muscle protein, a third genetic locus isoenzyme of carbonic anhydrase? Biochem Biophys Res Commun 76: 196–204

Maren TH (1967) Carbonic anhydrase: chemistry, physiology, and inhibition. Physiol Rev 47: 595–781

Maren TH, Rayburn CS, Lidell NE (1976) Inhibition by anions of human red cell carbonic anhydrase B: physiological and biochemical implications. Science 191: 469–472

Moynihan JB (1977) Carbonic anhydrase activity in mammalian skeletal and cardiac muscle. Biochem J 168: 567–569

Roughton FJW (1935) Recent work on carbon dioxide transport by the blood. Physiol Rev 15: 241–296

Zborowska-Sluis DT, L'Abbate A, Klassen GA (1974) Evidence of carbonic anhydrase activity in skeletal muscle: a role for facilitative carbon dioxide transport. Respir Physiol 21: 341–350

Zborowska-Sluis DT, L'Abbate A, Mildenberger RR, Klassen GA (1975) The effect of acetazolamide on myocardial carbon dioxide space. Respir Physiol 23: 311–316

CO_2 Transport and Carbonic Anhydrase in Kidney and Brain

Chairman: C. ALBERS

Significance of Carbonic Anhydrase for HCO_3^- Absorption and H^+ Secretion in Renal Tubules

E. FRÖMTER [1]

In the present paper I shall discuss some problems dealing with the role of carbonic anhydrase in renal bicarbonate absorption. I shall consider only the first part of the nephron, the so-called proximal tubule, because this is the site where most of the renal HCO_3^- absorption occurs. As observed in micropuncture experiments the proximal tubule absorbs 70%–80% of the filtered Na^+ ions and water, ~ 75% of the filtered Cl^- ions and ~ 95% of the filtered HCO_3^-, as well as practically all of the filtered glucose and amino acid molecules. Microscopically the wall of the proximal tubule consists of a typical single-layered epithelium of uniform, flat cells which are held together at their luminal end by quite leaky terminal bars. All substances which are reabsorbed by the cells enter the cells via the luminal cell membrane (brushborder) and leave the cells via the highly folded peritubular cell membrane. A great number of the transport mechanisms which are located in these cell membranes have been identified in recent years. Briefly we can say that the primum movens of almost all fluxes is a Na^+/K^+ pump located in the peritubular cell membrane which consumes energy by splitting ATP. This pump effects many different things. Firstly it brings about active transtubular Na^+ ion transport by exporting the Na^+ ions, which leak from the lumen into the cell, across the peritubular cell membrane into the interstitial space. Secondly it provides the energy for the absorption of glucose and amino acids – these substances are driven from lumen to cell by cotransport with Na^+ ions, accumulate in the cytoplasm and diffuse then into the interstitium along their concentration gradient – and thirdly it builds up the passive driving forces for the absorption of Cl^- ions and water. Other transport mechanisms, however, are less well defined. This is true for the transport of K^+, and the transport of weak acids and bases and it is also true for the HCO_3^- transport. With regard to HCO_3^- transport two main questions remain open: (1) Are bicarbonate ions absorbed as such or by means of H^+ secretion? and (2) Is HCO_3^- absorption (or respectively H^+ secretion) driven by a primary active pump or is it coupled to the active Na^+ transport as in the case of glucose and amino acids? While the first question is related to the present discussion of the role of carbonic anhydrase, the second question is not and will not be considered in detail here. The experimental evidence in favor of both Na^+-dependent (secondary active) HCO_3^- transport and Na^+-independent (primary active) HCO_3^- transport was recently discussed elsewhere [6].

---

1 Max-Planck-Institut für Biophysik, Kennedyallee 70, 6000 Frankfurt (Main) 70/FRG

1 The Classical View of the Role of Carbonic Anhydrase in Proximal Tubular Cells

The experimental evidence for the involvement of carbonic anhydrase in renal absorption of bicarbonate dates back to the 1940's when it was observed that administration of carbonic anhydrase inhibitors such as sulfanilamide or acetazolamide leads to the excretion of large amounts of bicarbonate-rich urine. Pitts and Alexander [14], followed by Berliner [1], concluded from this observation that bicarbonate ions must be disintegrated during the absorption process so that the interconversion of H_2CO_3 to CO_2 and back can become a rate-limiting step in the absorption process. This led to the development of a model concept according to which HCO_3^- ions are absorbed by means of secretion of H^+ ions from the cell into the tubular lumen. The addition of H^+ ions to the lumen fluid leads to the generation of carbonic acid (from HCO_3^-) which then splits into CO_2 and water. The luminal CO_2 pressure increases and CO_2 diffuses out of the lumen into the cell. Inside the cell it reacts with the OH^- ions, which were left behind by the H^+ ion pump, to form again HCO_3^-. This latter step was thought to be catalyzed by carbonic anhydrase. Inhibition by sulfanilamide or acetazolamide should then result in progressive accumulation of OH^- ions inside the cell. This results in a steeper adverse gradient which must be overcome by the H^+ pump (or the respective Na^+ ion/H^+ ion countertransport machinery) and reduces its efficiency or in other words reduces the supply of H^+ ions to the pump.

Later Maren [12] reexamined the concept quantitatively. He studied the inhibition kinetics of various carbonic anhydrase inhibitors on urinary HCO_3^- excretion and compared it with that of the isolated cytoplasmic enzyme. He found that the relative potency of the drugs in vivo and in vitro varied in parallel and that the kinetics both in vivo and in vitro conformed to a simple single-site inhibition model.

2 Indirect Evidence for Multiple-Site Actions of Carbonic Anhydrase Inhibitors

Indirect evidence against a single-site action of carbonic anhydrase inhibitors was obtained by Radtke et al. [15] who observed inhibition by acetazolamide of proximal tubular volume absorption even under conditions of bicarbonate-free media. Further indirect evidence came forth, when the inhibition kinetics of this drug were studied on the volume absorption of individual tubular loops rather than on the unrinary excretion of the entire kidney. Such studies were initially performed by Meng [13] and recently confirmed in our laboratory and extended to include ethoxzolamide and sulfanilamide. Plotting volume absorption versus the logarithm of inhibitor concentration, the data do not exhibit simple S-shaped curves but show a more flat relationship which points to the involvement of more than one binding site in the inhibition process. This is true of all inhibitors studied thus far. Since the renal cytoplasmic carbonic anhydrase cannot account for this behavior the question arises of whether there are other types of carbonic anhydrase present in tubular cells (possibly bound to cell membranes) or whether the inhibitors are not specific but inhibit also other steps in bicarbonate transport.

3 Evidence for "Luminal" Location of Carbonic Anhydrase

Assuming that all bicarbonate absorption was effected via secretion of H^+ ions Walser and Mudge [21] calculated that the uncatalyzed rate of carbonic acid dehydration in the tubular lumen would not suffice to achieve equilibrium between bicarbonate concentration, pH and carbon dioxide pressure as described by the Henderson-Hasselbalch equation and postulated an acid "disequilibrium pH" to build up in the proximal tubular lumen during HCO_3^- absorption. Since no such disequilibrium pH was found in direct intratubular pH measurements in normal rats it was postulated [16, 20] that the luminal fluid had direct access to the enzyme, and hence that some carbonic anhydrase must be located on the brushborder membrane itself. This notion was later supported [10, 22] or disputed [9, 17] on the basis of more or less conclusive evidence from various other experiments. The main supportive evidence is the biochemical observation of a specifically inhibitable carbonic anhydrase activity in isolated brushborder cell membrane fractions of kidney homogenates [22].

4 Evidence for Carbonic Anhydrase Activity in the Peritubular Cell Membrane

A significant carbonic anhydrase activity was also detected in isolated membrane fractions of the peributular cell membrane [22] and the same result was previously obtained with histochemical methods [11]. The functional significance of this finding can be assessed on the basis of electrophysiological experiments performed in our laboratory. With some restrictions, which I do not have time to discuss here in detail, it is possible to determine the transference number or relative permeability of a membrane for a given ion, by measuring the dependence of the cell membrane potential on the outside concentration of the ion in question. The principle is essentially identical to measuring the slope of an ion-selective electrode. Testing various normal ionic constituents of plasma as well as some foreign ions we have found that the peritubular cell membrane responds preferentially to changes in HCO_3^- concentration in the peritubular fluid. The typical time course of the potential change (see Fig. 1), as well as the slope of 27 mV per tenfold change of peritubular bicarbonate concentration, allows us to conclude that the peritubular cell membrane is highly permeable for bicarbonate buffer [7]. This finding closed a gap in our understanding of transtubular bicarbonate transport since thus far all interest focused on the transfer of bicarbonate across the luminal cell membrane, but little or nothing was known about the mechanism of transport across the peritubular cell membrane. In the following I shall report experiments from which we conclude that the passive bicarbonate permeability of the peritubular cell membrane can be blocked by carbonic anhydrase inhibitors.

One key observation with carbonic anhydrase inhibitors is already shown in Fig. 1. As can be seen, perfusion of the tubular lumen with acetazolamide leads to hyperpolarization of the cell and at the same time strongly reduces the potential response of the cell membrane to removal of HCO_3^- from the peritubular surface. This effect is fully reversible. In the presence of acetazolamide a tenfold change in HCO_3^- concentration produces only a potential change of 12 mV, as compared to

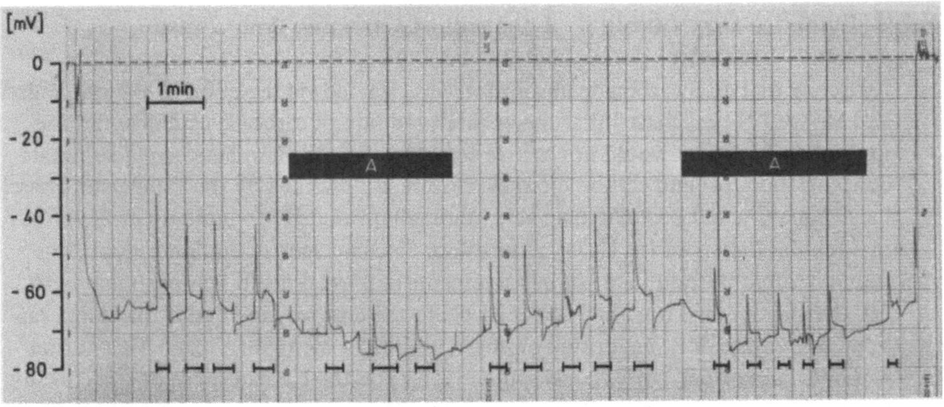

Fig. 1. Cell potential response to peritubular removal of HCO_3^- and effect of acetazolamide [7]. Trace record of cell potential measurement on a rat kidney proximal tubular cell. *Abscissa:* time as indicated, *ordinate:* potential difference across peritubular cell membrane in mV. During the light bars (⊢⊣) the peritubular blood capillaries were perfused with HCO_3^--free Ringer's solution. Note the typical time course of the potential response. During the heavy bar (*A*) the luminal perfusion was switched from HCO_3^- Ringer's to the same solution but containing 10^{-4} mol/l acetazolamide. Note that the cell hyperpolarizes and the potential response to peritubular bicarbonate removal decreases

27 mV under control conditions, indicating that the transference number or relative permeability of the peritubular cell membrane for bicarbonate is reduced. Whether or not this interpretation is correct can be tested by measuring the transference number (t_i) or relative permeability of other ions, because the sum of all transference numbers must be always 1.0, so that the transference numbers of other ions must increase when $t_{HCO_3^-}$ decreases. This consistency check was made and it was found that the transference number of K^+ increased under acetazolamide, as well as t_{Cl^-}, or respectively the permeability of Cl^- relative to that of SO_4^{2-} [8].

The main question now was, whether the decrease in $t_{HCO_3^-}$ reflects a decrease in the absolute HCO_3^- permeability of the peritubular cell membrane or reflects an increase in other ion permeabilities, such as K^+ permeability or Cl^- permeability. This question could be answered by resistance measurements. Although the presently available techniques do not allow us to measure the peritubular cell membrane resistance directly in the rat, we have obtained an estimate of the cell membrane resistance relative to control by two indirect methods: In the first approach we have passed current pulses transepithelially and we measured the resulting voltage deflection between lumen and interstitium (across the entire epithelium) and between cell and interstitium (across the peritubular cell membrane). It was observed that the transepithelial voltage deflections remained constant under acetazolamide, but that the deflections across the peritubular cell membrane increased by 21% ± 17% (n = 7) [5]. In the second approach we have investigated the effect of acetazolamide on the cell membrane depolarization induced by luminal perfusion with glucose Ringer's. It yielded a similar increase of ~20%, while again the transepithelial voltage deflections remained constant (I. Samaržija and E. Frömter, unpublished). The results from both experiments agree well and indicate that acetazolamide increases the peritubular cell membrane resistance slightly, Hence we conclude that acetazolamide decreases the absolute magnitude of the bicarbonate permeability of the peritubular cell membrane. However, since the change of the transference numbers was quite conspicuous, while the resistance change was rather small, it seems likely that besides, and simultaneously with, the reduction of bicarbonate permeability also other ion permeabilities increased, most likely the permeability of K^+. Such a double action of acetazolamide might also explain why the time course of the repolarization after bicarbonate removal is only slightly prolonged, whereas the magnitude of the initial depolarization decreases drastically.

The next question then was, whether the effects of acetazolamide on the peritubular ion permeabilities were direct drug effects, or were caused indirectly, such as for example by the acetazolamide-induced alkalinization of the cell. To answer this question properly is quite difficult. Our present knowledge of the problem, however, suggests that the increase in K^+ permeability may be an unspecific effect, while the decrease in HCO_3^- permeability appears to involve a specific action of the drug on the peritubular cell membrane. The first conclusion is based on experiments in which the cell membrane potential was measured in response to peritubular perfusions with bicarbonate containing Ringer's solution equilibrated with different CO_2 tensions. Decreasing CO_2 below normal and hence alkalinizing the cytoplasm hyperpolarized the cell membrane in direction toward the K^+-equilibrium potential while increasing CO_2 had the opposite effect. This observation and interpretation agrees also with a report from *Necturus* proximal tubule, where cell alkalinization was found to increase the relative K^+ permeability of the peritubular cell membrane [18].

The evidence that the inhibition of HCO_3^- permeability by acetazolamide may result from a direct effect of this drug on the peritubular cell membrane is based on studies with SITS. SITS (4-acetamido-4′-isothiocyano-2,2′-disulfonic stilbene) is a disulfonic stilbene derivative, which – in erythrocytes – binds covalently to the cell membrane without penetrating into the cell and inhibits the Cl^- exchange system [2]. In kidney tubules SITS was reported to inhibit specifically the HCO_3^- absorption [19] and since this effect occurs only if the drug is applied from the peritubular surface, it is likely that SITS does not penetrate into the tubular cells either.

When applied in 10^{-3} molar concentrations from the peritubular surface SITS decreased the cell potential response to peritubular HCO_3^- removal in the same way as acetazolamide, only the effect took longer to develop (Fig. 2). Similar results were recently obtained in *Necturus* proximal tubule [4]. This observation suggests that SITS acts in the same way as acetazolamide and, if SITS does not penetrate into the cell, that the common site of action is the peritubular cell membrane. Moreover, since it has been recently shown that SITS inhibits also carbonic anhydrase in vitro, both the cytoplasmic enzyme from red cells and the membrane-bound enzyme from dog kidneys (the half inhibition constants fall between 1 and 4 times 10^{-4} mol/l, Wistrand and Knuuttila, unpublished results), it would appear possible that both drugs, SITS and acetazolamide, act directly on the carbonic anhydrase activity of the peritubular cell membrane, which is associated with the HCO_3^-permeability of this membrane. It is interesting to note that similar conclusions were recently arrived at in studies on turtle bladder [3].

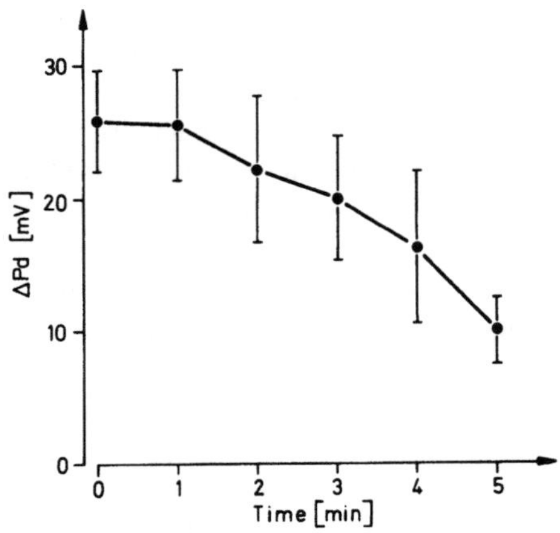

Fig. 2. Effect of SITS on potential response to HCO_3^- removal. *Abscissa:* time in minutes, *ordinate:* change of potential difference in mV in response to peritubular perfusion with HCO_3^- Ringer's solution. *Points* indicate mean value ± S.D. (n = 4). Experimental protocol as in Fig. 1, however, instead of liminal perfusion the drug (SITS, 10^{-3} mol/l) was applied continuously from time zero on in the peritubular superfusion fluid. Note that SITS reduces the potential response to HCO_3^- removal similarly as acetazolamide, but the effect takes longer to develop. (These data were obtained in collaboration with I. Samaržija)

In summary we conclude that carbonic anhydrase is involved in several distinct steps of bicarbonate absorption from rat kidney proximal tubular cells. Besides the classical cytoplasmic localization the enzyme is present in the brushborder membrane and in the peritubular cell membrane, where it facilitates active and passive HCO_3^- transfer, respectively. The accelerating action of carbonic anhydrase on bicarbonate transfer across these membranes and its inhibition by carbonic anhydrase inhibitors suggests that bicarbonate transport across both barriers involves a dehydration/rehydration reaction with separate transfer mechanisms for CO_2 and OH^-, (or respectively H^+ in opposite direction).

References

1. Berliner RW (1952) Renal secretion of potassium and hydrogen ions. Fed Proc 11: 695–700
2. Cabantchik ZE, Rothstein A (1972) The nature of the membrane sites controlling anion permeability of human red blood cells as determined by studies with disulfonic stilbene derivatives. J Membr Biol 10: 311–330
3. Cohen LH, Mueller A, Steinmetz PR (1978) Inhibition of the bicarbonate exit step in urinary acidification by a disulfonic stilbene. J Clin Invest 61: 981–986
4. Edelman A, Teulon J, Anagnostopoulos T (1978) The effect of a disulfonic acid stilbene on proximal cell membrane potential in Necturus kidney. Biochim Biophys Acta 514: 137–144
5. Frömter E (1977) Magnitude and significance of the paracellular shunt path in rat kidney proximal tubule. In: Kramer M, Lauterbach F (eds) Intestinal permeation. Excerpta Medica, Amsterdam, pp 166–178
6. Frömter E (1978) Primary and secondary active transport mechanisms in rat renal proximal tubule. In: Ullrich KJ, Vogel HG (eds) New aspects on renal function. Excerpta Medica, Amsterdam, pp 27–36
7. Frömter E, Sato K (1976) Electrical events in active H^+/HCO_3^- transport across rat kidney proximal tubular epithelium. In: Kasbekar DK, Sachs G, Rehm WS (eds) Gastric hydrogen ion secretion. Dekker, New York, pp 382–403
8. Frömter E, Sato K, Gessner K (1976) Electrical studies on the mechanism of H^+/HCO_3^- transport across rat kidney proximal tubule. In: Giovannetti S, Bonomini V, D'Amico G (eds) 6th Int Congr Nephrol. Karger, Basel, pp 108–112

9. Karlmark B, Danielson BG (1974) Titratable acid, P_{CO_2}, bicarbonate and ammonium ions along the rat proximal tubule. Acta Physiol Scand 91: 243–258
10. Lang F, Quehenberger P, Greger R, Oberleithner H (1978) Effect of benzolamide on luminal pH in proximal convoluted tubules of the rat kidney. Pflügers Arch 375: 39–43
11. Leder O (1969) Die intrazelluläre Verteilung der Carboanhydrase in der Niere von Ratten und Mäusen. Pflügers Arch 297: R54
12. Maren TH (1969) Renal carbonic anhydrase and the pharmacology of sulfonamide inhibitors. In: Hercken H (ed) Handb Exp Pharmacol, vol 24. Springer, Berlin Heidelberg New York, pp 195–256
13. Meng K (1972) Die am Nierentubulus wirksamen Diureticakonzentrationen. In: Hohenegger M (ed) Biochemische Aspekte der Nierenfunktion. Goldmann, München, pp 301–310
14. Pitts RF, Alexander RS (1945) The nature of the renal tubular mechanism for acidifying the urine. Am J Physiol 144: 239–254
15. Radtke HW, Rumrich G, Kinne-Saffran E, Ullrich KJ (1972) Dual action of acetazolamide and furosemide on proximal volume absorption in the rat kidney. Kidney Int 1: 100–105
16. Rector FC, Carter NW, Seldin DW (1965) The mechanism of bicarbonate reabsorption in the proximal and distal tubules of the kidney. J Clin Invest 44: 278–290
17. Sohtell M (1979) CO_2 along the proximal tubules in the rat kidney. Acta Physiol Scand 105: 146–155
18. Steels PS, Boulpaep EL (1976) Effect of pH on ionic conductances of the proximal tubule epithelium and the role of buffer permeability. Fed Proc 35: R465
19. Ullrich KJ, Capasso G, Rumrich G, Papavassiliou F, Klöss S (1977) Coupling between proximal tubular transport processes. Studies with ouabain, SITS and HCO_3^--free solutions. Pflügers Arch 368: 245–252
20. Vieira FL, Malnic G (1968) Hydrogen ion secretion by rat renal cortical tubules as studied by an antimony microelectrode. Am J Physiol 214: 710–718
21. Walser M, Mudge GH (1960) Renal excretory mechanisms. In: Comar CL, Bronner F (eds) Mineral metabolism. Academic Press, London New York, pp 287–336
22. Wistrand PJ, Kinne R (1977) Carbonic anhydrase activity of isolated brush border and basal-lateral membranes of renal tubular cells. Pflügers Arch 370: 121–126

Formation and Ionic Composition of Cerebrospinal Fluid: The Role of Carbonic Anhydrase

T.H. MAREN and B.P. VOGH [1]

It has been known for over 20 years that inhibition of carbonic anhydrase reduces the flow of cerebrospinal fluid (CSF), but until recently the mechanism underlying the effect was not known. The fact that CSF has virtually the same pH and HCO_3^- concentration as plasma appeared to lead away from conventional ideas regarding the role of this enzyme, although it was known that other neutral fluids, for example, the salt gland secretion of birds, are formed under its influence (reviewed in [5]).

The first hint in elucidation of the problem was the experiment of Fig. 1. Under appropriate conditions of perfusion and sampling, the dogfish, *Squalus acanthias,*

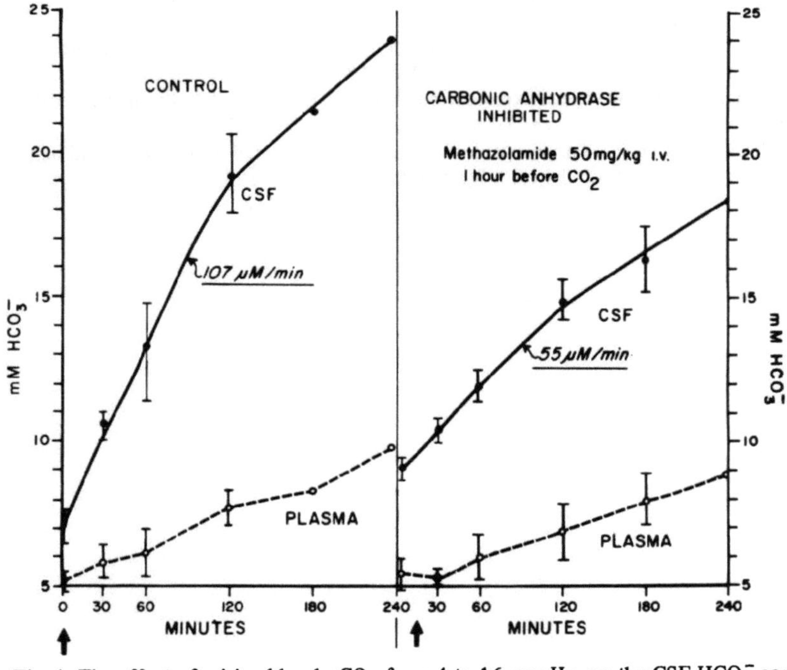

Fig. 1. The effect of raising blood pCO_2 from 4 to 16 mm Hg, on the CSF HCO_3^- concentration in the dogfish *S. acanthias.* At *arrow,* 2.5% CO_2 was added to cold oxygenated sea-water perfusing the gills [6]

1 Department of Pharmacology and Therapeutics, University of Florida, College of Medicine, Gainesville, Florida/USA

responded to hypercapnia by a fourfold increase in CSF HCO_3^- concentration; the rate of increase of CSF HCO_3^- was reduced to half by inhibition of carbonic anhydrase. This suggested that HCO_3^- reached the CSF by formation from CO_2, rather than by transfer of the ion from plasma to CSF [6]. It was soon shown that the same phenomenon could be elicited in the dog [10].

More important, perhaps, was the quantitative linking of HCO_3^- formation to entry of sodium and thus to fluid movement. It may be noted at the outset that the turnover constants for sodium entry to CSF, and for fluid turnover itself are the same, suggesting one-way passage of the ion, and indicating that sodium access is a measure of fluid formation [3]. Figure 2 shows the smoothed curves of the accession of isotopic Na^+, Cl^-, and HCO_3^- from plasma to CSF in the dogfish [6].

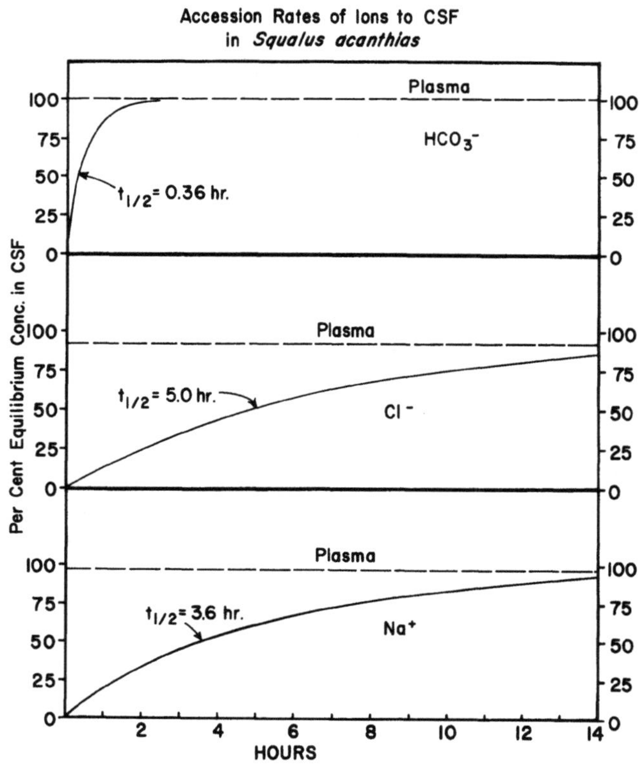

Fig. 2. Accession of isotopic ions to CSF in *S. acanthias*. Label injected at zero time. Smooth curves drawn to fit the observed rate constants. (From [6])

From these data the accession rate of each of the ions may be calculated, as shown in Table 1. The significant findings here are that the accession rate of Na^+ equals that of $Cl^- + HCO_3^-$, and that the concentration of HCO_3^- in newly formed fluid is some eight times that of plasma. The first of these findings suggests the validity

Table 1. Rates of accession of ions to CSF and their concentration in newly formed fluid

	1	2	3	4	5	6
	Plasma	CSF	k_{in}	t 1/2 to equil.	Accession rate Col. 1 x Col. 3	Conc. in secreted fluid [a]
	mM	mM	h^{-1}	h	mM/h	mM
Na^+	255	271	0.21	3.3	54	270
Cl^-	239	264	0.16	4.4	38	190
HCO_3^-	7.7	8.3	1.7 [b]	0.41	13	65
CO_2	0.26	0.26	51 [c]	–	–	–

a $= \dfrac{\text{Accession rate (Col. 5), mM/h x vol. CSF (1.2 ml)}}{\text{Formation of CSF (0.240 ml/h)}}$

b Calculated as if plasma HCO_3^- was the base of the process, to facilitate comparison with other ions. Entries in columns 4–6 of this row refer to kinetics of HCO_3^- in CSF, irrespective or origin

c Calculated as HCO_3^- formation from CO_2

of the method, particularly with regard to the use of $H^{14}CO_3$ injected into plasma, and its trace into the CSF. The presumption [6, 9] is that the isotope distributes in blood between HCO_3^- and CO_2 at their equilibrium concentrations, and the CO_2 which reaches the choroid plexus reacts there to form new labeled HCO_3^-, which is measured in the CSF. This idea is supported by the fact that fish pretreated with acetazolamide or methazolamide show about a 50% reduction in the appearance of labeled HCO_3^- in CSF. We emphasize that the equality of measured Na^+ to Cl^- + HCO_3^- accession in Table 1 suggests that the HCO_3^- found is not due to isotope exchange, but to formation of new ion. The second finding, of high HCO_3^- and low Cl^- concentration in newly formed fluid, further links the chemistry of CSF formation with that of aqueous humor and pancreatic fluid (Fig. 3). These data show that 24% of sodium entry is matched by HCO_3^-.

Similar experiments have been carried out in the cat (Fig. 4); we now show also the effect of carbonic anhydrase inhibition upon each ion. The turnover constants are used to yield the accession rates, as shown for Na^+ and Cl^- in Table 2. However, unlike for the fish (Table 1) the calculation of HCO_3^- accession in cat yielded a number higher than that for sodium, 3.5 mM/min. This shows that $H^{14}CO_3^-$ is reaching the CSF, in part, by exchange diffusion with plasma, or from rapid access to brain and then by "sink" effect into CSF. Thus we calculate the true HCO_3^- formation rate in CSF by Na^+ minus Cl^-, as shown in Table 2. This yields HCO_3^- concentration in new fluid of 62 mM, and shows that 37% of entering Na^+ is matched by HCO_3^-. On this basis, the fish and cat largely agree; the only difference is the large diffusional component for trace HCO_3^- in the cat. A possible basis for this difference is the much higher ratio of the volume of brain/CSF in the mammal (about 10) compared with the dogfish (about 2). This is an important topic for further work, linked to the relative rates of entry of HCO_3^- in CSF and brain.

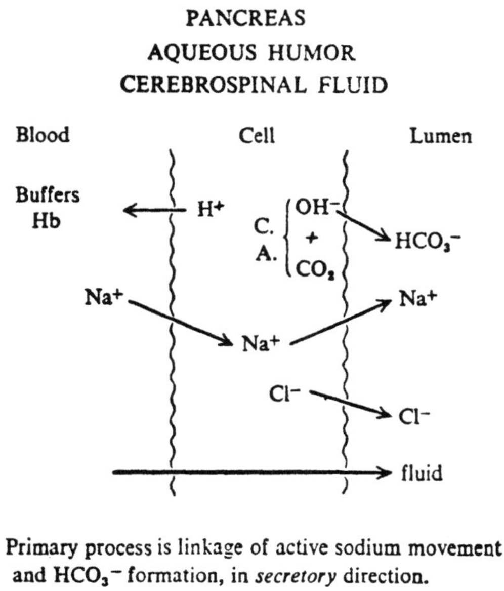

Fig. 3. Simplified model for secretory events in tissues which form HCO_3^- ion. The initial step is the protolysis of water, followed by spatial separation of H^+ and OH^- ions

Table 2. Accession of ions to CSF in cat

	1	2	3	4	5
	Plasma mM	k_{in} min^{-1}	Accession rate Col. 1 x Col. 2 mM/min	Conc. in new fluid [a] mM	Measured CSF conc. mM
Na^+, normal	147	0.016	2.4	164	158
CAI [b]		0.0076	1.1	157	
Cl^-, normal	115	0.013	1.5	104	134
CAI		0.0073	0.8	112	
HCO_3^-, normal					
as CO_2	0.95	0.94	0.9	62	22
as HCO_3^-	20	0.044			
CAI					
as CO_2	0.95	0.29	0.3	42	
as HCO_3^-	20	0.014			

a Col. 3 x fluid turnover time. The latter is given by the ventricular volume (1.4 ml) divided by the rate of CSF flow (0.020 ml/min) = 70 min. When carbonic anhydrase is inhibited the volume is unchanged and rate of formation is halved, whence fluid turnover time = 140 min

b Following 50 mg/kg acetazolamide or 30 mg/kg methazolamide, to inhibit completely carbonic anhydrase. The plasma and CSF concentrations of ions are not changed acutely following this treatment

Rate constants taken from Fig. 4. See [9] for original data

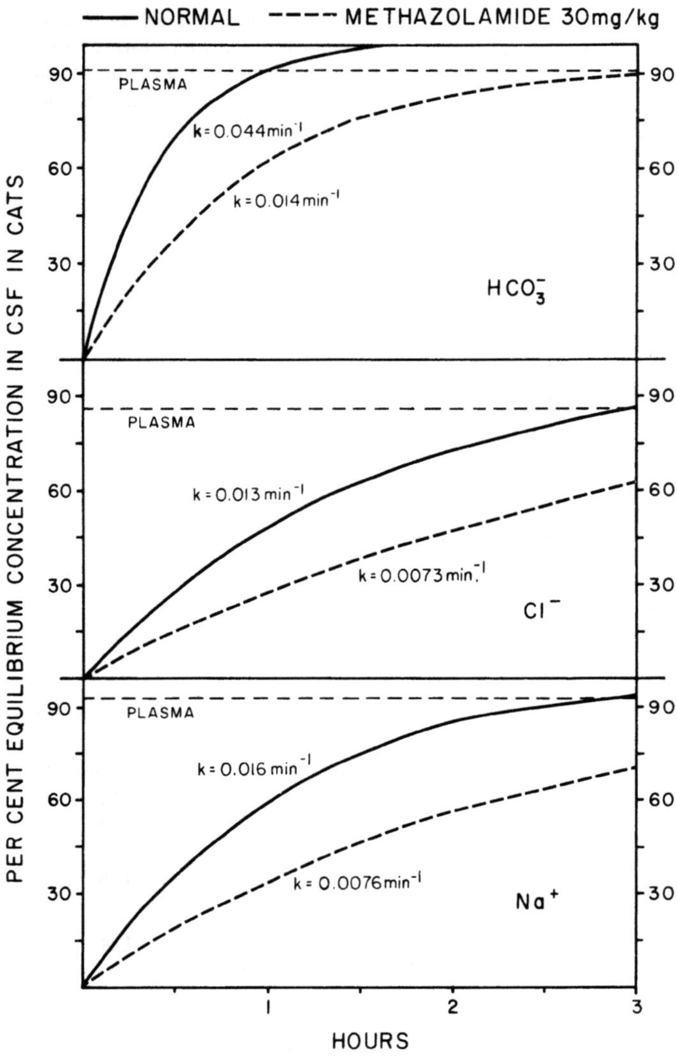

Fig. 4. Accession of isotopic ions to CSF in the cat. Label injected at zero time. *Horizontal lines* give plasma concentration of ion relative to CSF = 100. Drug injected intravenously 1 h before label. (From [9])

These data show that the effect of carbonic anhydrase inhibition in reducing CSF formation (reviewed in [5]; see also [8]) is a consequence of its effect in reducing sodium influx [4, 8], which in turn is due to reduction of HCO_3^- formation (Table 2). The data support the idea of a link between active sodium transport, dependent upon Na^+-K^+-ATPase [4] and HCO_3^- formation dependent on carbonic anhydrase. The precise nature of this link remains to be understood. In addition, there is a special connection between HCO_3^- formation and fluid movement as revealed by studies in the pancreas, that remains to be eludicated.

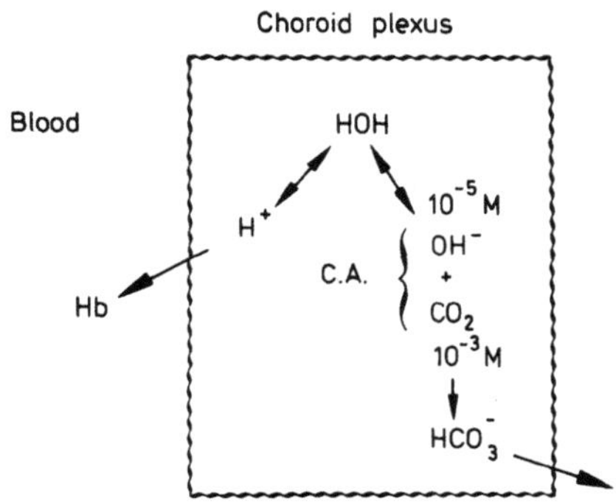

Fig. 5. Model showing chemical rates of HCO_3^- formation in CSF, based on data from cat [8]

$$V_{uncat} = k_2 \cdot OH^- \cdot CO_2$$
$$= 2 \times 10^4 \, M^{-1} s^{-1} \cdot 10^{-5} M \cdot 10^{-3} M$$
for 35 μL cell secretory volume
$$= 0.42 \, \mu mol/min$$

$$V_{cat} = \frac{k_{cat} \cdot E_0 \cdot S}{S + K_m}$$
$$= \frac{1 \times 10^6 s^{-1} \cdot 22 \mu M \cdot 1 mM}{1 mM + 14 mM}$$
for 35 μL cell volume
$$= 3300 \, \mu mol/min$$

The chemical basis for HCO_3^- accession is shown in model form in Fig. 5 where E_0 and the secretory volume are for cat choroid plexus. The numbers used are reasonably firm, except for the guessed $[OH^-]$ concentration at the site of formation, which enters into the calculation of the uncatalyzed rate. However, if this number were very much larger, the uncatalyzed rate would exceed the observed (see below); if it were much smaller there would be no gradient for OH^- or HCO_3^- across the membrane. As it turns out (Table 3), the calculated uncatalyzed rate from Fig. 5 agrees closely with the rate after complete inhibition of carbonic anhydrase [9]. Figure 5 shows that the calculated catalytic rate greatly exceeds the observed rate of 1.24 μmol/min (Table 3).

Table 3. Theoretical and observed rates of HCO_3^- production in cat CSF

Theory, maximum rates	μmol/min	Observed, in vivo [9]	μmol/min
Uncatalyzed	0.42	Total inhibition, i = 0.9999	
		30–50 mg/kg meth. or acet. = uncat.	0.40
Catalyzed	3300	Normal (total) rate	1.24
Catalyzed, at		Catalyzed rate: total – uncat.	0.84
i = EI/E = 0.9990	3	Methazolamide 10 mg/kg	
= 0.9997	1	i = 0.9985, observed	0.94
		Catalytic component	0.54

Table 3 shows the relations between theoretical (that is, calculated from the chemistry) and observed rates of HCO_3^- formation in CSF. The left frame gives the chemical rates from Fig. 5, and then the rates that would be expected at two levels of fractional inhibition (i) of the enzyme. The right frame gives the observed data, taken from Table 2 [9] and recent work on the dose-response relations using methazolamide [8]. HCO_3^- formation is given as μmol/min, and is directly obtained from the units of Table 2, mM/min, by incorporating the ventricular volume of CSF in the cat, 1.4 ml. The first line shows the agreement between the rate following total inhibition of the enzyme, and the calculated uncatalyzed rate. This suggests strongly that all of the HCO_3^- reaching the CSF is formed from CO_2, and that ionic transport as such is negligible. The second line gives the observed normal rate, and subtracting the uncatalyzed (or totally inhibited) rate gives the third line, the observed catalytic rate, 0.84 μmol/min. This is 3600 times less than the chemically calculated rate (left frame). Accordingly, when 10 mg/k g of methazolamide is given (i = 0.9985) and 0.15% of enzyme remains (fourth and fifth lines), the effect upon ion transport and flow is only partial [8].

The process that appears to limit HCO_3^- formation in CSF is blood flow to choroid plexus, and delivery of $HCO_3^- - CO_2$ to the tissue. Table 4 shows that if the total CO_2 were immediately available (in presence of blood carbonic anhydrase) to yield gaseous CO_2 for the hydroxylation reaction, the CSF HCO_3^- accession rate would be 2.3 μmol/min, about twice what is observed. In the inhibited state, the reaction $HCO_3^- \rightarrow CO_2$ in blood is slowed to 1.4 μmol/min, less than delivered, but still greater than the rate of HCO_3^- access in CSF [9].

Table 4. Rates of HCO_3^- formation in CSF, compared to calculated chemical rates, and delivery of HCO_3^-

	Chemically calculated rate [a]	Observed isotopic (Na minus Cl) rate [a]	Max rate from blood delivery of CO_2 [b]
		μmol/min	
Control	3300	1.24	2.3
Uncatalyzed or complete carbonic anhydrase inhibition	0.42	0.40	1.4 [c]

a See Table 3 and text
b Text and [9]
c Limiting event is $HCO_3^- \rightarrow CO_2$ in blood. See footnote 7 in [9]

Knowing the basal rates of HCO_3^- accession to CSF (Table 2), we may inquire how these may be changed by alterations in acid-base balance of the blood. The first two rows of Table 5 show HCO_3^- accession during normal acid-base balance, in control cats, and following carbonic anhydrase inhibition (from Table 2 above). In the next two rows (Table 5) we show what occurs in respiratory acidosis and metabolic alkalosis, based on the observed changes in concentration in CSF HCO_3^- [7], applied

Table 5. CSF HCO_3^- flux in cat (mM/h)

	In	Out
Normal [a]	53	53
Carbonic anhydrase inhibited [a]	17	17
Normal + pCO_2 rise from 42 to 65 mm Hg [b]	56	53
Normal + plasma HCO_3^- rise from 20 to 30 mM [b]	54	53

a From isotope data of Vogh and Maren [9]. See Table 2 above
b From concentration changes found by Nattie and Romer [7] applied to row 1 above

to the basal rate of row 1. Thus, elevation of pCO_2 causes a rise in CSF HCO_3^- of 3 mM/h, but this is a very small fraction of the basal rate. The effect of increasing plasma HCO_3^- is even less [7]. These data, together with the kinetic argument given above that all CSF HCO_3^- can be accounted for by formation from CO_2, suggest strongly that plasma HCO_3^- contributes very little (normally perhaps nothing) to CSF HCO_3^-.

We may inquire as to the HCO_3^- transport system in brain. There are no studies of isotopic accession of $HCO_3^- - CO_2$ to brain. The turnover constant for entrance of labeled sodium is the same as that for entrance to CSF [3]; however, there is a most significant difference in that acetazolamide and other drugs which reduce the entry of sodium into CSF, have no effect upon its access to brain [4]. It was concluded that whereas sodium entry to CSF represents the formation of new fluid, that to brain signifies diffusional exchange into a relatively stagnant fluid [4]. This is relevant to the HCO_3^- question, since we have seen that its entry into CSF is linked to that of sodium and fluid. In brain, as Fig. 6 shows, hypercapnia produces a remarkable elevation of HCO_3^- – the most rapid in the body and much greater than the effect upon CSF HCO_3^- [2]. Figure 6 also shows that acetazolamide yields a modest reduction in rate of brain HCO_3^- accumulation. In a similar study in the cat, 5% CO_2 caused a 7 mM rise in brain HCO_3^- in 20 min [1], accompanied by an equivalent fall in Cl^- concentration (H.R. Ahmad, personal communication).

Thus it appears that HCO_3^- accession in brain is based (as for CSF) on hydroxylation of CO_2, and that it subserves the vital role of maintaining brain pH within narrow limits in the face of changes in body CO_2. (Note from Fig. 6 the normal brain pH after 3 h of either hyper- or hypocapnia.) However, in brain the HCO_3^- formation is not accompanied by sodium or fluid movement, but exchange for chloride.

In conclusion, the reaction $CO_2 + OH \rightarrow HCO_3^-$ in the central nervous system subserves two important functions: Ion and fluid movement into CSF, and maintenance of pH of brain tissue.

Acknowledgment. This work was supported in part by National Institutes of Health grant EY 02227.

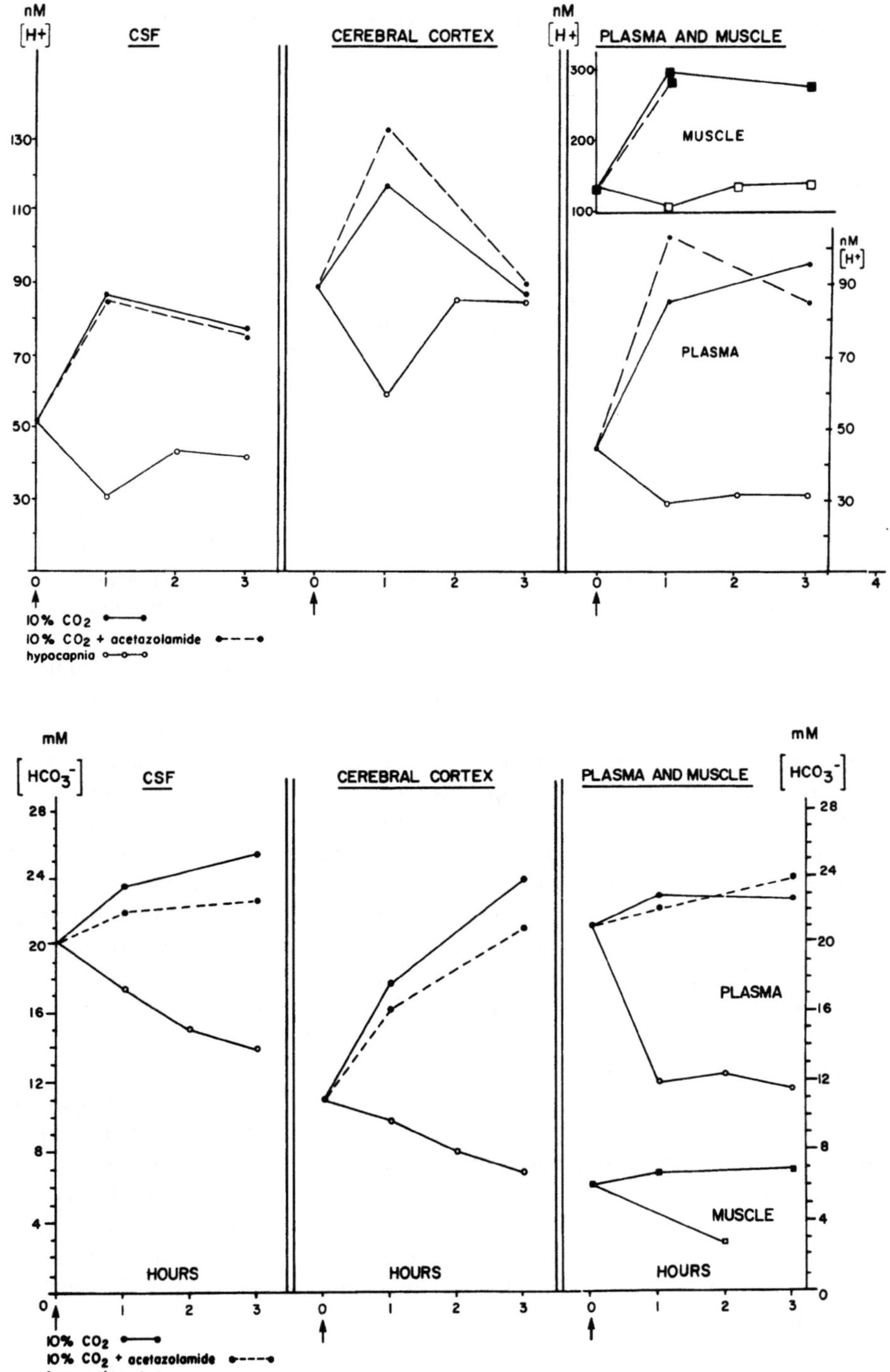

Fig. 6. Effects of hypercapnia (10% CO_2, arterial pCO_2 = 85 mm Hg) shown in *solid points,* and of hypocapnia (mechanical ventilation adjusted to arterial pCO_2 = 15 mm Hg) shown in *open points,* on HCO_3^- and H^+ concentration of CSF, cerebral cortex, plasma, and muscle of dogs. *Abscissae* show time from zero of altered pCO_2. *Dotted lines* show experiments in which dogs were pretreated with intravenous acetazolamide (100 mg/kg at 0 and 2 h) to inhibit completely carbonic anhydrase. Based on [2]

References

1. Ahmad HR, Loeschcke HH, Woidtke HH (1978) Three compartments model for the bicarbonate exchange of the brain extracellular fluid with blood and cells. In: Fitzgerald RS, Gautier H, Lahiri E (eds) The regulation of respiration during sleep and anesthesia. Plenum Press, New York, pp 195–209
2. Arieff AT, Kerian A, Massry SG, de Lima J (1976) Intracellular pH of brain: Alterations in acute respiratory acidosis and alkalosis. Am J Physiol 239: 804–812
3. Davson H (1967) Physiology of the cerebrospinal fluid. Churchill, London
4. Davson H, Segal MB (1970) The effects of some inhibitors and accelerators of sodium transport on the turnover of ^{22}Na in the cerebrospinal fluid and the brain. J Physiol 209: 131–153
5. Maren TH (1967) Carbonic anhydrase: Chemistry, physiology, and inhibition. Physiol Rev 47: 595–781
6. Maren TH (1972) Bicarbonate formation in the cerebrospinal fluid. Its role in sodium transport and pH regulation. Am J Physiol 222: 885–899
7. Nattie EE, Romer L (1978) CSF HCO_3^- regulation in isosmotic conditions: The role of brain pCO_2 and plasma HCO_3^-. Respir Physiol 33: 177–198
8. Vogh BP (1980) The relation of choroid plexus carbonic anhydrase activity to cerebrospinal fluid formation: study of three inhibitors in cat, with extrapolation to man. J Pharmacol Exp Ther (in press)
9. Vogh BP, Maren TH (1975) Sodium, chloride, and bicarbonate movement from plasma to cerebrospinal fluid in cats. Am J Physiol 228: 673–683
10. Wichser J, Kazemi H (1975) CSF bicarbonate regulation in respiratory acidosis and alkalosis. J Appl Physiol 38: 504–512

Autoregulation of CSF H^+ Activity

G.H. GURTNER, H.M. MAKHZOUMI, and R.J. TRAYSTMAN [1]

A few years ago we became interested in the acid base relationship between CSF and blood. Our initial interest was in the relatively large P_{CO_2} differences which could exist between CSF and blood during acidosis. We believe that these P_{CO_2} differences are a manifestation of the same phenomenon which we observed in the lungs (cf. Gurtner and Traystman, this vol) and felt that the P_{CO_2} differences might be important in CSF H^+ homeostasis. We developed a model in which the CSF H^+ activity was regulated passively with HCO_3^- and H^+ passively distributed between CSF responding to the P_{CO_2} difference and H^+ sensitive permeability changes of the blood brain barrier (Davies and Gurtner, 1973; Davies et al., 1973). A similar model was also developed involving H^+ sensitive changes in the blood brain barrier to Cl^- and HCO_3^- (Gurtner et al., 1974).

In order to test these models we endeavored to measure the permeability of the blood brain barrier to HCO_3^- and Cl^- at different blood and CSF H^+ activities; however, to measure the HCO_3^- permeability of the barrier accurately it was necessary to have an estimation of the HCO_3^- concentration in the freshly formed CSF. This could be achieved as part of the permeability measurement as follows:

Ventriculo-cisternal perfusions were carried out using a mock CSF containing no Cl^- save for trace amounts of $^{36}Cl^-$. The mock CSF had acetate as the major anion, trace amounts of ^{14}C inulin for the measurement of CSF production rate and varying amounts of HCO_3^-. We measured flux of unlabeled Cl^- into the CSF and flux of $^{36}Cl^-$ from CSF to blood. The flux of Cl^- was always greater than $^{36}Cl^-$ flux because of the presence of Cl^- in the freshly secreted CSF. An excess flux for Cl^- could be calculated by multiplying the difference in permeability between Cl^- and $^{36}Cl^-$ by the diffusion gradient for Cl^-.

The Cl^- concentration of the freshly secreted CSF was calculated by dividing the excess Cl^- flux by the CSF secretion rate. Since HCO_3^- and Cl^- are the major anions in both CSF and blood, and since both fluids are isosmotic, a good estimate of the HCO_3^- in freshly secreted CSF could be calculated by subtracting the Cl^- in the freshly secreted CSF from the sum of the serum Cl^- and HCO_3^-. The most surprising observation was that the HCO_3^- concentration of the freshly secreted CSF was very large and there was a strong correlation between the HCO_3^- concentration and the H^+ activity of the cisternal CSF (Makhzoumi et al., 1977, 1978). This is shown in Fig. 1. The permeability measurements are shown in Fig. 2. Note that HCO_3^- permeability

1 Depts. of Medicine and Environmental Health Sciences, The Johns Hopkins Medical Institutions, Baltimore, Maryland 21205/USA

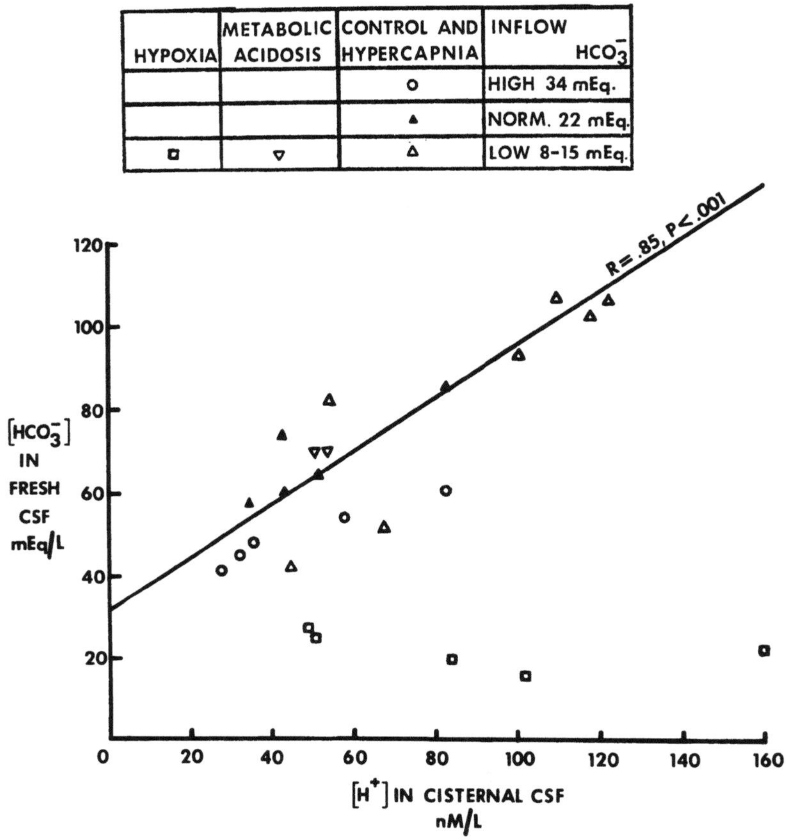

Fig. 1. The calculated HCO_3^- concentration of the freshly secreted CSF is plotted against the H^+ of the cisternal CSF. Data for different inflow HCO_3^- values are given. In most of the experiments the H^+ was changed by giving CO_2 in the inspired air. In one experiment, labeled metabolic acidosis, the serum H^+ activity was changed by infusion of Na^+ HCO_3^- at constant $PaCO_2$. In this experiment, the HCO_3^- concentration in the fresh CSF did not change. The relationship between the two variables during normoxia has a high correlation coefficient ($R = 0.85$, $P < 0.001$). This relationship during normoxia suggests that there is local regulation of the CSF H^+. Exposure of the animal to hypoxia markedly decreases the HCO_3^- in the freshly secreted fluid and abolishes the relationship between CSF H^+ activity and secreted HCO_3^-

is almost an order of magnitude larger than Cl^- permeability and does not change with blood pH; however, there appears to be a direct relationship between Cl^- permeability and blood H^+ activity.

The presence of an anomalously large constant permeability to HCO_3^- and the very high concentration of HCO_3^- in the freshly formed CSF did not appear consistent with our passive model and appeared more consistent with an active regulation of CSF H^+ activity mediated by HCO_3^- secretion.

Other experiments showed that the secretion of HCO_3^- was very sensitive to hypoxia (Fig. 1) and that the mechanism of HCO_3^- secretion was similar to that in the renal tubule in that the CSF K^+ level markedly influences the rate of HCO_3^-

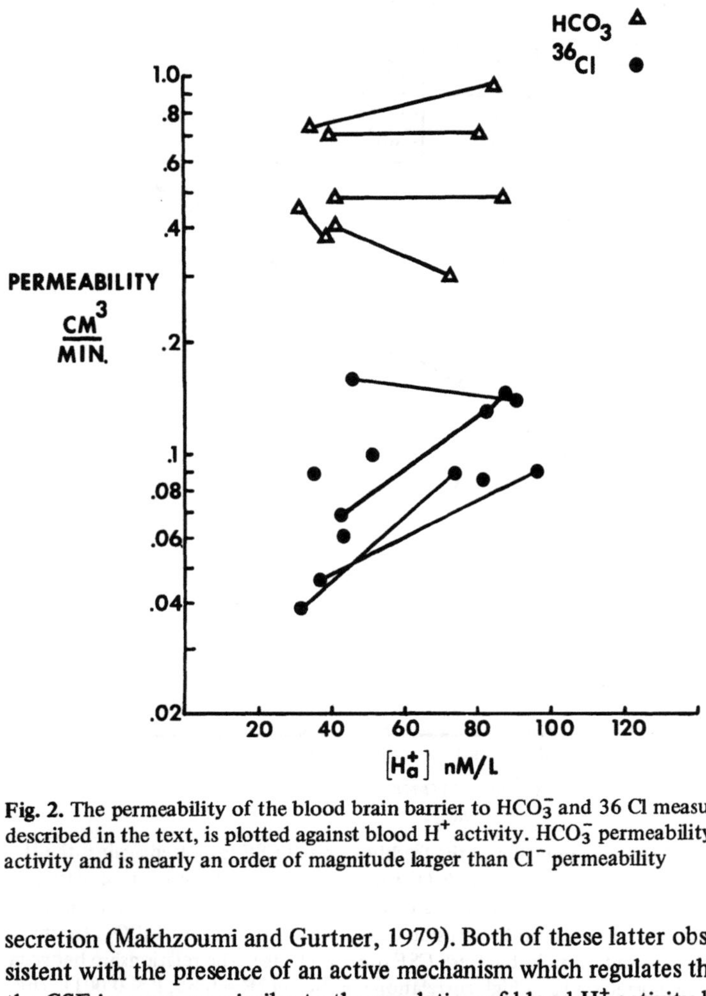

Fig. 2. The permeability of the blood brain barrier to HCO_3^- and 36 Cl measured by the method described in the text, is plotted against blood H^+ activity. HCO_3^- permeability is not related to H^+ activity and is nearly an order of magnitude larger than Cl^- permeability

secretion (Makhzoumi and Gurtner, 1979). Both of these latter observations are consistent with the presence of an active mechanism which regulates the H^+ activity of the CSF in a manner similar to the regulation of blood H^+ activity by the kidney.

References

Davies D, Gurtner GH (1973a) Cerebrospinal fluid acid-base balance and the Wien effect. J Appl Physiol 34: 249–254

Davies D, Fithgerald RS, Gurtner GH (1973b) Acid-base relationship between cerebral spinal fluid and blood during metabolc acidosis. J Appl Physiol 34: 243–248

Gurtner GH, Davies DG, Burns B (1974) Relationship between blood and extravascular pCO_2, (H^+) and (HCO_3^-) in different acid-base states in lung, CSF, and brain. In: Nahas G, Schafer K (eds) Carbon dioxide and metabolic regulations. Springer, Berlin Heidelberg New York

Makhzoumi HM, Gurtner GH (1979) The effect of CSF K^+ on the regulation of CSF H^+ activity. Fed Proc 38: 1147

Makhzoumi H, Traystman R, Sciuto AM, Burns B, Fitzgerald R, Gurtner G (1977) Local regulation of CSF HCO_3^-. Physiologist 20: 60

Makhzoumi, H, Traystman RJ, Burns B, Fitzgerald RS, Gurtner GH (1978) O_2 sensitive local regulation of CSF H^+ activity. In: Fitzgerald RS, Gautier H, Lahiri S (eds) Regulation of respiration in sleep and anesthesia. Plenum Press, New York, pp 185–193

Transients and Steady State of Chloride-Bicarbonate Relationships of Brain Extra-Cellular Fluid

H.H. LOESCHCKE and H.R. AHMAD [1]

Summary. In respiratory acidosis [HCO_3^-] in the extracellular fluid regularly increased and [Cl^-] decreased. The ratio of these changes closely approached 1. It is concluded that [HCO_3^-] in the extracellular fluid of the brain in respiratory acidosis increases because CO_2 reacts with the intracellular buffers and HCO_3^- exchanges from the intracellular to the extracellular phase in a 1:1 exchange against Cl^- similarly as in the HCO_3^--Cl^--shift between blood plasma and erythrocytes. The exchange is assumed to occur between glia cells and the extracellular fluid of the brain. All or the major part of the exchange takes place in less than the wash-in time of CO_2 which is limited by the local circulation. Acetazolamide prolonged and SITS diminished the time constants of pH, pCO_2 and Cl^-.

Introduction. The extracellular fluid (ECF) of the brain is in free exchange with the cerebrospinal fluid (CSF) (Katzman et al., 1968; Pappenheimer, 1967; Dermietzel, 1976). The ionic compositions of both fluids may therefore be assumed to be similar. This does not exclude that under certain conditions, as for example in hypoxia, gradients from ECF to CSF of hydrogen and other ions may be built up which could be temporary or also steady.

In a pioneer experiment Granholm and Pontén, 1969 investigated the "in vivo CO_2 dissociation curve" of CSF. They found that this curve in slope was similar to that in true rather than in isolated plasma or in a solution devoid of protein buffer as is CSF. The implication is that CSF (and ECF) in vivo must be in contact with buffer in a way comparable to plasma in vivo which exchanges ions with the interior of the erythrocytes, thus making the intracellular buffer available to the plasma especially if there is a load of CO_2 or other acids.

Such an ion exchange with ECF could be imagined on two lanes, either with blood or with brain cells. Obviously exchange occurs in both directions which led Ahmad et al. (1978) to what was called the three compartment hypothesis. Exchange of bicarbonate with blood was demonstrated by Leusen (1972); Leusen and Weyne (1973), by Kazemi et al. (1976) and by Ahmad et al. (1976). The role of an exchange with brain cells in respiratory acidosis was first demonstrated by Pannier et al. (1971) who let cats inhale CO_2 and observed increase of the bicarbonate concentration in CSF. To this experiment it might have been objected that in respiratory acidosis the bicarbonate concentration in the blood plasma rises and that this would cause a bicarbonate shift to ECF of the brain and to CSF. Pannier et al. (1971), however injected

1 Institut für Physiologie, Ruhr-Universität, 4630 Bochum/FRG

acid intravenously during such an experiment with the result that increase of the plasma [HCO_3^-] could be avoided or that even a decrease took place. Still, during inhalation of CO_2 a marked increase of [HCO_3^-] in CSF was observed and this experiment allowed them to exclude the blood plasma as the sole source of the [HCO_3^-] increase in CSF in respiratory acidosis. Similarly, and with the same result, in hypoxia the arterial bicarbonate concentration was kept constant by Severinghaus (1965). Recent experiments of Nattie and Romer (1978) could be interpreted in the same way as those of Pannier et al. (1971). Though it is obvious that in respiratory acidosis the main source of the bicarbonate appearing in CSF can be nowhere else than in the cells of the brain, it remains still open whether or not the changes in CSF and ECF would be similar and especially the questions arise what is the time course and what is the mechanism of such a bicarbonate exchange.

The shortcoming of the earlier experiments was that the concentrations were measured in CSF rather than in ECF. If fast exchanges between cells and ECF existed, this still would appear in CSF slowly because the volume of CSF is high compared with the adjacent ECF and the distribution of ions in CSF may occur mainly by diffusion especially in an anesthetized animal fixed to an operating desk which excludes movement of the head.

1 Methods

Ahmad et al. (1978) developed methods which may yield values much more representative for ECF and also are able to follow fast changes. Similar to the procedure of Loeschcke and Sugioka (1969) surface electrodes for pH and pCO_2 were used with a flat sensitive membrane attached to the intact brain surface and so measuring the parameters in a fluid film in direct contact with the extracellular spaces.

In the earlier experiments conventional CO_2 electrodes were used consisting of a glass electrode with a flat surface and a calomel reference, a silicon membrane and a bicarbonate buffer between membrane and electrode which also contained carbonic anhydrase. These electrodes had 90% reaction times of approximately 6 s. Later the electrode signal was passed through an antilog unit and a device which compensated for the attenuation diminishing the 90% time to less than 1 s (Luttmann et al., 1978).

In addition a chloride electrode was used consisting of a chlorinated silver plate which had gone through the procedure proposed by Marmont (1949) that is chlorination in m KCl, developing in photographic developper, washing, covering with a thin gelatine layer and finally hardening the gelatine by formaline. In connection with a calomel or silver-silver chloride reference electrode this electrode is reversible for halogen ions, and insensitive to changes of the cation concentrations including H^+. Since under the conditions of the experiments the other halogen ions are negligible, such an electrode is specific for chloride and gives the Nernst potential. The experts will realize that if pH, pCO_2 and [Cl^-] are measured simultaneously, much care must be taken to separate the circuits form each other.

The experiments were carried out in 21 cats of either sex and body weights between 1.5 and 3.1 kg, anesthetized with α-glucochloralose (40 mg/kg) and urethane (200 mg/kg). Arterial pressure, tidal volume, and expiratory pCO_2 were recorded with conventional methods. The body temperature was kept between 37.0° ± 0.2°C and the temperature at the brain surface was measured close to the electrode position and controlled by infrared radiation to 37° ± 0.1°C. The pH, pCO_2, and Cl^--electrodes were put on the brain surface. They were balanced in order to provide constant minimal force and pressure (approximately 1/2 pond and 0.06 pond·mm^{-2}) on the surface. The diameters of pH, pCO_2, and Cl^- electrodes were 5 mm. The electrodes were applied either to the parietal cortex or to the ventral surface of the medulla oblongata. By fixing a Plexiglas frame on the trephine hole a space was provided of at least 10 mm depth which filled with CSF. It was believed that this prevented appreciable loss of CO_2 from the brain surface.

The cat breathed oxygen in a semiclosed system without valves, circulated by a pump. The oxygen consumed was constantly replaced and CO_2 was absorbed. The tidal volume was recorded with a low inertia Plexiglas spirometer with an inductive transducer. In part of the experiments the cats were artificially ventilated with air by a pump.

CO_2 was added by injection of CO_2 gas to the inspiratory side. The flow was regulated by hand under observation of the infrared CO_2 recording in such a way that initially a high current was given which diminished so that a rectangular endtidal pCO_2 change was approached and maintained for 10 min or more. If the CO_2 flow was stopped quickly the system cleared itself from CO_2 by absorption. The on transients, the steady-state and the off transients of pH, pCO_2, $[Cl^-]$ and V_T could be observed under this regime.

2 Results

In Fig. 1 a typical experiment is shown where pH, pCO_2, and $[Cl^-]$ have been recorded on the surface of the brain cortex in an artificially ventilated anesthetized cat. By controlled addition of CO_2 to the inspiratory side endtidal pCO_2 was raised in an approximately rectangular fashion. pH, pCO_2, and $[Cl^-]$ on the brain surface – and it is assumed that this represents extracellular fluid of the brain – change in an approximately exponential function of the time approaching a steady-state level. The average time constants from similar experiments are for pH 40.3 s, for P_{CO_2} 40.4 s, and for Cl^- 30.5 s, which means they are similar for pH and pCO_2. In nine additional experiments on three spontaneously breathing cats the average time constants for pH_e on the ventral surface of the medulla oblongata was 42.9 s for the on transient and 39.6 s for the off transient. One of these cats was decerebrated (time constant 36.0 s). From pH and pCO_2 $[HCO_3^-]$ was calculated. The time constant was 30.7 s. This is similar to that of $[Cl^-]$ and shorter than the time constants of pH and pCO_2. Data are presented in Table 1.

In a similar series of experiments only two parameters were recorded simultaneously. The corresponding data of pH and $[Cl^-]$ or of pCO_2 and $[Cl^-]$ or of $[HCO_3^-]$ calculated from pH and pCO_2 were plotted against each other for the on and for the off transient in such a way that corresponding values were read and plotted for each 12 s as shown for $[HCO_3^-]$ and pCO_2 in Fig. 2. This yields a loop, and delays, if they occur, could be easily visualized because any delay would express itself by a divergence between on and off transient. If there is no area between the on and the off

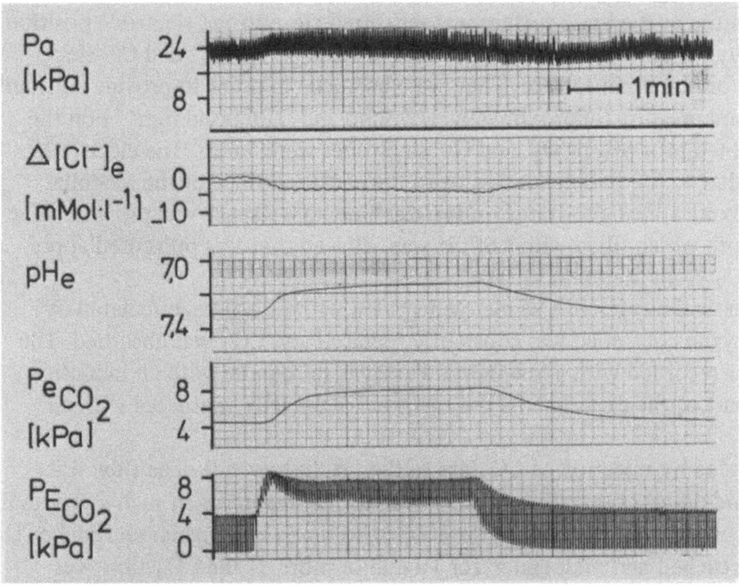

Fig. 1. Approximately rectangular change of endtidal pCO_2 in an artificially ventilated cat (19.12.78). Arterial pressure and extracellular $[Cl^-]$, pH and pCO_2 on the surface of the parietal cortex and expiratory pCO_2 are recorded

Table 1. Time constants τ of on-transient in approximately rectangular change of $pACO_2$ for pH_e, $peCO_2$, $[Cl^-]_e$ and $[HCO_3^-]_e$ in 14 experiments in 6 cats under constant artificial ventilation. Brain cortex. The probabilities of the zero hypothesis in group comparison are for pH_e against Cl_e^- 0.026, for pH_e against HCO_{3e}^- 0.055, for pCO_{2e} against Cl_e^- 0.026 and for pCO_2 against HCO_{3e}^- 0.060

Date	τ_{pH_e} s	τ_{peCO_2} s	$\tau_{Cl_e^-}$ s	$\tau_{HCO_{3e}^-}$ s
18.1.78	33	–	32	–
	31	–	28.5	–
22.9.78	42	40	55.2	30
	36	48	40.8	50.4
19.12.78	29	33.6	33	33
	36	48	36	48
3.1.79	45.6	45	22.6	36
	50.4	49.2	24	26.4
	37.2	36	25.2	19.2
	20.6	20.6	14.4	12
1.3.79	57.6	44.4	–	19.2
	57.6	39.6	–	33
12.3.79	43.0	–	26.0	–
	45.8	–	28.4	–
$\bar{x}$	40.3	40.4	30.5	30.7
$\pm s_m$	± 2.8	± 2.8	± 3.0	± 3.9
Number of exp.	14	10	12	10
Number of cats	6	4	5	4

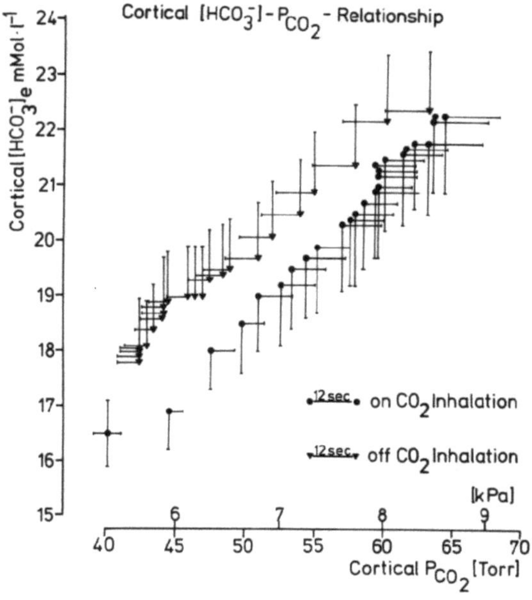

Fig. 2. In 6 experiments as in Fig. 1 cortical extracellular [HCO_3^-] is calculated from pH and pCO_2 every 12 s after approximately rectangular change of $PACO_2$ and plotted against the cortical extracellular pCO_2. The on-transient starts at the lower left corner. There is a small hysteresis between on- and off transients. This plot may also be considered as representing the in vivo HCO_3^--pCO_2 relationship in brain extracellular fluid. The *bars* represent the standard error. (From Ahmad et al., 1978, courtesy of Plenum Press, New York and London)

plots this indicates simultaneity of the two parameters. In fact usually there is some hysteresis indicating that the recording of [Cl^-] often follows pH and pCO_2 with a short delay.

The time course of [HCO_3^-] and [Cl^-] are compared in Fig. 3 in an on-transient of a $PACO_2$ step. They seem to correspond well enough (a small delay of pCO_2 against pH was compensated in the calculation of [HCO_3^-]).

From Fig. 4 showing averages of six experiments it may be seen that the increase of [HCO_3^-] is regularly correlated to a decrease of [Cl^-] and that the ratio of these two changes considering the standard errors is very close to one. In most experiments, however, the chloride concentration does not completely return to the control level.

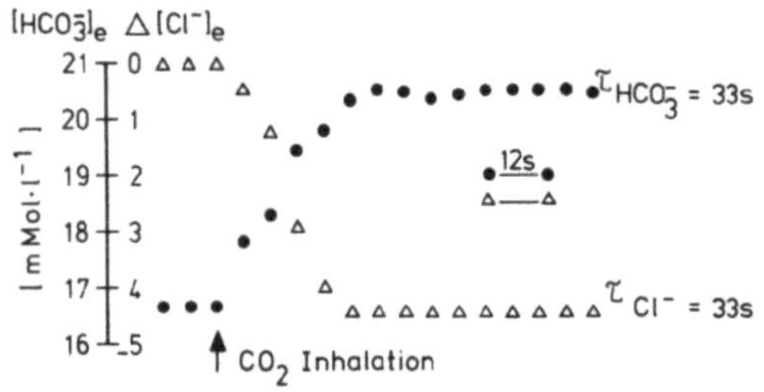

Fig. 3. From a recording as in Fig. 1 brain extracellular [Cl^-] and also brain extracellular [HCO_3^-] are plotted against time during an approximately rectangular step of $pACO_2$. In the calculation of [HCO_3^-] from pH and pCO_2 a small delay of pCO_2 against pH was considered as a technical error probably due to an interference with the local circulation under the CO_2 electrode and therefore eliminated. The cat is artificially ventilated

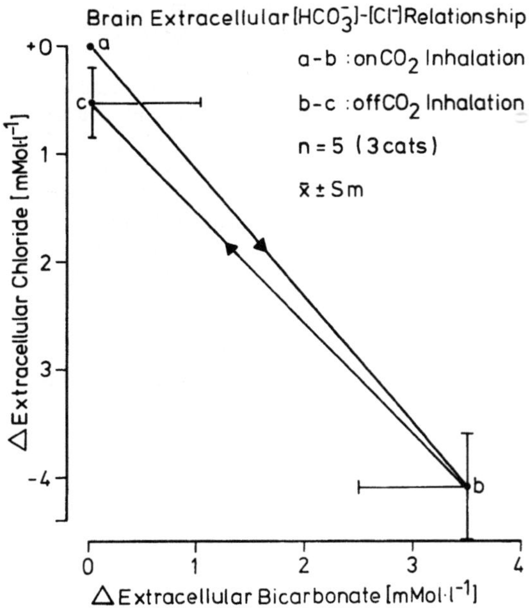

Fig. 4. The average change of extracellular $[Cl^-]$ was plotted against the average change of HCO_3^- in 5 experiments on 3 cats after a step change of $pACO_2$. The ratio of the changes is approximately 1:1. The chloride concentration at the end of the experiment (*c*) did not completely return to the control value before inducing the respiratory acidosis (*a*)

In erythrocytes the anion exchange can be blocked for example by SITS (4 acetamino-4′-isothiocyanostilbene-2,2′-disulfonic acid). In such a case no HCO_3^- could pass from the cell to the ECF. The time constants of pH in ECF and of pCO_2 in ECF following a rise of $pACO_2$ should then be shorter because less CO_2 is bound by the cell and because the CO_2 transport is limited by the circulation. SITS was dissolved in natural CSF and pH was titrated with NaOH to CSF pH equilibrated with 5% CO_2 and after a slight elevation of the electrode injected to the brain surface at the electrode site and after 5 min the electrode again was brought into contact with the brain. Table 2 represents data on the time constants of pH before and after application of SITS and the expected effect becomes evident, with the exception of the first experiment of the series (12.1.79) which is probably technically less good, as may be indicated by the higher scattering of the data. The difference in a group comparison is still significant ($p = 0.02$).

In preliminary experiments acetazolamide was applied similarly. In the four experiments in two cats the time constant of pH_e was prolonged.

3 Discussion

The results of this series of experiments are based on new techniques. Before drawing conclusions the reliability must be discussed for steady and for transient states separately. The pH electrode is conventional and there is no doubt that it can be calibrated with high accuracy and that it also follows fast changes. CO_2 electrodes such as membrane-covered glass electrodes are known to be reliable for steady-state comparisons, they are, however, slow because diffusion is involved in the measuring process. The electrode used in the critical experiments was an electrode of the fast type of Lutt-

Table 2. Time constants τ of on-transients in approximately rectangular changes of $pACO_2$ for pH_e before (control) and after application of SITS in 18 (16) experiments in 6 cats under constant artificial ventilation. Brain cortex

Date	Control	SITS
	τ_{pH_e}	τ_{pH_e}
	s	s
12.1.79	72	24
Left hemisphere	36	67
		55
Right hemisphere	43	
	74	96
	72	
	48	
31.1.79	48	25
	43	24
13.2.79	55	35
	55	38
		31
		34
		26
		26
1.3.79	58	46
	58	48
8.3.79	38	29
	43	31
12.3.79	43	31
	46	34
$\bar{x}$	52.0	38.9
$\pm s_m$	3.04	4.34
Number of exp.	16	18
Number of cats	6	6

mann et al. (1978) as described in Methods. The 90% response of this electrode system was reached in less than 1 s. Since the time constants observed in the present experiments were in the range of 12–96 s, the speed of the electrode is amply sufficient for correct recording. In other experiments the classical type of a CO_2 electrode was used, however with carbonic anhydrase added to the fluid between membrane and glass electrode. This electrode reached 90% deflection in 6 s which still is good enough to record the changes occurring. The Cl^- electrode is a silver-silver chloride electrode (see Methods). This electrode was tested for reproducibility and for selectivity. As long as the temperature was constant and as long as the measurement was in protein-free solutions the electrode was reproducible. It, as the other electrodes used, is dependent on temperature. Much care therefore was taken to control the temperature at the site of the measurement by using a thermocouple and radiant heat.

The chloride electrode starts drifting in fluids containing proteins in an uncontrollable way. CSF and ECF, however, are almost free of proteins and in the experiments described the drift even in a period of hours was minimal and needed no consideration in the short experiments described here. Of course any bleeding had to be avoided. The gelatine cover is an impediment to diffusion but it is thin and the time for 90% deflection of the electrode in a rectangular [Cl^-] change was determined to be 7 s or less.

The next question to be discussed is whether or not the fluid investigated is extracellular fluid. The use of flat electrodes to the surface of the brain for investigation of extracellular fluid, of course, is based on the assumption that the fluid film under the electrodes is in contact and exchange with extracellular fluid rather than with CSF. This is only possible if the electrode touches the surface of the brain with a pressure high enough to exclude free exchange with CSF, but on the other hand this pressure should be light in order not to disturb the local circulation under the electrode. The experimental arguments that a fluid representing ECF rather than CSF is measured are (1) that pH changed (usually slightly to the alkaline side) if the electrode was lifted up from the surface and (2) that the time constants of all parameters if CO_2 was inhaled were short, much shorter than in CSF (as measured with a lifted electrode), and that they were reproducible in the single animal and similar between animals. Furthermore the observation that in the best experiments there is only small hysteresis in on- and off-transient loops between parameters may be interpreted as meaning that a fluid representative of extracellular fluid is measured.

In ECF the protein content is negligible. This is a most favorable condition for an experiment on ion exchange in respiratory acidosis because in such a fluid without exchange no appreciable increase of [HCO_3^-] would occur if pCO_2 is elevated. The complete increase of [HCO_3^-] therefore is due to the exchange. Also the full decrease of [Cl^-] is due to exchange. This means that the exchange ratio may well be determined in such an experiment without consideration of other sources or sinks of chloride or bicarbonate. It must be kept in mind, however, that ECF exchanges with cells and also with blood. From experiments of Pannier et al. (1971) it becomes clear that the exchange in respiratory acidosis is predominantly with cells, but the blood component is difficult to assess. The statement from the experiments presented must therefore be that in respiratory acidosis there is an exchange of HCO_3^- against Cl^- between ECF and cells and blood but predominantly cells. The total exchange ratio is close to one and this is probably still true for the major component (cells). It may be different for the exchange with blood, but this has to be elaborated by other experiments.

The ion exchange is a fast exchange because there is no appreciable lag between pCO_2, [HCO_3^-] and [Cl^-] and the time constants for the anions were even faster than those for pH and pCO_2.

The time constants determined in this kind of experiments are not the time constants of the ion exchange itself. Their main determining factor is the wash in or wash out time of CO_2 in a tissue containing buffer. The observed time constants therefore represent an upper limit of the time constants of the ion exchange itself. The latter have to be assumed to be shorter. Also it must be taken into account that the step of arterial pCO_2 was not completely rectangular. This again would cause an error in the sense that the real time constants are shorter than those determined.

A fast exchange of HCO_3^- was demonstrated with a similar but less elaborated method on the surface of the choroid plexus in cats by Loeschcke and Sugioka (1969) and exchange against chloride was assumed. This obviously was an exchange between blood and CSF while in the present investigation the exchange between cells and ECF is predominant. Bicarbonate exchanges which were thought to be slow because they were measured in the bulk of CSF were described by several authors as reviewed by Leusen (1972), Leusen and Weyne (1973) and more recently by Ahmad et al. (1976, 1978), Kazemi et al. (1976) and by Nattie and Romer (1978). Only in the latter papers was the role of Cl^- acknowledged. It is the first time in the present experiments that an exchange ratio of approximately one was determined.

Which cells could be made responsible for the anion exchange? In the brain, except for capillaries, there is nothing but neurons and glial cells. The glia cells provide the bulk of the cell mass. This is why they are the first candidates for the exchange of Cl^- against HCO_3^- ions. Also the permeabilities of the neuronal membrane for anions are probably too low to allow the exchange observed even if it is kept in mind that only a relatively small number of ions must be exchanged to reach equilibrium because the extracellular space is quite narrow (approximately 200 Å). The properties of glial cells are not well known. Their membrane potential is high and reflects the potassium concentration ratio between inside and outside (Kuffler and Nichols, 1966). This does not exclude that anions are exchanged. Recent views concerning the erythrocytes assume that there is a carrier (probably a protein or peptide) which mediates an electroneutral one-to-one exchange of HCO_3^- against Cl^- or other anions. The few and incomplete data about sodium and chloride concentrations seem to indicate that these are relatively high inside the glia cell (Koch et al., 1962; McIlwain and Bachelard, 1971), and also the fluxes, if abnormal concentration differences are created, are high (McIlwain and Bachelard, 1971).

In erythrocytes the anion exchange may be blocked by disulfonic stilbene derivatives like SITS. In a complete blockade both compartments in a respiratory acidosis would be equilibrated separately. The equilibration of the extracellular phase would need less CO_2 transport because of the negligible protein buffer concentration and the decrease of pH would be expected to be faster than in the control. In the experiments presented here the equilibrium of the intracellular phase evades observation.

A shorter time constant for pH would be expected and this was the case in the experiment. There should be no change in extracellular HCO_3^- if the exchange is blocked completely. These results indicate that SITS also blocks the anion exchange between glia cells and extracellular fluid of the brain.

Glia cells are rich in carbonic anhydrase (Giacobini, 1962; Maren, 1967) which accelerates the hydration and conversion of CO_2 into bicarbonate ion. This conversion can be prolonged by blocking the carbonic anhydrase. This would mean that the appearance of HCO_3^- in the cell is delayed, and therefore the exchange. The initial decrease of extracellular pH should be accelerated, but there should be a late phase of ion exchange which may even reduce the hydrogen ion concentration outside after a while. The HCO_3^- increase outside should be sluggish. Indeed the experiment shows a retardation of the decrease of extracellular pH in respiratory acidosis.

Acknowledgments. The skilled help of Mrs. Renate Hildebrandt, Mr. Hans H. Woidtke and Mr. Hashim Shams is appreciated.

References

Ahmad HR, Berndt J, Loeschcke HH (1976) Bicarbonate exchange between blood, brain extracellular fluid and brain cells at maintaned pCO_2. In: Acid base homeostasis of the brain extracellular fluid and the respiratory control system. Thieme, Stuttgart, pp 19–27

Ahmad HR, Loeschcke HH, Woidtke HH (1978) Three compartments model for the bicarbonate exchange of the brain extracellular fluid with blood and cells. In: Fitzgerald RS, Gautier H, Lahiri S (eds) Regulation of respiration during sleep and anaesthesia. Plenum Press, New York, pp 195–209

Dermietzel R (1976) Central chemosensitivity, morphological studies. In: Acid base homeostasis of the brain extracellular fluid and the respiratory control system. Thieme, Stuttgart, pp 52–65

Giacobini E (1962) A cytochemical study of the localization of carbonic anhydrase in the nervous system. J Neurochem 9: 169–177

Granholm L, Pontén, U (1969) The in vivo CO_2 buffer curve of the intracellular space of cat cerebral cortex. Acta Neurol Scand 45: 493–501

Katzman R, Graziani L, Ginsburg S (1968) Cation exchange in blood, brain and CSF. In: Lajtha A, Ford DH (eds) Brain barrier systems, vol 29, Progress in brain research. Elsevier Publ Comp, Amsterdam London New York, pp 283–294

Kazemi H, Weyne J, van Leuven F, Leusen I (1976) The CSF HCO_3 increase in hypercapnia relationship to HCO_3, glutamate, glutamine and NH_3 in brain. Respir Physiol 28: 387–401

Koch A, Ranck JB Jr, Newman BL (1962) Ionic content of neuroglia. Exp Neurol 6: 186–200

Kuffler SW, Nichols JG (1966) The physiology of neuroglial cells. Ergebn Physiol 57: 1–90

Leusen I (1972) Regulation of CSF composition with reference to breathing. Physiol Rev 52: 1–56

Leusen I, Weyne J (1972) Aspects of the regulation of the acid-base status in CSF and brain. Bull Physiopathol Respir 9: 635–646

Loeschcke HH, Sugioka K (1969) PH of cerebrospinal fluid in the cisterna magna and on the surface of the choroid plexus of the 4th ventricle and its effect on ventilation in experimental disturbances of acid-base balance. Transients and steady states. Pflügers Arch 312: 161–188

Luttmann A, Mückenhoff K, Loeschcke HH (1978) Fast measurement of the CO_2 partial pressure in gases and fluids. Pflügers Arch 375: 279–288

Maren TH (1967) Carbonic anhydrase: chemistry, physiology and inhibition. Physiol Rev 47: 595–781

Marmont G (1949) Studies on the axon membrane. J Cell Comp Physiol 34: 351–382

McIlwain H, Bachelard HS (1971) Biochemistry and the central nervous system. Churchill Livingstone, Edinburgh London

Nattie EE, Romer L (1978) The role of chloride and other anions in cerebrospinal fluid bicarbonate regulation. In: Fitzgerald RS, Gautier H, Lahiri S (eds) The regulation of respiration during sleep and anaesthesia. Plenum Press, New York, pp 211–218

Pannier JL, Weyne J, Leusen I (1971) The CSF/blood potential and the regulation of the bicarbonate concentration of CSF during acidosis in the cat. Life Sci 10: 287–300

Pappenheimer JR (1967) The ionic composition of cerebral extracellular fluid and its relation to control of breathing. Harvey Lect 61: 71–94

Severinghaus JW (1965) Electrochemical gradients for hydrogen and bicarbonate ion across the blood-CSF barrier in response to acidbase balance changes. In: Brooks Ch McC, Kao FF, Lloyd BB (eds) Cerebrospinal fluid and the regulation of ventilation. Blackwell Sci Publ, Oxford, pp 247–258

Subject Index

Acetazolamide
see also Carbonic anhydrase, inhibition
binding in lungs 340
inhibition constant 295, 413
permeation through cell membranes 341, 401
Angiotensin
carbamino adducts 106
Anion exchanger
removal of HCO_3^- 78
Antibodies
chicken erythrocyte carbonic anhydrase 402
Association constant
of oxygen
influence of CO_2 122, 125

Bicarbonate
absorption in kidney tubules 419, 420, 423, 424, 438
concentration
in brain 433
extracellular fluid 446
in cerebrospinal fluid 426–428, 432, 436, 440
conductivity 8
diffusion 27, 36, 40, 51
disequilibrium, lung space 349
exchange
brain CSF and cells 440
brain extracellular fluid and cells 433, 440, 446
choroid plexus 447
red cells and plasma 285, 292
formation rate in CSF 432, 433, 437
inhibition constant for carbonic anhydrase 33
permeability
peritubular cell membrane 421, 422, 423
red cell membrane 180, 286, 288
placental transfer 386, 390, 393
exchange against lactate 394
secretion
into cerebrospinal fluid 437, 438
self-exchange in red cells 285, 288
Blood
brain barrier, H^+ sensitivity 436
Bohr effect
of hemoglobin 356
influence of CO_2 85, 124, 128
speed of
influence of carbonic anhydrase 372
Bradykinin
carbamino adducts 106
Brain
cerebrospinal fluid 426, 436, 439, 446
formation of 426, 427
formation during carbonic anhydrase inhibition 427, 430
ionic composition of 426
role of Na^+ and HCO_3^- transport for its production 430
cortex, parietal 441
extracellular fluid 439, 441, 446
glial cells 447

Carbamate
α-amino groups 68, 72, 75, 86, 122
^{13}C magnetic resonance spectroscopy 90, 102, 106
determination of 75, 89, 97, 102
ε-amino groups 76, 91
equilibrium constants 96, 119, 128
equilibrium reactions 76, 89
formation
bird hemoglobin 85
cat hemoglobin 114
crocodile hemoglobin 86
electrostatic aspects 96
fetal hemoglobin, human 86
fish hemoglobin 84
peptide hormones
equilibrium constants 109
quantitation by ^{13}C NMR 106
sheep hemoglobin 86, 116
oxygen-linked 67, 84, 96, 114, 122, 127
physiological significance 86, 93
X-ray crystallography 68, 70, 85, 119

Carbamic acid
pK 89
Carbaminohemoglobin 67, 75, 84, 89, 115, 122
Carbonate
diffusion 51
Carbonic acid
dehydration rate constant 8, 9, 11
dissociation equilibrium and kinetics 8, 10, 11
Carbonic anhydrase
acceleration (activity) factor
dependence on enzyme concentration 32
in skeletal muscle 413
pH dependence 31, 134, 135, 182
red cell lysate 181
red cells 177, 180, 302, 368
active site, ionizing group 211, 213, 231, 246
activity in fetal blood
consequence for placental CO_2 exchange 382
activity measurements by ^{18}O depletion from CO_2 254
anion binding 31–33, 134, 136, 138, 162, 195
binding of chloride 262
kinetic rates 248, 268
pH dependence 242, 244, 265, 270
binding of phosphate
pH dependence 242
binding of sulfate
pH dependence 238, 243
buffer mediated proton transfer 212, 231, 254
^{13}C magnetic resonance spectroscopy 213, 226
catalytic activity
human blood 177, 180, 302, 368
CO_2 hydration parameters 209
cobalt derivative 152, 228, 273, 277
dependence of hydration activity on pH 207
efficiency of catalytic step
activation of CO_2 221, 222, 234
dependence on enzyme concentration 216
intramolecular proton transfer 212, 217, 231, 232, 258
human isozymes B and C
turnover number 206, 214, 230
independence of hydration activity on pH 238
Carbonic anhydrase, inhibition by
acetazolamide 292, 295, 308, 326, 329, 336, 340, 368, 401, 407, 413, 420, 428, 433, 444
physiological consequences 366
anions 31–33, 134, 162, 195, 238, 246, 262
benzolamide 308, 326
ethoxzolamide 414, 420
imidazole 184, 200, 209, 210, 213
methazolamide 292, 296, 427, 428, 431
sulfanilamide 420
sulfate anions 238
sulfonamides (crystal structure) 196
isozymes B and C 134, 154, 157, 184, 206, 214, 230, 414, 415
covalent modification 208, 213
differences in chloride interaction 265
hydration activity for CO_2 207, 213
kinetic properties
Michaelis constant 30, 412
turnover number 30, 32, 133, 414, 415
localization
by autoradiography 399, 400
by histochemistry 333, 406
by immunocytochemistry 399, 402
magnetic resonance spectroscopy 151, 226, 262, 273
manganese derivative 273
mechanism of action 133, 151, 202, 212, 216, 234, 249, 280
covalent modification of active sites 208, 213
critique of "hydroxide hypothesis" 202, 239, 280
"hydroxide hypothesis" 207, 216, 230
"imidazole hypothesis" 281
metal ion function 151, 216, 218, 226, 280
"neutral net charge hypothesis" 244, 247, 250
proton transfer 136, 137, 148, 213, 232, 250, 259
rate limiting steps 212, 230
metal ion-substituted forms 141, 143, 151, 194, 226, 274, 280
metal substituted forms
proton relaxation measurement 151, 218
models for water binding 218, 219
nitrophenylacetate hydrolysis 273
orientation of acetate groups 276
nucleophilic nature of OH^- 213, 220, 223, 239

Carbonic anhydrase
  protonation reaction 152, 213, 217, 232
  rate of CO_2 activation 234
  rate of CO_2 hydration 207, 235, 236, 238
  rate of proton transfer 212, 231, 259
  rate of water exchange
    effect of buffers 258
  structure
    active site structure 139, 140, 151, 192
    amino acid composition 154, 155
    amino acid sequence 154, 157, 159
    crystal structure 184
    histidine residues liganded to the zinc 136, 152, 162, 191
    secondary structure 190
    side chains 191
    tertiary structure 186
  in various tissues
    bone 403
    brain 402
    cardiac muscle 412
    chorioallantoid membrane 403
    connective tissue 403
    endothelial cells 313, 342
    fetal red cells 300, 302
    gastric mucosa 401
    glial cells 447
    kidney brush border cell membrane 407, 421, 424
    kidney peritubular cell membrane 407, 423, 424
    kidney proximal tubular cells 18, 419, 420
    kidney tubular cell cytoplasma 401, 403, 420, 424
    liver 313
    lung 18, 308, 313, 331, 332, 336, 339, 408
    lung endothelial cell membrane 332, 339, 341
    nuclei 399, 403
    osteoclasts 403
    pancreas 401, 403
    placental tissues 408
    plasma 313
    red cells 18, 177, 285, 291, 299, 302, 368, 401–403
    shell gland mucosa 401, 403
    skeletal muscle 18, 313, 399, 401–403, 408, 413
    skeletal red muscle 412
    skeletal white muscle 412
  various types
    A isozyme 145

Carbonic anhydrase, various types
    B, C isozymes 134, 154, 157, 184, 206, 214, 230, 414, 415
    muscle enzyme 414, 415
  water splitting 223, 232
  zinc ion 151, 188, 192, 194
Cerebrospinal fluid
  see Brain, cerebrospinal fluid
Chloride
  concentration
    of cerebrospinal fluid 428
    of extracellular fluid of the brain 441, 446
  electrodes 321, 440, 445
  transport to CSF 428
Chloride-bicarbonate exchange 285, 299, 303, 306, 326, 433, 439, 446, 447
  inhibition of 286, 288, 296, 444
  pH dependence of 288
CO_2
  activation by carbonic anhydrase 221, 222
  diffusion constant 16, 27, 37, 40, 58
  dissociation curve "in vivo"
    of cerebrospinal fluid 439
    of fetal and maternal blood 384
  exchange
    during exercise 62, 63, 304, 356, 358
    in the lung 304, 306, 356
    in the placenta 304, 379, 385
  excretion in the lung 315, 333, 334, 337, 356, 359
  hydration velocity
    determination of 5, 7, 178, 180, 411
    rate constant 7, 9, 11, 28, 133
    in red cells 177, 178, 304
  partial pressure gradients in lung capillaries
    effect of carbonic anhydrase inhibition 368, 370
  permeability of red cells 297
  reactions in water 3, 11, 15, 23, 50, 133
    equilibrium constants 4, 8, 11
    kinetic constants 5, 7, 8, 11, 133
  solubility 3, 27
  transport in man 367
  uptake by red cells 177, 292, 299, 304, 306

Diffusion limitation
  CO_2 transfer, lung 359

Electron density maps
  CO_2 difference 70, 72
Electrostatic stabilization
  of hemoglobin by N-terminal valines 97

Ethoxzolamide
see also Carbonic anhydrase, inhibition
inhibition constant 414
Eye
aqueous humor formation 428

Facilitated diffusion of CO_2
in buffer solutions 36, 43, 52
effect of
buffers 49, 52, 62
carbonic anhydrase 15, 18, 23, 29, 37, 43, 49, 58
diffusion potentials 40, 49, 54
path length 15, 27, 62
P_{CO2} 29, 37, 58, 61
relaxation time of proton exchange reaction 47
spatial distribution of carbonic anhydrase in the layer 15, 18, 20
in HCO_3^-/CO_3^{--} solutions 15, 23, 51, 52
mechanisms of 36, 39, 46, 47, 51, 52, 62
in muscle 58, 61, 410, 413
of oxygen 42
in protein solutions 36, 39, 43, 53
Facilitated diffusion of H^+ 36, 39, 43, 47, 52
Feto-maternal concentration gradient
lactate and bicarbonate 393

Glucagon
carbamino adducts 106

H^+
conductivity 8
passive transport mechanism 36, 39, 46, 52, 136, 137
secretion
in bone and egg shell 403, 404
in kidney proximal tubule 419
transfer at carbonic anhydrase active site
see Carbonic anhydrase, mechanism of action
transfer between red cells and plasma 306, 324, 326, 332
Haldane effect
hemoglobin 95
Hansson's histochemical carbonic anhydrase test 399, 400, 406
Häusler's histochemical carbonic anhydrase test 399, 406
Hemoglobin
anion binding 70, 98, 115, 116
Bohr effect
influence of CO_2 85, 95, 124, 128
Hemoglobin, carbamate
determination 75, 89, 97, 102
formation 67, 75, 84, 89, 102, 115, 122
X-ray crystallography 67, 97
earthworm 37, 38, 42
first association constant of oxygen
influence of CO_2 122, 126
N-terminal groups 67, 72, 75, 84, 97, 114, 122, 129
Hydration
of carboxyl compounds 10
of CO_2: see CO_2 hydration velocity
Hypercapnia 433

Jacobs-Stewart cycle 286, 307, 312

Lactate
permeation across red cell membrane 326
permeation mechanisms in red cell membranes 329
placental transfer 396
Lung
active CO_2 elimination 343
CO_2 transfer 315, 333, 334, 337, 356, 359
diffusion limitation 359
diffusing capacity, O_2
rest and exercise 362
diffusing capacity, CO_2
rest and exercise 362
intracapillary CO_2 gradient 353

Medulla oblongata 441
Metabolic alkalosis 432
Myoglobin 36, 43

Non-ionic diffusion
of lactic acid 329

Oxygen-18 depletion from CO_2 177, 180, 291
Oxygen affinity of hemoglobin
regulation by CO_2 114, 120, 122, 124

Pancreas
pancreatic fluid formation 428
pCO_2
oscillations in alveolar gas and arterial blood 332
changes
pulmonary postcapillary blood 315, 354
carbamino CO_2 and Bohr effect 319
effect of carbonic anhydrase 316
differences
alveolar to blood
charged membrane hypothesis 343, 353

pCO_2 differences
H$^+$ disequilibrium 319, 348, 354, 356
unequal $\dot{V}/\dot{Q}$ distribution 349
between alveolus and endcapillary blood 351, 356, 358, 436
between cerebrospinal fluid and blood 436
umbilical vein to uterine vein 381
disequilibrium in blood 304, 306, 321, 356
electrodes 58
response times of 321, 440, 445
surface of 440, 444
of extracellular fluid of the brain 441
oscillations in alveolar gas and arterial blood 332
Peptide hormones
physiological implications of carbamate formation 110
Permeability for
bicarbonate in the blood brain barrier 436
chloride in the blood brain barrier 436
pH
in brain tissue 433
of cerebrospinal fluid 426, 436
disequilibrium
in blood 306, 312, 319, 331, 339, 348, 356
in kidney tubule 421
electrodes
in stopped flow apparatus 411
surface 440, 444
equilibration in blood 306, 321
of extracellular fluid of the brain 441
Placenta
bicarbonate transfer 386, 390
CO_2 exchange 304, 385
species differences 379
CO_2-lactic acid exchange 397
functional efficacy for gas exchange 380
lactate transfer
facilitated transport 396
shunts and CO_2 exchange 382
uneven distribution of blood flow 383
water movement 386
Postcapillary pCO_2 changes 306, 310, 336, 356
Postcapillary pH changes 306, 310, 312, 331, 332, 337, 342, 356
Pressure-jump-technique 5
Protein
carbamate reaction 89
diffusion
rotational 36, 38, 42, 45
translational 36, 38, 40, 41, 45

Rapid reaction
continuous flow measurements 178, 285, 300, 322
stopped flow measurements 286, 410
Respiratory acidosis 432, 439, 446

SITS 286, 327, 423, 444, 447
Sodium
transport to CSF 427, 428, 433
Sulfate
permeability 326

List of Participants

AHMAD, H.R., Bochum/FRG
ALBERS, C., Regensburg/FRG
ARNONE, A., Iowa City, Iowa/USA
BARNIKOL, W.K.R., Mainz/FRG
BARTELS, H., Hannover/FRG
BAUER, C., Regensburg/FRG
BAUMANN, R., Hannover/FRG
BERTINI, I., Florence/Italy
BÖNING, D., Hannover/FRG
BREEPOEL, P.M., Nijmegen/Netherlands
BURKHARD, O., Mainz/FRG
BURSAUX, E., Suresnes/France
COLEMAN, J.E., New Haven, Connecticut/USA
CRANDALL, E.D., Los Angeles, California/USA
DE KONING, J., Nijmegen/Netherlands
EFFROS, R.M., Torrance, California/USA
FORSTER, R.E., Philadelphia, Pennsylvania/USA
FRÖMTER, E., Frankfurt/FRG
GAY, C.V., State University of Pennsylvania, Pennsylvania/USA
GROS, G., Essen/FRG
GURD, F.R.N., Bloomington, Indiana/USA
GURD, R.S., Bloomington, Indiana/USA
GURTNER, G.H., Baltimore, Maryland/USA
HENDERSON, L.E., Frederick, Maryland/USA
HILL, E.P., La Jolla, California/USA
HOLLAND, R.A.B., Kensington, New South Wales/Australia
KANNAN, K.K., Bombay/India
KHALIFAH, R.G., Charlottesville, Virginia/USA
KLOCKE, R.A., Buffalo, New York/USA
KNOCHE, W., Göttingen/FRG
KOENIG, S.H., Yorktown Heights, New York/USA
LEDER, O., Freiburg/FRG
LINDSKOG, S., Umea/Sweden
LOESCHCKE, H.H., Bochum/FRG
LUCHINAT, C., Florence/Italy
LUTTMANN, A., Bochum/FRG
MAREN, T.H., Gainesville, Florida/USA
MELDON, J.H., Medford, Massachusetts/USA
MERGNER, G., Baltimore, Maryland/USA
MERGNER, W.J., Baltimore, Maryland/USA
MOLL, W., Regensburg/FRG
PERRELLA, M., Milan/Italy
PESANDO, J.M., Boston, Massachusetts/USA
PIIPER, J., Göttingen/FRG
PONTE, J., Bristol/GB
POYART, C., Suresnes/France
PURVES, M.J., Bristol/GB
QUINN, J.A., Philadelphia, Pennsylvania/USA
RISPENS, P., Groningen/Netherlands
SCHEID, P., Göttingen/FRG
SCHULTZ, J.S., Ann Arbor, Michigan/USA
SILVERMAN, D.N., Gainesville, Florida/USA
TASHIAN, R.E., Ann Arbor, Michigan/USA
WARD, R.L., Livermore, California/USA
WOOLLEY, P., Berlin (West)
ZIJLSTRA, W.G., Groningen/Netherlands
ZOCK, J.P., Groningen/Netherlands